SURGE TECTONICS:
A NEW HYPOTHESIS OF GLOBAL GEODYNAMICS

Solid Earth Sciences Library

Volume 9

The titles published in this series are listed at the end of this volume.

Surge Tectonics: A New Hypothesis of Global Geodynamics

by

ARTHUR A. MEYERHOFF†

former International Geological Consultant, Tulsa, OK, U.S.A.

IRFAN TANER

Consulting Geologist, Tulsa, OK, U.S.A.

A. E. L. MORRIS

Consulting Geologist, Los Angeles, CA, U.S.A.

W. B. AGOCS

Consulting Geologist, Allentown, PA, U.S.A.

M. KAMEN-KAYE

Consulting Geologist, Cambridge, MA, U.S.A.

M. I. BHAT

Wadia Institute of Himalayan Geology, Dehra Dun, India

N. C. SMOOT

U.S. Naval Oceanographic Office, Stennis Space Center, MS, U.S.A.

DONG R. CHOI

Consulting Geologist, Higgins, A.C.T. Australia

edited by

DONNA MEYERHOFF HULL

The University of Texas at Dallas, U.S.A.

KLUWER ACADEMIC PUBLISHERS

DORDRECHT / BOSTON / LONDON

A C.I.P. Catalogue record for this book is available from the Library of Congress

ISBN 0-7923-4156-2

Published by Kluwer Academic Publishers,
P.O. Box 17, 3300 AA Dordrecht, The Netherlands.

Kluwer Academic Publishers incorporates
the publishing programmes of
D. Reidel, Martinus Nijhoff, Dr W. Junk and MTP Press.

Sold and distributed in the U.S.A. and Canada
by Kluwer Academic Publishers,
101 Philip Drive, Norwell, MA 02061, U.S.A.

In all other countries, sold and distributed
by Kluwer Academic Publishers Group,
P.O. Box 322, 3300 AH Dordrecht, The Netherlands.

Printed on acid-free paper

Printed in the Netherlands

DEDICATION

Dedicated to Howard A. Meyerhoff, father of the senior author and former professor of geology at Smith College, to V.V. Beloussov, formerly of the Russian Academy of Sciences, and to Alexander Baird, formerly of Pomona College, California; three distinguished geologists who greatly influenced the career of the senior author throughout his life.

TABLE OF CONTENTS

PREFACE
TECTONICS AND PHYSICS

Geology, although rooted in the laws of physics, rarely has been taught in a manner designed to stress the relations between the laws and theorems of physics and the postulates of geology. The same is true of geophysics, whose specialties (seismology, gravimetry, magnetics, magnetotellurics) deal only with the laws that govern them, and not with those that govern geology's postulates. The branch of geology and geophysics called tectonophysics is not a formalized discipline or subdiscipline, and, therefore, has no formal laws or theorems of its own. Although many recent books claim to be textbooks in tectonophysics, they are not; they are books designed to explain one hypothesis, just as the present book is designed to explain one hypothesis. The textbook that comes closest to being a textbook of tectonophysics is Peter J. Wyllie's (1971) book, *The Dynamic Earth*.

Teachers, students, and practitioners of geology since the very beginning of earth-science teaching have avoided the development of a rigorous (but not rigid) scientific approach to tectonics, largely because we earth scientists have not fully understood the origin of the features with which we are dealing. This fact is not at all surprising when one considers that the database for hypotheses and theories of tectonics, particularly before 1960, has been limited to a small part of the exposed land area on the Earth's surface.

Some unfortunate attitudes have developed within and concerning geology and geologists, largely because geology has not been a rigorous science in the strict meaning of that expression. Many geologists consider geology a science separate from physics because it has its own data sets that distinguish it from physics and other fields. Yet many specialists in fields other than geology often refer to it as an "inexact" science. It is no accident that there is no Nobel Prize in the geological sciences. (It is ironic that the monies for the Nobel Prizes came from sales of the petroleum produced from the Nobel-owned oil fields near Baku along the western margin of the Caspian Sea, fields discovered after 1885 by the Nobel's own geologist, Hjalmar Sjögren [Owen, 1975]!) Nor is this negative attitude toward geologists limited to physicists. The same attitude is common among engineers, and even some chemists and biologists. Such attitudes have become less common, however, since the advent of plate tectonics, which has provided a more quantitative basis to the geological sciences.

Geologists have pointed out that many of the papers published in physics and engineering were quantitative studies that, where applied to Mother Nature, had very little meaning. Most of us, as geologists, pride ourselves on being "practical" scientists rather than the theoretical idealists that others, especially physicists, might be. Be that as it may, it was--and still is--perfectly true that many geologists have two or less years of university-level physics, chemistry, and mathematics. Geologists usually do not have an engineering background except for those who have had petroleum or civil engineering courses.

This is not to say that there never have been scientists who have tried to quantify the earth sciences, especially geology, in some degree. Nor are we saying that quantification can be done easily. What we are saying is that, in order to quantify certain aspects of the earth sciences, structural geology and tectonics in particular, we geologists must make the fullest possible use of the laws of physics, so that, if we are to make any postulates of our own, they will be firmly rooted in those laws. Moreover, as a theoretical science, geology must become wedded thoroughly to classical physics for future progress to be made on a solid foundation. The late Prof. Chester Ray Longwell of Yale and Stanford Universities tried to instill in each student, beginning with the freshman year, a commitment to learning the laws of physics. His was a lone voice in the wilderness.

We do take note of the fact that many geologists and geophysicists have tried to instill physical principles in their students and to use them in their own published works. Some of the many names that come to mind include E.M. Anderson, E.V. Artyushkov, M.A. Biot, Hugo Benioff, Walter H. Bucher, L.M. Cathles, J. Goguel, David T. Griggs, Ross Gunn, W. Hafner, J.M. Handin, M. King Hubbert, J.A. Jacobs, J.C. Jaeger, Sir Harold Jeffreys, Konrad B. Krauskopf, R.A. Lyttleton, Gordon J.F. MacDonald, John D. Moody, H. Ode, Hans Ramberg, J.G. Ramsey, R.D. Russell, W.W. Rubey, Adrian E. Scheidegger, R.B. Smith, the Weertmans, and the early work of J. Tuzo Wilson. However, we do not agree with the recent statement by J. Tuzo Wilson that "The Earth is so modest and quiet a fragment of the universe that laws of classical physics can be assumed to control its behavior, but it has not been easy to relate specific laws to geological observations. The explanation is believed to be that due to changes in conditions with depth the layers do not all follow the same laws and their modes of behavior differ. This complicates their interactions" (Wilson, 1990, p. 6611). Rather, we are convinced that the classical laws of physics do indeed find expressions in familiar geological structures under all conditions of pressure, temperature, and depth. Prof. Hans Ramberg has proved this in many tectonic situations (e.g., Ramberg, 1952, 1955, 1960, 1962, 1963a,b, 1967, 1968a,b, 1972a,b, 1973, 1981, 1985, 1986, 1989, and 1991).

We hope that the contents of this book will move geology and tectonics even farther towards a goal of integrating physical laws with our global geodynamics hypotheses.

Arthur A. Meyerhoff
Written Summer, 1994
Deceased Sept. 18, 1994

Editor's Postscript

This book and the geodynamic hypothesis that it presents, surge tectonics, are the culmination of a lifetime of geological thinking by Arthur A. Meyerhoff, ideas shared and enhanced by his co-authors and colleagues. Although his opposition to the plate-tectonics hypothesis is well-known among earth scientists who started their careers in the 1950s through 1970s, most younger geologists only know of his objections second- or third-hand. Many of the reasons for his objections have been forgotten, dismissed as unimportant, or are believed to be solved, despite the fact that many of his original objections, published in his hallmark paper in 1974 (Meyerhoff and Meyerhoff, 1974), still remain unanswered today. Some of the "old" questions are elucidated once again in this text and new questions are asked on the basis of new data collected by more advanced technology during the last two decades. The answers to all of these questions, he believed, lay in the concept of surge tectonics.

As plate tectonics grew in popularity and acceptance, it was commonly thought that Art Meyerhoff had fallen "behind the times". Nothing could have been further from the truth. He devoured current literature, subscribing to no less than 40 geological and geophysical journals in addition to receiving a steady stream of books on every conceivable aspect of geology. Those books and journals did not simply go unread onto a shelf. He took copious amounts of notes, continually compiling information on paper and especially in his mind. His habit of packing papers with references has been overwhelming to some of his reviewers, but the habit was well-acquired. He knew that standing in opposition to the majority would require extra effort and knowledge on his part. Admittedly, he did find delight some days in being the "devil's advocate" and he was an ardent iconoclast. But he did not endure years of criticism for the delight of it; he did so because he did not see 100% infallible logic in the concept of plate tectonics. The hypothesis simply was not able to explain a great deal of geological evidence that he had amassed over the years. During the 1970s, at a conference where he expounded upon the flaws of that hypothesis, one of his fellow earth scientists gave him a sharp rebuke, pointing out that if he didn't like the plate tectonics hypothesis, perhaps he ought to come up with something better. It was a comment never forgotten by my father, Art Meyerhoff, and to this end he devoted the last ten years of his life.

His earliest ideas about asthenosphere flow and the "surge" of magma developed during conversations with his fellow geologist and father, Howard A. Meyerhoff, during the early 1970s. Their ideas focused on the origin of island arcs, and were presented in a joint paper in 1977 (Meyerhoff and Meyerhoff, 1977). These ideas then rested for a few years, until the mid- to late-1980s, when A.A. Meyerhoff again took up the subject of asthenosphere surge, this time in earnest. As he developed his hypothesis, which he termed "surge tectonics," he reached out to hundreds of earth scientists, hoping for new ideas and feedback on his own thoughts. Many replied and many did not; many were enthusiastic and encouraging, others silent or sardonic. Familiar with mixed reviews, he continued forward, gathering data, testing his ideas, modifying and strengthening them and always asking more questions of himself and his hypothesis. A few earth scientists (including the co-authors of this text), excited by his ideas, joined him in the development of surge tectonics. Finally the time came to take their ideas "public." In 1990, he and Irfan Taner attended a conference in Washington, D.C., entitled, "New Concepts in Global Tectonics" (organized by S. Chatterjee and N. Hotton, III). There, he presented the surge-tectonics hypothesis for the first time.

Unfortunately, however, time and health were not on Art Meyerhoff's side, and the symposium in Washington was followed shortly thereafter by the discovery of lung cancer. Not one to be stopped by untimely surgery, he brought his journals to the hospital and continued to read, write, and think about geology during his recovery. In the next few years, he finished three papers for the New Concepts in Global Tectonics Symposium volume, then began working on several new papers, which included a compendium on the surge-tectonics evolution of southeastern Asia (Meyerhoff, in press), a paper on global paleobiogeography of the Phanerozoic (Meyerhoff et al., in press; GSA Special Paper 189), short discussions of the Chicxulub proposed impact site (with C. Officer) and a re-evaluation of the interpretations of seafloor bathymetric data with N.C. Smoot (Smoot and Meyerhoff, 1995). He also sought, and acquired, a contract to write a book on the surge-tectonics hypothesis. Always energized by thinking and the discovery of new ideas, he worked and wrote steadily until his death in late 1994.

The production of this book has therefore taken a different path than originally envisioned, and has been completed regrettably without the senior author's presence and the mass of detailed information that he carried in his mind. The hypothesis itself has been published and is described here once more with specific examples. Sections written by the co-authors before and after Art Meyerhoff's death are combined with his writings and with oral and written communications concerning the surge-tectonics hypothesis which we received during his lifetime. It has been an unexpected challenge in the editor's life, and together with the co-authors, I hope that we have served my father's memory and ideas well with this publication.

Mantle diapirism, bivergent foldbelts, ridge-parallel flow--each of these concepts can be found in the "mainstream" geological literature today. Each of these concepts is part of the foundation of the surge-tectonics hypothesis. The concept is not so far afield as some might think, but it is, undoubtedly, a different approach. It is not flawless. Art Meyerhoff did not believe that his hypothesis was flawless; he was working on it, developing it, and hoped to have the chance to do so for years to come. He encouraged his colleagues to continue thinking about the hypothesis and wanted them to continue to improve it with their own data and ideas. If he were here today, he would ask no less of you, the reader. He also understood the attraction to the hypothesis of plate tectonics, and knew well why it was so popular. His debates grew not out of the questions that plate tectonics can answer, but out of the questions it leaves unanswered.

Among the scientists my father admired most was Sir Harold Jeffreys, who began his 1974 paper entitled "Theoretical Aspects of Continental Drift" (AAPG Memoir 23)with the following quotation (p. 395):

"Forty million school books can't be wrong," Grant said after a little.
"Can't they?"
"Well, can they?"
"I used to think so, but I'm not so sure nowadays."
Josephine Tey, *The Daughter of Time*, p . 92

Perhaps we should never be too sure.

Donna Meyerhoff Hull
April 28, 1996

ACKNOWLEDGMENTS

The authors wish to acknowledge the following persons for their critiques of parts of the following publication, assistance in acquiring literature, maps, charts, and data, and for thoughts and ideas concerning the surge-tectonics hypothesis. Their time, patience, and willingness to be involved in this effort will always be appreciated. They include: K. Alam, H.E.F. Amundsen, V. Anfiloff, E. Atalik, D.L. Baars, R.N. Bergantino, A. Boucot, A. Bowsher, H. Duque Caro, S.E. Cebull, R.E. Chapman, S. Chatterjee, J.K. Davidson, S.B. Devine, J.M. Dickins, A.T. Donnelly, A.K. Dubey, Ch. Ducloz, R. Freitag, A.C. Grant, P.E. Gretener, Mrs. Fu Jing, Y. Fujita, N. Guler-Quazir, A.K. Hamilton, C.W. Hatten, J.R. Hertzler, N. Hotton III, M. Ilich, MS. Kashfi, T.T. Khoo, K. Kis, F.G. Koch, F. Kottlowski, K. Krauskopf, H. Le Grand, D. Love, P.D. Lowman, A. Lowrie, W.D. Lowry, B.D.A. and M. Mantura, Martin, J.C. Meyerhoff, R.D. Meyerhoff, P. Miles, M.T. Moussa, W. Muehlberger, S. Nagumo, W.W. Olive, V. Raiverman, H. Ramberg, D. Rigassi, T.S.M. Ranneft, M. Schalk, V.M. Seiders, R. Shanker, D.R. Shawe, W. Stannage, K.M. Storetvedt, P.M. Sychev, B.K. Tan, C. Teichert, W.A. Thomas, G.A. Thompson, P. Trurnit, T.A. Tuezov, G.B. Udintsev, Lee Woodward, C.T. Wrucke, W.L. Youngquist and B. Xingbi. As noted in earlier papers of this hypothesis, our acknowledgment of these colleagues in no way implies agreement with our views, although it is encouraging that some do. A special note of thanks to Barbara M. Christy of the Library of Congress for her moral support and for her help in finding many, many publications over a twenty-year period. We thank Kathryn L. Meyerhoff, Ernestine Voyles, Sally Reid, and Joyce Moon for contributing so much time and energy in the preparation of the manuscript and figures. Enormous thanks to Stuart and Jean Jenness of Ottawa, Canada and to Robert Hull and Suchran Taner for their encouragement and help in the long, final hours of editing.

Because of the death of the senior author in the early stages of the preparation of this book, the remaining authors selected some excerpts and figures from previous publications by A.A. Meyerhoff and others. Reproduction permission is gratefully acknowledged from the following sources and publishers:

> Excerpts and figures reprinted from "Origin of midocean ridges," "Reykjanes Ridge: quantitative determinations from magnetic anomalies," and "Surge tectonics: a new hypothesis of Earth dynamics," by A. A. Meyerhoff and others, *In, New Concepts in Global Tectonics*, 1992, S. Chatterjee and N. Hotton III, eds.,with permission from **Texas Tech University Press**, Lubbock, Texas, USA.

> Excerpts and figures reprinted from *Journal of Southeastern Asian Earth Sciences*,"Surge-tectonic Evolution of Southeastern Asia: A Geohydrodynamics Approach" by A. A. Meyerhoff, in press (1996), with permission from **Elsevier Science Ltd**, The Boulevard, Langford Lane, Kidlington 0X5 1GB, UK.

Finally, our thanks to Kluwer Academic Publishers for the publication of this book. In doing so, they recognize that new hypotheses based on scientifically sound data deserve and, for the healthy progress of science, need to be heard and discussed, whether they flow with the mainstream of thought, or take a new and controversial approach.

CHAPTER 1

WHY A NEW HYPOTHESIS?

1.1 Introduction

Before 1962, the year in which H.H. Hess revived and revised Arthur Holmes's (1931) concept of seafloor spreading (which also was proposed by Ampferer [1941]), the geology and geophysics departments of the world taught several geodynamics hypotheses. These hypotheses stimulated lively discussions and resulted in the publication of a highly diversified spectrum of ideas. After Hess's version of seafloor spreading was published, diversity in geodynamics thinking began to wane, and outside of Asia and Eastern Europe, had all but vanished by the end of 1963. Most of these earlier and now contrary concepts are no longer taught or are presented briefly as old, outdated ideas in light of the new hypothesis of plate tectonics. This fact should generate concern among all scientists, for historically the rigorous testing of ideas effectively ceases in intellectual environments dominated by a single concept. Furthermore, it is the belief of these authors that as intensive geotectonic research has vastly increased the database for Earth-dynamic studies, plate tectonics has not adequately and completely explained the geology of many regions of the world.

The purpose of this book is to present a comprehensive and internally consistent hypothesis of global tectonics, an hypothesis that we call *surge tectonics*. The need for such an hypothesis has become apparent as new sets of geological and geophysical data were found impossible to explain within the framework of the plate tectonics hypothesis (see Chapter 2). Because each of these unexplained data sets consists of information gathered from all parts of the Earth, the inability of plate tectonics---or any other hypothesis, for that matter---to explain them fully indicates that current tectonic hypotheses can no longer be regarded as adequate explanations of Earth dynamics. Our own research, therefore, concentrated on finding an hypothesis that explains *all* sets of structural, tectonic, and related data.

Today, geologists and geophysicists tend to treat pieces of the Earth's crust like a roomful of furniture, objects that can be pushed around at will into whatever configuration is required to satisfy a particular model. Unfortunately, the Earth's crust is not so easily manipulated, particularly if one is faithful to physical laws as well as the geologic data. Instead, the rigid crust and uppermost mantle form a massive interlocking mosaic, which we call the lithosphere. Surfaces along which the lithosphere can move freely with minimal resistance simply do not exist, except where the upper surface of the lithosphere comes into contact with the atmosphere.

To complicate matters, a huge body of evidence has accumulated to show that this lithosphere mosaic is in a state of *equiplanar tangential stress* (e.g., Denham et al., 1979). That is, compressive stress is ubiquitous in the lithosphere; moreover it is tangential and directed approximately equally in all directions of the compass, in accord with Newton's Third Law of Motion. This fact alone means that, for one part of the mosaicwork to move laterally (and tangentially), *all* parts must shift in order to accommodate the movement of the one part (Garfunkel, 1966). One of several convincing proofs of this involves the classical *hole-in-the plate-problem* of architecture and architectural engineering (Inglis, 1913; Griffith, 1920; Pöschl, 1921; Donnell, 1941; Stevenson, 1945; Timoshenko and Goodier, 1951; Murrell (1964a, 1964b); Jaeger and Cook, 1979; McCartan and Gettings, 1991). Within any body (or plate) subjected to equiplanar tangential stress (e.g., compression), stresses in all directions are approximately equal and opposite, in accord with Newton's Third Law, unless there is a flaw (hole) in the body. Wherever a flaw, or 'hole', is present, the compressive stresses must of necessity (Third Law of Motion) be oriented at right angles to the flaw, or hole. A well-known example is the San Andreas fault zone in California, where the axis of maximum compressive stress is everywhere at right angles to the fault trend (Mount and Suppe, 1987; Zoback et al., 1987).

We also know from geological field mapping that objects within the lithosphere mosaic *are* moved substantial distances, both vertically and laterally. However, the argument that large lithosphere plates, each 50 to 200 km thick, each extending for thousands of kilometers in all directions, and each weighing incalculable tons, can be moved freely and systematically about the Earth's surface defies all physical laws and common sense. Strictly lateral tangential movements are out of the question to explain the observed lateral and vertical motions that have been mapped in the field. To accommodate these *visible*, measurable, large lateral movements, rock bodies within the lithosphere mosaic must be able to move. To do this requires (1) upward (vertical) motion of rock bodies to positions of least resistance, followed by (2) lateral outward motions of the newly freed bodies on the upper lithosphere surface where the stresses required for lateral movements are far less than those required *within* the lithosphere. To accomplish the observed motions---which are not confined to relatively narrow mobile belts but occur everywhere within the lithospheric plates---a geodynamic explanation other than conventional plate tectonics and any other existing geodynamic hypothesis is required.

Surge tectonics is a new hypothesis which proposes that the Earth acts like a hydraulic press. The containment vessel for this press is an interconnected network of magma chambers and channels in the lithosphere; the fluid in the chambers is magma from the asthenosphere; and the trigger mechanism, or press, is episodic collapse of the lithosphere into the asthenosphere along points of weakness. Three interdependent and interacting processes are involved: (1) lateral flow of fluid, or semifluid magma through the interconnected channels; (2) cooling of the Earth causing contraction, which contributes to tectogenesis; and (3) the Earth's rotation. Surge tectonics draws on well-known physical laws, especially those related to the laws of motion, gravity, and fluid dynamics.

In the pages that follow, we first review the major concepts of Earth dynamics that have dominated scientific thought over the last two centuries. Chapter 2 contains a discussion of several types of data used in tectonic studies, and describes numerous data sets that are largely unexplained by most tectonic hypotheses, thus setting the foundation for a new global hypothesis. Chapter 3 presents this hypothesis, which we call "surge tectonics." Chapter 4 gives examples of surge channels in different tectonic settings, and Chapter 5 presents an application of the surge-tectonics hypothesis to southeastern Asia. Finally, Chapter 6 discusses the origin of flood basalts and magma floods in the context of surge tectonics.

1.2 Former and Current Concepts of Earth Dynamics

1.2.1 GENERAL

For nearly four centuries, inquisitive minds have pondered the origin and development of the Earth, as well as the forces and processes that shape it. Several hypotheses of the Earth's origin have caught the popular fancy---particularly during the last 150 years---each holding sway for a time until scientific advances led to a newer and better concept. In our century, several concepts have attracted a sizable following, but none more passionately than the plate-tectonics hypothesis that emerged during the 1960s and is now widely accepted by the earth-science community. We review briefly the major Earth-dynamics hypotheses that have been advanced over the last few decades.

1.2.2 CONTRACTION

Cooling of the Earth, or contraction, is the oldest scientific hypothesis of Earth dynamics, having been proposed late in the 16th century by Giordano Bruno, who lived from 1548 to 1600. He was burned at the stake for insisting that the Earth orbited the sun. His classic analogy between a shrivelled apple and a cooling Earth was immortalized in a diagram published by Descartes (1644; see Dennis, 1982). Élie de Beaumont (1831) in Europe and Dana (1847) in North America, however, seem to have been the first to apply the concept to specific foldbelts as an explanation of tectogenesis. Charles Davison (1887) and Sir George Darwin (1887) were the first to work out the mathematical theory of contraction, an hypothesis that remained popular until the 1950s.

Two types of contraction hypotheses were developed. The first was the *elastic instability* form of the hypothesis, according to which the lithosphere, as the Earth cools, buckles into a series of folds, much like the surface of the shrivelled apple of Bruno (*in* Descartes, 1644). Jeffreys (1970) showed that such folds are unstable, and that failure would not be by folding, or buckling, but by fracturing. Thus the second form of the contraction hypothesis was born, the *fracture-contraction hypothesis* (Jeffreys, 1970; Meyerhoff et al., 1972).

Contraction of the Earth may not be solely the result of simple cooling of a sphere, as Jeffreys (1970) has stated. Gravitational differentiation also may be an important factor (Scheidegger, 1963). MacDonald (1959, 1963, 1965a) has stressed the fact that the Jeffreys (1970) contraction model fails to account for the radioactive decay which must have attended the differentiation of the Earth.

MacDonald (1959, 1963) argued that, because of the widespread presence in the primitive Earth of radioactive elements, the Earth must have expanded during much of its history. MacDonald (1963) reasoned that, during the first part of its history, decay of the radioactive elements would have caused heat production to exceed heat loss, thereby causing the Earth to expand. With continuing decay, the two would equalize until, during the last 1.1 Ga (billion years) or so, heat loss would exceed heat production, and contraction would take place. This contraction model, although very different from the Jeffreys (1970) model in which contraction of the Earth takes place from its inception, also is a fracture-contraction model.

MacDonald (1963) showed that, during the gentle contraction of the last 1.1 Ga, thermal stresses would be released. In addition, because radioactive materials extend more deeply into the mantle beneath ocean basins than beneath continents, thermal stresses will concentrate at the boundaries separating the oceanic and continental mantle materials. Such concentrations could lead to the development of deep faults such as the Benioff zones.

Even though MacDonald (1963) answered many of the growing objections to the contraction hypothesis, the hypothesis fell from grace. One obvious reason for this is that the idea of ocean-floor spreading was riveting the attention of increasing numbers of earth scientists. Another reason was the widespread conviction that any loss of heat attributable to Earth cooling was more than offset by the radioactive heat generated by uranium, thorium, and potassium (Urey, 1952). A third reason was the observation that the amount of measured foreshortening in foldbelts is far greater than the amount that contraction can account for. MacDonald's (1959, 1963) model accounts for the second reason, but the third reason---measured foreshortening exceeds anything that contraction can produce---still, in our judgment, is valid. If contraction does take place, another mechanism must produce the foldbelts. Regardless, many geophysicists (e.g., Stacey, 1981; Lyttleton, 1982) still regard contraction as an ongoing process within the Earth.

A contracting Earth is an extremely attractive model for tectonic processes, because---in theory at least---it can provide directly for tangential compression at the Earth's surface. However, contraction as the sole cause of tectogenesis is highly unlikely for many reasons, most of which were discussed by Scheidegger (1963) and Bott (1971). Not the least of these is the fact that in neither the contraction envisioned by Jeffreys (1970) nor that described by MacDonald (1963) can all of the true shortening in mobile belts be accounted for. However, if weak zones---surge channels---containing magma are present in the lithosphere, contraction can play a role much different than that usually attributed to it.

1.2.3 Mantle Convection

So many papers favorable to mantle convection have been published that many earth scientists regard its existence as established fact. However, a critical review of the literature reveals that the basis of its existence is almost entirely theoretical. Little physical evidence is known to support the postulate of *large* mantle convection cells---whether mantle-wide or in the upper mantle only---and many data are now available that suggest that large mantle convection cells do not exist. Undoubtedly, as MacDonald (1965b) has pointed out, small-scale convection occurs in the mantle, but there is no longer significant support in factual data for the kinds of convection cells that propel large lithosphere plates and give rise to the world's foldbelts.

As an hypothesis in geology, the use of convection cells seems to have appeared during the early 19th century to explain the distribution of active volcanism at the Earth's surface (e.g., Hopkins, 1839). Convection gained little acceptance, however, before Vening Meinesz's (1933) classic paper on mantle-wide convection to explain the genesis of foldbelts and the great negative gravity anomalies beneath oceanic trenches. Subsequently, through mathematical treatments by Pekeris (1936) and Hales (1936) and a laboratory experiment by Griggs (1939), the concept became respccted and well entrenched. With the advent of plate tectonics, mantle convection received a major boost, because it is the only mechanism known that might propel lithosphere plates. Plate tectonics suggests that midocean ridges and few other great rift zones (e.g., the East African rift) are the sites of mantle upwelling, whereas the deep-sea trenches and island arcs are the sites of mantle consumption (Hess, 1962; Dietz, 1961; Girdler, 1965; Orowan, 1965; Runcorn, 1965; Oxburgh and Turcotte, 1968). This greatly simplified image of the Earth's large-scale convection pattern still persists.

Sir Harold Jeffreys (1959) opposed mantle convection on several grounds, including the theoretical argument that the Lomnitz Law (Lomnitz, 1957, 1962) provided a self-damping mechanism that effectively eliminated large-scale mantle convection. Orowan (1967), among several, attacked the Lomnitz Law, and argued that it could not be applied to large-scale phenomena. Jeffreys and Crampin (1970) then formulated the Jeffreys-Lomnitz Law, according to which large-scale damping does eliminate large active convection cells. Subsequent theoretical work by Strick (1984) demonstrated the probable validity of Jeffreys and Crampin's (1970) conclusions.

Thus most arguments that are pro or con mantle convection are theoretical. However, theory is not a substitute for field data. Hence, the recent appearance of new and pertinent data sets bearing on the problem may lead to a more definitive conclusion concerning the merits of convection. These data sets, which indicate that large-scale mantle convection is highly unlikely, are reviewed briefly.

1. Residual geoid height maps from many studies were thought to show a regular pattern of geoid highs and lows with a characteristic 3,000-4,000-km wave length. This regular pattern and a seeming correlation with topography (bathymetry)---but only in a few places---were interpreted as evidence for the existence of mantle convection cells (Watts et al., 1985). Subsequently, Sandwell and Renkin (1988) demonstrated that the seemingly regular geoid pattern is an artifact caused by the removal of spherical harmonic coefficients through degree and order 10. Sandwell and Renkin (1988) found instead a correlation between geoid height and seafloor depth.

In our own work, we found that each surface-tectonic feature studied by us reflects the presence of a major structure beneath, in either the crust, mantle, or both. (We demonstrate the correctness of this statement in this book.) This observation that deep structure is reflected everywhere in surface phenomena at least in the more than 100 large tectonic features that we examined---indicates to us that, if convection cells are active in the upper mantle, their presence must be reflected at the surface. Hence, Sandwell and Renkin's (1988) study constitutes strong evidence that large active mantle convection cells do not exist.

2. A second data set is the recent identification of abundant seismic reflections in the upper mantle to a depth of at least 80 km (Flack and Warner, 1990; Posgay et al., 1990). Earlier, Mooney and Brocher (1987) rejected the opinion expressed by several workers that good seismic reflections can be obtained from the upper mantle. They rejected the notion on the grounds that so many high-quality reflection studies had failed to identify such reflectors that it was not likely that they could have been overlooked. Posgay et al. (1990), using a low-pass 4-32 Hz filter, found high-quality reflections to a depth of at least 24 s (75 to 80 km) beneath the Pannonian basin of Hungary. Good reflectors had also been found in the upper mantle to depths of 60 to 80 km (18-24 s) by other workers in western Europe (e.g., Dohr, 1970; Flack and Warner, 1990). Thus the failure to find upper-mantle reflections turns out to be an artifact caused by the use of too narrow a seismic filter band (12-32 Hz).

We do note, however, that the depth of penetration reported to date---80 km---(e.g., Posgay et al., 1990) and the lack of sufficient data from the upper mantle preclude a judgment at this time regarding the ability of deep reflection surveys to identify the upper parts of convection cells. Suffice it to say for now that the several surveys that we have seen contain no information suggestive of upper-mantle convection patterns. Our point is that this method holds some promise for future investigations of the structure of the asthenosphere and upper mantle.

3. Modelling experiments on the effects of a downgoing plate (slab) beneath Benioff zones

by Christensen and Yuen (1984) indicate that the plate should depress the 670-km discontinuity by 50 to 200 km. To test this, Revenaugh and Jordan (1989) applied waveform-inversion techniques to a very large number of long-period, reverberative-interval seismograms with western Pacific paths. Their study demonstrated that nowhere in the western Pacific is the 670-km discontinuity depressed more than about 30 km. They wrote (p. 5811) that "Much greater topography, *if it exists*, should be easily detectable by our methods" (emphasis added). Thus another technique fails to provide evidence for active mantle convection.

4. The most conclusive technique now in use is that of seismotomography (Dziewonski and Anderson, 1984; Woodhouse and Dziewonski, 1984; Dziewonski and Woodhouse, 1987; Grand, 1987). Using only two-dimensional, plan-view images, Olson et al. (1990) claimed that seismotomographic techniques had proved the existence of well-organized convection patterns. Dziewonski and Woodhouse (1987), as well as the other references just cited, provided excellent three-dimensional seismotomographic data, including numerous cross sections. These cross sections *were not utilized* by Olson et al. (1990). The cross sections provide no evidence for convection-cell geometry of any type, neither the type referred to by Oxburgh and Turcotte (1968 and subsequent papers) nor the type of Bercovici et al. (1989). This is especially true for the upper-mantle region above 670 km. We consider the seismotomographic three-dimensional results by Anderson, Dziewonski, Grand, Woodhouse, and several others as nearly conclusive.

5. Finally, we present a theoretical but practical reason why mantle convection is unlikely. The suggestion is not original with us, but is one that requires further study.

In a convection cell 670 km deep, a plate moving downward beneath a Benioff zone must enter and pass through a zone of increasing density and, therefore, more compact crystal structure. At the same time, the Earth's circumference at each deeper level is smaller. Thus a "lighter" mass from a region of Earth circumference 40,000 km must penetrate a "denser" region whose circumference (at a depth of 670 km) is only 35,800 km, or 11% less than the surface circumference. At some later stage in the convection process, the reverse process must take place, with "heavier" or "denser" material rising beneath the midocean ridges to propel the plates toward their "subduction" zones. This simplified, but widely accepted, "model" implies that the downward-moving plate undergoes phase changes and, therefore, increases in density that are more or less proportional to their depth and changing circumference. The upward-moving process is one that is fairly simple to envision, but the downward motions accompanied by reduced circumference, increased density, and more compact crystallinity---all acting in perfect concert---are not too believable. However, now that data are available---especially seismotomographic data---that suggest that convection cells are not present in the upper mantle, it may soon be unnecessary to discuss the pros and cons of convection on such a theoretical level.

1.2.4 Earth Expansion

This hypothesis seems to have originated with Rothpletz (1903) and Tammann (1903). It had very few adherents (e.g., Joly, 1925; Hilgenberg, 1933) until the 1950s when Egyed (1956) and Dicke (1957) revived it. Today the hypothesis has staunch supporters in such notables as Carey (1976) and Owen (1976). A major problem with the concept is that it provides no mechanism for lateral compression. Another unsatisfactory feature is that the Earth must expand in a highly controlled way to account for the distribution of continents and ocean basins. Moreover, no mechanism is known for the rapid expansion necessary during Phanerozoic time. Finally, MacDonald (1963) has shown that, whereas expansion probably was important during the first three eons of Earth history, it was rather minor and almost certainly is not taking place today.

1.2.5 Vertical Tectonics

Vertical-tectonics hypotheses attribute folding and thrusting to gravity sliding from the tops and flanks of vertically rising fault blocks, structural arches, mantle diapirs, and like phenomena. Consequently, the term *gravity tectonics* commonly is used in place of vertical tectonics (de Jong and Scholten, 1973). North (1965) published the best and most complete review of gravity (vertical) tectonics. Variants include the *oscillation* hypothesis (Haarmann, 1930), the very similar *undation* hypothesis (van Bemmelen, 1931), the *blister* hypothesis (Rich, 1951), and the *oceanization* hypothesis (Tetyayev, 1934; Beloussov, 1948). Van Bemmelen (1976) has attempted to fit the undation hypothesis into the framework of plate tectonics with only moderate success. Oceanization concepts differ from those of oscillation and undation in that, in oceanization, continental crust is engulfed by mantle basalts that convert continental crust into oceanic crust (Beloussov, 1980, 1981, 1989). No such resorption process exists in the oscillation-undation hypotheses. Today, vertical-tectonics hypotheses are not widely accepted, except in the Soviet Union and Eastern Europe, where Beloussov's (1980, 1981) ideas still command a broad following.

1.2.6 Zonal Rotation

Jardetzky (1935, 1954) proposed that Earth rotation produces massive east-west displacements between adjacent crustal blocks. An example is the geographic relationship between North and South America. In Jardetzky's (1954) scheme, differential rotational lag between the "heavier" Northern Hemisphere and the "lighter" Southern Hemisphere caused the Northern Hemisphere to lag behind the Southern, thus accounting for the fact that South America lies so far east of North America. Differential shear between the two continents created the young (geologically) east-west foldbelts of the Caribbean region. Because all foldbelts are the products of differential shear in Jardetzky's (1935) hypothesis, large shifts in the rotational pole through time are necessary. Unfortunately, Jardetzky's ideas were published in an obscure Balkans journal in the French language, and the hypothesis never received the attention that it deserved.

1.2.7 Continental Drift, Polar Wandering

These hypotheses are well known because of their close relationship with the plate-tectonics hypothesis. Alexander von Humboldt (1801) seems to have been the first earth scientist to write of continental drift, mainly to explain similarities in the shapes of coastlines. Goodacre (1991) referred to two authors (Thomas Dick in 1838 and Antonio Snider-Pellegrini in 1858) who suggested continental drift in their papers. Pickering's (1907) is one of the earliest modern discussions. Then Taylor (1910) published his *Polarflucht* concept shortly afterward. Wegener (1912, 1915) actively pursued the concept of continental drift. Mobilistic concepts had been in existence for many years in Europe, thanks largely to the development of nappe theory, and, therefore, Wegener's (1915) ideas fell on some fertile ground. Suess (1885) was an enthusiastic supporter of large-scale horizontal movements in tectogenesis, as were Heim (1878), Lugéon

(1902), Argand (1924), Daly (1925), Collet (1927), and many others. Argand (1924), Daly (1925), and Collet (1927) are especially noteworthy for their energetic support of the concept of drifting continents. In the Southern Hemisphere, where apparent similarities among the several continents dominated much of geological thought, Wegener's (1915) ideas received almost universal support (e.g., du Toit, 1937).

The lack of an obvious mechanism, however, led to the near-total rejection of the hypothesis outside of the Southern Hemisphere. Wegener's (1915) idea was that the continental sial of the continents "ploughed through" an "ocean" of mafic rocks. This concept of less dense rock moving through more dense rock seemed so highly improbable that most geoscientists regarded it as a fatal flaw.

Polar wandering, commonly associated with continental movements, persists as a viable concept, partly because it seems to explain some otherwise anomalous paleoclimatic and paleontologic data, and partly because it can be neither proved nor disproved. The latter is true because of the great uncertainties involved in describing the Earth's internal behavior.

Two forms of polar wandering have been proposed. In *true polar wandering*, the Earth's rotational axis actually shifts through time. In *apparent polar wandering*, the lithosphere slides intact over the asthenosphere, so that the rotational pole appears to wander, when in fact it is the lithosphere that is "wandering."

The concept apparently originated with the Comte de Buffon, who lived from 1707 to 1788, and appeared in one of the 44 volumes of his remarkable *Histoire Naturelle* (Munk and MacDonald, 1960). The concept gained a great deal of popularity (e.g., Peano, 1895; Volterra, 1895; Barrell, 1914), despite Darwin's (1887) very convincing arguments for the Earth's essential rigidity (based on research by Lord Kelvin). Just after World War II, paleomagnetic research suggested that continental motions might have taken place (e.g., Graham, 1949), and the hypothesis became popular once more. Gold (1955---the famous "Gold's beetle" paper) demonstrated that polar wandering not only is physically feasible but is to be expected, provided that the anelastic properties of the mantle proposed by Bondi and Gold (1955) from the damping of the Chandler wobble are actually present (Munk and MacDonald, 1960). Munk and MacDonald (1960) showed that such conditions cannot be proved and, in fact, are unlikely. Despite an attempt by Goldreich and Toomre (1969) to revive the hypothesis, it cannot be regarded as anything but speculation. In fact, apparent polar wandering---the sliding of the lithosphere over the asthenosphere---is so fraught with physical problems that it can be regarded as impossible (Munk and MacDonald, 1960).

A major obstacle to the hypothesis is the discovery of deep continental roots. Seismic-velocity and heat-flow studies (e.g., MacDonald, 1963; Jordan, 1975, 1978) show that the asthenosphere is very thin or absent beneath Precambrian nuclei, and that these nuclei have very deep roots, from 250 to 500 km (Fig. 1.1). This conclusion is supported by a large volume of data, including neodymium and strontium isotope studies (Wasserburg and DePaolo, 1979), as well as by strontium isotope studies in young volcanic rocks (Brooks et al., 1976). The absence, or near-absence, of a low-velocity asthenosphere below the Precambrian nuclei led Lowman (1985, 1986) to propose that the continents are fixed and, if sea-floor spreading takes place, it is limited to oceanic regions.

1.2.8 Seafloor Spreading and Plate Tectonics

The currently popular and widely accepted plate-tectonics hypothesis (Dietz, 1961; Hess, 1962;

Morgan, 1968) evolved from the Holmes-type (1931) convection model. It took root initially from interpretations of linear magnetic anomaly patterns on midocean ridges (Vine and Matthews, 1963; Morley and Larochelle, 1964), and developed explosively after Wilson's (1965) predicted "transform faults" seemed to be confirmed by seismic studies (Sykes, 1967). In general, the hypothesis proposes the following.

Convecting mantle rises beneath, and thereby sustains, the midocean ridges. In this way, new crust forms at the ridge crests where it divides, with part going in one direction and part in the opposite direction. Movement is orthogonal to the ridge. Thus crustal age increases away from the ridge crests. The linear magnetic anomalies associated with the ridges represent alternate times of "normal" and "reversed" polarity of the Earth's magnetic field. Finally, the oceanic crust is consumed (subducted) in the Earth's Benioff zones. In this way, the Earth maintains a constant radius by consuming as much crust as it creates.

However, continental nuclei have deep roots. Figure 1.1 shows that, in the Canadian shield, these roots extend downward at least to 500 km; an asthenosphere apparently is absent (MacDonald, 1963; Jordan, 1975, 1978; Grand, 1987; Lerner-Lam and Jordan, 1987). The existence of such deep roots has been vigorously challenged (e.g., Anderson, 1987a), but Figure 1.1 speaks for itself. Because deep roots are present, Lowman (1985, 1986) proposed a new sea-floor spreading model in which the continents are fixed.

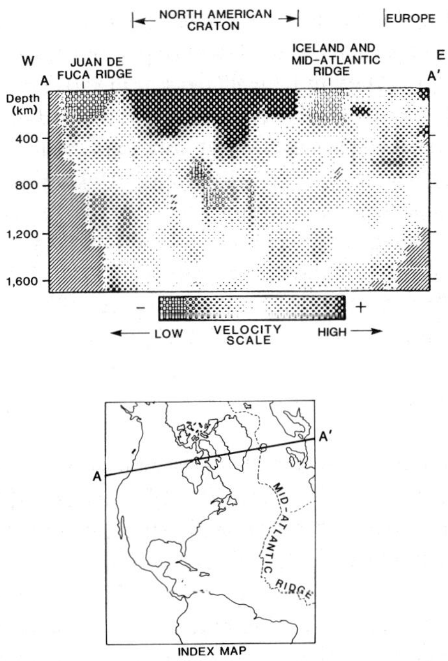

Fig. 1.1. Southwest-northeast seismotomographic cross section showing velocity structure across the North American craton and North Atlantic Ocean. High-velocity (i.e., colder) lithosphere, shown in dark tones, underlies the Canadian shield to depths of 250 to 500 km. The section shows that, as noted by Jordan (1975, 1978), continents have deep mantle roots and the asthenosphere low-velocity zone below them is either extremely thin or absent, suggesting that continents are fixed and immovable (Lowman, 1985, 1986). Note: velocity scale above 320 km is -3% to + 3%; from 320 to 405 km, it is -1.5% to +1.5%; and below 405 km, it is -0.9% to +0.9%. From Grand (1987).

1.2.9 Tectonostratigraphic Terranes

In recent years, an hypothesis reminiscent of the sliding lithosphere has gained a measure of popularity, the concept of *tectonostratigraphic terranes*. This hypothesis holds that continental accretion may take place in part by the rafting of discrete tectonic blocks, or terranes, from regions that are thousands of kilometers distant. Each of these terranes is characterized by a distinctive stratigraphy, paleontology, and, in some cases, tectonic style.

The concept is not new. Those familiar with Alpine geology will recognize at once the *facies-belt concept*, according to which each nappe, or group of nappes, has a unique stratigraphy and is separated from correlative stratigraphic sequences in adjacent nappes by a major fault, or fault zone. The concept originated with Ch. Lory (1860) and was summarized by Bertrand (1897). The only new idea in the tectonostratigraphic-terrane concept is the idea that terranes can be carried from one side of an ocean to another (Wilson, 1966, 1967).

Support for long-distance terrane transport (other than by thrusting) is derived mainly from paleomagnetic studies. Saul (1986) and Newton (1988) have shown that the paleontological support for paleomagnetics claimed by some workers (e.g., Monger, 1984) is poorly founded. Northrop and Meyerhoff (1963) long ago showed the uncertainties of paleomagnetic data. In more recent years, scores of detailed geological field studies have been published in which field and paleomagnetic data clash (e.g., Laubscher, 1975; Donovan and Meyerhoff, 1982; McDowell et al., 1984; Seiders, 1988; and many more). More recently, Butler et al. (1989) and Storetvedt (1992) were able to explain the contrary paleomagnetic data by vertical and horizontal rotations that do not require long-distance transport.

Sharp contrasts in tectonic styles and lithofacies on which the terrane concept is based were explained by Bertrand (1897), and more recently by Aleinikoff et al. (1987) and Hansen (1988). Silberling et al. (1987) were able to show that all Western Cordilleran terranes fit neatly into an alpine-type facies-belt model like that described for the Alps by Bertrand (1897). Aleinikoff et al. (1987), on the basis of lead-isotope studies from Alaska, concluded that most of the terranes which they studied formed adjacent to North America. Our conclusion is that long-distance transport, beyond that which is provided by nappe emplacement, may not be necessary to explain the tectonostratigraphic terranes of alpinotype foldbelts.

1.2.10 Wedge Tectonics

Since the "new global tectonics" (plate tectonics) revolution, some earth scientists have begun to see shortcomings in plate tectonic theory. Wedge tectonics is based on the interactions among compression, tension and shear forces on the crust. It is thought to provide predictable structural and tectonic trends (Pratsch, 1982).

1.2.11 Plate Tectonics with Fixed Continents

Lowman (1985, 1986) refuted continental drift based on several lines of new evidence, including the presence of deep continental lithospheric roots (Fig. 1.1), the absence of a low-velocity zone under shields, the absence of hot-spot trails on continents, contradictions among paleomagnetic data, and other reasons. Lowman's model incorporates fixed continents with several of the basic precepts of plate tectonics, including seafloor spreading, transform faulting, and subduction. Opening of the Atlantic Ocean by seafloor spreading is explained in the Lowman model by very slow subduction along the passive margins of the Atlantic.

1.2.12 Zipper Tectonics (Spiral Tectonics)

Trurnit (1988) proposed a new geodynamic model in which he attempted to integrate gravitational forces (expansion and contraction), the centrifugal force of the spinning Earth, thermal convection and tidal forces (Trurnit, 1988, 1991). The model is known as spiral tectonics or zipper tectonics. In this model, the stress field moves eastward and the oceans open and close following the "zip fastener principle" as the continents move westward. The Wilson Cycle is thus completed in this model in 400 to 500 m.y., forming a spiral shape, which then reverses in a N-S direction.

1.2.13 Viscous Flow Model

The viscous flow model of Keith (1993) is also intended to fulfill the shortcomings of plate tectonics by providing an alternative and promoting critical reexamination. This model suggests that the radioactive elements are concentrated in continental rocks and depleted in mid-ocean ridge basalt and in xenoliths of the upper mantle. The continental masses are rooted, and together with their "keel" they function as a heat source, insulator and boundary layer. Thus mantle upwelling occurs under continental masses and flows toward the mid-ocean ridges, opposite to the convection direction proposed by plate tectonics. Convection cell boundaries are the locations of subduction zones. In this model, no continental drift or mid-ocean spreading takes place.

1.3 Conclusion

Although each concept discussed here has strong points in its favor, each has one or more serious flaws inasmuch as they rely largely on the plate-tectonic model and/or on mantle convection, both of which also have flaws. None of the existing models has, so far, been successful in integrating all data sets and explaining all geologic features and dynamic events on the Earth. In view of the deficiencies of all existing hypotheses of earth dynamics, a new hypothesis is needed that can explain the data sets and geologic features while adhering to all physical laws. We believe that our hypothesis, surge tectonics, has that capability.

CHAPTER 2

UNRAVELING EARTH HISTORY: TECTONIC DATA SETS

In this chapter, we discuss types of data sets available for discerning Earth tectonics todays that we feel have not been adequately explained by current tectonic hypotheses.

2.1 Data Availability

From the beginning, the plate tectonic model was based mainly on data from rock exposures on land. Even today, a viable global-tectonics hypothesis has to be based principally on land data. However, information from the deep ocean floors also is essential. When plate tectonics was first elucidated during the 1960s, less than 0.0001% of the deep ocean had been explored and less than 20% of the land area had been mapped in meaningful detail. Even today, only a tiny percentage---possibly between 3 and 5%---of the deep ocean basins has been explored in any kind of detail, and not much more than 25 to 30% of the land area can be said to be truly known

A difficulty during the 1960s and 1970s, when most plate-tectonic concepts were formulated and accepted, was that with rare exceptions technologies for studying the ocean floors were few, and most of them were prohibitively expensive. Thus plate tectonics began with essentially no oceanic data and far fewer land data than are available now. Bathymetric information was crude and accurate underwater mapping methods were not developed. Gravimetric and magnetic studies, although feasible and relatively inexpensive, were only marginally useful without detailed bathymetry and reliable information on rock compositions, densities, and magnetic properties. Heat-flow data were scarce and their potential importance was unappreciated. Teleseismic information was abundant, but only along travel paths between recording stations, the number of which were woefully inadequate. Refraction-seismic data were few, and the cost of obtaining new refraction lines was expensive. Moreover, refraction data at best are difficult to interpret. The best tool available, reflection seismology, would have involved massive expense. Moreover, for reflection data to be really useful, recording times would have to be doubled or even tripled below the normal 5 to 6 seconds recorded by the petroleum companies who were the principal users of this technology. In addition, all recording equipment then available would require major redesigning to record times of 12 seconds or more. Interpretations of such long paper records, already a cumbersome manual process, would become next to impossible. Regardless of these difficulties, however, had reflection seismology been used from the inception of plate tectonics, many of its misconceptions and erroneous assumptions might have been avoided.

2.2 New Data Acquisition

2.2.1 Submersibles and Deep-Sea Drilling

To explore the rocks of the ocean floors, the Deep Sea Drilling Project, initially a national scientific project of the United States, was conceived. Beginning in late 1968, deep-sea samples were collected systemically for the first time by deep-sea drilling from the research vessel, *Glomar Challenger*. (A brief drilling program had been conducted with the M. V. *Caldrill I* during 1965 in the shallower water of the Blake Plateau just east of Florida to determine whether deep-sea drilling was even feasible [JOIDES, 1965].) In some areas, such as deep-sea trenches and the rift valleys atop midocean ridges, drilling information was supplemented by direct observations from deep-sea submersible vehicles, or bathyscaphes. Such research continues today with the use of one- to three-man submersibles and the increasingly popular Remotely Operated Vehicles (ROVs).

Deep-sea drilling was limited by several factors. One of these, the inability to hold the ship's position for extended periods during drilling operations, was soon overcome by dynamic positioning technology. Drilling results have been hampered seriously in some cases by the inability to reenter holes and to orient cores. Then and today, operations have been further impeded by the inability of drill bits to penetrate and recover some rock types encountered, especially chert and other hard rocks. Because of these shortcomings, it was impossible except in rare cases to know (1) whether true basement had been reached with the drill bit and (2), therefore, the age of the oldest rock at a given drillsite. (3) Moreover, it was not possible in most cases to determine whether the deepest basalt penetrated was a flow or a sill. In many places where contacts with the overlying sediments were recovered and observed to be baked, the basalt still was called basement (e.g., Shipboard Scientific Party, 1972, p. 401). (4) Finally, the magnetization directions of the rocks penetrated could not be determined for the first 5½ years of the Deep Sea Drilling Project until 331 sites had been drilled (mid-1974), and even then the cores were "semioriented" (Hall and Ryall, 1977). Deep sea drilling continues today, and data acquisition improves as new technology advances.

2.2.2 Sonography

Another technique developed to study the deep ocean floors is sonography, a technique that uses side-scanning sonar to obtain visual images of the ocean floor. The first side-scanning sonar was developed in the United Kingdom in the late 1960s. This is the GLORIA (Geological LOng-Range Inclined Asdic) system (Rusby, 1970; Rusby and Somers, 1977). The imagery of the first GLORIA system was very poor, and the images obtained were very difficult to interpret. Hence, sonography was not used widely until a clearer and more easily interpreted image was perfected in the mid-1970s, the GLORIA II system (Somers et al., 1978). Unfortunately for earth science, sonography did not come into its own until the mid-1980s after GLORIA II had proved itself and improved techniques had been developed (SeaMARC I, SeaMARC II [Sea Mapping And Remote Characterization]; Blackinton et al., 1983; Reed and Hussong, 1989). Although the images of the structures observed on the midocean ridges show features that were very different from those predicted by plate tectonics, the structures were given plate-tectonic explanations, however unrealistic.

2.2.3 Accurate Bathymetry

The Seabeam technique of producing highly accurate three-dimensional bathymetric maps is another method developed and used for deep-sea exploration (Glenn, 1970; Rénard and Allenou, 1979; de Moustier and Kleinrock, 1986). The Seabeam bathymetric survey system is a multibeam echo sounder that produces near-real-time, high-resolution, contoured swath charts of the sea floor to its maximum depth of about 12 km. The widths of the swaths range from two thirds to twice the water depth at any given place. The multibeam sonar picks up most features with a relief of 10 m or more. If coupled with good navigation, Seabeam can map about 1,200 km^2 of sea floor per day with a contour interval of 10 to 20 m. However, most of the long faults, fractures, and fissures that characterize midocean ridges, for example, have reliefs of less than 10 m. As a consequence, to study the tectonics of a given segment of ocean floor, Seabeam and sonography must be used together.

The U. S. Navy has collected bathymetric data for decades, but until recently most of these data were classified and not available to the scientific community. A full discussion of the evolution and use of available bathymetric data in earlier years (1960s and 1970s) is presented subsequently in this chapter in "Ocean Floor Bathymetry". Smoot and Meyerhoff (1995) give detailed accounts of early developments in bathymetric data collection and interpretation and show how bathymetric data were forced to fit preconceived configurations.

2.2.4 Seismotomography

Technologies other than those developed for deep-ocean floor exploration have also produced extremely important data sets for developing a new hypothesis of global dynamics. One of the most useful---and certainly one of the most exciting---new techniques developed in this century is the technique of seismic tomography, or seismotomography. This method utilizes teleseismic data to produce three-dimensional images of the Earth's interior. The velocities of P and S waves passing through the Earth's interior can now be measured by more than 1,000 recording stations at the Earth's surface (Woodhouse and Dziewonski, 1984; Dziewonski and Woodhouse, 1987; Anderson et al., 1992a). The velocity information is digitized along each ray path and then adjusted in position to match the corresponding velocities recorded on intersecting ray paths. Thus three-dimensional images of the Earth's interior are constructed. These images actually show within fairly narrow limits of error the distribution of high- to low-velocity zones, or bodies, inside the Earth. The high-velocity zones, or bodies, are assumed to be "cold," whereas those having a low velocity are assumed to be "warm." In this way, seismotomography reveals the overall thermal structure of the Earth's interior.

Seismotomographic images can be prepared for a wide range of scales. One may, for example, examine only a selected part of the Earth's interior, as demonstrated by Liu Futian et al. (1989), Liu Jianhua et al. (1989), Sun Ruomei et al. (1991), and Anderson et al. (1992a). Liu Jianhua et al. (1989) and Liu Futian et al. (1989) mapped the upper 1,100 km of the Earth beneath all of China and showed its thermal structure in great detail. Seismotomographic studies have produced results that are largely unexplained by plate tectonics and, in fact, the findings to date were unexpected (Anderson et al., 1992a).

2.2.5 Space Geodesy

The term "space geodesy" refers to techniques developed for making precise position measurements among sites on the Earth's surface that may be separated by distances up to 12,000 km. Four of these techniques are VLBI (Very Long Baseline Interferometry), SLR (Satellite Laser Ranging), GPS (Global Positioning System), and DORIS (Doppler Orbitography and Radiopositioning Integrated by Satellite). Laser ranging to the moon (LRM) is a variety of SLR. All of the techniques are based on developments in space research, especially radio astronomy and satellite tracking (Cazenave et al., 1992; Gordon and Stein, 1992). One stated purpose of these techniques is to determine the velocities and movement directions of the plates of plate tectonics.

The VLBI technique depends on the precise timing of radio noise from compact extragalactic sources (e.g., quasars) recorded by a network of radio telescopes. Analysis of the noise permits estimation of the location of a particular radio telescope site in the essentially inertial frame defined by the radio-telescope noise sources. In a typical VLBI geodetic experiment, two to seven radio telescopes record radio noise from one source at a time, ultimately recording noise from 5 to 15 sources. The noise recorded at different times at one site is correlated with the noise records from other sites. On the basis of the correlations, delays and delay rates among the radio telescopes can be determined, as well as site locations, the Earth's orientation, and the azimuths to the radio sources. Repeat experiments allow determinations of the relative motions of each site.

SLR is based on the round-trip times of laser pulses reflected off a satellite, especially the LAGEOS (LAser GEOdynamics Satellite) launched in 1976. This satellite orbits at approximately 5,900 km. The distance (range) to the satellite is half of the product of the speed of light and the total elapsed time between emission of a laser pulse and its reception time at the same tracking site. The distance calculations from many sites define the satellite's trajectory as well as the three-dimensional framework of the tracking stations. Repeated experiments define the sense of motion of each tracking station. Because the satellite orbit moves with the Earth's center of mass, site motions are referred to a dynamically defined and nearly inertial reference center that moves with the center of mass.

GPS uses a "constellation" of several NAVSTAR (NAVigation Satellite Time And Ranging) satellites which are 20,000 km above the Earth's surface in 12-hr orbits. Each satellite broadcasts a message giving the time and the satellite's position. The distance between the satellite and a GPS receiver is calculated from the transit time of the message from the satellite to the receiver. Because multiple satellites are tracked simultaneously, the location of the receiver can be calculated to within a few meters by trilateration. Because most of the sources of error affect all elements of the GPS system at the same times, analysis of the data gives results with a precision of a few centimeters. The resulting position changes are estimated to be sufficiently precise that, within a few years of the startup of a GPS system, tectonic motions should be detectable. This method is less expensive than the others and is being used increasingly.

DORIS utilizes the Doppler effect, which is a widely used method of tracking satellites. The system is based on a worldwide network of ground beacons transmitting on two frequencies, a satellite equipped with a Doppler receiver, and a ground control system that controls and monitors the satellite's orbit. DORIS is one of several new or modified space-geodesy systems now in the planning or experimental phase.

We noted above that one of the stated purposes of the space-geodetic systems is to determine the velocities and directions of movement of the plates postulated in plate tectonics. Despite incautious claims to the contrary (e.g., Argus and Gordon, 1990; Gordon and Stein, 1992), movements of large continental blocks have not been proved (e.g., Lowman, 1991). In fact, most of the space-geodetic results have shown no definite pattern and, in fact, have been very contradictory (e.g., Ma et al., 1989). Some authors have proposed a variety of *ad hoc* hypotheses to explain the contradictions, especially those that do not confirm the widely accepted Minster and Jordan (1978) model of plate motions (e.g., Harrison and Douglas, 1990; Fallon and Dillinger, 1992; Gordon and Stein, 1992). In one example, after finding no tensile motions associated with the East African rift, the investigators felt forced to postulate that plate motions there must be episodic (Asfaw et al., 1992). In contrast, Gordon and Stein (1992) maintain that plate motions are steady if ". . .averaged over a few years. . .." Space geodesy should prove to be an extremely useful tool for several tasks, including the monitoring of lateral and vertical motions predicted in tectonic models.

2.2.6 Satellite Photography

Satellite photography has enabled scientists to see the component parts of the solar system in ways never dreamed of only a few decades ago. The cameras and other instruments aboard satellites have provided whole new data sets for study, not just of the Earth but also other planets and their satellites.

Photographs of the Earth, for example, have revealed the presence in all tectonic belts, especially the active ones, of belt-parallel systems of faults, fractures, and fissures essentially identical with those found with sonographs of the midocean ridges. Identical belt-parallel fault-fracture-fissure systems are present on both active rift systems, such as the East African Rift (e.g., Mohr, 1976) and the Dead Sea Rift (Foose, 1985), and foldbelts, such as the California Coast Ranges (Johnson and Page, 1976) and the Alps, Apennines, and Dinaric Alps (Foose, 1977, 1985).

Nearly identical belt-parallel fault-fracture-fissure systems have been observed by the authors in photographs taken of Mars and its moon, Phobos; of three Jovian moons, Europa, Ganymede, and Io; of the Saturnian moons, Enceladus and Tethys; and of the Uranian moons, Ariel, Miranda, Oberon, and Titania.

2.2.7 Satellite Radar Altimetry

Satellite radar altimetry was intended to measure the shape of the ocean's surface. However, in 1973 when this radar altimetry was tried from the Skylab, it was discovered that this method is also able to map the ocean floors in addition to ocean surface, currents, eddies, tides and fronts (Lowman, 1994). Satellite radar altimetry depends on microwaves that do not penetrate an electrical conductor, and radar return is from the surface of the water. The mean sea surface over areas of mass deficiency shows a depression of a few meters and over areas of mass excess shows elevations of the same magnitude. Thus mean sea surface or geoid is a subdued replica of the ocean-floor topography. The principle behind this is that the positive topography tends to pull the ocean toward it into depressions and the surrounding negative topography results in elevation of the sea surface.

After an initial experiment of Radar Altimetry aboard the Skylab in 1973, the first satellite radar altimetry survey was conducted in 1978 by Seasat. The Seasat mission lasted about four months. The collected data were used to map the physical geoid, bathymetric features of the ocean, and gravity of the world. The maps generated from this initial Seasat radar altimetry data (e.g., Haxby, 1987) reveal many previously unmapped features (seamounts, guyots, ridges and fracture zones) and a bathymetry strikingly different than previously published bathymetric maps (e.g., Heezen and Tharp, 1977; GEBCO, 1978-1982; National Geographic, 1992).

Satellite Radar Altimetry surveys have been carried out by other satellites: U. S. Navy GEOSAT (1985-1990), European Earth Remote Sensing Satellite (1991-Present), and Japanese Earth Resources Satellite (1992-Present). All U. S. Navy GEOSAT data with ground-track spacing averaging 4 km have recently been declassified.

2.2.8 Radar Mapping of Venus

Because of Venus's perpetual cloud cover, an approach other than satellite photography had to be developed for mapping the surface of that planet. A low- to medium-resolution radar altimeter was developed for the Pioneer Venus orbiter mission begun in 1978 (Pettengill, 1977; Pettengill et al., 1979). The maps from that mission were published shortly afterward (U. S. Geological Survey, 1981, 1984). Subsequently, scientists of the former U.S.S.R. mapped parts of the Venusian surface with side-looking, synthetic-aperture radar during late 1983 and early 1984 (the Venera-15 and 16 missions; Barsukov et al., 1986; Sukhanov et al., 1989; U. S./U.S.S.R. Joint Working Group, 1989). Resolution of surface features was in the 1- to 3-km range and many individual structures were clearly recognizable.

High-resolution mapping of Venus with the Magellan spacecraft began in September 1990. For the first time, very detailed information was acquired from the surface of a planet very nearly the same size as the Earth. Because of the many physical similarities between Venus and the Earth, it had been confidently expected that the surface of Venus would reflect clearly the same plate-tectonic features that had been interpreted on the Earth. Instead, a very different tectonic style appeared to have been discovered in which plate tectonics plays no obvious role (Solomon et al., 1991). Instead, the tectonic style is very much like that described for the Earth in this book, namely, the style of surge tectonics.

During each day of operations, Magellan acquires 7.3 orbits of imagery. Each orbit records a strip 20 km wide that extends 17,000 km from the Venusian north pole to below 70° south latitude. The near-polar orbit (85.5° inclination) is elliptical, ranging in elevation above the planet's surface from 294 km at the periapsis (which is at 9.9° N) to 2,225 km at the apoapsis (north polar region). Resolution of individual surface features ranges from 120 m to about 300 m, depending on the elevation of the spacecraft (Pettengill et al., 1991, Tyler et al., 1991).

Magellan carries only one scientific instrument, a 12.6-cm-wavelength radar system that performs three operations. The first is the taking of radar images through a side-looking synthetic aperture; the second is nadir-directed altimetry (a near-vertical incidence angle); and the third is thermal emission radiometry. The side-looking synthetic aperture radar equipment and the altimeter are active sensors that measure the planet's surface reflectivity in oblique and near-vertical views respectively. Thermal emissions from Venus at microwave frequencies are measured by sampling the radar receiver output at times when there are no radar echoes present.

The radar operates through a narrow beam-width high-gain antenna whose angles of incidence (from a normal to the Venusian surface) range from about 15° near the pole to 50° at periapsis. Altimetry is mapped by brief diversions of the radar energy to a broad beam-width horn antenna that is aimed roughly toward the nadir.

The radar images obtained of the Venusian surface provided many surprises to Magellan's audience, but were not a surprise to the authors of this book. They so closely resemble the surge channels of our surge-tectonics hypothesis that we consider the radargraphs (radar images) to be a dramatic confirmation of the surge-tectonic hypothesis.

2.2.9 Other Techniques

Of the many techniques used in geology, none has been more effective than field geology, or "walking the outcrop." Yet for many decades, the man in the field has had to map large areas of igneous and metamorphic rocks whose protoliths and ages were a total mystery. Both of these seemingly insuperable difficulties have been partly resolved with the development of new techniques in geochronology and the acquisition of voluminous data on the geochemistry of igneous and metamorphic rocks. It is now possible to obtain at least minimum ages on many rock types, and geochemical comparisons between unmetamorphosed and metamorphosed rock types have revealed the protoliths of many metamorphic rocks. However, many geochronological data conflict with one another and with other data so that such results must be regarded cautiously. Moreover, not all geochemical analyses give a unique protolith solution, a fact that has not always been appreciated. Despite their shortcomings, these methods---and the very fact of their existence---have made the field geologist's task infinitely more rewarding. The once widespread assumption that all metamorphic rocks are Precambrian is now known to be as erroneous as the equally widespread assumption that all Precambrian rocks are metamorphosed. The once seemingly ubiquitous "Lipalian Interval" no longer enjoys even the status of a ghost.

2.3 Data Sets Unexplained in Current Tectonic Models: Foundation for a New Hypothesis

2.3.1 Linear Structures

Sonographs of the midocean ridges reveal the presence everywhere of long, linear, ridge-parallel faults, fractures, and fissures. The ridge-parallel sets of faults, fractures, and fissures are not restricted to the crestal regions of the ridges, but extend down the ridge flanks to levels where the sediments of the adjacent abyssal-plain basins lap onto the ridges (Laughton and Searle, 1979). Because several of the midocean ridges extend into adjacent continents (e.g., East Pacific Rise; Heezen, 1960; Meyerhoff et al., 1992a, 1992b), we extended our study of the ridge-parallel fault, fracture, and fissure systems to embrace all tectonic belts within the continental regions.

So that there will be no misunderstanding, it is necessary to define here our use of the term *tectonic belt*. In general, a tectonic belt is *any* structural megafeature developed at the Earth's surface above what we call a surge channel. Thus a tectonic belt includes the full spectrum of linear tectonic features known on Earth. In continental regions, these include continental rifts, strike-slip fault zones, germanotype and alpinotype foldbelts, and continental volcanic arcs. They also include such linear cross-strike features as the Colorado Mineral Belt and the Lower Yangzi Valley plutonic-volcanic belt. In oceanic regions, tectonic belts include midocean ridges, "aseismic ridges," linear island and seamount chains, and oceanic island arcs.

Tectonic belt-parallel systems of faults, fractures, and fissures were found in every tectonic belt examined, whether a continental rift, a strike-slip fault zone, or a foldbelt. Examples include the Western Cordillera of the United States (Fig. 2.1; Stewart, 1978) and, at a smaller scale within the same tectonic province, the California Coast Ranges-San Andreas fault zone (Fig. 2.2; Johnson and Page, 1976). Other examples are the East African Rift system

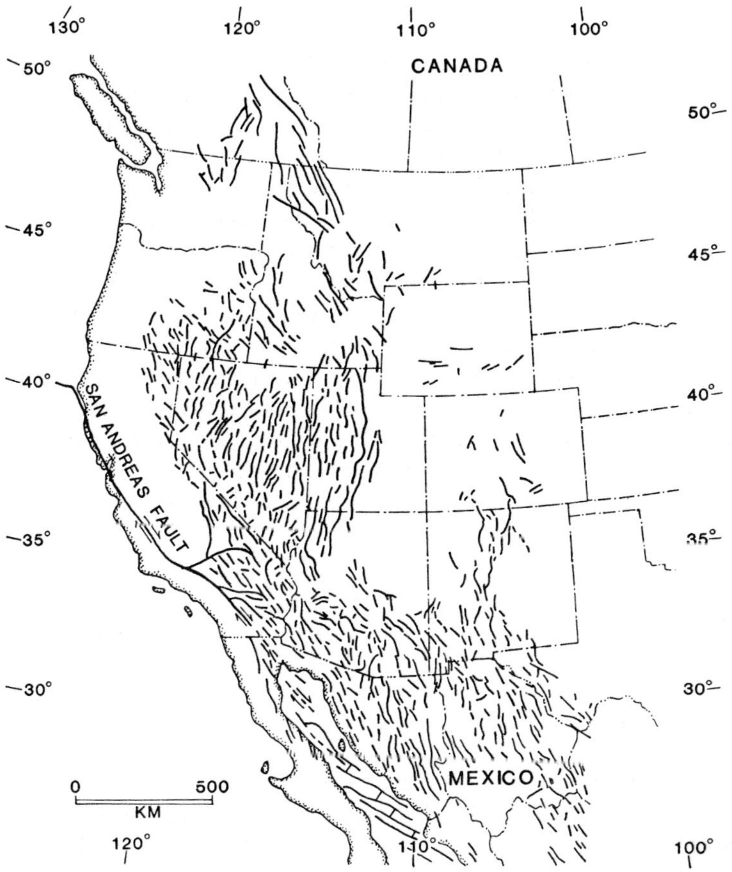

Fig. 2.1. Fault pattern of part of the Western Cordillera. Shown are normal and strike-slip faults. This pattern, following Stoke's Law, shows that the predominant flow direction beneath the Western Cordillera of the U.S. is north-south to north-northwest-south-southeast. Compare with Figures 2.2, 2.3, 2.4, 2.5, 2.6, 3.28 and 3.34. From Stewart (1978).

(Fig. 2.3; Mohr, 1976, 1987), the Rhine Graben (Fig. 2.4; Illies and Greiner, 1979), the Front Range of New Mexico, Colorado, and Wyoming (Fig. 2.5; Eaton, 1986), and the Reelfoot Graben beneath the Mississippi Embayment (Figs. 2.6, 2.7; Fletcher et al., 1978; Hildenbrand, 1982). Together, these systems involve a huge body of data that are not well explained in plate tectonics, and with rare exceptions, have not been addressed. The fact that faults, fractures, and fissures parallel the strike of each tectonic belt indicates, as a simple consequence of Stoke's Law (see Appendix), that each of these belts has been, or is underlain by a mobile body that moves parallel with the tectonic belt. Thus the primary motions producing these systems of faults, fractures, and fissures are not at right angles to the tectonic belts (e.g., Cox and Hart, 1986).

Linear evaporite trends and many types of linear basins originate in half-grabens, grabens, and compression-produced topographic (synclinal) lows, and generally are explained as a consequence of tension or compression. However, all linear basins and all oval basins (e.g.,

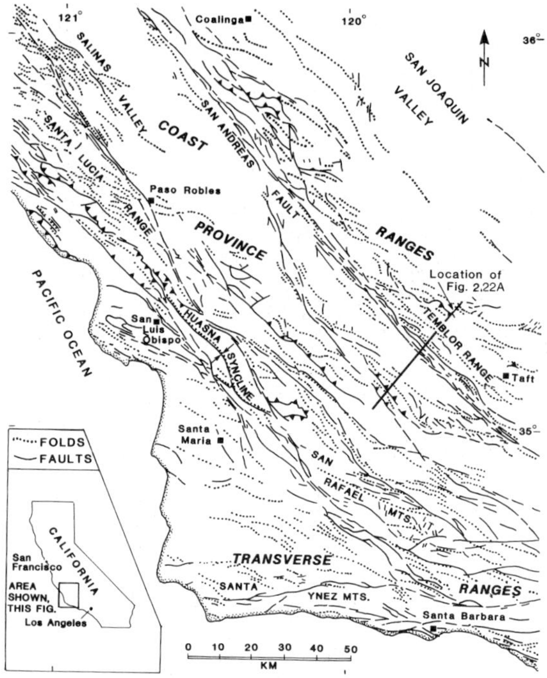

Fig. 2.2. Map of part of the California Coast Ranges showing the north-northwest-south-southeast trends of the faults and fold axes. Following Stoke's Law, this map shows that flow beneath the region shown on the map parallels the fault and fold-axis directions. Compare this fault and fold-axis pattern with the patterns shown on Figures 2.1, 2.3, 2.4, 2.5, 2.6, 3.28, and 3.34. Note that the fault and flow-axis pattern directly overlies the postulated surge channel of Figure 2.20. From Johnson and Page (1976).

Paris basin, Williston basin, Illinois basin, Moscow basin, Sichuan basin), both on cratons/platforms and in less stable regions such as rifts and foldbelts, are underlain by lenses of 7.0-7.8 km/s material.

Linear valleys and mountain systems commonly can be explained as inherited from the strike of underlying older structures. Mountains that are transverse to regional structure, however, pose bigger problems (e.g., the California Transverse Ranges, the Uinta Mountains of the Rocky Mountains, the Wichita and Arbuckle Mountains of the United States Great Plains). Similarly, long, straight river courses across regional strike do not always have an obvious explanation. Examples include the lower courses of the Mississippi River (Mississippi Embayment), the St. Lawrence River, and the Yangzi River. These linear to curvilinear valleys are underlain by lenses of 7.0 to 7.8 km/s material at the Mohorovičić discontinuity.

2.3.2 Lithosphere Diapirs and Lithosphere Magma Chambers

Ever since the publication of Wegmann's (1930, 1935) pioneer papers on the topic, mantle diapirism has been invoked increasingly as a mechanism for generating or promoting

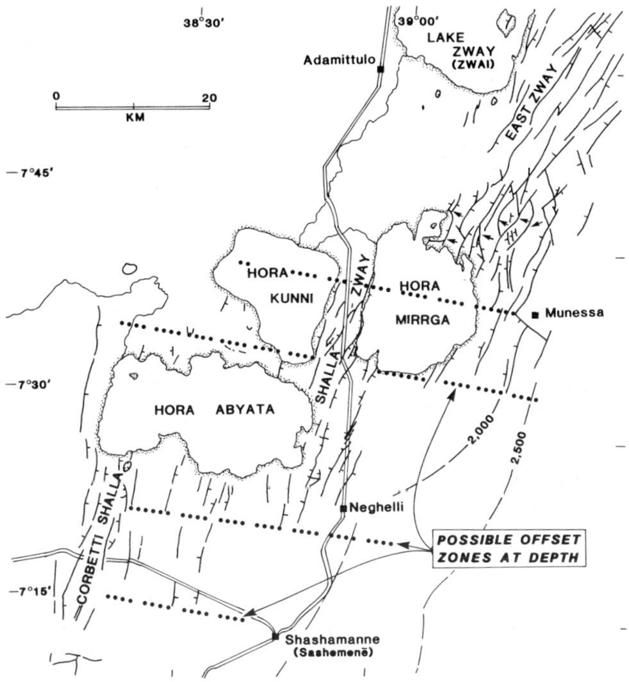

Fig. 2.3. East African rift valley, Ethiopia. Note the linear faults, fractures, and fissures parallel with the north-northeast-striking rift. These form belts of fractures that, along strike, are offset (offsets are indicated by dotted lines at right angles to the rift valley). We call these fracture-band offsets. In our opinion, they originate in the same way as offsets of ridge-parallel fractures (Fig. 3.28) along the ridge-transverse fracture zones (transform faults) that cut at approximately right angles across the midocean ridges (see Meyerhoff et al., 1992a), forming one end member of a series of related structures--fracture band offsets (=offsets along transform faults), channel overlaps (overlapping spreading centers), eddylike structures and vortex structures (these have no name in plate tectonics). Figures 2.4, 3.9, 3.11, 3.34, and 4.21-4.22 illustrate in part this gradational series. Modified from Mohr (1987).

tectogenesis. Van Bemmelen (1933) and Glangeaud (1957, 1959), for example, favored mantle diapirism and subsequent lateral sliding and/or compression for creating the structures of the Mediterranean Sea region. Mantle-diapirism hypotheses have found favor at different times with many geologists (e.g., Maxwell, 1968) for explaining the structural evolution of the Mediterranean belt, and indeed still do (e.g., Krebs, 1975; Locardi, 1988).

The evidence adduced for extensive lithosphere diapirism is now formidable (Beloussov, 1980, 1981, 1989, 1992). Shallow magma chambers are ubiquitous beneath active tectonic belts, whether they be rift zones, streamline (strike-slip) fault zones, or foldbelts. Some rift-valley examples of shallow magma chambers or diapirs include the East African Rift system (Tongue et al., 1992), the Red Sea Graben (Gaulier et al., 1988), the Rhine Graben (Jacoby, 1984), the Baykal Rift (Krylov et al., 1979; Sychev, 1985), the Rio Grande Rift (Sanford and Einarsson, 1982), Iceland (Sanford and Einarsson, 1982), the Hetao-Yinchuan Graben (Liu Jianhua et al., 1989), the Fen Wei (Wei He) Graben (Liu Jianhua et al., 1989), and many, many more. Streamline (strike-slip) fault zone examples include the San Andreas Fault zone (Lachenbruch and Sass, 1980), the Dead Sea Fault zone (Bayer et al., 1989), the Alpine Fault

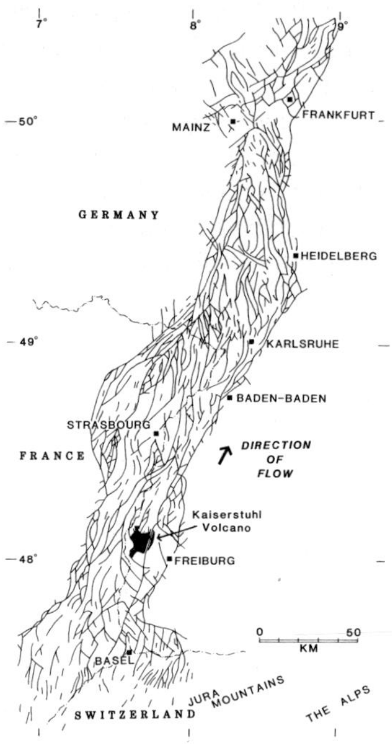

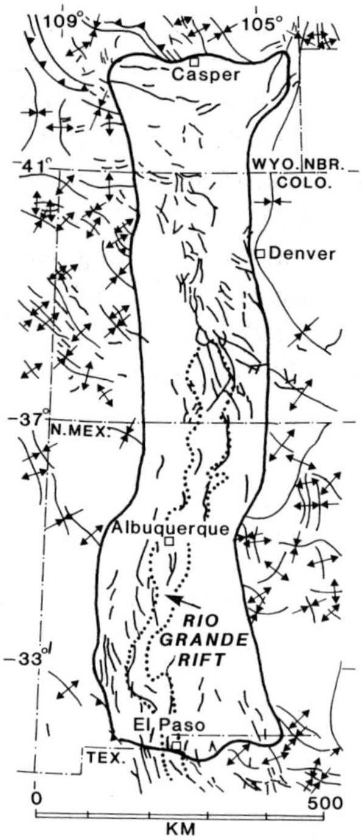

Fig. 2.4. (Left) Fault and fracture patterns of the Rhine graben, France and Germany. From Illies and Greiner (1979). Horsetail patterns point predominantly from south-southwest toward the north-northeast in the direction of flow within the underlying surge channel.

Fig. 2.5. (Right) Southern Rocky Mountains structures superimposed on Laramide structures (after Eaton, 1986). The Southern Rocky Mountains structures strike north-south unlike the Late Cretaceous-middle to late Eocene Laramide structures that strike northwest-southeast. The Southern Rockies uplifts are largely Oligocene and younger. One purpose of this figure is to show the important role of surge channels in creating and sustaining both uplifts and depressions. In this case, one surge channel simultaneously created (1) the north-south uplift that extends from El Paso, Texas, to Casper, Wyoming, and (2) the Rio Grande rift.

(Pandey, 1991), the Queen Charlotte Fault zone (Dehler and Clowes, 1988), and many more. One problem with finding examples of magma chambers at shallow depths along streamline (strike-slip) fault zones is that these zones have not been studied in the same way as rifts and foldbelts. Hence the discovery of shallow melt, or potential melt, zones beneath streamline fault zones has been largely serendipitous. Examples of shallow diapirs beneath foldbelts are also abundant. A few examples include the California Coast Ranges (Blümling and Prodehl, 1983; Blümling et al., 1985), the Alps (Miller et al., 1982; Mueller, 1983), the Dinaric Alps (Dragašević, 1974), the Himalaya and Qinghai-Xizang (Tibet) Plateau (Wu Gongjian et al., 1991), the Yunnan Himalaya (Liu Jianhua et al., 1989), the Japan Arc (Kushiro, 1987), and the Pyrenees (Velasque et al., 1989).

Almost since the beginning of the plate tectonics era, geophysicists such as Lliboutry (1971), Bonini et al. (1973), and many others have pointed out the important role that diapirism must play in any scheme of Earth dynamics. Despite this, mantle diapirism and related upwelling processes received little consideration *as intrinsic parts of plate tectonics* until Dewey (1988a) recognized their possible importance throughout the Alpide-Mediterranean and parts of the Circum-Pacific tectonic belts. Dewey's (1988a) explanations, however, do not account for coexisting states of compression and tension, as the field data from many areas required (e.g., Alboran Sea; Platt and Vissers, 1989). In contrast, surge tectonics *requires* the simultaneous formation of side-by-side compressional and tensile regimes during tectogenesis. Table 2.1, based on random sampling of some recent literature, shows how widespread the idea of mantle diapirism and upwelling has become. More than 50% of the examples listed are alpinotype or germanotype foldbelts; the remainder are tensile belts. Our point is that, whereas mantle diapirism may have a place in the tensile regimes of plate tectonics, it cannot be accommodated in the compressional regimes.

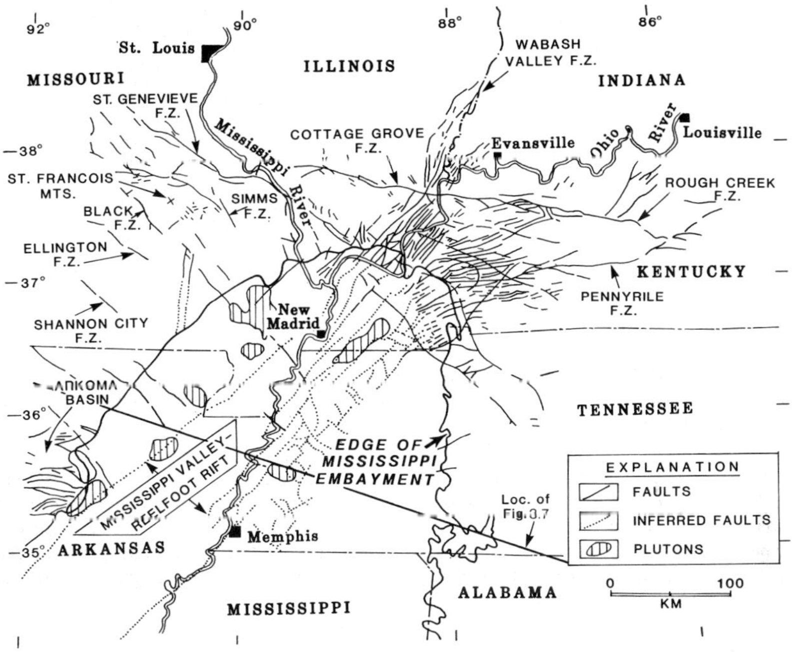

Fig. 2.6. Upper Mississippi embayment. This map shows the linear fault, fracture, and fissure system mapped in the subsurface and at the surface. Note that Stoke's Law again applies. The principal fault zones are shown by name. The fault patterns suggest that a surge channel beneath the upper Mississippi embayment trifurcates, with one branch following the Simms-St. Genevieve fault zone, another the Wabash Valley fault zone, and the last the Pennyrile and Rough Creek fault zones. The positions of several subsurface plutons are also shown. Compiled from many sources (see Figure 2.7).

2.3.3 MAGMA CHAMBER-RELATED PHENOMENA

Lithosphere magma chambers and related asthenosphere upwellings form zones of reduced seismic velocity, the low-velocity zones of the literature. Commonly a large magma chamber forms close to the mantle-crust boundary, followed by the formation at still higher levels in the middle to upper crust of smaller magma chambers whose sizes decrease upward, thereby forming a "Christmas Tree" structure as described by Corry (1988) for sill complexes in the

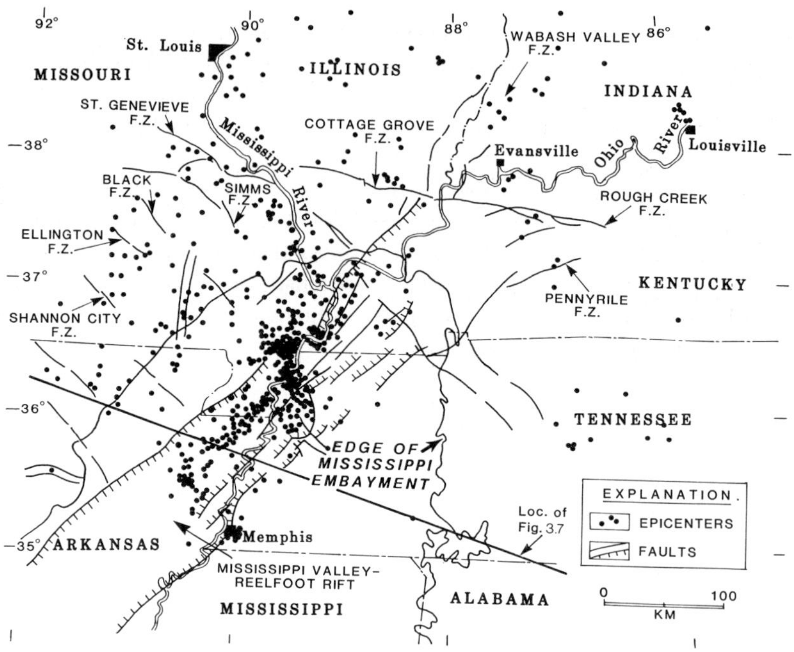

Fig. 2.7. Upper Mississippi embayment. The epicenters are plotted with relation to the principal fault zones. Site of the New Madrid earthquakes of 1811-1812 is shown. Compiled from Fletcher et al. (1978), Hildenbrand (1982), and Hinze and Braile (1988). The purpose of this illustration and Figure 2.6 is to show that the long linear Mississippi Valley's course and that of the associated Mississippi embayment are controlled by a surge-channel system. In short, surge channels generate topographic depressions, or basins.

upper crust (see Fig. 2.8). The large magma chamber close to the mantle-crust boundary is pod-shaped (see Fig. 2.9), and is referred to in the literature by various names---lenses, lenticles, lozenges, pillows, rift pillows, pods, shear pods, anastomosing networks of shear zones, and so forth---terms that show the lack of knowledge of their origin (s). These lenses, for many years, have been termed layers of "anomalous upper mantle" or, conversely, "anomalous lower crust." Although they have been observed most commonly at the mantle-crust boundary, such lenses do occur in some tectonic belts in the middle to upper crust (Finlayson et al., 1989; Thompson and McCarthy, 1990; Figs. 2.10, 2.11).

The lens at the mantle-crust boundary typically has a P-wave velocity in the 7.0-7.8 km/s range (Revelle, 1958; Ewing and Ewing, 1959; Le Pichon et al., 1965; Taylor et al., 1980; Mooney et al., 1983). Many of them contain a low-velocity zone (5.4-6.6 km/s) near their centers (Prodehl, 1970, 1979). Beneath continents and many parts of the ocean basins, these lenses are typically between 100 and 500 km wide, most commonly in the 150-250-km range (e.g., Taylor et al., 1980; Mooney et al., 1983). Where not present at the mantle-crust boundary, they pinch out laterally into a thin but nearly omnipresent zone with a velocity range of 6.9 to 7.9 km/s (Mooney and Brocher, 1987; Mooney and Braile, 1989).

An identical but much larger lens occupies the crust-mantle boundary zone beneath midocean ridges (Ewing and Ewing, 1959; Le Pichon et al., 1965), where they were first discovered by Revelle (1958). Here, beneath the midocean ridges, the lenses are typically 1,000 to 3,000 km across and they occupy the full 65,000-km length of the midocean ridge system.

Because these lenses pinch out laterally from the centers of the midocean ridges, they

Table 2.1. Random sample of literature proposing mantle diapirs. From Meyerhoff et al. (1992b).

Type of mobile belt Name of mobile belt	*Author(s) (Year)*	*Diapirs inferred mainly on basis of:* Geology	Geophysics	Both	Theory
Alpinotype Foldbelt					
Yenisey-Sayan-Baykal	Sheynmann (1968)	x			
Rif, Betic Cordillera, Alboran Sea	Bonini et al. (1973)		x		
Northern Appalachians	Ramberg (1973)	x			x
California Coast Ranges	Maxwell (1974)	x			
Alps	Smith and Woodcock (1982)				x
Massif Central (France)	Nicolas et al. (1987)			x	
Rhenish Massif (Germany)	Witt and Seck (1987)			x	
Alps to the Himalaya	Dewey (1988)				x
Apennines, Tyrrhenian Sea	Locardi (1988)			x	
Carpathians, Pannonian Basin	Ádam et al. (1989)		x		
Western Cordillera of North America	Hamilton (1989)				x
Honshu (Japan)	Hirahara et al. (1989)		x		
Rif, Betic Cordillera, Alboran Sea	Platt and Vissers (1989)	x			
Pyrenees	Valasque et al. (1989				x
Continental and Oceanic Rifts					
Mid-Atlantic Ridge	Ramberg (1973)		x		x
Zabargad Island (Red Sea)	Bonatti et al. (1981)	x			
Red Sea	Nicolas (1987)	x			
Oman, Red Sea	Boudier and Nicolas (1988)	x			
Gulf of Aden, Red Sea	Isaev (1987)			x	
Base of continental slope, west of Iberian Peninsula	Boillot et al. (1989)			x	
Rhine Graben	Clauss et al. (1989)		x		
Iceland	White (1989)		x		
East Pacific Rise	Winterer et al. (1989)		x		
Aseismic Ridges					
Hawaiian Ridge, Cape Verde Rise	White (1989)		x		
Theoretical Models					
Foldbelts and rifts	Ramberg (1973)				x
Foldbelts and rifts	Beloussov (1974)				x
Cratonal basins	Artyushkov and Baer (1986)				x
Island arcs and rifts	Turcotte (1989)				x
Flood basalt, backarc basins	White and McKenzie (1989)				x

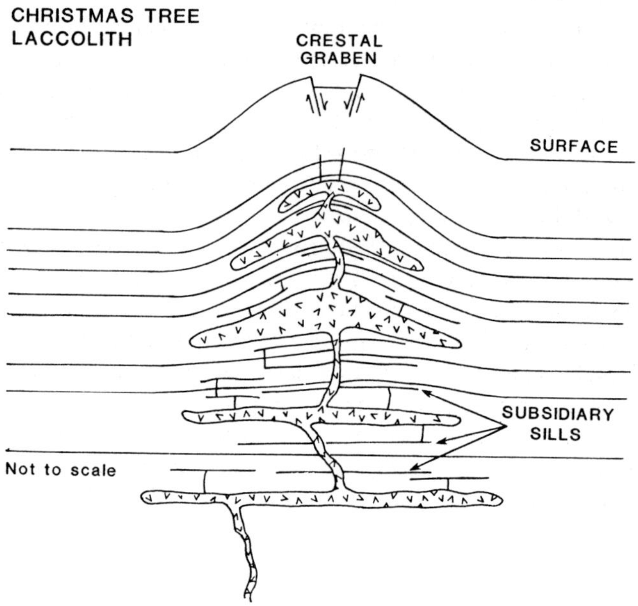

Fig. 2.8. Corry (1988) published this figure, and called it a "Christmas tree laccolith." It differs from Bridgewater et al.'s (1974) model in that the Christmas tree structure is a series of laccoliths, one above the other.

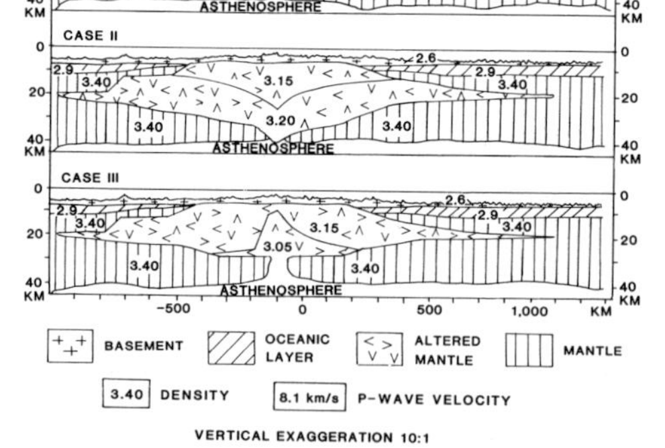

Fig. 2.9. Three possible crustal models across the Mid-Atlantic Ridge (between 27° and 47° N latitude). All three satisfy the gravity data--at the top of the figure--and the seismic-refraction data (Le Pichon et al., 1965). The numbers in the 2.5 to 3.4 range are density values; numbers 7.3 and 8.1 are P-wave velocity values. We regard this figure as a "type locality" for midocean-ridge surge channels. After Talwani et al. (1965).

were at first perceived as an obstacle to the newly formulated hypothesis of sea-floor spreading (Hess, 1962; Dietz, 1961), but were soon provided with an explanation conforming to plate tectonic models. The problem, as it was perceived, was that, if the "anomalous mantle" lens formed at the midocean ridge crests (as it had to do, in sea-floor spreading), then some process had to remove the 7.0-7.8-km/s material as the oceanic crust moved away from the midocean ridge crest toward its laterally coeval and subparallel subduction zones. Two speculative solutions to the problem were suggested and, to the best of our knowledge, were accepted without benefit of additional research.

The first solution was proposed by Drake and Nafe, 1968, p. 185): "Velocity-depth data indicate that velocities in the range 7.2-7.7 km/sec are almost completely absent in the deep ocean basins away from ridges or prominent seamounts and under the low-lying continental shields, but are present in all other regions to some degree. The material in this velocity range must be derived from the mantle but is of lower density than normal. If, as is suggested by the data, it is of a transient nature, its appearance and disappearance may be related to the changes in elevation associated with tectonic activity." Elsewhere in the same paper, Drake and Nafe (1968, p. 184) wrote that oceanic crustal layer 3 (the lowest oceanic crustal layer which overlies the Mohorovičić discontinuity) "...would receive permanent additions of rock with the properties

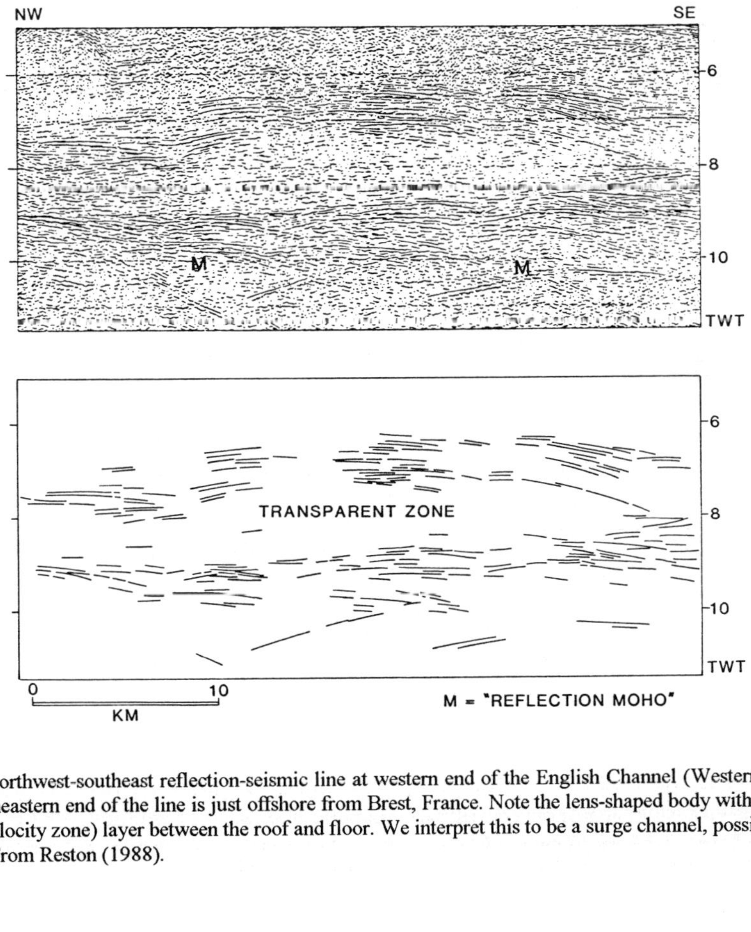

Fig. 2.10. Northwest-southeast reflection-seismic line at western end of the English Channel (Western Approaches basin); southeastern end of the line is just offshore from Brest, France. Note the lens-shaped body with a transparent layer (low-velocity zone) layer between the roof and floor. We interpret this to be a surge channel, possibly including some melt. From Reston (1988).

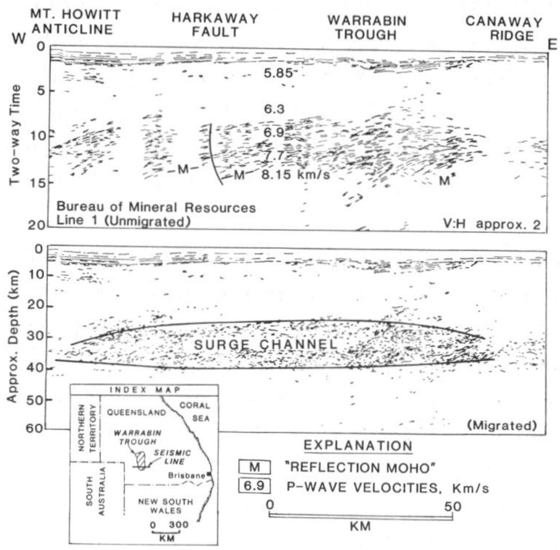

Fig. 2.11. West-east reflection seismic line (tracing) across the Warrabin trough, Eromanga basin, southwestern Queensland. P-wave velocities are shown. The lens, which we interpret to be a former surge channel of Paleozoic age, has reflectors throughout. Unlike in Figure 3.20, a transparent zone is lacking. This indicates that the contents of the channel have cooled and that the channel is no longer active. From Finlayson et al. (1989).

of gabbro, and a 7.2- to 7.7-km/sec layer would first develop and then vanish. In this view, the principal contribution of the 7.2- to 7.7-km/sec layer is to increase total thickness and, through isostatic adjustment, to increase surface elevation during the orogenic process, and then to disappear with an accompanying reduction in thickness and elevation."

Referring to the 7.2- to 7.7-km/s layer as low-velocity mantle, Vogt et al. (1969) wrote that, "The occurrence of low-velocity mantle under [the midocean ridge] crest could well be a steady-state phenomenon. That is, it may be constantly created under the axis and converted to normal mantle under the flanks" (Vogt et al., 1969, p. 593). In further explanation, Vogt et al. (1969, p. 593) wrote that the low-velocity mantle "...under the ridge axis is most likely an ultrabasic crystal slush through which basaltic fluids must rise to feed the growing layers 2 and 3.... This slush then probably solidifies and becomes 'normal' mantle as it withdraws from the axis." This second explanation is no more satisfactory than that proposed by Drake and Nafe (1968).

The problem has not been researched further, to the best of our knowledge, and remains unsolved. The problem is crucial, because these lenses are found under every type of tectonic belt. Under the continents, for example, long linear lenses of 7.0-7.8-km/s material underlie all rifts (e.g., Mooney et al., 1983), all streamline (strike-slip) fault zones (e.g., Mooney and Weaver, 1989), and all foldbelts (e.g., Drake and Nafe, 1968; Taylor et al., 1980; Marillier et al., 1989). Under the ocean basins, identical lenses underlie the midocean ridges (e.g., Ewing and Ewing, 1959; Le Pichon et al., 1965), linear island and seamount chains (e.g., Drake and Nafe, 1968), and other aseismic oceanic ridges (Goslin and Sibuet, 1975). *The lenses are a common denominator for all tectonic belts* and, therefore, cannot be transient features, as maintained by Drake and Nafe (1968). Nor can the material that forms them become "normal" mantle as it withdraws from the axis of each tectonic belt, as suggested by Vogt et al. (1969). In plate tectonics, the midocean ridges are the only tectonic belts from which rock materials (i.e., the new crust formed at the axes of midocean ridges) can withdraw. If Vogt et al. (1969) are right, then a second explanation must be developed to explain the presence of identical lenses in other types of tectonic belts.

In many foldbelts, the "anomalous" lenses have been deformed together with the shallower rocks (e.g., Figs. 2.13-2.15). Where a foldbelt has been found to be deformed bilaterally (i.e., the belt is bivergent), one side of the belt is said to have a zone of "backthrusts" (e.g., Coward and Smallwood, 1984; Mattauer and Collot, 1986), although few of the "backthrusts" exhibit the criteria for backthrusts (Rich, 1951). In this work, we demonstrate that *all* foldbelts are bilateral (i.e., bivergent), an observation made long ago by Kober (1925), Vening Meinesz (1934), and many others. We call these bivergent foldbelts *kobergens*, a concept that we define and explain in detail in the following chapter. Examination of our many figures illustrating bivergent foldbelts reveals at once that such features combine the effects of compression and tension (Fig. 2.15). Along the two flanks of the foldbelt are folds, thrusts, and nappes whose vergence on one flank is the opposite of that on the other flank. Between the two flanks is a zone of tension. Thus compression and tension act together, side by side, in belts hundreds, even thousands, of kilometers long. Consequently, the seemingly contradictory evidence for stress regimes noted by workers in foldbelts (e.g., Dewey, 1988a; Platt and Vissers, 1989) is not at all contradictory but is an inevitable consequence of tectogenesis. The literature on bivergent foldbelts dates at least to Suess (1885) and has increased steadily to the present (e.g., Platt and Vissers, 1989).

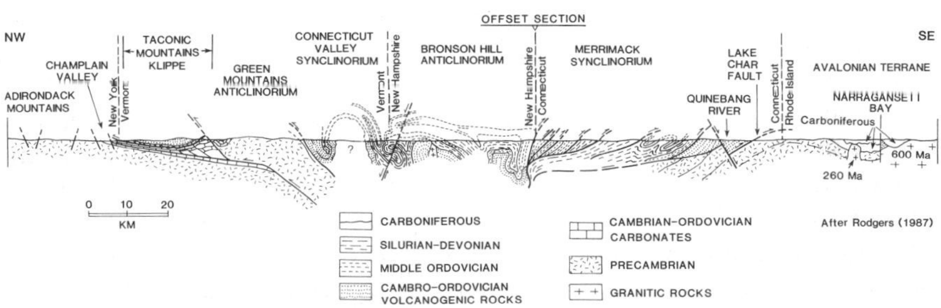

Fig. 2.12. Northwest-southeast structural cross section of the northern Appalachians, New England. This section is located very close to the Taylor et al. (1980) section in Figure 3.13. Note the bivergent symmetry centered on the gneiss domes of the Bronson Hill anticlinorium. Gneiss domes, consisting of metamorphosed igneous rocks--in large part granitic batholiths--are characteristic of the cores, or root zones, of alpinotype foldbelts. After Rodgers (1987).

2.3.4 Tectonostratigraphic Terranes

The kobergen and the lithotectonic (tectonostratigraphic) terrane concepts are inextricably intertwined. This is illustrated by the fact that where a bivergent foldbelt is present, one flank is assumed to be the zone of primary vergence, whereas the other flank is assumed to be a zone

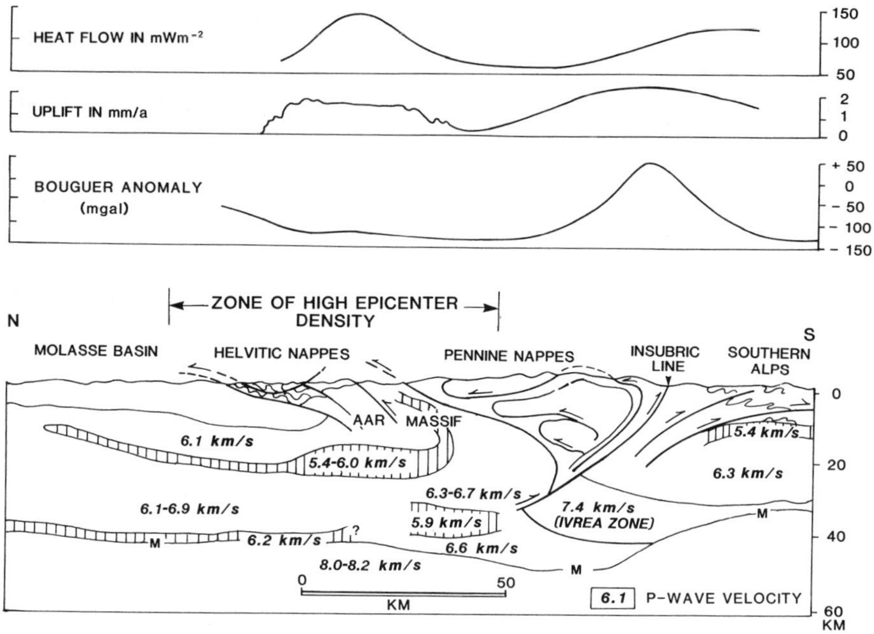

Fig. 2.13. North-south structural cross section of the kobergen of the western Alps. Based on surface geology and several types of geophysical studies. Heat flow, rates of uplift, and Bouguer gravity also are shown. Compare with Figure 2.15. The low-velocity zones shown are in the crust. Note that one is believed to crop out (Miller et al., 1982; Mueller, 1983). As we interpret the section, the Ivrea zone is part of a Late Jurassic-Eocene surge channel that was partly destroyed by Late Cretaceous-early Tertiary orogeny and split into two parts, one north and one south of the Insubric Line. The latter is a *Verschluckungszone*, whose origin is discussed in the text (Figs. 2.23-2.24 and 3.16-3.17). This section was compiled from data published by Cermák and Rybach (1979), Miller et al. (1982), Mueller (1983), Bayer et al. (1989) and Frei et al. (1989).

of back-thrusting (e.g., Dewey, 1988a). Therefore each major thrust-fault zone in every kobergen, regardless of the flank the thrust zone is on, separates one lithofacies assemblage (lithotectonic terrane) from another. Thus each thrust-fault zone that separates one lithotectonic terrane from another is of equal rank. This is not true of backthrusts (Rich, 1951).

Since Lory (1860) proposed the lithofacies belt concept, many descriptions of these Alpine belts have been published for the Jura, the Alps, the Carpathians, and the Dinaric Alps. A definitive and still indispensable paper was published by Bertrand (1897). Trümpy (1960) published another useful paper reviewing the entire topic of lithofacies belts in the central and western Alps. The concept seems to have reached North America via Brönnimann and Pardo (1956), who applied it successfully to the Cuban alpinotype belt. The thrust-bounded *facies structural zones* of Cuba have been the subject of a large literature (e.g., Khudoley, 1967; Khudoley and Meyerhoff, 1971; Pszczólkowski, 1987; Iturralde-Vinent, 1988). Wilson (1966, 1967) may have been the first to introduce the concept to the Western American Cordillera. The lithofacies-belt concept flourished there, where a large group of authors postulated that some of these thrust-bounded lithofacies belts (called tectonostratigraphic terranes or lithotectonic terranes) have been "rafted in" to the Western Cordillera for distances of many thousands of kilometers (e.g., Jones et al., 1977). Careful work by some geologists, however,

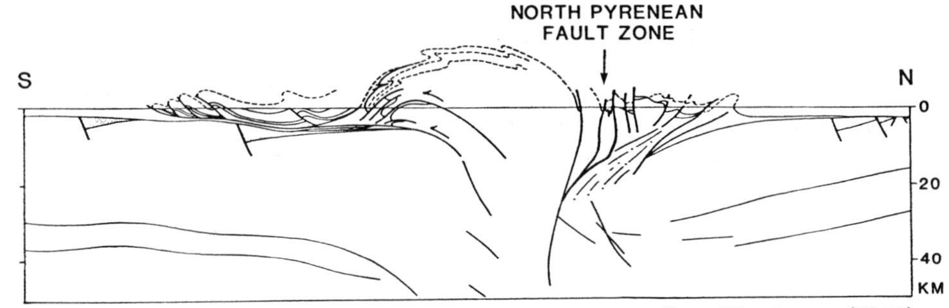

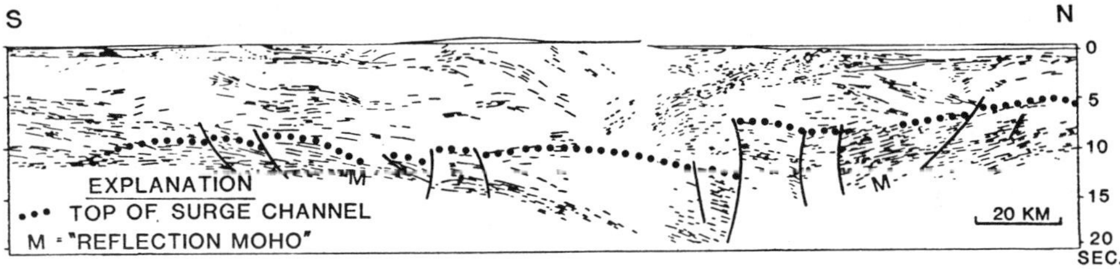

Fig. 2.14. South-north structural cross section of the Pyrenees, border between France and Spain. This section, although based on reflection-seismic data, is supported by detailed field mapping. The lower section shows the quality of the reflections. Note that the North Pyrenean fault (upper section) lies at the crest of a kobergen structure. The line of large dots in the lower profile outlines what we believe to be the surge channel lens beneath the Pyrenean kobergen. It has been strongly faulted (offsets up to 5 km); even the reflection Moho is offset.

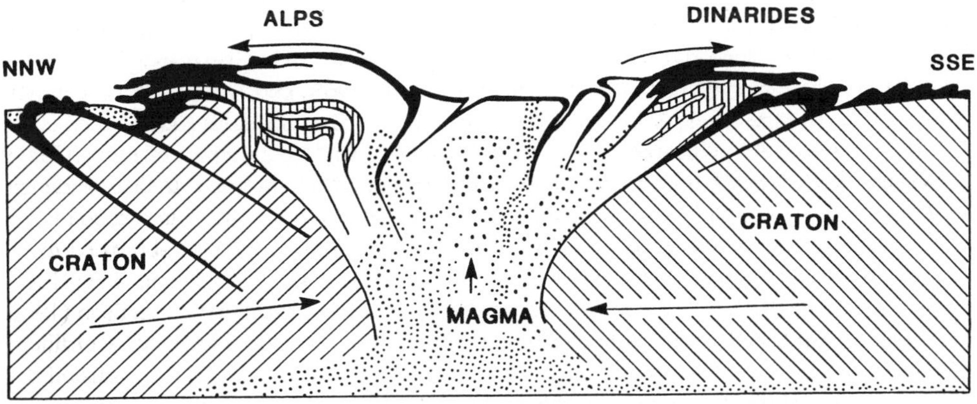

Fig. 2.15. Kober's (1925) mechanism--compression of a magma chamber (equivalent to surge channel)--for bivergent foldbelt deformation. This fan structure is what we call a kobergen and is the core of surge tectonics.

indicates that the lithofacies belts of the Western Cordillera are essentially in place and have not been transported long distances except possibly by thrusting (e.g., Saul, 1986; Hansen, 1988; Newton, 1988; Butler et al., 1989; Irving and Archibald, 1990; Storetvedt, 1990). An eloquent summary of this old European concept, including the reasons why the present lithotectonic terrane concept should be abandoned, was published by Şengör (1990).

2.3.5 Stretching Lineations

Many examples have been given in the literature of stretching lineations in foldbelts that *parallel* the foldbelt and of similar lineations that are *orthogonal* to it (Benn et al., 1992). Pêcher (1991), like Brunel (1986) before him, found during his studies in the Himalayas that stretch lineations in the fault gouges associated with thrust sheets parallel the direction of thrusting, whereas, away from fault gouges, both vertically and laterally, the stretch lineations parallel the trend of the tectonic belt. Other well known examples of belt parallel stretching lineations are found in the Dabie Shan (Suo and Zhou, 1992) and the Tian Shan (Makeyeva et al., 1992). Many workers have attempted to explain this apparent paradox by attributing the foldbelt-parallel lineations to wrench, or strike-slip faulting (e.g., Harland et al., 1992; Suo and Zhou, 1992). However, if one considers the foldbelt-parallel lineations to be the result of foldbelt-parallel flow beneath the tectonic belt and the foldbelt-orthogonal lineations to be younger, produced during the tectogenesis of the foldbelt, then the difficulty disappears. Makeyeva et al. (1992) also recognized the fact that subcrustal, belt parallel flow is a more likely cause. Thus the two sets of lineations are a direct consequence of Stoke's Law (Appendix). The fact should be noted that most stretch lineations in a foldbelt parallel the fault-fracture-fissure systems mapped from satellite photography. We return to this topic in our presentation of the surge-tectonics hypothesis.

2.3.6 Segmentation

Segmentation is a fundamental property of every type of tectonic belt, being present in midocean ridges (Lonsdale, 1982; Macdonald and Fox, 1983; White, 1984, 1989), linear island and seamount chains (Detrick and Watts, 1979; Lonsdale, 1988; Watts et al., 1988), other aseismic oceanic ridges (Detrick and Watts, 1979; Meyerhoff et al., 1992b), island arcs (Marsh and Carmichael, 1974; Beck and Christensen, 1991; Taylor et al., 1991), germanotype foldbelts (Hamilton, 1988), alpinotype foldbelts (Thompson et al., 1968; Leo, 1991), continental rifts (Mohr, 1987; Ebinger et al., 1989a, 1989b), streamline (strike-slip) fault zones (Allen et al., 1989; Rockwell, 1989; Kneupfer, 1989; Embley and Wilson, 1992), zones of normal faults (Crone and Haller, 1989; dePolo et al., 1989), and tectonic belts observed on several planets and some of their satellite moons (Yamaji, 1991). White (1984) referred to the segmentation on midocean ridges as *pinch-and-swell geometry*. The average lengths of the principal and most obvious midocean-ridge segments range from 50 to 300 km (Macdonald et al., 1988), with an average range of 80 to 200 km. However, detailed studies show that the range can be as great as 15 to 500 km. A thorough examination of several long ridge segments reveals the presence of at least two lower orders of segmentation (Macdonald et al., 1988; Winterer et al., 1989). Gordin (1989) has found at least one higher order of segmentation in the South Atlantic

Ocean and possibly elsewhere. The number of lower orders is unknown, but must be considerable, possibly extending to the scale of individual lava flows (e.g., Ballard and Moore, 1977; Yamagishi, 1985).

Whatever the scale and order of segmentation, each segment and subsegment is separated from the adjacent segment or subsegment by a topographically lower zone that strikes at right angles to the ridge. The major segments, whether those observed by Macdonald et al. (1988) or the higher order segments reported by Gordin (1989), are separated by *ridge-transverse fault zones* (the *transform faults* of plate tectonics). The lower order segments are separated by less prominent ridge-transverse depressions variously called *accommodation zones, transfer zones, devals*, and so forth (Ebinger et al., 1989a, 1989b; Winterer et al., 1989). High-order segments, where not separated by ridge-transverse fracture zones, commonly are separated by eddy-like structures which, in plate-tectonic terminology, are called *overlapping spreading centers*. (In a subsequent section we propose a very different origin for these features.)

Segmentation in all other types of tectonic belts shows the same diversity as that observed on midocean ridges, as well as a complex subdivision into several orders (Meyerhoff et al., 1992a, 1992b). On Figure 2.16, we show examples. Ebinger (1989) and Ebinger et al. (1989a) have noted that the segmentation observed in the East African rift system is similar to that reported along the midocean ridges (Macdonald and Fox, 1983; Macdonald et al., 1988). Yet nowhere could we find comparative studies of segmentation patterns in different types of tectonic belts or large-scale landforms. Apparently the pinch-and-swell geometry that characterizes every tectonic belt that we studied has not been recognized.

2.3.7 RIFT ZONE TO VERSCHLUCKUNGSZONE

A very large group of linear structures is present in foldbelts. These include such features as *troughs*, *rifts*, *strike-slip fault zones*, *suture zones*, *lines*, *median tectonic lines*, *Narbenzonen*, *Scheitelzonen*, and *Verschluckungszonen*. Examples of troughs and rifts include the Okinawa trough, the Taiwan longitudinal valley (Fig. 2.17), and the Kamchatka-Olyutor trough or graben. Strike-slip zones and sutures include the North Pyrenean fault (Fig. 2.14), the North Anatolia fault (Fig. 2.18), the Armorican shear zone, the Taurus-Zagros suture (Fig. 2.19), the Indus-Yarlung suture, the Philippine fault zone (Fig. 2.17), the San Andreas fault (Figs. 2.2, 2.21-2.22), and the Brevard shear zone (Fig. 2.23). Lines and median tectonic lines, including *Verschluckungszonen*, *Narbenzonen*, and *Scheitelzonen*, are widespread (Ampferer, 1906; Ampferer and Hammer, 1911). Examples are the median tectonic lines of Japan and New Zealand (Figs. 2.23, 2.24), the Insubric line (Fig. 2.13), the Periadriatic line (Fig. 2.23), the Balanton line (Fig. 2.23), and many others (Fig. 2.23).

In our research, we found that these linear features form an unbroken genetic series from rifts and grabens at one end of the spectrum to alpinotype foldbelts at the other. The literature on such linear features is very large, yet attempts to interrelate them are few, especially since 1955. With rare exceptions in fact (e.g., Mueller, 1983), the extremely useful and well-established concept of *Verschluckungszone* has all but vanished. In a tectonic belt, strike-slip zones begin to form when compressive stress predominates. Strike-slip zones in alpinotype foldbelts commonly grade into *Verschluckungszonen*, which is translated into English as *engulfment zones* (Meyerhoff et al., 1992b). Closely related but not synonymous are the terms

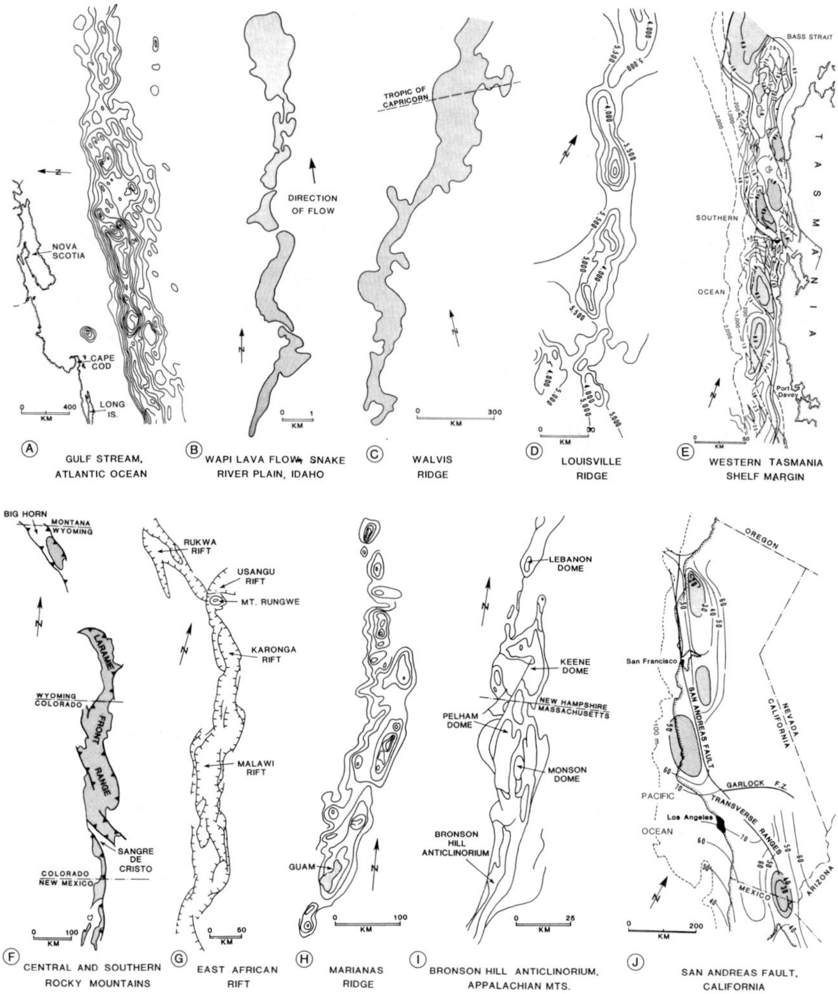

Fig. 2.16. Segmentation and sinuosity flow patterns in tectonic belts. **A** = Gulf Stream; **B** = recent lava flow; **C** = submarine aseismic ridge; **D** = linear island and seamount chain; **E** = shelf-margin structures; **F** = germanotype or block-faulted foldbelt; **G** = continental rift valley; **H** = island (magmatic) arc; **I** = root zone of Paleozoic alpinotype foldbelt; and **J** = strike-slip fault system. From Meyerhoff, in press.

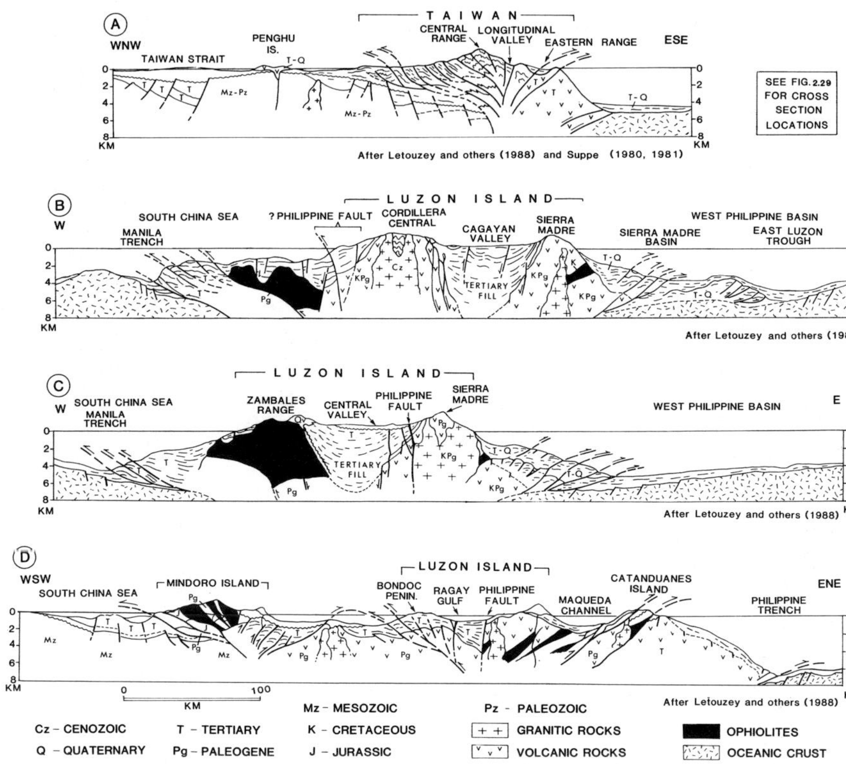

Fig. 2.17. Four west-to-east structural cross sections of Taiwan and the Philippine arc system (from Létouzey et al., 1988). The sections are based on reflection-seismic and field-geology data. The following should be noted: (1) the Benioff zone (in each cross section, at the eastern side of Taiwan or Luzon) has a kobergen just behind it, demonstrating the influence of the Earth's rotation; (2) a major strike-slip fault system lies at the crest of the kobergens that underlie the volcanic arc–the longitudinal valley of Taiwan and the Philippine fault of Luzon; (3) the kobergen beneath northern Luzon bifurcates to form a separate kobergen beneath Mindoro and Palawan Islands (Fig. 2.29). Locations of cross sections appear on Figure 2.29.

root zones (Heim, 1922; Bucher, 1933; see Boyer and Elliott [1982] for an up-to-date discussion) and *steep belts* (Milnes, 1974).

2.3.8 Hydrothermal Manifestations

These are well known on land, where they form large natural geyser-steam fields in various tectonic regimes. Examples from the extensional regime include Iceland, the Great Basin, the Yellowstone area of the Western North American Cordillera, and the "black smokers" of several midocean ridges---the East Pacific Rise, the Juan de Fuca Ridge, the Galapagos rift, and the Mid-Atlantic Ridge. Examples from transtensional and transpressional regimes include New Zealand, the Lau basin (Fig. 2.25), and northern California. Examples from compressive belts include the Lhasa block of Tibet, the Apennines, Japan, and to some degree New Zealand and northern California. Our point is that almost identical thermal phenomena appear in a complete spectrum of tectonic conditions, from tensional to compressional. Contrary to widespread belief (e.g., Von Damm, 1990), we do not think that hydrothermal phenomena are simply "...a consequence of the emplacement of hot rock at divergent plate boundaries..." (Von Damm, p. 173). Some hydrothermal features have been explained on a case-by-case basis, but no hypothesis has been able to provide a single mechanism to explain all occurrences, whatever the stress state.

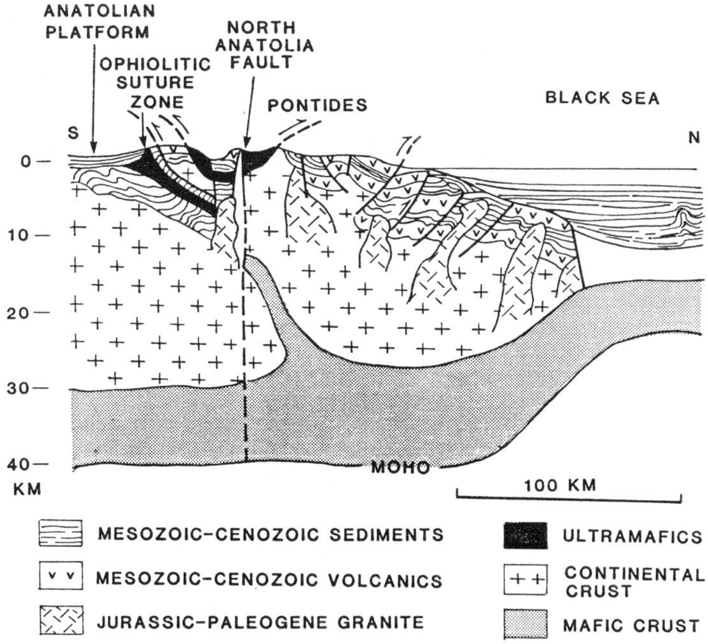

Fig. 2.18. South-north structural cross section across the North Anatolia fault (from Létouzey et al., 1977). As in the case of the San Andreas fault, the master regional strike-slip fault lies at the center of a kobergen. This figure shows that the San Andreas is not an isolated example. Compare with Figures 2.13, 2.14, 2.17, and 2.22.

2.3.9 Reticulate Pattern of High Heat-Flow Bands

The plate-tectonics hypothesis predicts simple heat-flow patterns around the Earth. It postulates a broad band of high heat flow beneath the full length of the midocean rift system, and parallel bands of high and low heat flow along the Benioff zones. In the Circum-Pacific, the low-heat flow band is postulated to lie seaward from the deep-sea trench, whereas the parallel high heat-flow band is postulated to underlie the adjacent volcanic arc continentward from the trench. Intraplate regions are predicted to have low heat flow.

Figure 2.26 shows the pattern actually observed. It is apparent that it is unlike the pattern predicted by the plate tectonics hypothesis. Exceptions to the predicted pattern have been explained by the "hot spot" phenomenon which are hot asthenosphere upwellings or diapirs ("mantle plumes"), fixed in time and space in the asthenosphere, and over which the lithosphere plates move, leaving "hot-spot trails" behind them. Thus a given "hot spot trail" should show an age progression from one end to the other. However, good age progressions are very rare (Fig. 2.27), and dated rows of mantle diapirs that have little or no age progression (Fig. 2.28) form a large majority. (It is worth mentioning that almost all rows of mantle diapirs that show an age progression show it to be from west to east.)

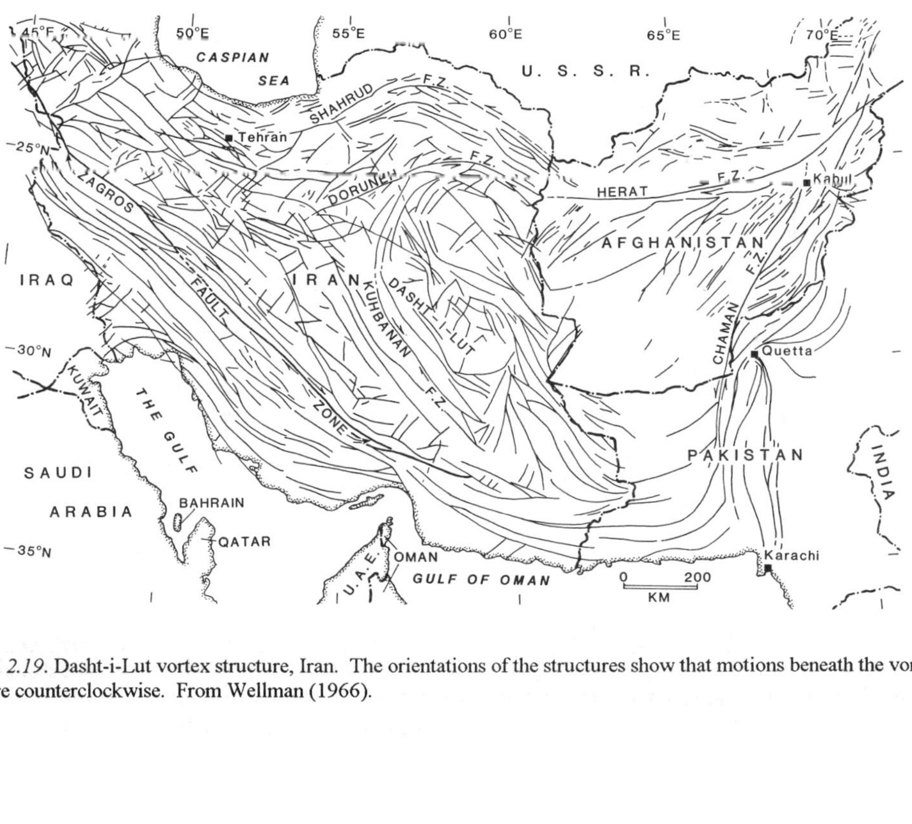

Fig. 2.19. Dasht-i-Lut vortex structure, Iran. The orientations of the structures show that motions beneath the vortex were counterclockwise. From Wellman (1966).

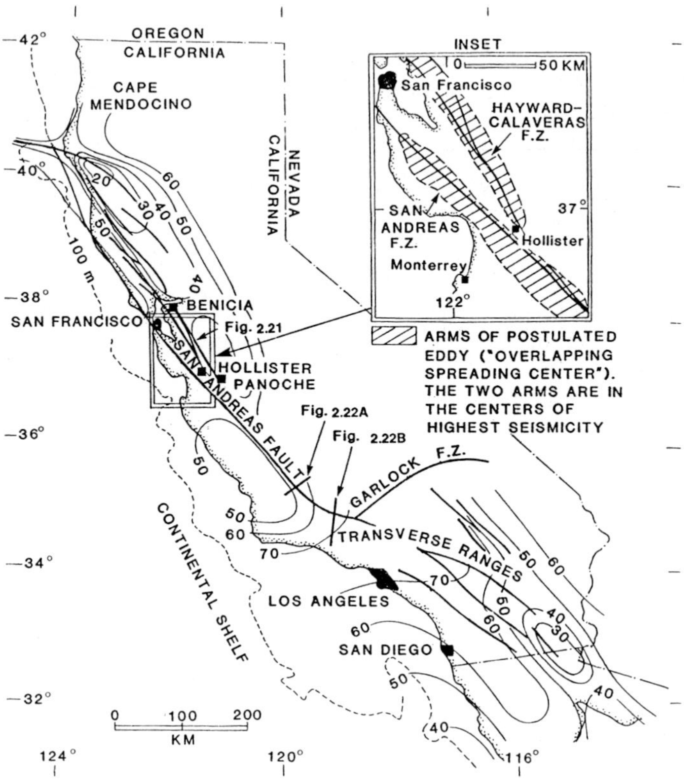

Fig. 2.20. Structural contour map of the top of the California Coast Ranges northwest-southeast-trending surge-channel complex. Zandt and Furlong (1982), from whose work this map was taken, interpreted the contour values to represent lithosphere thickness. However, the shape, dimensions, and orientation parallel with California's fold and fault system (Figure 2.2) indicate that the linear feature is a surge channel. As we point out in the text, this linear feature also is associated with Cenozoic volcanic rocks, elevated heat flow, high seismicity, a row of hot springs, 7.0-7.8 km/s material, and other phenomena generated by surge-channel activity. The depths shown are to the top of the main magma chamber, not to the top of the shallower chambers that are shown on Figure 2.21. The location of Figures 2.21 and 2.22 are shown. Inset map: this is a postulated eddy in the San Francisco Bay region where the zone of most intense seismic activity is transferred from the San Andreas fault to the Hayward-Calaveras fault (Smith, 1978).

Figure 2.26 is at present a preliminary map subject to considerable revision in some regions because insufficient heat-flow measurements have been made to draw a more accurate map. However, as a brief examination of the map shows, even substantial revisions in such areas as Siberia, South America, and parts of the Pacific basin will not seriously alter the reticulate pattern that is shown here.

2.3.10 Microearthquake Bands

Modern ideas on the Earth's epicenter belts are based in large part on the classic Gutenberg and Richter (1949, 1954) hypocenter catalogues and accompanying maps. Almost all earthquakes shown by Gutenberg and Richter (1954) are in the 5.5-9.0 (Gutenberg scale) magnitude range. The belts lie along island arcs, the crests of midocean ridges, continental rift zones, and wrench-

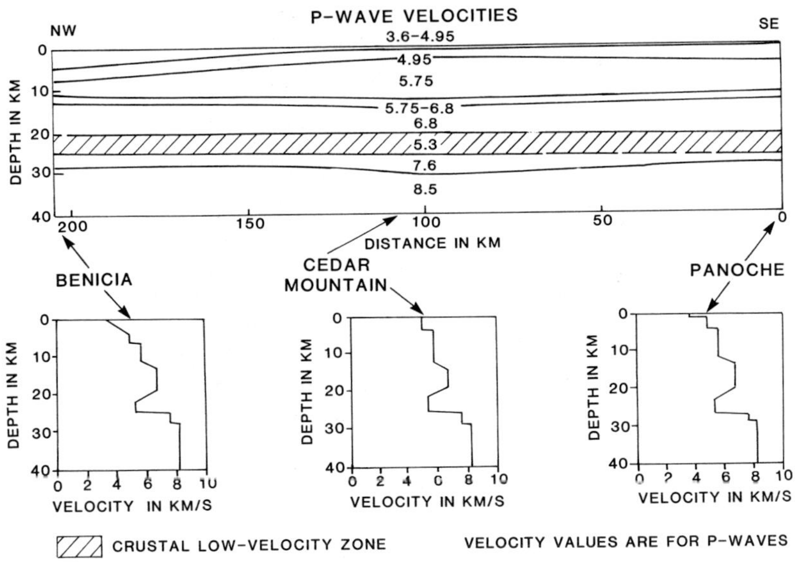

Fig. 2.21. Velocity-depth interpretation for a northwest-southeast reflection-seismic line through the Diablo Range, central California, from Benicia to Panoche. The location of the line is shown on Figure 2.20. *Top*: numbers indicate the P-wave velocities in km/s. Note the low-velocity zone associated with 7.6-km/s material below 20 km. *Bottom*: one-dimensional velocity-depth functions for selected profiles. The Benicia profile is at the northwestern end of the refraction line, the Panoche profile at the southeastern end, and the Cedar Mountain profile in the middle. The purpose of this figure is to demonstrate that a low-velocity zone associated with 7.0-7.8-km/s material at depth does underlie the Coast Ranges. From Blümling and Prodehl (1983).

fault zones. This Gutenberg and Richter "image" of the distribution of epicenter beds has become a cornerstone of plate tectonics.

Today earthquakes of magnitude 1.0 to 2.0 can be detected, provided that the proper recording instrumentation is available. However, earthquakes of such small sizes can be produced by a variety of events ranging from quarry blasts to street traffic; thus, many have no tectonic significance. However, almost all earthquakes in the 3.5 to 5.5 range *are* tectonically significant, and their records are (1) adding important volumes of information to the Earth's seismicity data base and (2) greatly modifying the Gutenberg and Richter (1954) model.

The routine recording of these smaller earthquakes in areas where good instrumentation is available (e.g., Canada, United States) is revealing the presence of belts of seismicity that were essentially unknown to Gutenberg and Richter (1949, 1954). Where adequate data on the smaller magnitude earthquakes are present, they are seen to form a pattern of reticulate bands of seismicity, *which coincide with the high heat-flow bands shown on Figure 2.26* (Stegena, 1990). Stegena (1990) has noted that these facts are unexplained by current tectonic hypotheses.

2.3.11 Diffuse Plate Boundaries

The concept of "diffuse plate boundaries" has been introduced to explain the presence of wide

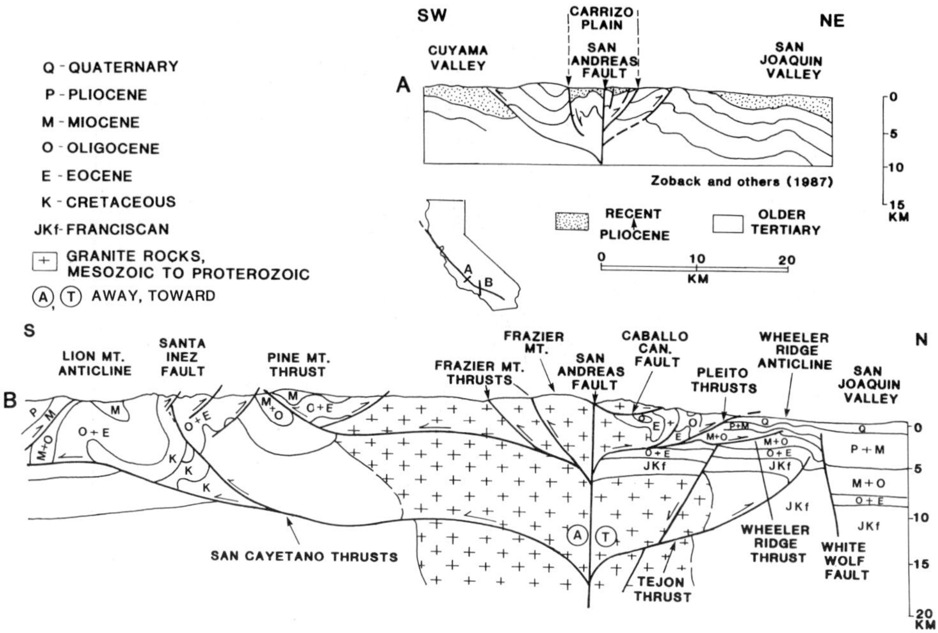

Fig. 2.22. Two structural cross sections across the San Andreas fault, central California. In all structural sections across this fault the regional strike-slip fault lies at the center of a developed, or developing kobergen. This implies that the regional master fault is at or very close to the top of a surge channel, or former surge channel. See the text for detailed discussion. The locations of these cross sections are on Figures 2.2 and 2.20. Compare this figure with Figures 2.17b and 2.17c (Philippines fault) and 19 (North Anatolia fault). Figure 2.13 (Insubric Line of the Alps) shows a related feature. The geometric relation between regional strike-slip faults and kobergen mechanics means that much of what has been written about strike-slip zones needs to be reevaluated. Based on Namson and Davis (1988).

(up to several hundred kilometers across) bands of shallow seismicity. A well-known example is the San Andreas fault zone where a width on the order of 300 km has been postulated to explain (1) the broad band of shallow diffuse seismicity of that region (Mooney and Weaver, 1989) and (2) the fact that measured offsets along the San Andreas fault are far short of some plate-tectonic predictions. Barrows et al. (1985) found that a "diffuse plate boundary" was essential in southern California where they showed that lateral movements along the San Andreas fault in Los Angeles County from Jurassic time to the present could not have exceeded 21 km and, along the whole zone of San Andreas-parallel faults, 102 km. Although the postulate of a "diffuse plate boundary" has proved useful in regions where plate boundaries are thought to be present, it does not explain the many bands of shallow seismicity that are far from plate boundaries. In actual fact, these "diffuse plate boundaries" are parts of the reticulate network of microseismicity discussed in the preceding section, and underlie bands of elevated heat flow.

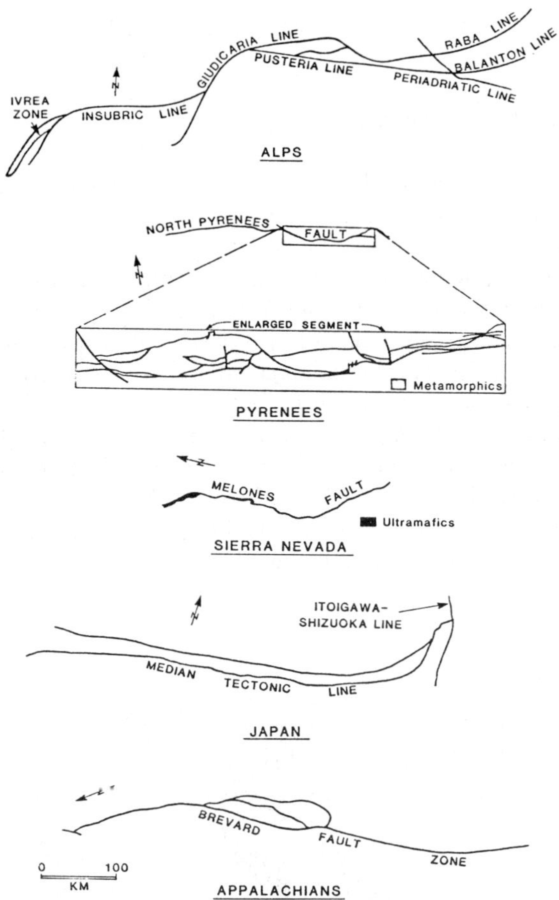

Fig. 2.23. Traces of several median tectonic lines and their associated ultramafic and/or metamorphic rocks. We interpret these to be the Narben (scars) of *Verschluckungszonen* (engulfment zones) that formerly were straight strike-slip (streamline) fault zones but which, as a consequence of compression during circumferential shortening, have become twisted and deformed. From Meyerhoff et al. (1992b).

2.3.12 Linear Anorogenic Belts

There is essentially no explanation, except on an *ad hoc* basis, for linear rows of anorogenic "within-plate" plutons, volcanoes, volcanic fields, and related geothermal phenomena. Linear anorogenic igneous provinces abound worldwide, and display a great diversity in styles and compositions. Examples include the Upper Rhine and other peri-Alpine Tertiary igneous centers (Brousse and Bellon, 1983), the West African Permian through Holocene anorogenic magmatic province (Wright et al., 1985), the Syrian-Arabian Tertiary volcanic province (Camp and Roobol, 1992), the Tertiary Southern Cordilleran basaltic andesite province of the southwestern United States and northwestern Mexico (Cameron et al., 1989), and many others. Asian examples include the Permian rift province that extends from the Panjal Traps of India to the Emeishan Basalt of southwestern China (Zhang Yunxiang et al., 1990), the Late Jurassic-Tertiary Lower Yangzi plutonic-volcanic province (Liu Xiangpei et al., 1988), and the Cenozoic

Mongolian-Transbaikalian Trap province of southern Siberia and Mongolia (Litvinovskiy et al., 1989).

2.3.13 Continental-Margin Phenomena

Meyerhoff et al. (1992b) suggested, on the basis of very preliminary data, that active surge channels might underlie the continental margins near the continental shelf/slope break. After that paper was written, large volumes of evidence were amassed to support that tentative conclusion. Continental margins with a few rare exceptions are associated with a belt of elevated heat flow (Reiter and Jessop, 1985; Bodri and Jessop, 1989; Shanker, 1991); an associated belt of microearthquakes (Adams and Basham, 1991; Hasegawa, 1991; Berrocal et al., 1993); a relative gravity positive trend (a gravity "high"; Anfiloff et al., 1976; Bureau of Mineral Resources, 1976; Wellman, 1979; Anfiloff, 1988; Lawver et al., 1993); a structural arch (Lehner and Ruiter, 1977; Goldfinger et al., 1991; Wannesson et al., 1991); lenses of 7.0-7.8 km/s material close to the Mohorovičić discontinuity (Trehu et al., 1989; White and McKenzie, 1989; Rosendahl et al., 1991; Olafsson et al., 1992); and a high-conductivity magnetotelluric or electrical anomaly (Roussel and Lesquer, 1991; Jones, 1992). In addition, segmentation like that in tectonic belts is common (Goldfinger et al., 1991; Moore et al., 1992),

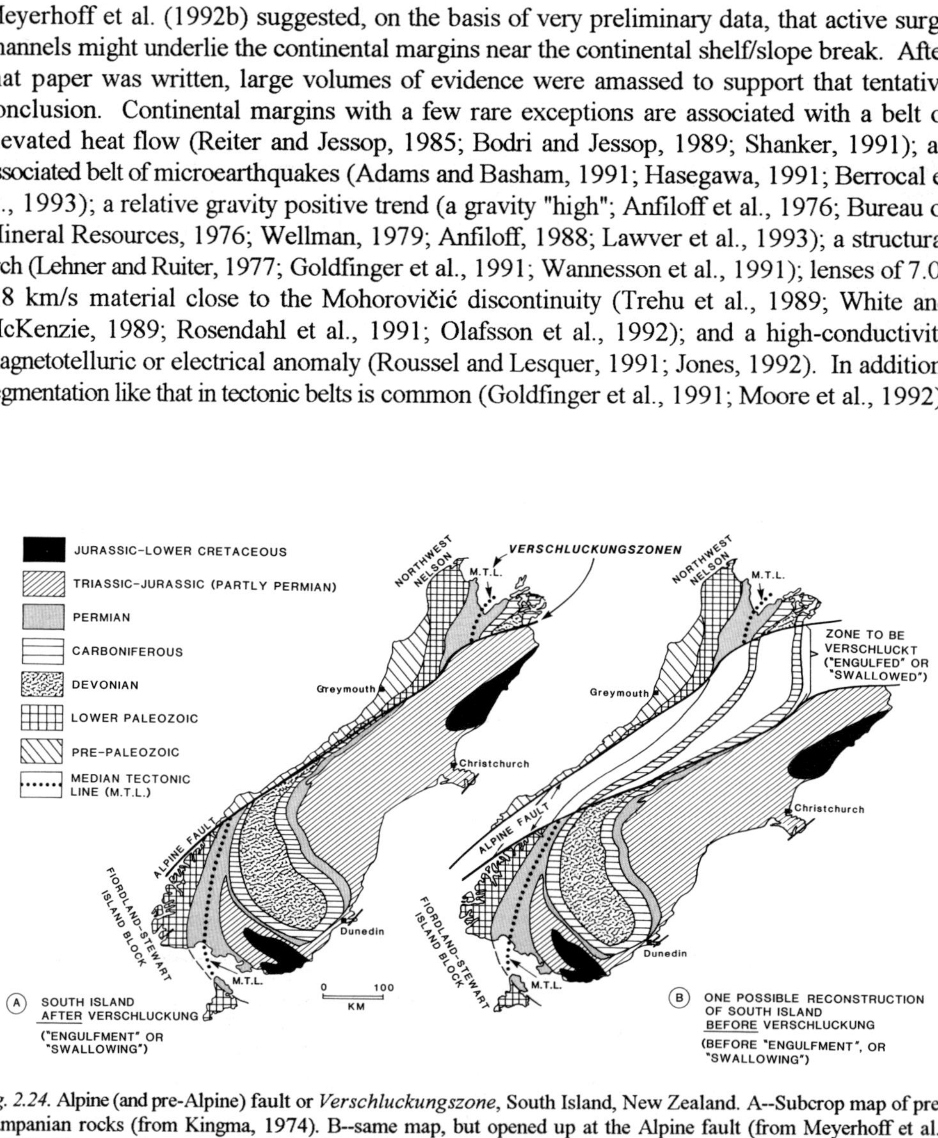

Fig. 2.24. Alpine (and pre-Alpine) fault or *Verschluckungszone*, South Island, New Zealand. A--Subcrop map of pre-Campanian rocks (from Kingma, 1974). B--same map, but opened up at the Alpine fault (from Meyerhoff et al., 1992b). Between the two sides of the fault is shown the minimum amount of surface area that was engulfed (*Verschluckert*) by the underlying surge channel during the middle Cretaceous Rangitata orogeny. We have shown the missing surface area along the Alpine fault, or a pre-Alpine fault at the same location, as the *Verschluckungszone*; that zone originally was the Median Tectonic Line (M.T.L.). Our points are that (1) a very large surface area was engulfed along this *Verschluckungszone* in South Island; (2) much of this zone's surface trace now coincides with the Alpine fault; (3) the *Verschluckungszone* explanation obviates the need for large strike-slip movements along the Alpine fault; and (4) the *Verschluckung* process is a logical mechanism for circumferential shortening in the lithosphere.

as are grabens and half-grabens with and without thick evaporite sequences (Iliffe et al., 1991; Artyushkov, 1992). Cloos and Dietz (1954), Martin (1975), and Cox (1989) noted that many of the world's coastlines are associated with rugged escarpments and rising topography in regions such as Brazil, Greenland, the Canadian Arctic, Scandinavia, South Africa, Mozambique, India, the Red Sea, and the Gulf of Aden. One unusual example of a surge channel along a continental-shelf margin is the Taiwan-Sinzi Ridge that connects Taiwan with Japan (Emery and Niino, 1968; Létouzey and Kimura, 1985).

2.3.14 Vortex Structures

Vortex structures have been known to exist at least since the mid-1920s (Li Siguang, 1929, 1939, 1973), but have been almost universally ignored. Descriptions of vortex structures have been published not only by Li Siguang (1929, 1939, 1973), but also by Wellman (1966), Huang Qinghua (1980), Neev and Hall (1984), Neev et al. (1985), Rousseau (1987, 1992), Westaway (1990), Zeng Zuoxun (1990), Meyerhoff et al. (1992b) and Leybourne et al. (1995). Vortices have been variably interpreted by geologists as median massifs, orogen collapse structures, overlapping spreading centers, pull-apart basins and microplates. A surge tectonics mechanism for the formation of vortex structures is presented in Chapter 3.

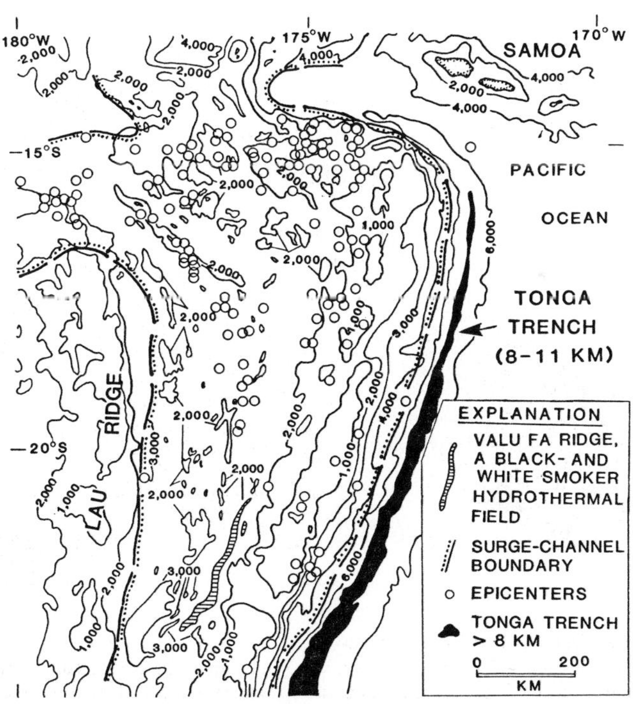

Fig. 2.25. Tonga Ridge, Lau Ridge, and Lau basin, southwestern Pacific; epicenters are shown. Note that the epicenters tend to cluster in rows along the tops of sea-floor ridges. This fact indicates that (1) motions beneath the Lau basin parallel the basin axis, (2) streamline shears underlie these linear sea-floor ridges, and (3) subsurface flow, therefore, is laminar. Some of these ridges (e.g., Valu Fa Ridge) contain rows of black smokers (from NAUTILAU Group, 1990). Location of the figure is on Figure 2.29.

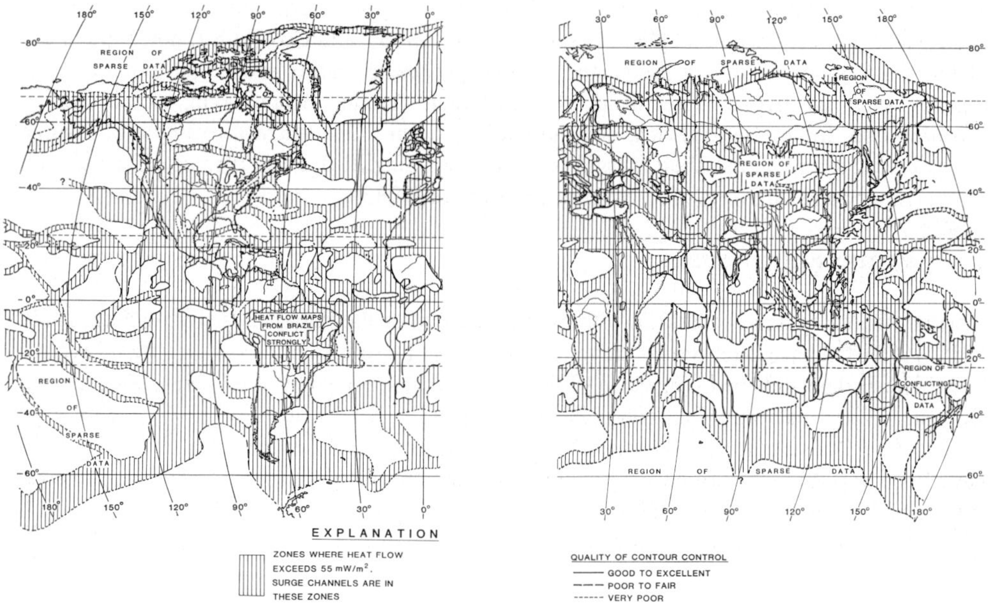

Fig. 2.26. Preliminary world map showing bands of high heat (>55 mW/m^2). All active surge channels lie within the bands of high heat flow. Control is weak in parts of Asia, North America, South America, and the Pacific basin. Compiled from more than 500 sources, the most important of which are listed in Meyerhoff et al. (1992b). From Meyerhoff et al. (1992b).

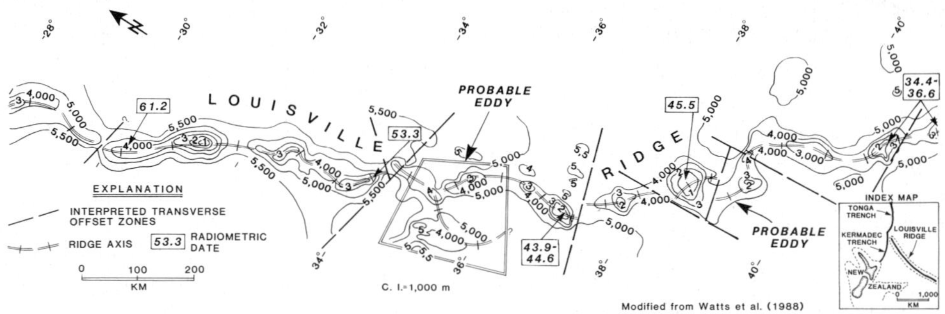

Fig. 2.27. Louisville Ridge, southwestern Pacific, the location is shown on Figure 2.29. Note the sinuous pattern and eddylike features. This is a typical breakout channel, studded with volcanoes in the forms of seamounts and guyots. This class of surge channel commonly (1) shows an age progression oldest-to-youngest from west to east, never the reverse; and (2) originates at its western end at the cusp where two volcanic arcs intersect (see Fig. 2.29). These ridges originate only on the western sides of ocean basins. This fact and the age progression from west to east--of some, but not all breakout channels--demonstrate that the "hot spot" theory should be reevaluated.

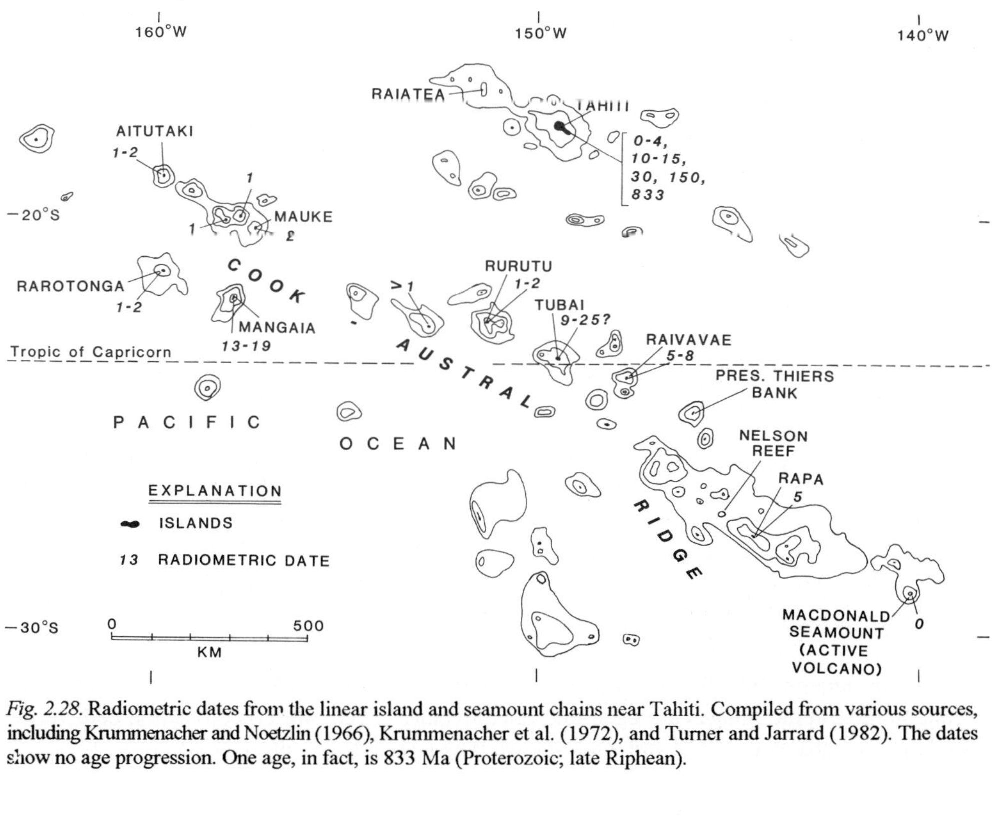

Fig. 2.28. Radiometric dates from the linear island and seamount chains near Tahiti. Compiled from various sources, including Krummenacher and Noetzlin (1966), Krummenacher et al. (1972), and Turner and Jarrard (1982). The dates show no age progression. One age, in fact, is 833 Ma (Proterozoic; late Riphean).

2.3.15 Magma Floods

Magma floods, e.g., flood basalts, have been treated in much of the geological literature as a special geological problem (e.g., White and McKenzie, 1989). Yoder (1988) has pointed out that basaltic floods are the most important rock type at the Earth's surface, because they include not only the isolated flood-basalt provinces observed on the continents, but also the basalts of the midocean ridges, the adjacent ocean floors, and oceanic seamounts, guyots, and islands. Thus magma floods cover approximately 68% of the Earth's surface.

In Chapter 6 it is shown that, in addition to flood-basalt provinces, flood-andesite (Petterson et al., 1992), flood-trachyte, flood-dacite, and flood-rhyolite provinces are also major features of the Earth's surface. Examples of flood-rhyolite provinces include the Patagonia "Gondwana magmatic suite" (including the huge Jurassic Chon Aike Group), the Cordillera Occidental (Mexico), the Central Queensland volcanic province, the Lord Howe Rise, the Primor'ye volcanic belt, the Okhotsk-Chukotka volcanic belt, and many more. Flood-dacite/flood-rhyolite provinces grade petrographically, geochemically, and isotopically into flood-basalt provinces. More important, flood-basalt to flood-rhyolite provinces show an unbroken chemical and petrographic spectrum, with midocean ridge basalts at one end and continental magmatic-arc basalts at the other. This fact has been demonstrated in part in a recent paper by Dupré et al. (1994). These statements are contrary to claims that each tectonic environment has a distinctive geochemical signature (e.g., Pearce, 1983; Thompson et al., 1983). In Chapter 6, we describe numerous magma-flood provinces and their geochemistry, and then discuss the origin of magma floods in the context of our new hypothesis, surge tectonics.

2.3.16 Seismotomography and Convection

Because plate-tectonic models require that upwelling occurs beneath midocean ridges and downwelling occurs beneath volcanic-arc trenches, a well-organized convection-cell pattern was anticipated in the mantle, and seismotomographic images were expected to provide the proof. In addition, those images were expected to show the extent of the convection cells as well as solid evidence for the existence of mantle plumes, or hot spots (Woodhouse and Dziewonski, 1984; Dziewonski and Woodhouse, 1987). Interestingly, seismotomographic images have not provided conclusive evidence for organized convection in the mantle and mantle plumes do not appear (Anderson et al., 1992a).

The seismotomographic results could have been anticipated on the basis of other factors. O'Keefe (1965) showed that, if mantle convection occurs, the continents must be concentrated either at the poles or along the equator, because the existing equatorial bulge has to be dynamically sustained. Jeffreys (1959, 1970, 1976) went a step farther, showing that convection cannot take place in the mantle because it is a self-damping process in accordance with the Lomnitz Law (Lomnitz, 1957; Jeffreys and Crampin, 1970). Sandwell and Renkin (1988) found no expression in the geoid of convection-cell geometry. Zindler and Hart (1986) wrote that convection is not occurring because the compositions of the igneous rocks and xenoliths reaching the Earth's surface from the mantle are too heterogeneous, and form no systematic distribution pattern. It is no longer necessary, however, to depend on these arguments against mantle convection. As pointed out by Anderson et al. (1992a), the seismotomographic images are clear and unequivocal. There are no convection cells of the Rayleigh-Bénard type---or of any other known type (Anderson et al., 1992a).

Another important result of seismotomographic studies has been to prove that the roots of continental cratons are deep. Anderson et al. (1992a, 1992b) argued that roots deeper than about 200-250 km do not exist, because the tectonic plates could not move if the roots extend to the base of, or below, the asthenosphere---up to 400 km or more according to Jordan (1975, 1978) and Lowman (1985, 1986). MacDonald (1963, 1964, 1965b), on the basis of heat-flow and gravity studies, already had shown that the roots must be close to 500 km. Grand (1987) confirmed it, showing on the basis of a seismotomographic study of North America, that cratonic roots extend to between 400 and 500 km, well below the base of the asthenosphere (Fig. 1.1). Russo and Silver (1995) have expanded on this idea to include South America. They also show that subduction is not occurring on the west coast of South America, and suggest flow of the mantle around the northern and southern parts of that continent.

On the basis of data indicating the presence of deep continental roots, coupled with seismic-refraction information, gravity studies, and geochemical data, Lowman (1985, 1986) concluded that the cratons of the continents are fixed; now all seismotomographic studies amply confirm MacDonald's (1965b), Jordan's (1975, 1978), and Lowman's (1985, 1986) conclusions.

2.3.17 Antipodal Positions of Continents and Ocean Basins

Arldt (1907) appears to have been among the first to recognize the fact that the continents are almost wholly antipodal to ocean basins, a fact which means that the Earth was partly or entirely molten early in its history. Harrison (1968) demonstrated that the antipodal relation cannot be a matter of chance or coincidence. The reason for the antipodal positions is almost certainly a response to the Earth's rotation according to which masses on the Earth's surface should be distributed more or less axisymmetrically for rotational stability (Munk and MacDonald, 1960, 1975). Axisymmetry can be attained only under largely molten conditions. The idea of antipodal positions may be rooted in a map published by Gregory (1899, 1901), but the concept goes back even further, as it is a demonstration of Green's (1875, 1887) tetrahedral theory of the Earth (Bucher, 1933). Regardless, the antipodal positions of the continents and ocean basins show that the Earth's continents must have reached their present positions in a molten Earth very early in the Earth's history (Bucher, 1933; Wilson, 1954; Harrison, 1968).

2.3.18 World Evaporite Distribution

Meyerhoff (1970) found that 95% of the world's evaporites, middle Proterozoic through recent, lie in areas which today receive less than 1,000 mm/a of rainfall; i.e., almost all evaporites of the past lie in *today's* dry-wind belts. This is not possible unless the Earth's wind-circulation patterns have remained approximately the same through time. The wind-circulation pattern could have been the same through time only if the present positions of continents and ocean basins have remained similar since middle Proterozoic or earlier time.

2.3.19 Preferential Eastward Flow of the Asthenosphere

Meyerhoff et al. (1992b) showed that the igneous activity of numerous tectonic belts shifts eastward during any given tectonic cycle and, in some places, from one tectonic cycle to the next. The cause of eastward flow is the Earth's rotation, which causes differential lag between the lithosphere and the strictosphere, producing an eastward "squeezing" of the asthenosphere between them (Meyerhoff and Meyerhoff, 1977; Nishimura, 1986). (The strictosphere, as defined by Bucher [1956], is the hard shell underlying the asthenosphere. The uppermost layer of the strictosphere has a P-wave velocity range of 8.1-8.7 km/s). Chi and Luo (1992) recently have stressed the potentially great importance of rotation and tidal forces in lithosphere deformation. Exceptions to this eastward rule exist and likely are due in each case to local geological factors. The exceptions include shifts westward, northward, or southward. However, westward shifts are minuscule in contrast to eastward shifts.

Many examples of purely eastward migration are known from the eastern part of Asia. The most dramatic are the eastward-facing island arcs of the western Pacific Basin. Meyerhoff and Meyerhoff (1977) interpreted these arcs as having been created by eastward surges of the asthenosphere from beneath Asia toward the Pacific. Such surges would explain why all western Pacific arcs have backarc basins and why all of them have an eastward component of direction. Similar mechanisms were proposed long ago by Bucher (1947), Meyerhoff (1954), and Hamilton (1963) to explain the formation of the eastward-facing Lesser Antilles and Scotia arcs. Meyerhoff and Meyerhoff (1977) wrote that the same asthenosphere surge that produced the backarc basins of the western Pacific could also explain the *absence* of backarc basins along the western continental margins of North and South America, where all volcanic arcs are continental rather than oceanic.

Although the Meyerhoff and Meyerhoff (1977) asthenosphere-surge model is theoretical, concrete examples from Asia of eastward migration of igneous phenomena abound and give substance to the theory. These include:

(1) Cusps of oceanic volcanic arcs;

(a) the Hawaiian-Emperor chain, Figure 2.29;

(b) the Shatsky Rise. This feature originates from asthenosphere surge at the Bonin Ridge which passes through the cusp formed between the Izu and Bonin trenches, continues via the Uyeda Ridge/Nelson Guyot complex, forms the Shatsky Rise and finally becomes the Chinook Fracture Zone (Smoot, 1995a), a trans-Pacific feature. A later feature also arises in that cusp in the form of the Ogasawara Plateau/Michelson Ridge to form a K-structure (see Chapter 3 for definition). According Smoot and King (1993), the Dutton Ridge forms at the East Mariana Ridge, passes through the segmented Mariana Trench, and continues trans-Pacific as the Mendocino Fracture Zone which is parallel to the Chinook Fracture Zone; Smoot (1991, 1994a, b) has also proposed that the Magellan Seamounts form from surge at the East Mariana Ridge, which crosses through a more southerly bulge in the segmented Mariana Trench, and passes southeasterly as part of the lineation including the Marshall-Gilbert Seamounts, the Louisville Ridge, and the Eltanin Fracture Zone; and

(c) the Carolina Seamount chain. This feature proceeds from a cusp between the Yap and Mariana arcs;

(2) Numerous examples from Eastern Asia (Nishimura, 1984, 1985, 1986; Fig. 2.30); see also Chapter 5 for discussion of the surge-tectonics evolution of southeastern Asia;

(3) Okhotsk-Chukotka volcanic belt (Figs. 2.19, 2.20; Zonenshain et al., 1990);

(4) Primor'ye volcanic belt (Figs. 2.29, 2.30; Solov'yev, 1966);

(5) Kamchatka volcanic belt (Fig. 2.29; Marsh, 1979);

(6) Japan arc (Fig. 2.29; Utada, 1980; Banno and Nakajima, 1992); (7) Korean Peninsula (Fig. 2.29; Kim and Lee, 1983; Lee, 1981; Choi, 1993);

(7) Southeastern China (Figs. 2.30, 2.31, 5.2; Grabau, 1931; Li Siguang [J. S. Lee], 1939; Wang Liankiu et al., 1980; Yang Zunyi et al., 1986; Teng, 1992);

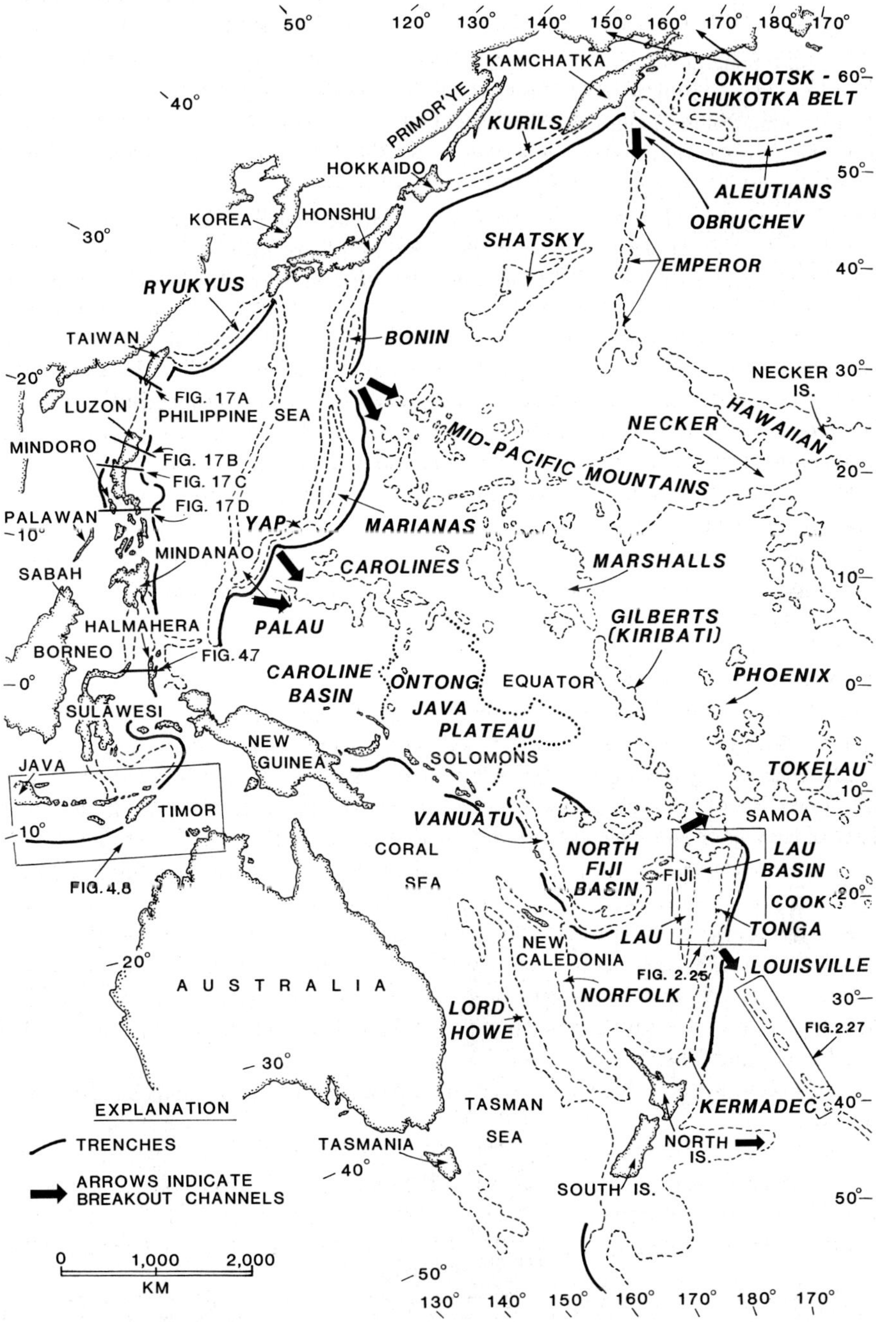

Fig. 2.29. Sketch map of the western Pacific Ocean. This map shows that all major linear island and seamount chains originate either (1) at the cusp where two island-arc Benioff zones intersect or (2) from a linear chain that originates at such a cusp. The locations of Figures 2.25, 2.27, 4.7, 4.8, and 2.17A-17D are shown. Modified from Meyerhoff et al. (1992b).

(8) Emeishan Basalt (Fig. 2.18, Panjal, Abor, and related traps there; Zhang Yunxiang et al., 1990); and

(9) Ordos basin (Zhang Buchun et al., 1985; Wang Jingming, 1987).

The eastward-migrating igneous/magmatic belts listed above are but a few of many of the eastward-migrating features of eastern Asia. The fact should be noted that tectonic activity, not just magmatic activity, also shifts eastward in many places in time. Some 54 such eastward-migrating features are listed in Table 2.2. A well-known example is the eastward shift of the Japanese geosynclinal system as described by Kimura (1973) and many other Japanese workers. Figure 2.30 also shows eastward migration of major sedimentary basins. These are just a few of the phenomena; eastward migration of asthenosphere flow requires explanation.

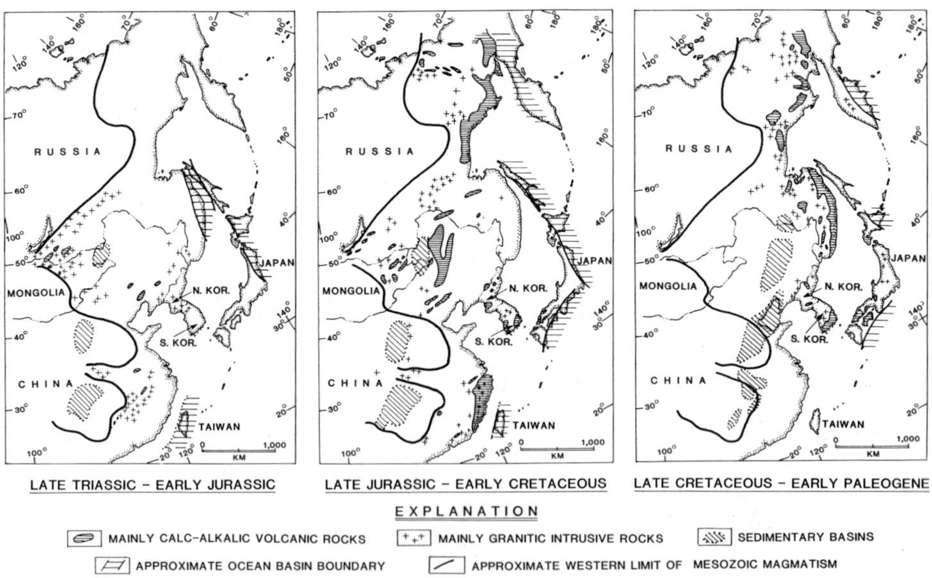

Fig. 2.30. Late Triassic through early Paleocene igneous activity, eastern Asia. The map shows the steady eastward shift of igneous activity in eastern Asia during the Indosinian, Yanshanian, and Himalayan tectogeneses. From Nishimura (1985).

Table 2.2. Evidence for eastward movements within the lithosphere (surge channels) and asthenosphere. From Meyerhoff et al. (1992b; figure numbers have been updated to this publication).

Phenomenon Example(s)	Age	Comment(s)	Reference(s)
1. *Volcanic arc with backarc basin; most arcs face east*		Earth's rotation *promotes* backarc basin development	Meyerhoff and Meyerhoff (1972b, 1974)
a. Western Pacific arcs	Mainly Phanerozoic	Fig. 2.29	Wilson (1954)
b. Lesser Antilles arc	Late Cretaceous-present	Fig. 3.11	Bucher (1947); H. Meyerhoff (1954); North (1965); Meyerhoff and Meyerhoff (1972a)
c. Scotia arc	Cenozoic	---	Hamilton (1963)
d. Seychelles arc	Proterozoic-early Paleozoic	Extinct arc	Kamen-Kaye and Meyerhoff (1980)
2. *Volcanic arc without backarc basin, most arcs face west*		Earth's rotation *inhibits* backarc basin development	Meyerhoff and Meyerhoff (1972b, 1974)
a. Eastern Pacific arcs	Mainly Phanerozoic	Examples: Middle America and Peru-Chile trenches	Wilson (1954)
3. *Stack-up of tectonic phenomena only along the western sides of Benioff zones*			
a. Eastern Pacific fracture zones	?Precambrian	Examples: Clipperton, Clarion, Murray, Mendocino, Easter Island, and others Fzs	Menard (1960); Meyerhoff and Meyerhoff (1972b,1974)
b. Massive asthenosphere upwelling and/or diapirism	Precambrian	i. Backarc basins of the western Pacific; Fig. 4.5	Meyerhoff et al. (1992b); see Woodhouse and Dziewonski, (1984)
		ii. East Pacific Rise: Fig.4.5	Ditto
4. *Feeder channels (see text);; part of the broad category of ridges called aseismic ridges*			
a. Walvis Ridge	Probably Permian-Cenozoic	These ridges branch *only* from the eastern flanks of midocean ridges; Fig. 4.1	Meyerhoff et al. (1992b)
b. Others: Cape Verde Rise, Cocos Ridge, Nazca Ridge, Galapagos rift, and others	----	Ditto	Meyerhoff et al. (1992b)
5. *Overflow, or breakout channels (see text); part of the broad category of submarine ridges called "aseismic ridges"*			
a. Hawaiian Ridge	Late Cenozoic	Fig. 2.29; these ridges branch *only* from eastern sides of volcanic arcs and continents; Fig. 4.3	Jackson (1976); Shaw et al. (1980)
b. Louisville Ridge	Maastrichtian to Quarternary	Fig. 2.27	Watts et al. (1988)
c. New England Seamount chain	Neocomian-Miocene	----	Houghton et al. (1978)
6. *North-south plutonic belt; all or part of the belt shifts eastward through time*			
a. Coast Range Batholith, British Columbia and southeastern Alaska	Cretaceous-middle Eocene	----	Gehrels and McClelland (1988)
b. Sierra Nevada and Great Basin granitic plutons	Cretaceous-Oligocene	----	Coney and Reynolds (1977); Bateman (1983)
c. Baja California andwestern Mexico granitic plutons	Cretaceous-middle Eocene	Fig. 3.12	Clark et al. (1982)
d. Granitic plutons of Colombian-Ecuadorian Andes	Cretaceous-middle Eocene	Fig. 4.17	Mégard (1987)
e. Peruvian Andes granites	Cretaceous-Tertiary	----	Cobbing and Pitcher (1983)
f. Chilean Andes granites	Cretaceous-Quaternary	----	Aguirre (1983)
g. Lachlan foldbelt granites of southeastern Australia	Silurian-Carboniferous	Lachlan foldbelt is in Tasman foldbelt	White and Chappell (1983)

Table 2.2. Continued.

Phenomenon Example(s)	Age	Comment(s)	Reference(s)
h. Tasman foldbelt (overall)	Paleozoic-Triassic	----	Leitch and Schiebner (1987)
i. Southeastern China	Late Jurassic-Cretaceous	----	Wang Liankui et al. (1980)
j. Korean granites	Cretaceous	----	Kim and Lee (1983)
k. Japanese granites	Late Cretaceous-Neogene	----	Utada (1980)
l. Urals granites	Paleozoic	----	Nalivkin (1973)
m. Southern Appalachians	Paleozoic	----	Fullagar and Butler (1979)
n. West Africa			
7. *Overall west-to-east belt; eastern tip propagates eastward*			
a. Middle and lower Yangzi Valley granites and mineral deposits	Early-Jurrasic-middle Cretaceous	----	Zhu Ming (1989)
8. *North-south volcanic belt; all or part of the belt shifts eastward through time*			
a. Cascade Range	Late Eocene-Recent	High Cascades are built on and east of older Cascades	Taylor (1990)
b. Argentine Andes and eastward	Triass-cmiddle Cretaceous	Shift is northeastward torward southern Brazil	Uliana et al. (1989)
c. New Zealand	Cenozoic	----	Stern and Davey (1989)
d. Southeastern China	Late Jurassic-Tertiary	NNE-striking Neocathaysian zone	Yang Zunyi et al. (1986)
e. Panxi rift, southwestern China	Early Permian (west) to Late Permian (east)	Figs. 2.31, 3.15	Zhang Yungxiang . et al. (1990)
f. East African rift	Early Miocene-Recent	Figs. 2.4, 3.37	Baker (1987)
9. *NW to SE, W to E, and SW to NE volcanic belt; eastern tip propagates eastward*			
a. Young volcanics of British Columbia Coast Ranges	Miocene-Pliocene	----	Bevier (1989)
b. Snake River Plain-Yellowstone caldera	Miocene-Recent	----	Sears et al. (1990)
c. Southern British Columbia-northern Great Basin	Late Eocene-early Miocene	Overall NW to SE shift	Armstrong (1978)
d. Southern Great Basin	Cenozoic	Overall SW to NE shift	Leeman and Fitton (1989)
e. Colorado Plateau	Cenozoic	Overall SW to NE shift at 2.9 cm/yr	Tanaka et al. (1986)
f. Cenozoic silicic volcanics of Tasman foldbelt	Maastrichtian Recent	Mafic volcanics have no pattern	Pilger (1982)
g. Tyrrhenian Sea, Mediterranean Sea	Miocene-Quaternary	----	Locardi (1988)
10. *NW to SE, W to E, and SW to NE rift, eastern tip propagates eastward*			
a. Hetao-Yinchuan rift, Ningxia and Shaanxi, China	Late Eocene-Quaternary	Fig. 2.31	Zhang Buchun et al. (1985)
b. Fenwei rift, Shaanxi and and Shanxi, China	Late Eocene-Quaternary	Fig. 2.31	Wang Jingming (1987)
c. Baykal rift, south-central Siberia	Eocene-Pliocene	Figs. 3.6, 4.20	Logatchev and Florensov (1978)
11. *North-south foldbelts migrating eastward*			
a. U.S. Western Cordillera	Devonian-middle Eocene	----	Roberts et al. (1965); Verrall (1989)
b. Lesser Antilles	Middle Cretaceous-Recent	Aves Ridge is Cretaceous, Volcanic Caribbees, younger	Meyerhoff and Meyerhoff (1972a)
12. *North-south depocenters migrating eastward*			
a. Izumi trough, Honshu, Japan	Late Cretaceous	----	Miyata (1990)
b. New Zealand	Cenozoic	----	Stern and Davey (1989)
c. Southeastern Arabian Peninsula, Oman	Late Proterozoic-Late Cretaceous	----	Ries and Shackleton (1990)
13. *Surface structures showing eastward-flow directional features*			
a. Rhine graben fault pattern, Germany	Eocene-Recent	Fig. 2.5. Horsetail structures point northeastward	Illies and Greiner (1979)
b. Ankara mélange, central Turkey	Late Cenozoic	Horsetail structures point northeastwar	Norman (1984)

2.3.20 Seismic Studies of Oceanic Crust

Worzel (1965) seems to have been the first to recognize another major puzzle. He noted that, no matter where crustal thickness is measured in the deep ocean basins, it is everywhere approximately the same, that is, in the range of 5-7 km. Vogt et al. (1969) noted the great variation in assumed sea-floor spreading rates on different midocean ridges, yet the oceanic-layer "...thickness is always the same. Production of this layer by branches of the mid-oceanic ridge must keep close pace with the crustal-spreading rate, and any theory for [this] layer must account for this fact" (Vogt et al., 1969, p. 593). Increasing numbers of studies of oceanic

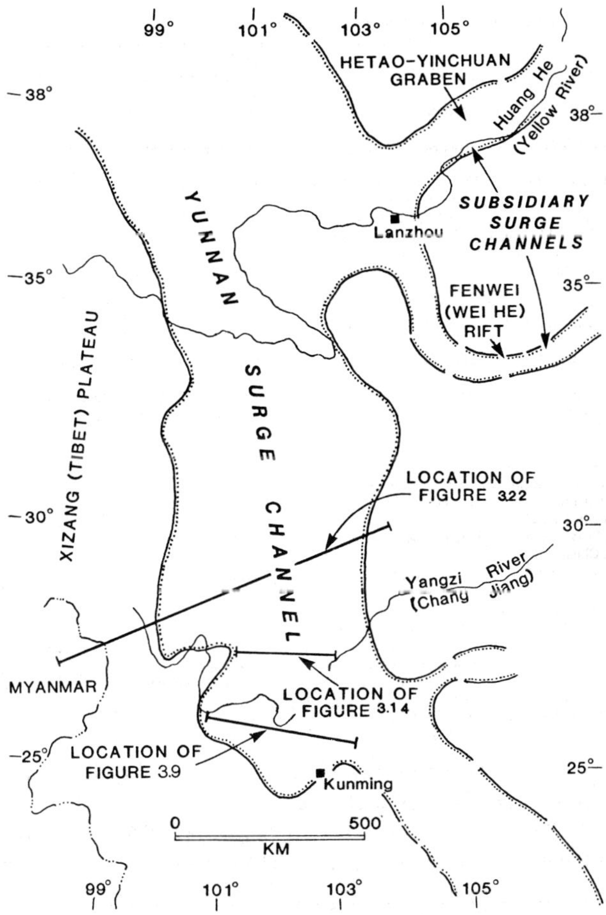

Fig. 2.31. Seismotomographic map (at approximately 50 km depth) of surge channels in southwestern China. The principal channel, a continental trunk channel, is the north-south surge channel that underlies and sustains the eastern margin of the Qinghai-Tibet Plateau. The Yunnan surge channel portrayed here underlies the kobergen shown on Figure 3.22. Locations of Figures 3.9, 3.14, and 3.22 are shown. Three continental rifts (Hetao-Yinchuan, Fenwei, and unnamed) branch *eastward* from the Yunnan channel; these are continental breakout channels as defined in the text and illustrated on Figure 3.39. They constitute further proof of the importance of the earth's rotation in surge-channel and asthenosphere flow. This figure also demonstrates that foldbelts and rifts have a common origin. Compiled from seismotomographic maps published by Liu Jianhua et al. (1989). From Meyerhoff et al. (1992b).

crustal thicknesses by seismic-refraction and seismic reflection techniques have shown that this statement still holds true. However, the plate tectonics hypothesis has no explanation for nearly constant thickness of the oceanic crust and no attempts have been made, to the best of our knowledge, to provide an explanation.

2.3.21 Ocean-Floor Bathymetry

By the end of the 1950s, Navy bathymetric data was being collected on the DEWLINE surveys using a portable fathometer and ships of opportunity (to collect harbor data). Deep ocean data in the form of sounding tracks were received at the Navy Hydrographic Office (HYDRO) and cut up into their appropriate areas to be filed in letter boxes. It is here that Bruce Heezen formulated the idea of a circumglobal midocean ridge system based on data collected in this manner from the *Challenger*, *Meteor*, and *Verma*. However, prior to the 1960s, depths and ships' tracks were simply plotted onto one overlay and contoured as the need arose, with little regard for accuracy. In fact, shoals and shallows were all that mattered, as they represented navigational hazards. Before 1960, all of the data collected by the Navy were unclassified for use by the surface fleet.

As technological advances improved the resolution of bathymetric data, it became evident that older techniques were introducing large errors into the data. First, the sonar system introduced artificial slope migration. In the early days, ensonification was in a 65° cone. The hydrophone collected the first return from within that cone, and the fathogram recorded the two-way travel time. For the Navy collectors that beam width became 9°, and this lessened the slope migration features only to the accuracy of that system combined with the accuracy of the navigation system. Surveys in the early 1960s used long range navigation (LORAN-C; J. Rubisch, pers. comm., 1993) as the horizontal positioning control. Used with the 9° cone sonar collector, these surveys were quite accurate, and the charts are still in use today. Surveys outside the LORAN-C area had little value because they were using dead reckoning as the horizontal positioning control for the unclassified surveys.

Second, older techniques created what is known as the "sonar bottom." In essence, the sonar bottom is an artifact of the sound velocity used to collect the bottom data. If an artificial number, such as 4800 feet per second is used, then the two-way travel time of the sonar ping is modified. Most sound velocities are in excess of 5000 ft/sec. When multiplying that number by the one-way travel time, the seafloor level is artificially raised, hence creating the "sonar bottom."

In 1962, armed with new technology, HYDRO became the Naval Oceanographic Office (NAVOCEANO). With it came an explosion in seafloor data collection. Its presence was equally appreciated by academicians, who began using the data extensively. B. Heezen and M. Tharp were known to bring grocery carts to NAVOCEANO to transport volumes of documents down the halls for copying while working on an Office of Naval Research contract (Dick Murchison, pers. comm., 1993). New surveying departments were formed, and numerous individuals were hired to staff new projects. One of these projects was the Ocean Survey Program (OSP) Division. The mission of this group was to collect classified bathymetric and geophysical data in various deep-ocean basins, progressing from north to south at a predetermined line spacing. The Navy Navigation Satellite System was orbited to help navigation, long - range navigation (LORAN-C) nets were emplaced, an inertial navigation system (SINS) to help dampen the effects of gravity was added to the survey platforms, and dedicated surveys were begun in earnest.

The use of bathymetry outside the auspices of the Navy took major strides forward during this time frame. As a historical adjunct, Allen Lowrie (pers. comm., 1995) of NAVOCEANO's Geophysical/Acoustic Data Base Division, was working for Heezen at Columbia University's Lamont-Doherty Geological Observatory at this time. Lowrie notes that Heezen used pieces of bathymetric data in the North Atlantic that were obtained from NAVOCEANO. Heezen had "discovered" the Mid-Atlantic Ridge, and felt that fracture zones were connected compression features across the ocean basin. He also believed that the Earth was not expanding, and was particularly drawn to the concept of seafloor spreading. With this concept in mind, Heezen aligned his fractures perpendicular to that azimuth (the Mid-Atlantic Ridge). This introduced the straight-line fracture zones in the North Atlantic that are readily apparent on many seafloor maps (e.g., any of the Heezen-Tharp maps).

By 1965-1966, several monumental events happened to change forever the face of unclassified bathymetry and bottom contour maps. At this time, most employees of NAVOCEANO were shorebased, working with contour maps. Two of these individuals, Joe Gilg and Fred Sorensen, were making point, or custom built, charts of specified areas where the data could be more closely scrutinized before inclusion into the data set. That idea was a revolution in itself because the old method of contouring was to simply draw lines and give particular attention to the shallows. Also at this time, the multibeam sonar collector, Sonar Array Sounding System (SASS) was put into use on three of the NAVOCEANO hulls (the USNS *Bowditch*, *Dutton*, and *Michelson*, affectionately known as the "big grey pigs"). The operational date was 12 July 1967 (R. Jones, pers. comm., 1993). This was the first use of a system that, when employed with swath mapping techniques, provided a total bottom contour image (Glenn, 1970). There was no need to guess any more about actual features because of the 1°x1° beam width sonar collection process which employed a fan-of-sound and a series of hydrophone receivers. That fact alone established the SASS as the world's premier sonar collector to date. It also reduced the artificial slope migration figure appreciably. However, the SASS data were totally classified.

In the late 1960s the Navy decided to compile atlases of the northern hemisphere seafloor bathymetry. Scripps Institution of Oceanography was awarded a grant to compile the North Pacific. NAVOCEANO sent individuals to help in that effort, which resulted in the publication of the North Pacific maps in 1971. Interestingly, the Mapmaker Seamounts which were discovered during this compilation were later shown by NAVOCEANO to be the tops of the parallel ridges of the passing Mendocino and Surveyor Fracture Zones (Mammerickx and Smith, 1985). Other discrepancies between academic and NAVOCEANO interpretations were noted by NAVOCEANO workers as well (J. Gilg, pers. comm., 1992).

The Navy's World Bathymetry Group (WBG) was assigned the job of compiling the bathymetric data for the North Atlantic, which was partially classified. Since the WBG was an unknown entity, an advisory group of experts from universities and oceanographic institutions was appointed to oversee their endeavors. That panel included B. Heezen, R.L. Fisher of Scripps, B. Hersey from the Office of Naval Research, and E. Uchupi from Woods Hole Oceanographic Institute. During this time, the WBG and its advisors found that many of the contours did not agree with the pre-conceived ideas proposed by the plate-tectonics model; for example, N-S trends are present where E-W trends theoretically should be present in the North Atlantic. F. Sorensen, who cross-trained with OSP in 1967, realized that the fracture zones were oriented 90° to the tectonic fabric, and took this knowledge back to the WBG. He also noted bends in the fracture zones, and a 30 nautical mile spacing of transverse faults in the midocean ridges in many areas, as well as the lack of these faults in other areas (F. Sorensen, pers. comm., 1990; Dick Murchison, pers. comm., 1995). The information sent the WBG into intensive discussion, disagreement, and discord (F. Marchant, pers. comm., 1994), but few, if any, hypotheses were altered to fit the real data.

In 1974, NAVOCEANO was split again and the WBG reorganized. At this time B. Bergantino joined the group to enhance their ability to show three dimensional data of the seafloor. In this effort, Bergantino published numerous maps in which he tried to portray the third dimension uniformly, keeping the distortion of shapes to a minimum and the location of features as close as possible to reality. Interestingly, Bergantino's maps, in which he used methods by Raisz, Robinson and Thrower to produce planimetrically correct perspectives, were derived from the same WBG data that Heezen and Tharp used. Bergantino's more correct use of cartographic principles allowed the quality of his map to supersede the Heezen and Tharp map by virtue of better analyses and less vertical exaggeration. These maps have been virtually ignored by the community since they were completed.

Figure 2.32a shows a bathymetric map of a part of the central North Atlantic Ocean, taken from the GEBCO (1975-1982) charts. Figure 2.32b is a NAVOCEANO physiographic diagram of exactly the same area based on single-beam sonar data, and compiled by R. N. Bergantino during 1973 and 1974. The two maps are compiled from nearly the same random track data. Significant differences between the two figures are evident (see figure captions for Figs. 2.32a and 2.32b). Further analyses and comparisons of the Bergantino and GEBCO maps have been published recently by Smoot and Meyerhoff (1995), who show that in light of more recent data, Bergantino is the more correct in his interpretation of the Atlantic Ocean seafloor bathymetry than the authors of the GEBCO and the National Geographic maps. As an example of errors in the latter map, the 1995 version of the National Geographic Atlantic Ocean Map labels the Faraday and Maxwell fracture zones, even though these do not exist bathymetrically (Smoot, 1989). As Smoot (1989, 1990b) and Smoot and Meyerhoff (1995) wrote, the increasing quantity of data from the U. S. Navy Oceanographic Office's multibeam sonar charts and GEOSAT satellite radar altimeter surveys will undoubtedly improve future tectonic models. As an example, Figures 2.33a and 2.33b show how information collected by the GEOSAT satellite altimeter can be used. From March 1985 to October 1986, the GEOSAT satellite measured the marine geoid height to determine the gravity geoid (Fig. 2.33a). The GEOSAT orbited once every 100 minutes at a track spacing of 3 km or less, collecting information between 72°N and 72°S latitude. The tracks were manually edited to remove obvious spikes and noise, as well as misadjusted tracks and segments over polar ice. The base for Figure 2.33b was produced by applying a two-dimensional high-pass filter. The filter extremes were set at 70-125 nautical miles. With certain exceptions, the structural trends correlate well with the bathymetry. The concept has already been ground-truthed with satisfactory results using the less-accurate Seasat altimetry data (Smoot and Meyerhoff, 1995). These methods will help in determining large-scale events in the southern ocean basins, where very little bathymetry exists.

2.3.22 Tensile Stress Characterizes the Midocean Ridge Regimes

Fundamental to the plate-tectonic hypothesis is the requirement that ridges are dominated by tension. However, new CDP reflection-seismic data summarized on Figure 2.34 show that this is not true. Antipov et al. (1990) have demonstrated that a zone of thrust faults, 300-400 km wide, flanks both sides of the Mid-Atlantic Ridge. These data indicate that the bilateral thrusting of the midocean ridges is a worldwide phenomenon (Meyerhoff et al., 1992a).

2.3.23 Oceanic Basement

Many geologists assume that the basalt found at the base of deep-sea drillholes is basement. We have examined in detail the first 429 drillsites of the Deep Sea Drilling Project (Legs 1-54), and found the following. Of these 429 sites, only 165 reached basalt. Some in fact penetrated more than one basalt. All but 12 of the 165 basalt penetrations were called basement including 19 sites where the upper contact of the basalt with the sediments was baked. In 33 sites, the upper contact was depositional. The contacts at 101 sites were *not* recovered, yet all 101 were called depositional contacts. Table 2.3 summarizes the basalt data from the first 54 legs of the Deep Sea Drilling Project.

The practice of calling baked contacts basement was explained during Leg 9 in the Pacific Ocean. The purpose of this leg was to determine the ages of the linear magnetic anomalies of the eastern Pacific. At Site 79 on a magnetic anomaly believed (on the basis of theory) to be 21 Ma, a baked contact was penetrated at total depth of the drillhole. Because the shipboard scientists needed a depositional contact, they navigated the ship along the strike of the magnetic anomaly to a second location, Site 80. Again a baked contact was reached, but beneath sediments of the exact same age as those at Site 79. Since the paleontological age was very close to the basement age predicted by theory, the Shipboard Scientific Party (1972, p. 401) declared the basalt sill to be basement, stating "The baked sediment...overlying the basalt is included in the same foraminiferal and nannofossil zones as those at the base of Site 79 [where a baked contact also was found]. Because these two sites are on [strike], such age equivalence is in accord with the concepts of sea floor spreading and although the basalt encountered at Site 80 is a sill, suggests that the silling process does not invalidate the basal sediment age as an indicator of basement age." We conclude that the "basements" defined by the Deep Sea Drilling Project are not always reliable and should not be used as irrefutable evidence of sea-floor spreading or plate tectonics.

2.3.24 Increasing Age of Oceanic Crust Away from the Midocean-Ridge Crest

Another imperative of the plate-tectonics hypothesis is that the oceanic crust becomes older from the midocean ridge crests toward the adjacent continents. However, Paleozoic and Proterozoic igneous rocks were recovered in cores during Legs 37 and 43 (Reynolds and Clay, 1977; Houghton et al., 1979), both on the Mid-Atlantic Ridge and the Bermuda Rise. Not one of these occurrences of ancient rocks was mentioned in either the Cruise Site Reports or Cruise Synthesis Reports. Reynolds and Clay (1977), reporting on a Proterozoic date (635 ± 102 Ma) near the crest of the Mid-Atlantic Ridge, wrote that the age had to be wrong because, on the basis of magnetic anomalies, the site could not contain rocks older than about 10 Ma. Here is a classic example of a result being used to reject its own research.

Figure 2.35 shows age data from the Atlantic Ocean from Legs 1 through 54 of the Deep Sea Drilling Project. Although the number of sites is limited, the data show that the simple assumption of increasing age with increasing distance from the midocean ridge crest is not

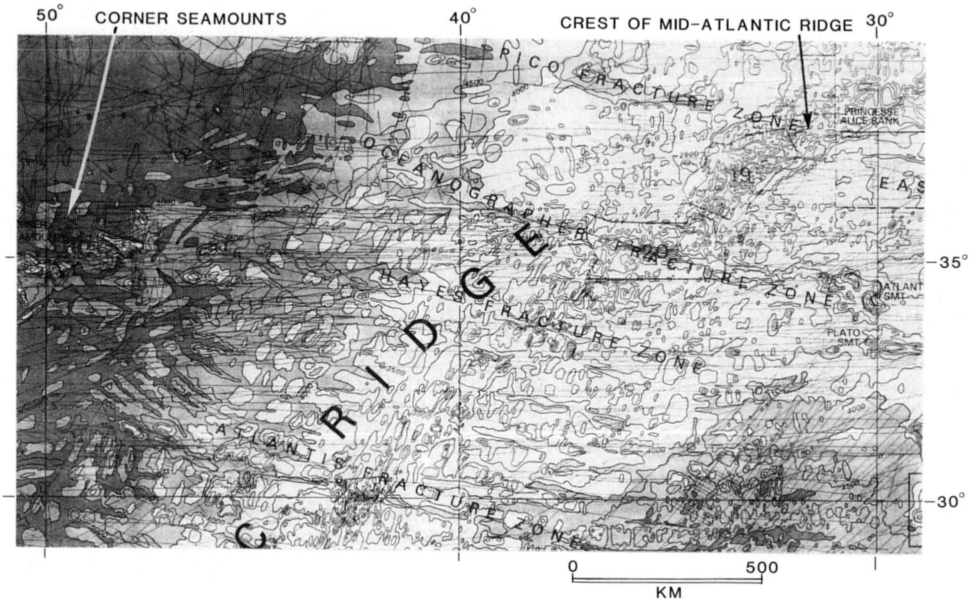

Fig. 2.32a. Bathymetric map, central North Atlantic Ocean, approximately 29° W to 51° W longitude, 29° N to 52° N latitude, from GEBCO (1975-1982). This map is based on bathymetry compiled from soundings from many sources and methods that date back more than 100 years. Sources do include at least some of the same random track data used to compile Figure 2.32b. More detailed comments appear in the caption for Figure 2.32b. From Smoot and Meyerhoff (1995).

always accurate. If dredge samples are included, the whole concept must be called to serious question.

2.3.25 Mid-Atlantic Ridge Geology North of 37° N Latitude

In the area bounded by 37° and 45°30' N latitude and by 19° and 35 °W longitude, an area greater than 1,800,000 km^2, thousands of outcrops of Proterozoic and Paleozoic rocks appear to be present. This conclusion is based on the fact that nearly every dredge haul from uplifted blocks in the region contains old rocks, many of them granitic (Aumento and Loncarevic, 1969). Aumento and Loncarevic (p. 13) wrote that dredge hauls in the Bald Mountain region just west of the Mid-Atlantic Ridge crest at 45° N latitude consistently recovered about 74% silicic rocks, and commented that this is "...a remarkable phenomenon...." The dismissal of such hauls as evidence of "glacial erratics" and "ship ballast" (Heezen et al., 1959) can no longer be given credence. Moreover, the number of pre-Cretaceous rock localities on the Mid-Atlantic Ridge is so great that such localities, extending from the Arctic Ocean through the South Atlantic Ocean, can no longer be disregarded. The Bald Mountain locality is a case in point (Aumento

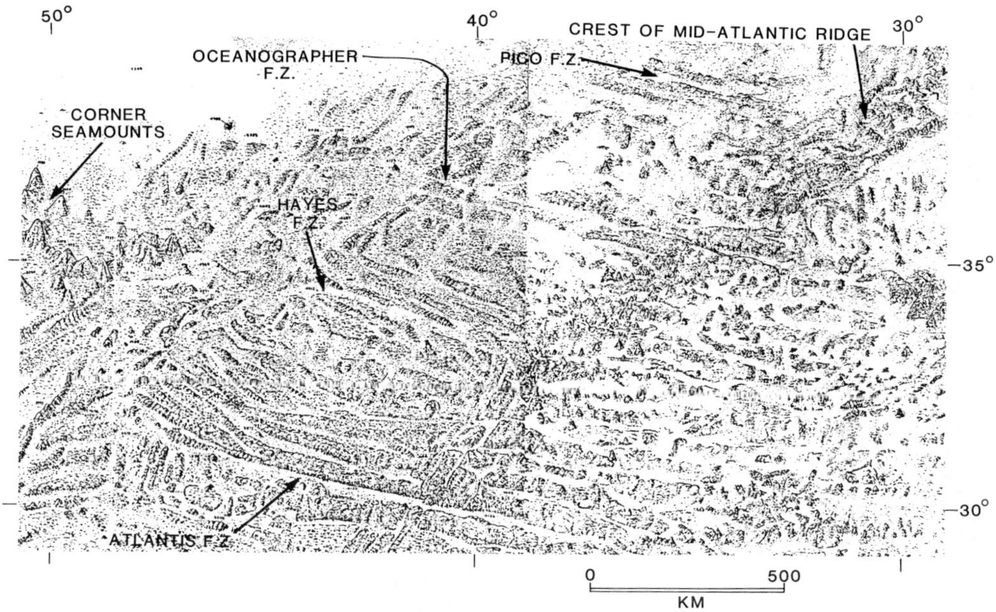

Fig. 2.32b. Physiographic diagram, central North Atlantic Ocean, same area as Fig. 2.32a. Source: Bergantino (1973), U. S. Naval Oceanographic Office, World Chart Compilation Group, NAVOCEANO. There are more than 400 discrepancies between the bathymetric features shown on Figure 2.32b and those shown on Figure 2.32a. Figure 2.32b is based on a single method--ship-of-opportunity single-beam sonar charts prepared by the World Chart Compilation Group of NAVOCEANO. Consequently, Figure 2.32b is far more accurate. A few significant differences should be noted: (I) On Figure 2.32b, the Pico fracture zone (f.z.) is 300 km long, but is forced on Figure 2.32a through a distance of 1,100 km. (ii) On Figure 2.32b, the Hayes f.z. is 550 km long, but has been forced for about 1,000 km on Figure 2.32a. The Hayes f.z. bifurcates in two places on Figure 2.32b; it does not on Figure 2.32a. (iii) The Atlantis f.z. is shown totally differently on the two figures. (iv) Neither the Pico nor the Hayes f.z. crosses the crest of the Mid-Atlantic Ridge. The Pico f.z. does not even reach the crest. (v) A large S-shaped structure, produced by five parallel ridges and depressions, dominates the topography just west of and up to the crest of the Mid-Atlantic Ridge on Figure 2.32b between the Atlantis and Oceanographer f.z.'s. There is no hint of this huge structure on Figure 2.32a. (vi) East-west structures extend almost uniformly west of the Mid-Atlantic Ridge crest for a distance of 600-700 km, where they are cut off by northeast-southwest ridges and troughs. There is almost no hint of this major morphotectonic change on Figure 2.32a. (vii) East of the Mid-Atlantic Ridge crest, through-going fracture zones are scarce (only the Atlantis and Oceanographer f.z.'s). In contrast, at least eight through-going fractures are shown on Figure 2.32a. (viii) The near-total difference between the topography east and west of the ridge crest is striking. Additional analyses of these maps and others were published recently by Smoot and Meyerhoff (1995). From Smoot and Meyerhoff (1995).

and Loncarevic, 1969). Centered at 45°13' N latitude, 28°52' W longitude, it is 13 km long, 3.5 km high, and up to 5 km wide, with an estimated volume of 80 km^3 (hardly a glacial erratic or ship's ballast!). It consists of granitic and silicic metamorphic rocks ranging in age from 1,690 to 1,500 Ma (Wanless et al., 1968), and is intruded by 787-Ma mafic dikes or sills.

Along the eastern flank of the Mid-Atlantic Ridge north of the Azores, along the northern flank of the Azores Ridge (Azores-Gorringe Bank area), and along the scarps facing the King's

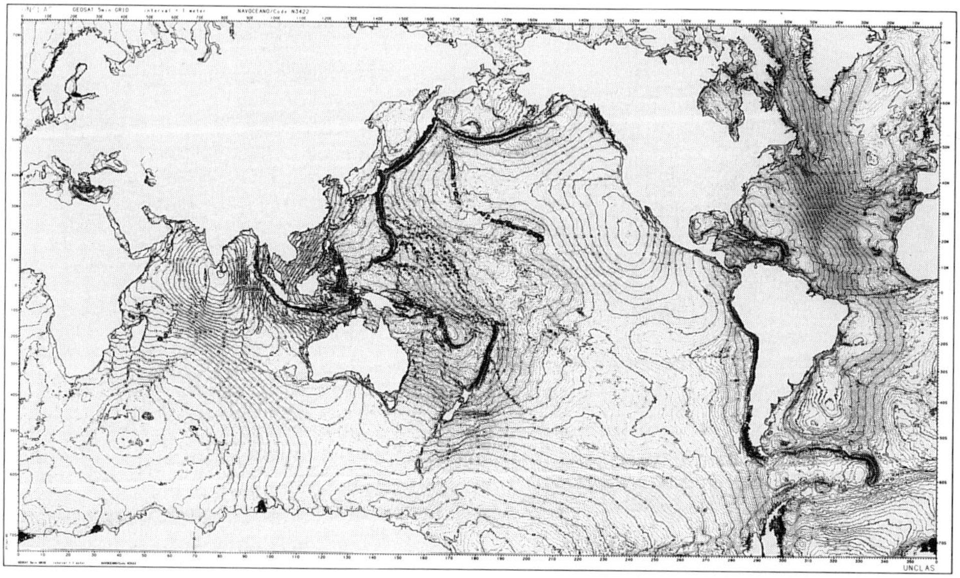

Fig. 2.33a. Gravity geoid based on 5-cm GEOSAT altimetry data showing the regions of upwelling (positive gravity geoid) and downwelling (negative gravity geoid). By applying a high-pass filter, Pacific Ocean structural trends become visible (see Figure 2.33b).

Trough (44° N, 22° W), fossiliferous Paleozoic sedimentary rocks have been recovered during dredging operations, and even on fishing nets, since 1883 or earlier (Furon, 1949; Meyerhoff, 1981a). The *Talisman* (1883) collections were reported by Furon (1949). The more recent collections from fishing nets are in the Sedgwick Museum, Cambridge University. C. P. Hughes (Cambridge University, written and oral communs., 1978-1980) has identified many of them. One of his collections from the southern escarpment of the King's Trough contains the trilobite *Triarthus* aff. *T. spinosus* and the graptolite *Climacograptus typicalis* (Meyerhoff, 1981a), both of them being taxa from North America. Sparker profiles of the sea bottom suggest that many of the specimens are from outcrops in low-lying cuestas (C. P. Hughes, oral commun., June 1980). The faunas in every collection to date include only ostracods, trilobites, graptolites, worm trails, and similar faunal remains that would be expected on a Paleozoic ocean floor. Shallow-water taxa have not so far been recovered. The Sedgwick Museum had a good collection of deep Atlantic Ocean Paleozoic faunas on display in the 1980s.

2.3.26 Mid-Atlantic Ridge Geology of the Equatorial Atlantic

Melson et al. (1972), studying St. Peter and Paul's Rocks at the crest of the Mid-Atlantic Ridge just north of the equator, found an 835-Ma peridotite stock associated with other rocks giving 350-, 450-, and 2,000-Ma ages. In the same locality, middle Miocene sediments and Quaternary basalt flows are present. At 2°15.5' N latitude, 150 km north of the St. Peter and

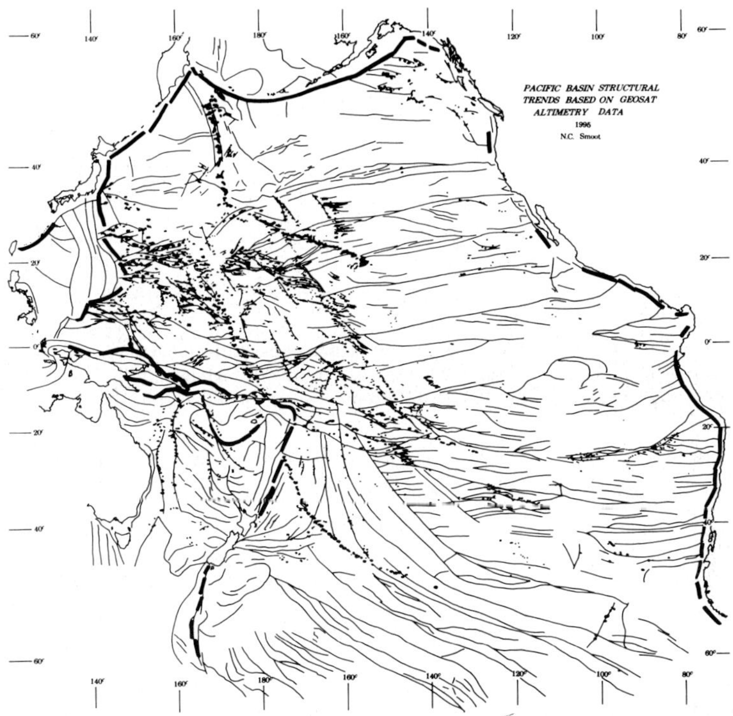

Fig. 2.33b. Structural trends of the Pacific Basin. Trends are related to the height above sealevel measured by the satellite. The East Pacific Rise has been filtered out of this data set. Compare with Figure 2.33a.

Paul's locality, Udintsev et al. (1992) found a section of sedimentary breccia containing a sequence of Maastrichtian through Holocene marine strata, the breccia itself is Pliocene. These and an additional Eocene locality farther south are only 20 to 30 km east of the central rift valley.

Timofeyev et al. (1992) and Udintsev and Kurentsova (1993) sampled many other localities in the equatorial Atlantic straddling the Mid-Atlantic Ridge (0°-5° N latitude, 25°-35° W longitude), and found continental flood basalts, sericitic mica phyllite, quartzite, carbonaceous shale, brown coal, and many other rocks derived from continental sources, as well as midocean-ridge-type and eugeosynclinal alpinotype rocks. Late Cretaceous-Eocene samples are common. West of the central rift is a 200-300-km-wide plateau that dips gently westward into the Guyana Basin northeast of Brazil; east of the rift valley, the plateau remains elevated and merges into the Sierra Leone Rise west of Africa. The plateau is capped with sedimentary rocks 400-1,200 m thick, sections which do not thin away from the midocean ridge crest. Timofeyev et al. (1992) and Udintsev and Kurentsova (1993) concluded that sea-floor spreading in this >550,000-km^2 area is not possible.

2.3.27 Linear Magnetic Anomalies

Magnetic field reversal during generation of new oceanic crust is the only hypothesis that has

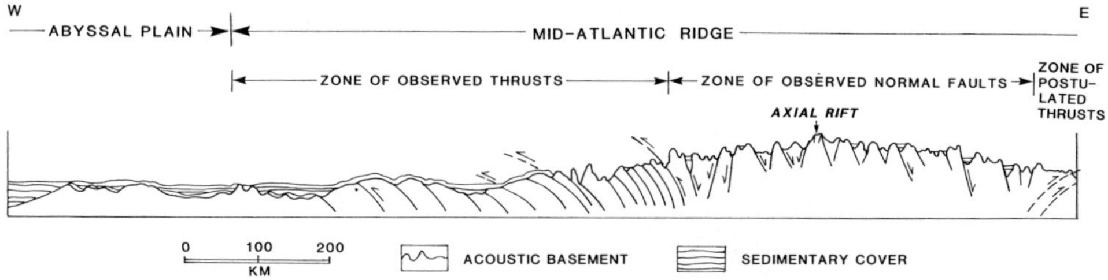

Fig. 2.34. This is a west-to-east schematic structural interpretation of the Mid-Atlantic Ridge between 13° and 14° N latitude. The interpretation, from Antipov et al. (1990), is based on several high-quality CDP reflection-seismic lines. Note the bivergent (kobergen) structure that proves the compressive origin of the Mid-Atlantic Ridge at this latitude.

been used to explain the alternate positive and negative magnetic anomalies that characterize much---but by no means all---of the world's midocean ridges. This hypothesis was first proposed by L. W. Morley in June, 1963 (Morley and Larochelle, 1964) and shortly thereafter by Vine and Matthews (1963). Agocs et al. (1992) presented an alternative interpretation, specifically, that the anomalies are a consequence of magnetic susceptibility contrasts. It should be noted that the authors of the earlier hypothesis did not transform the data to the pole (recalculated to a common latitude), and the hypothesis, therefore, is speculative. In contrast, Agocs et al. (1992) made a truly quantitative study with controlled modelling parameters.

In the Agocs et al. (1992) scheme, the anomalies are not caused by alternate strips of positively and negatively magnetized mafic rock, but by alternate strips of more positive and less positive (or, conversely, more negative and less negative) mafic rock. This type of magnetic susceptibility contrast model was first proposed by Luyendyk and Melson (1967). These authors, as well as Opdyke and Hekinian (1967), reported magnetization values from oceanic rocks that are more than adequate to produce the observed anomalies by magnetic susceptibility

Table 2.3. Basement penetration summary, Legs 1-54 (Sites 1-429), Deep Sea Drilling Project, August 1968-June 1977. From Meyerhoff, in press.

	Total Sites Drilled	As a % of 429	As a % of 165
TOTAL SITES DRILLED	429	100	--
- Basalt not penetrated	264	62	--
- Basalt penetrated	165	38	100
BASALT DATA			
- Total number of sediment-basalt contacts <u>not seen</u>	101	24	61
- Sediment-basalt contact data recovered	64	15	39
- Contact is depositional	33	8	20
- Contact is baked	31	7	19
- Total number of baked contacts that were called "basement"	19	4	12

Source: DSDP volumes 1-54, 1969=1980.

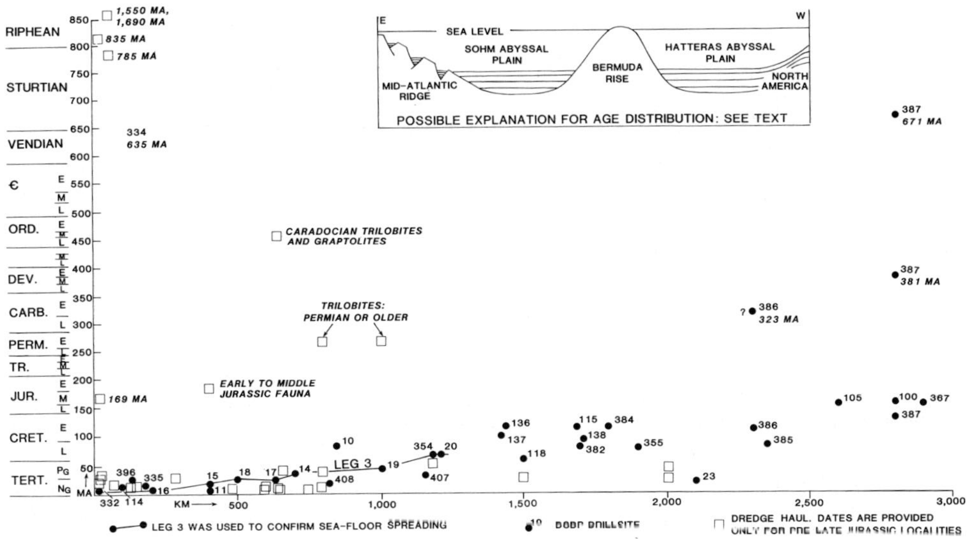

Fig. 2.35. This is a plot of rock age vs. distance from the crest of the Mid-Atlantic Ridge. The figure is designed to show (to scale) rocks of all ages, whether from drill holes of the Deep Sea Drilling Project or dredge hauls. An east-west cross section of the western North Atlantic basin is inserted to provide an explanation for the pre-Jurassic rocks. Pre-Jurassic rocks so far have come mainly from the Mid-Atlantic Ridge and Bermuda Rise. Sources of the age data are listed in Meyerhoff et al. (1992a). The inset is highly schematic.

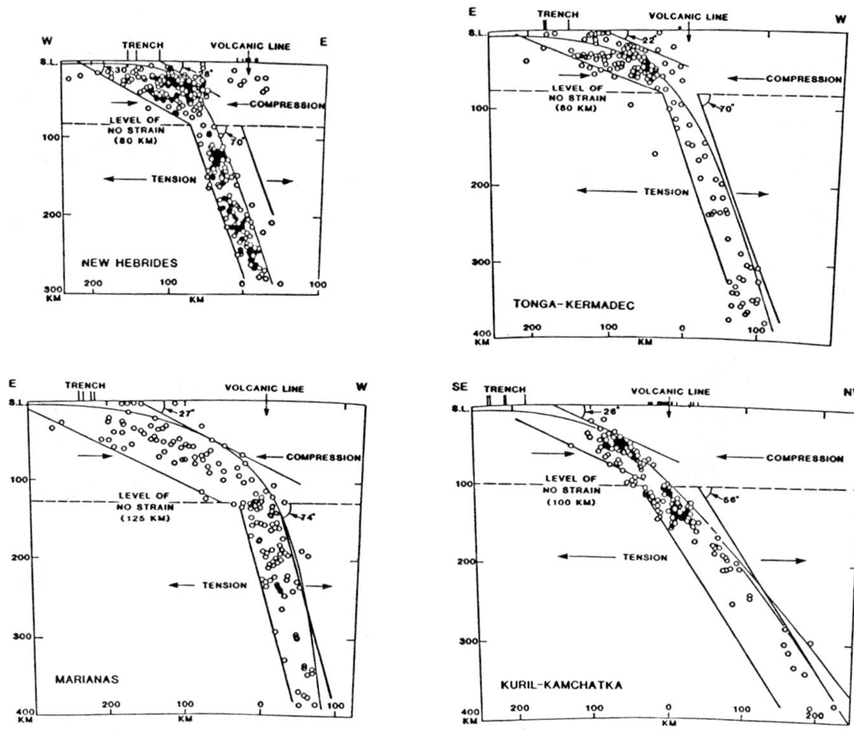

Fig. 2.36. Sections showing hypocenter distribution beneath four western Pacific volcanic arcs (from Isacks and Barazangi, 1977). Each section displays distinct lithosphere and strictosphere Benioff zones, each with its own dip angle. For comparison, we have retained--in each section--the curved line drawn by Isacks and Barazangi (1977) to represent the shape of the Benioff zone.

contrasts. The linearity of the anomalies in some parts of the midocean ridge system is explained in surge tectonics as a natural consequence of ridge-parallel flow of mafic magma in channels beneath the midocean ridges.

2.3.28 Paleomagnetism

Paleomagnetism studies during the late 1940s and early 1950s (e.g., Graham, 1949) revived interest in the hypotheses of continental drift and polar wandering. By the mid-1960s, several books on paleomagnetism had garnered widespread attention (e.g., Nagata, 1953, 1961; Blackett, 1956; Irving, 1964). Continuing research in the field has produced several additional textbooks (e.g., Parkinson, 1983).

Paleomagnetic techniques and results have been plagued with uncertainties from the outset. Most of these were reviewed by Rezanov (1962, 1968), Northrop and Meyerhoff (1963), and Meyerhoff (1970). Despite these uncertainties, paleomagnetism's proponents have continued to amass data on the magnetic properties of rocks---declinations, inclinations, intensities, and so forth---and on the basis of their studies, have proposed horizontal translations of landmasses for distances up to 7,000 km and more (e.g., Drewry et al., 1974). In some cases, these move-

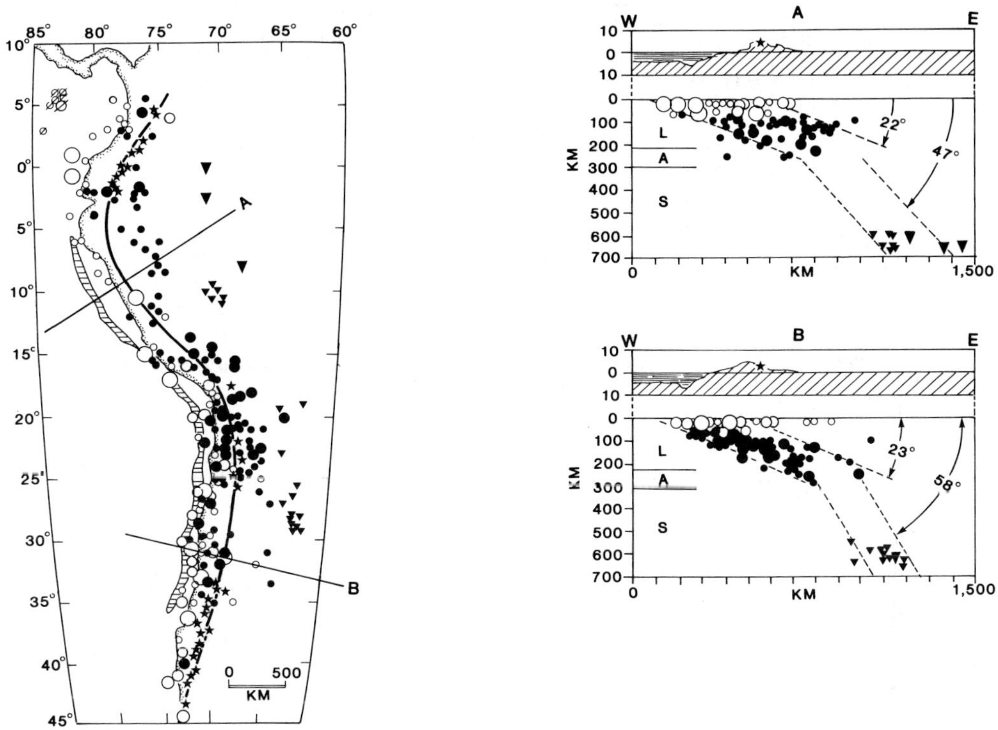

Fig. 2.37. East-west cross sections across Peru-Chile Trench by Benioff (1954). Hypocenters are shown. The change in dip from zone L (lithosphere) to zone S (strictosphere) takes place in zone A (asthenosphere). Thus the asthenosphere is the *level of no strain* in Davison's (1887) contraction theory, and the change in dip from 22°-23° in the lithosphere to 47°-58° in the strictosphere follows the Navier-Coulomb maximum shear-stress theory. Hence the Earth is contracting.

ments have not been compatible with field data (e.g., Donovan and Meyerhoff, 1982; Parnell, 1982; Saul, 1986; Hansen, 1988; Newton, 1988; Butler et al., 1989; Irving and Archibald, 1990), and hence two new hypotheses that explain the paleomagnetic data in full have been developed, one by Butler et al. (1989) and the other by Storetvedt (1990). Butler et al. (1989) explain paleomagnetic results by vertical block rotations and tilts, Storetvedt (1990) by horizontal block rotations. Either of the two hypotheses, or a combination of the two, can explain the paleomagnetic data, and long-distance translations of thousands of kilometers are not necessarily required.

2.3.29 Benioff Zones

We believe that Benioff zones, another *sine qua non* of the plate-tectonics hypothesis, have been misinterpreted in that hypothesis. They generally are depicted as 100-km-thick slabs entering the lithosphere at a shallow angle near the Earth's surface and gradually curving around to an angle between 60° and 75° to the Earth's surface (Isacks and Barazangi, 1977; Fig. 2.36).

This configuration does not exist as Benioff (1949, 1954), their discoverer, proved (Figs. 2.37, 2.38). Benioff showed that a Benioff zone consists of two separate segments, one shallow and one deep. The shallow segment between the Earth's surface and a depth of about 150 km dips between 22° and 45°; the deep segment below 150 km dips between 45° and 75°. The average dip of the shallow segment is 30°; the average dip of the deep segment is 60°. Benioff (1954) even found that the two segments in some places are offset 100-200 km from one another! Isacks and Barazangi (1977) found exactly what Benioff (1949, 1954) did but, in accordance with plate tectonic models, depicted the zone as a single curving segment (compare Figs. 2.36, 2.37). In an analysis of all of the world's Benioff zones (Smoot, 1993; 1994a; in press; in review), it appears that deep earthquakes are disconnected from shallow ones. Very few intermediate earthquakes exist. Smoot (1993, 1994a, in press, in review) observed that deep earthquakes lie landward of aseismic, buoyant features located at the trench segment junctures, and that these sites are also the loci of clustered surficial volcanoes. This suggests that these phenomena are interrelated, possibly in the form of surge channel constrictions (Smoot, 1993, 1994a; in press; in review).

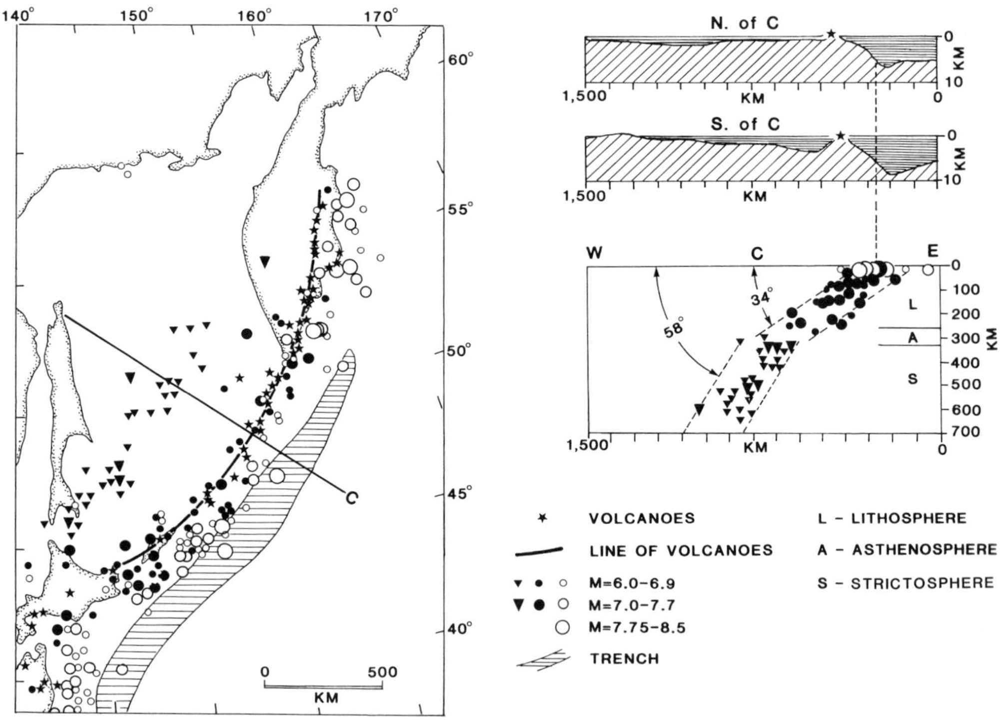

Fig. 2.38. Northwest-southeast cross section across Kuril-Kamchatka Trench by Benioff (1954). Hypocenters are shown. The change in dip from zone L (lithosphere) to zone S (strictosphere) takes place in zone A (asthenosphere). Thus the asthenosphere is the *level of no strain* in Davison's (1887) contraction theory, and the change in dip from 34° in the lithosphere to 58° in the strictosphere follows the Navier-Coulomb maximum shear-stress theory. Hence the Earth is contracting.

2.3.30 Mechanism

All tectonic hypotheses have always been plagued by the need for a mechanism. Many workers have written about these mechanisms; two good summaries are by Wyllie (1971) and Cox and Hart (1986). Cox and Hart (p. 337) wrote that the likely mechanism is convection, whether the plates ride as passive passengers above moving mantle or whether the plates are themselves part of the active convection process. Closely related is an ongoing debate about the size of convection cells: is the whole mantle involved or only that part above the 670-km discontinuity? Theoretical treatments have been published by Turcotte and Oxburgh (1969, 1976), Lomnitz (1974), Cathles (1975), and Stevenson and Turner (1979).

What energy sources are available to generate convection? Wyllie (1971, p. 353-354) argued that the radioactive decay of U, Th, and K provides the only adequate energy source. Lliboutry (1969) proposed ridge-push in which rising magma pushes the plates apart at the ridge crests. Hales (1969) proposed that gravitational forces provide the energy. Cox and Hart (1986) opined that all of these mechanisms together maintain convection. In contrast, Olson et al. (1990) stated somewhat arbitrarily that slab subduction "...applies a torque on the viscous mantle and provides the main driving force for convection..." (p. 209). The problem is still unresolved, as concluded by Scheidegger (1982).

2.4 Conclusion

We have outlined briefly the new sets of geological and geophysical data acquired through new and old research technologies---data sets that are very difficult to reconcile with current tectonic models. A very high priority in earth science, therefore, must be to find a hypothesis that *works*.

CHAPTER 3

SURGE TECTONICS

3.1 Introduction

Surge tectonics is a new hypothesis quite unlike previously proposed hypotheses, although many of its component parts are based on ideas long known. We believe the hypothesis provides a comprehensive and internally consistent explanation of all tectonic phenomena without the necessity of making unsupported assumptions or ad hoc explanations. We have found nothing that surge tectonics cannot explain in a simpler way than other tectonic hypotheses. Surge tectonics draws on well-known physical laws, especially those related to Newton's laws of motion and gravity. Fluid dynamics plays an important role in surge tectonics. (For more information on the laws we utilize, those mentioned in the text are defined in the Appendix; those wishing more detail are referred to two standard physics textbooks by Sears et al. [1974] and Blatt [1983]. An excellent state-of-the-art fluid-dynamics text is that by Tritton [1988]).

Surge tectonics is based on the concept that the lithosphere contains a worldwide network of deformable magma chambers (*surge channels*) in which partial magma melt is in motion (*active surge channels*) or was in motion at some time in the past (*inactive surge channels*). These surge channels play the role of the holes in the "hole-in-the-plate" problem ("elliptical hole problem") of civil and construction engineering, industrial engineering, and rock mechanics (Inglis, 1913; Donnell, 1941; Murrell, 1964a, 1964b; Jaeger and Cook, 1979; McCartan and Gettings, 1991). The presence of surge channels means that all of the compressive stress in the lithosphere is oriented at right angles to their walls. As this compressive stress increases during a given geotectonic cycle, it eventually ruptures the channels that are deformed bilaterally into ***kobergens*** (Fig. 2.15). Kobergens were named by Meyerhoff et al. (1992b) in honor of the Austrian geologist, Leopold Kober (1921, 1925, 1928). Kober observed that Alpide foldbelts have been bilaterally deformed with the northern ranges vergent toward Europe and the southern toward Africa (see Fig. 12 in Meyerhoff et al., 1992b). Thus, bilaterally deformed foldbelts in surge-tectonics terminology are called kobergens.

Surge tectonics involves three separate but interdependent and interacting processes. The first process is the contraction or cooling of the Earth. The second is the lateral flow of fluid, or semifluid, magma through a network of interconnected magma channels in the lithosphere. We call these surge channels for reasons that will become apparent. The third process is the Earth's rotation. This process involves differential lag between the lithosphere and the strictosphere (the hard mantle beneath the asthenosphere and lower crust), and its effects---eastward shifts---already discussed (Table 2.3). Because lithosphere compression caused by cooling is the mechanism that propels the lateral flow of fluid, or semifluid, magma through surge channels, we discuss first the velocity structure of the Earth's lithosphere and underlying layers, then the contraction hypothesis (Earth cooling), and then the effects of contraction on the Earth's outer shells.

Table 3.1. Seismic interpretation of crust and upper mantle. Principal sources are Press (1966) and Iyer and Hitchcock (1989). From Meyerhoff et al. (1992b).

Shell	Thickness (km)	Depth to base (km)	P-wave volocity range (km/s))	Comments regarding P-wave range
Lithosphere mantle	60-180	60-180	0-8.45	Includes water through upper
Asthenosphere	160-40	220	7.85-8.05	Values are for the soft interior
Strictosphere	----	----	8.1-8.7	Top of shell only; upper "contact" transition a

3.2 Velocity Structure of the Earth's Outer Shells

3.2.1 Basic Framework

The Earth's outer shells (Fig. 3.1, Table 3.1) consist of a "hard" lithosphere above a "soft" asthenosphere (Dana, 1896; Barrell, 1914, Daly, 1940). The interpreted seismic structure of these two shells is given in Table 3.1, which is based on Press (1966) and Iyer and Hitchcock (1989). The asthenosphere overlies another hard shell that Bucher (1956, p. 1295) named the strictosphere. Its seismic characteristics (at least near the upper surface of the strictosphere) also are given on Table 3.1. Of these shells, the lithosphere is especially important because visible tectonic effects provide the principal clues to the origin of tectogenesis. It has not been too long, since the seismic structure of the crust and upper mantle became sufficiently well imaged to permit more than just educated guesses about it.

Before 1958, when Revelle (1958) discovered a layer of material with a velocity of 7.3 km/s on the southern part of the East Pacific Rise, the lithosphere was perceived as consisting of 6.6-km/s crust overlying directly the 8.1-km/s mantle. What Revelle (1958) discovered was a lens or high-velocity crust, or low-velocity mantle, with a P-wave velocity of 7.0-7.8 km/s separating the "normal" crust above from the "normal" mantle below. Similar lenses were found almost everywhere beneath the midocean ridge system (e.g., Ewing and Ewing, 1959; Talwani et al., 1965). By 1982, a similar lens of 7.0-7.8-km/s material was found beneath most of the Earth's rifts (Figs. 2.9, 3.2-3.8). Mooney et al. (1983) suggested that such lenses are a characteristic of extensional tectonic belts.

During the 1970s, refraction shooting in the northern Appalachians discovered a similar 7.0-km/s lens under the Acadian (Devonian) foldbelt (Taylor et al., 1980; Marrilier et al., 1989; and Fig. 3.13). It was not long before identical lenses beneath foldbelts were identified in many parts of the world (Figs. 2.13, 2.18, 2.21, 3.9-3.14). Good images of these lenses were recorded on reflection-seismic lines used for deep continental and oceanic tectonic studies (Figs. 2.10, 2.11, 2.14, 3.15; Nelson, 1988). Figures 2.10 and 2.11, from the English Channel and southwestern Queensland respectively, are particularly good images of two of these lenses near the base of the continental crust.

Mooney and Braile (1989), summarizing the present state of knowledge of the structure of the Earth's crust, inserted a 7.0-7.8-km/s lower crustal layer between the 6.6-km/s sialic crust above, and the 8.1-km/s mantle below. They showed the layer to be absent in places. In general, it is present beneath cratons, platforms, foldbelts, rift systems, and wrench-fault zones. Under cratonic areas the layer is 7-15 km thick; under tectonic belts it is thicker, ranging from 10-25 km. In many places the 7.0-7.8-km/s velocity is gradational with mantle velocities (> 7.9 km/s). In these places, the Mohorovičić discontinuity is not a true discontinuity but a transition zone several kilometers thick (Mooney and Braile, 1989). Areas were found also beneath ancient platforms and shields where the 7.0-7.8-km/s layer is up to 30 km thick. Examples include the Baltic Shield (3.5; Luosto et al., 1989) and parts of the Canadian Shield (Braile, 1989).

The preceding suggests that the 7.0-7.8-km/s layer is distributed rather randomly, and that its thickness in a given area is a result of random processes. These conclusions are almost unavoidable if one tries to explain the distribution and thickness with any of the Earth-dynamics hypotheses published during the last century. In surge tectonics this apparent randomness of distribution and thickness is in part predictable.

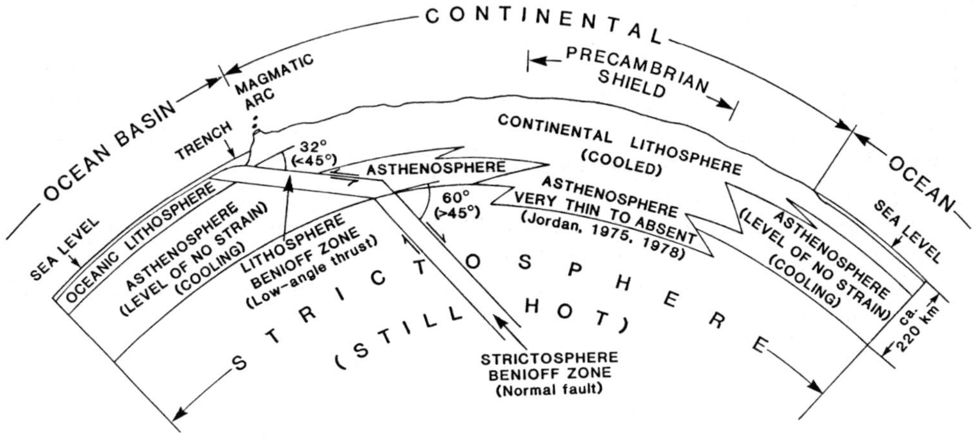

Fig. 3.1. Idealized cross section through two ocean basins, a continent, and a Benioff zone. The section shows that (1) the asthenosphere wedges out beneath continental nuclei, (2) continental nuclei are rooted to the deep mantle (strictosphere), and (3) the Benioff zone consists of two segments, a low-dipping thrust zone in the lithosphere (which is under compression) and a steeply-dipping normal fault zone in the strictosphere (which is under tension). The dip angles of the two fault segments are predicted by the Navier-Coulomb maximum shear-stress theory (Scheidegger and Wilson, 1950; Jaeger, 1962; Jaeger and Cook, 1979), a fact which demonstrates that the strictosphere is still contracting (MacDonald, 1963). From Meyerhoff et al. (1992b).

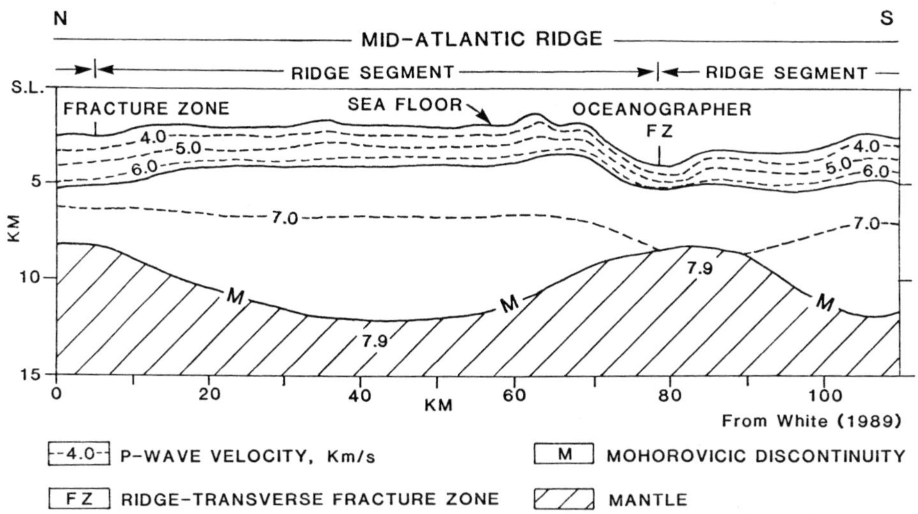

Fig. 3.2. North-south refraction-seismic section showing the velocity structure of the Mid-Atlantic Ridge between 40° and 37°N latitude. Note how the 7.0-7.8-km/s layer thins essentially to 0 thickness where large fracture zones transect the midocean ridge. The purpose of this figure is to show the morphological similarity between the longitudinal section of a midocean ridge and that of a lava tube (Figure 3.32). Figures 3.2 and 3.32, in our opinion, provide an excellent reminder that physical laws operate at all scales (another example is on Figure 3.28). We suggest that the ridge segments between the major fracture zones are mechanical equivalents of the basalt pillows between transverse spreading cracks (constrictions) of Figure 3.32. If the analogy is correct, this constitutes further evidence that flow in the surge channels is ridge-parallel.

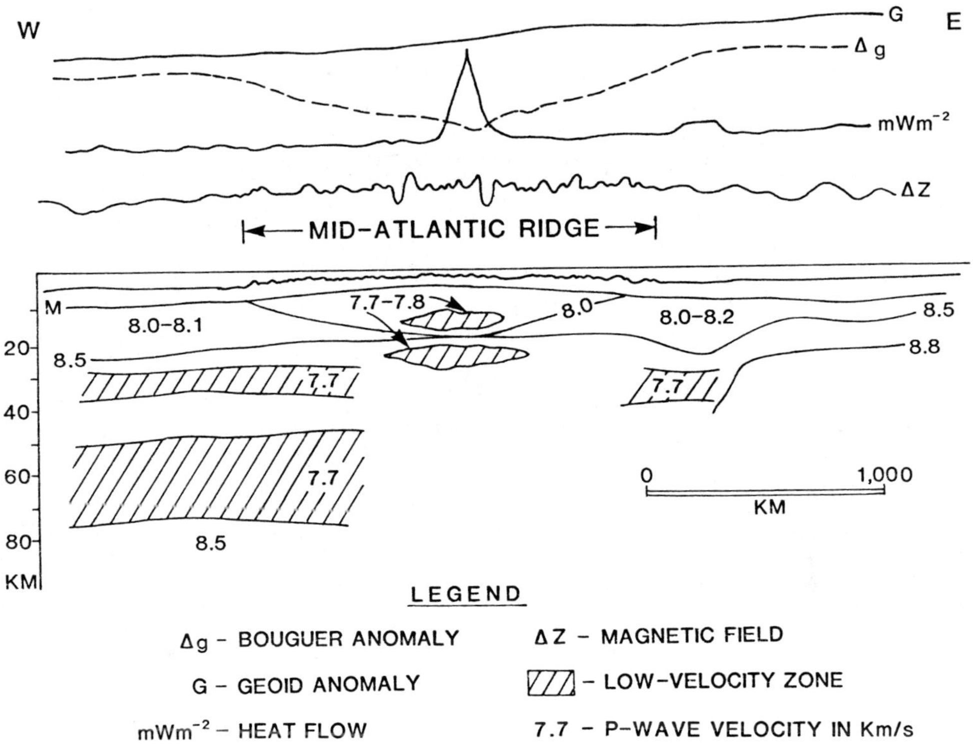

Fig. 3.3. West-east refraction-seismic section across the Mid-Atlantic Ridge at 11°S latitude. Geoid, gravity, heat flow, and magnetic profiles also are shown. Note the close similarity between the gravity profiles of Figures 2.9 and 3.3. Figure 3.3 shows that the internal structure of a midocean ridge is much more complex than Figure 2.9 suggests. The low-velocity zone at 75-50 km is the asthenosphere; that at 37-28 km is the main surge channel; two small low-velocity channels underlie the ridge axis in the 30-12 km depth range. Together these low-velocity zones form a large surge-channel complex (from Pavlenkova, 1989).

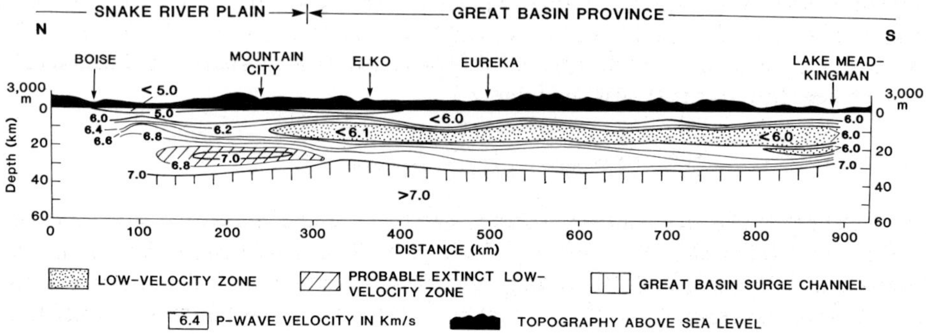

Fig. 3.4. North-south refraction-seismic profile of the eastern Great Basin, western North American Cordillera. Location shown on Figures 4.10 and 4.11. The numbers are P-wave velocities in kilometers per second. A prominent feature of the figure is the 2-6 km thick low-velocity zone that underlies much of the Great Basin at a depth of 10 to 12 km. At the northern end of the profile is a lens of 7.0-km/s material that probably is an extinct late Proterozoic-Paleozoic lens in a surge-channel complex (see text). Its dimensions (180 x 12 km) are comparable with those of the Lake Baykal surge channel of Figure 3.6. Note that the "mother" surge channel, which lies under the whole of the Great Basin (with P-wave velocity of 7.0 km/s) is generally below 30 km. From Prodehl (1970, 1979).

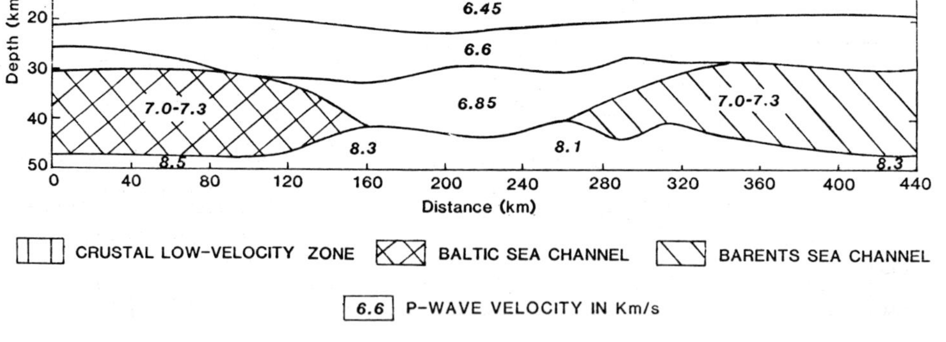

Fig. 3.5. Southwest-northeast refraction-seismic profile across the northern margin of the Baltic Shield. Two active surge channels at 30-50 km depth are shown, one in the north beneath a branch of the middle Paleozoic Caledonides and one in the south that underlies the Gulf of Bothnia. The location of Figure 3.5 is shown on Figure 4.13. Modified from Luosto et al. (1989).

The origin of the 7.0-7.8-km/s layer almost certainly is closely related to that of the high reflectivities of the lower crust as described and illustrated by Klemperer et. al., (1986), Klemperer (1987), Goodwin and Thompson (1988), Thompson and McCarthy (1990), and others (see Figs. 2.10, 2.11, 2.14, 3.15). We emphasize the fact that a lens (or lenses) or 7.0-7.8-km/s material underlies all of the microearthquake bands that we studied, and therefore, all of the high heat-flow bands shown on Figure 2.26.

3.2.2 Continents Have Deep Roots

An important aspect of upper mantle and crustal structure is that continental cratons have deep crustal roots (MacDonald, 1963; Jordan, 1975, 1978; Woodhouse and Dziewonski, 1984; Dziewonski and Woodhouse, 1987; Grand, 1987; Russo and Silver, 1995; Gossler and Kind, 1996). Contrary to general belief (e.g., Anderson, 1987a), continental roots are fixed to the strictosphere (Lowman, 1985, 1986). This conclusion is supported by large and increasing volumes of data, including neodymium and strontium studies of crustal rocks (Wasserburg and DePaolo, 1979). The absence, or near-absence, of a low-velocity asthenosphere beneath ancient cratons led Lowman (1985, 1986) to propose an Earth-dynamics hypothesis of sea-floor spreading between fixed continents. In this hypothesis, if sea-floor spreading takes place, it is restricted to suboceanic regions. Thus, the deep roots of continents are a major obstacle to any hypothesis requiring continental movements (Lowman, 1985, 1986; K.B. Krauskopf, pers. comm., 1990).

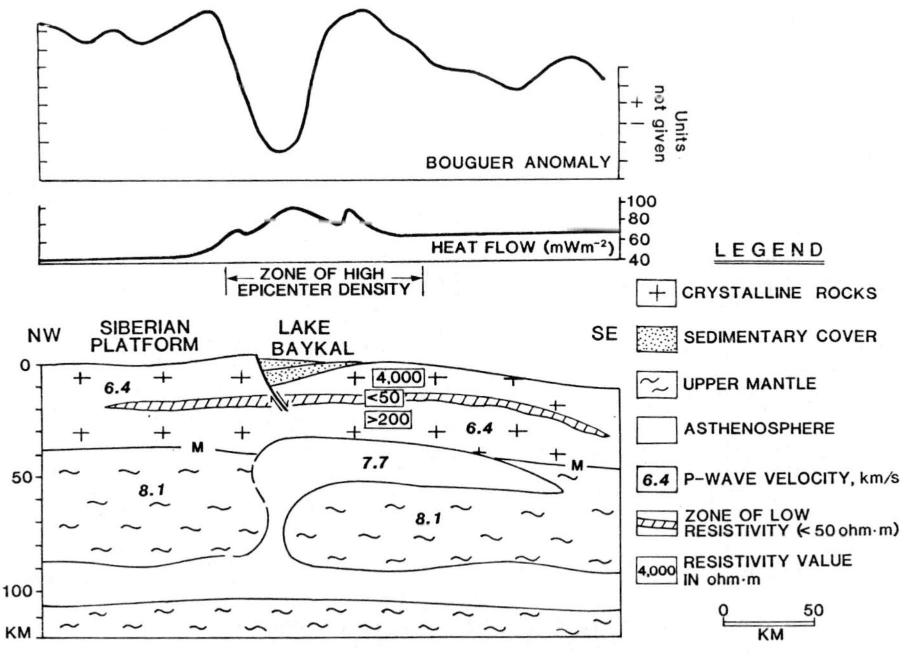

Fig. 3.6. Northwest-southeast refraction-seismic section across the Lake Baykal surge channel, south-central Siberia (location of Lake Baykal shown on Figures 4.20, 5.1). Magnetotelluric, heat-flow, and gravity data are also shown. The 170-km width and 20-km height of the channel are typical of several types of oceanic and continental surge channels. Compiled from Golenetskiy and Misharina (1978), Krylov et al. (1979), Zakharova (1980), Sychev (1985), and Popov (1987).

An example of deep continental roots is presented in Figure 1.1, a seismotomographic cross section of North America. The dark shading beneath the Canadian Shield shows a root extending to 400-450 km (Grand, 1987). Similar deep roots are seen beneath parts of all of the Earth's ancient cratons. In places, however, lenses of 7.0-7.8-km/s material containing low-velocity zones (Fig. 3.5) are present (e.g., Baltic Shield; Luosto et al., 1989). Such lenses containing low-velocity layers postdate the establishment of the deep cratonic roots, as we show in subsequent sections.

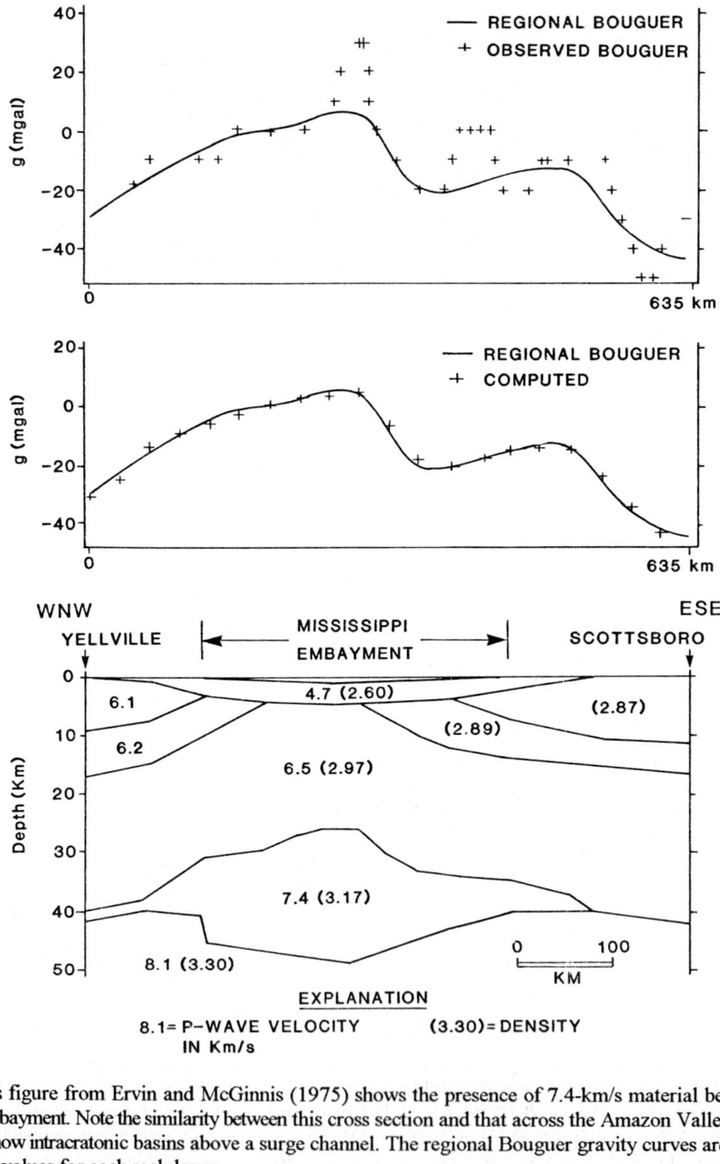

Fig. 3.7. This figure from Ervin and McGinnis (1975) shows the presence of 7.4-km/s material beneath the upper Mississippi embayment. Note the similarity between this cross section and that across the Amazon Valley on Figure 3.8. Both figures show intracratonic basins above a surge channel. The regional Bouguer gravity curves are shown, as well as the density values for each rock layer.

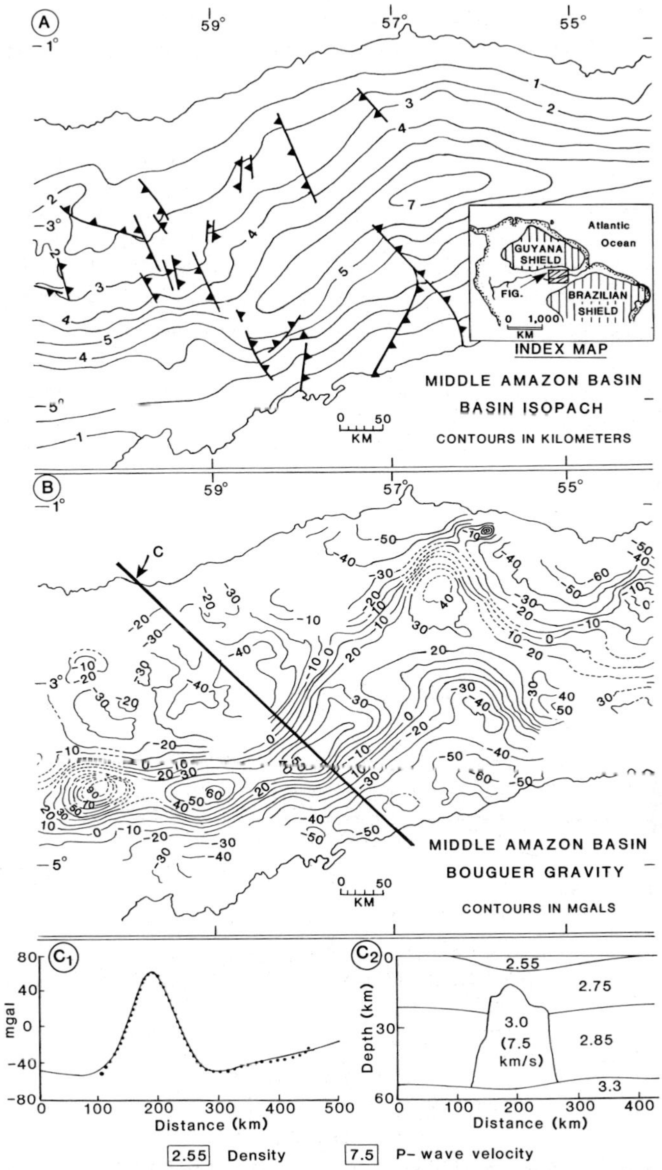

Fig. 3.8. Middle Amazon basin, Brazil; basin isopach map, Bouguer gravity map, and cross section based on gravity data (from Nunn and Aires, 1988). Compare this figure with Figure 3.7. The basin depocenter and an east-west positive Bouguer anomaly coincide almost exactly. The source of the anomaly is a lens of 7.0-7.5-km/s material, probably mafic rock, the remnant of a Paleozoic surge channel (see text). The minimum lifespan (according to radiometric dates from mafic rocks in the basin) of this channel is 425 Ma. This figure shows the genetic relationship between a surge channel and intracratonic basin formation.

3.3 Contraction

3.3.1 General

A discussion of the history and concept of contraction have been presented in Chapter 1 of this book. It will not be repeated here, but the interested reader is encouraged to review that discussion.

3.3.2 Contraction Skepticism

Many workers today either doubt that contraction is taking place or fail to see why the possibility should even be considered. Bott (1971, p. 270), expressing a common opinion, wrote that because of the success of plate tectonics in producing foldbelts, contraction now "...is irrelevant to tectonic problems." Two reviewers of the paper by Meyerhoff et al. (1992b) also expressed doubts that contraction can be taking place. However, one of them, K.B. Krauskopf (pers. comm., 1990), conceded that "...too little is known about what goes on in the [Earth's] interior for any definite statement to be made." He noted that MacDonald's (1959, 1963, 1965a) models could easily be as sound today as they were in 1965 because "...not much more is known today..." about the concentration of radioactive elements in the Earth's interior.

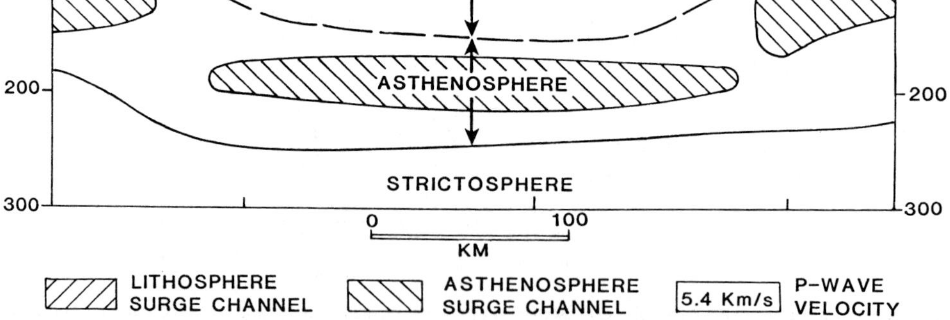

Fig. 3.9. West-east seismotomographic cross section of part of the Yunnan surge channel shown on Figure 2.31 (see also Figure 3.22). Note the presence of at least one shallow low-velocity channel above the main surge channel. This is an example of a surge-channel complex. Location is shown on Figure 2.31. After Liu Jianhua et al. (1989).

3.3.3 Evidence For a Differentiated, Cooled Earth

The evidence is straightforward. The most salient facts follow.

1. The Earth includes several concentric shells, which are explicable only if the Earth differentiated efficiently and at a much higher temperature than today.

2. The outermost of these shells may be the oceanic crust whose thickness ranges from about 4-7 km. This crust is characterized by relatively constant thickness and fairly uniform seismic properties. Both Worzel (1965) and Vogt et al. (1969) observed that if the plate-tectonic explanation of ocean-crust generation is correct, it is a truly remarkable process that produces such a uniform layer in all ocean basins regardless of the spreading rate----1.2 cm/yr or 60 cm/yr. This uniformity is explained, however, if the oceanic crust is the outermost of the Earth's concentric shells. There are other explanations, one of which is discussed later.

3. Mehnert (1969), among several, noted that the further back one looks into the geological record, the greater is the abundance of mafic rocks. This is explained if the lithosphere has been thickening through time by cooling, as Mehnert (1969) suggested.

4. Miyashiro et al. (1982), reporting on studies of the Earth's metamorphic rocks, noted that Precambrian rocks show the highest geothermal gradients and that geothermal gradients of younger rocks generally decrease to the present time.

5. A convincing evidence that huge segments of the lithosphere have been and are being engulfed by tangential compression is the existence of the previously discussed *Verschluckungszonen* (swallowing or engulfment zones) of Ampferer (1906) and Ampferer and Hammer (1911). In places along such zones, whole metamorphic and igneous belts that are characteristic of parts of a given foldbelt simply disappear for hundreds of kilometers along strike (e.g., Alps: Mueller, 1983; Kyushy-Shikoku foldbelt: Miyashiro, 1973; New Zealand

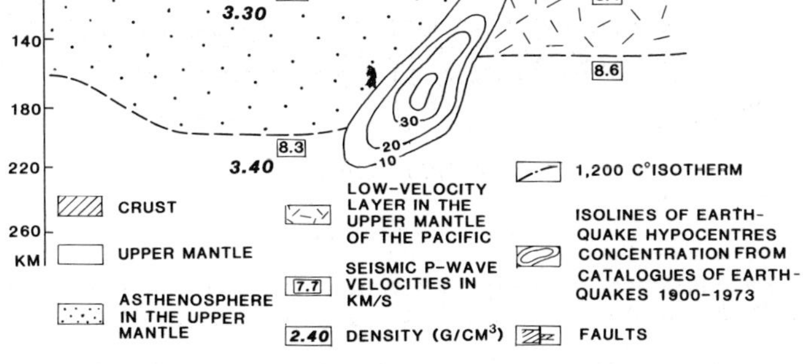

Fig. 3.10. Velocity, density, and thermal structure of a typical western Pacific volcanic arc and backarc basin, based on refraction-seismic, gravity, and heat-flow data. From Rodnikov (1988). Note the very thick (150 km) section of asthenosphere, which is dammed behind the lithosphere Benioff zone as a result of the Earth's rotation (see Figure 4.5).

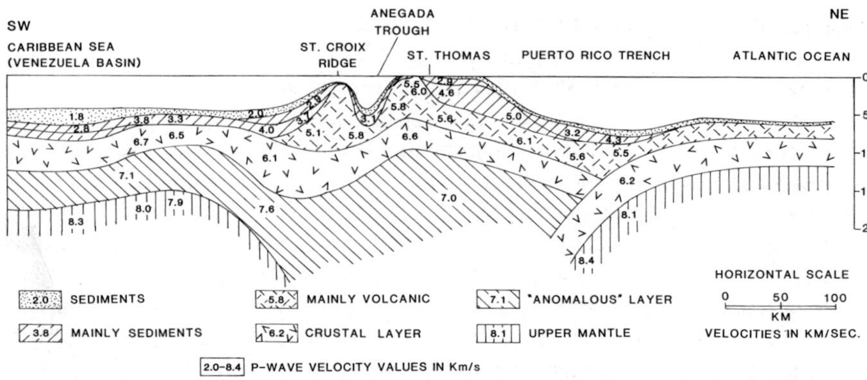

Fig. 3.11. Southwest-to-northeast refraction-seismic cross section of the northern part of the Lesser Antilles arc (from Officer et al., 1959). This figure is included to show the similarity between the Caribbean island arc system and those of the western Pacific. See Table 4.2.

Alps: Kingma, 1974; southern California Transverse Ranges: Humphreys et al., 1984). Figures 2.23-2.24 and 3.16-3.17 illustrate the characteristics of typical *Verschluckungszonen*. Although Mueller (1983), Humphreys et al. (1984), and other workers considered these features to be former subduction zones, this interpretation is difficult to defend because all of these zones, regardless of age, are near-vertical bodies (1) reach only the top or middle of the asthenosphere (150 to 250 km deep) and (2) do not deviate more than 10° to 25° from the vertical (Mueller, 1983; Humphreys et al., 1984).

6. The antipodal positions of the continents and ocean basins mean that Earth passed through a molten or near-molten phase (Arldt, 1907; Bucher, 1933; Wilson, 1954; Harrison, 1968). Such antipodal relations are unlikely to be a matter of chance or coincidence (Harrison, 1968).

7. Theory (Jeffreys, 1970) and laboratory experiment (Bucher, 1956) showed that heated spheres cool by rupture along great circles. Remnants of two such great circles (as defined by hypocenters at the base of the asthenosphere) are active today: the Circum-Pacific and Tethys-Mediterranean fold systems. The importance of Bucher's (1956) experiment to contraction theory, in which he reproduced the great circles, is little appreciated.

8. As Earth cooled, it solidified from the surface downward. Because stress states in cooled and uncooled parts are necessarily opposite one another, compression above and tension below, the two parts must be separated by a surface or zone that Davison (1887) called the level of no strain (Fig. 3.1). We, as did Wilson (1954), equate the cooled layer with the lithosphere (Fig. 3.1). The uncooled part below is what Bucher (1956) called the strictosphere. Thus, as

originally proposed by Scheidegger and Wilson (1950), Davison's (1887) level of no strain must be the asthenosphere, or a zone of no strain across which the change in stress states is gradual (Fig. 3.1). Only in a cooling Earth, which approximates a closed thermal system, can an asthenosphere form.

9. Continued cooling deepens the asthenosphere and the upper surface of the strictosphere. The stresses accumulated through cooling are relieved episodically by rupture along the great-circle fractures that are the Earth's cooling cracks or the Benioff zones of current literature. Because the lithosphere is being compressed and the strictosphere subjected to tension, the mechanics of rupture should follow the Navier-Coulomb maximum shear-stress theory (Jaeger, 1962). Accordingly, the lithosphere Benioff zone must dip less than 45° to a tangent to the Earth's surface (in actual fact, it dips 22° to 44°; Figs. 2.36, 3.1). In contrast, the strictosphere Benioff zone must dip more than 45° to a tangent to the Earth's surface (50° to 75°; Figs. 2.36, 3.1). Benioff (1949, 1954) discovered the change in Benioff-zone dip from lithosphere to strictosphere, but Scheidegger and Wilson (1950) recognized these dips as an expression of the Navier-Coulomb maximum shear-stress theory (Figs. 2.36, 3.1). The dip values of the lithosphere and strictosphere Benioff zones confirm that the Earth is a cooling body. (Ritsema [1957, 1960], working independently, also discovered the abrupt dip changes in the dip of the Benioff zone with increasing depth.)

An important fact concerning the Benioff zone is that the two segments, one in the lithosphere and the other in the strictosphere, do not necessarily form a single, continuous zone as depicted in diagrams and cross sections (e.g., Figs. 2.36, 3.1). Benioff (1949, 1954) found

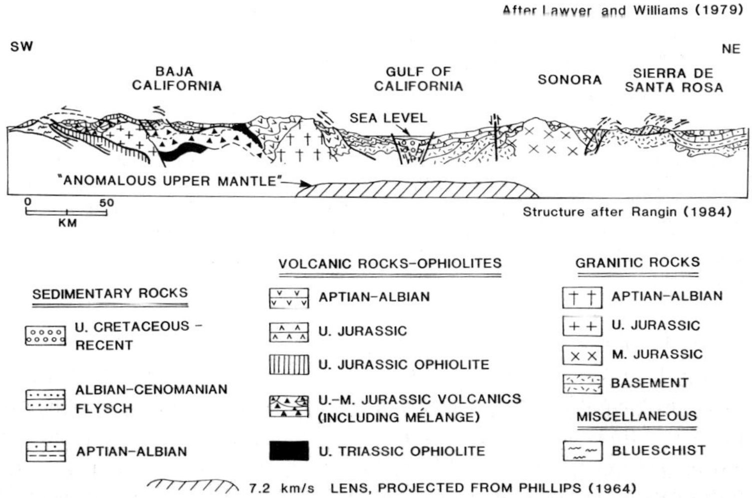

Fig. 3.12. Southwest-northeast structural cross section across the Gulf of California. A surge channel (anomalous upper mantle) and the heat flow also are shown. The location is shown on Figure 4.10.

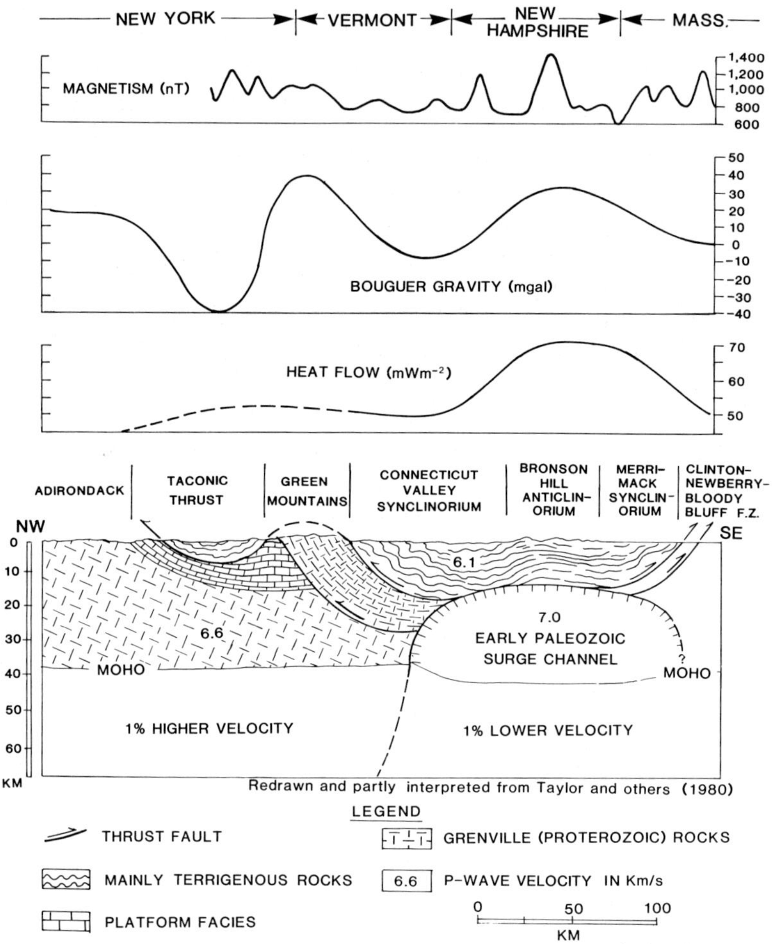

Fig. 3.13. Northwest-southeast refraction-seismic profile across the Appalachian Mountains of the northeastern United States. The right side of the section is the shoreline of the Atlantic Ocean. (The location is approximately the same as the reflection profile shown on Figure 3.21.) This figure shows the early Paleozoic Acadian (Middle Devonian) kobergen centered on the Bronson Hill anticlinorium. Heat-flow, gravity, and magnetic data are also shown. Compiled from Taylor et al. (1980), Hildenbrand et al. (1982), Zietz (1982), and McCartan and Architzel (1988). From Meyerhoff et al. (1992b).

that the two segments of the Benioff zone, instead of joining near the base of the asthenosphere, may be offset for distances of 100 to 200 km (Fig. 3.18). In some places such as the Lesser Antilles arc, a strictosphere Benioff zone may not even be present below the lithosphere Benioff zone (e.g., Lesser Antilles and Scotia volcanic arcs). These facts are explained in our surge-tectonics hypothesis but not by other hypotheses. In fact, all detailed modern studies of hypocenter distribution in Benioff zones show the same clear division into a shallow, gently dipping lithosphere Benioff zone and a deeper, steeply dipping strictosphere Benioff zone (Fig. 2.36; Grow, 1973; Isacks and Barazangi, 1977).

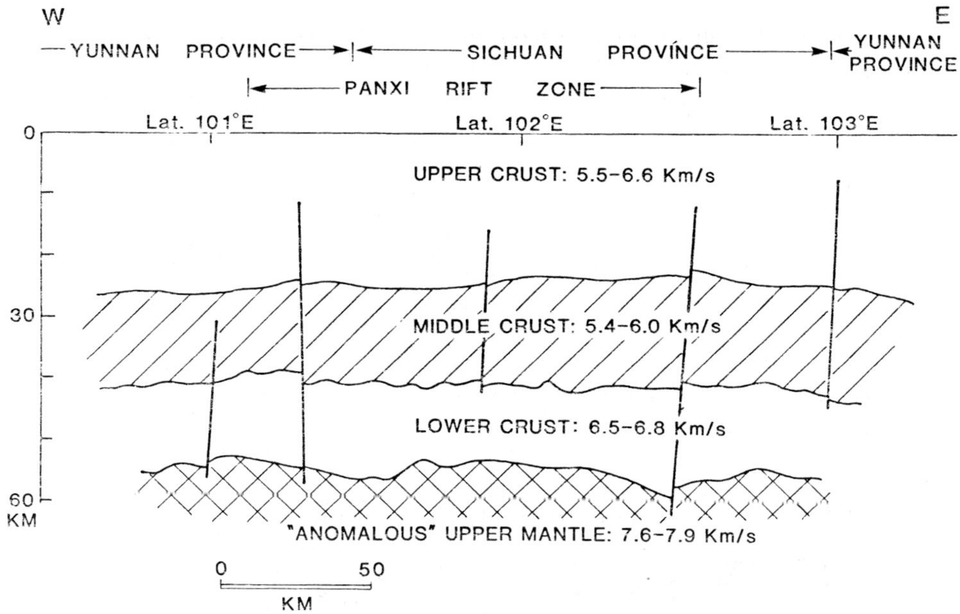

Fig. 3.14. West-east seismotomographic section across the Permian Panxi rift in southernmost Sichuan Province, southwest China. The main surge channel is below 50 km. This is the same channel which on Figure 3.9 is in the 60-30 km depth zone. Comparison of this figure with Figure 3.9 shows how much the depth of a surge channel can change in a short distance. Location of Figure 3.14 is shown on Figure 2.31. From Liu Jianhua et al. (1989).

Ritsema's (1957, 1960) focal-mechanism studies of shallow, intermediate, and deep earthquakes showed that the Benioff-zone dip in the lithosphere is only half of its dip in the strictosphere. An even more significant discovery made by Ritsema (1957, 1960), although he attached little importance to it, was the revelation that earthquake foci above 0.03R (approximately 180 km depth) show mainly compression, whereas those below 0.03R show mainly tension. Most earthquakes above and below 0.03R, as Scheidegger (1963) also noted, have a strike-slip component. Thus Ritsema's findings lend support to Scheidegger and Wilson's (1950) interpretation of the Benioff zone as a manifestation of the Navier-Coulomb maximum shear-stress theory.

10. Computer simulations of possible Earth thermal histories (MacDonald, 1959, 1963; Jacobs, 1961; Reynolds et al., 1966), using a broad spectrum of assumed initial temperatures and chemical compositions, show that the Earth is cooling (see also Stacey, 1981; Lyttleton, 1982). The fact that Earth's Benioff zones still are active earthquake-generating zones provides strong support for this conclusion. Perhaps the strongest support comes from the Basalt Volcanism Study Project (1981) report by 101 petrologists, mineralogists, and petrographers, who wrote that the repeated extrusion of basalt to the Earth's surface through its history is proof of the Earth's long history of cooling.

11. Finally, the existence of *Verschluckungszonen* in the lithosphere and upper mantle also constitutes evidence that the Earth is actively cooling. *Verschluckungszonen* are interpreted by us to be large masses of lithosphere and upper mantle that were downbuckled into the upper

mantle during tectogenesis as the lithosphere readjusted its shape to fit the underlying, cooling strictosphere (Figs. 2.24, 3.16-3.17). If this interpretation is correct, the existence of *Verschluckungszonen* may constitute proof that the Earth has been cooling to the present day. We discuss *Verschluckungszonen* in a subsequent section.

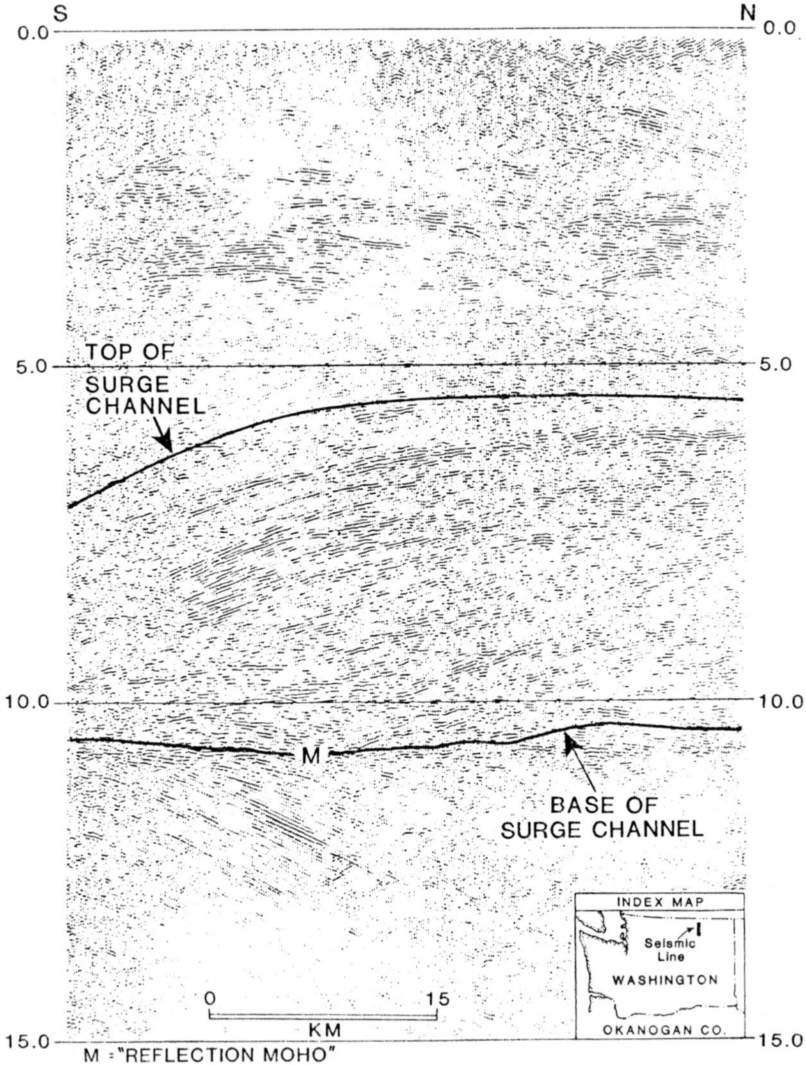

Fig. 3.15. South-to-north reflection-seismic line through Okanogan dome, northeastern Washington (from Nelson, 1988). The Okanogan dome is a metamorphic core complex, whose origin is related closely to that of mantled gneiss domes (Figures 2.12, 3.13, 3.21). This structure coincides exactly with the presence of a lens-shaped, highly reflective structure in the lower crust. We show this figure to emphasize the close genetic relations among alpinotype foldbelts (Figures 2.13, 2.14, 2.16, 3.12, 3.13, 3.14, 4.17), germanotype foldbelts (2.16, 4.15, 4.17, 4.19), backarc basins and island arcs (Figures 2.16-2.17, 3.10-3.11), rifts (Figures 2.10, 2.11, 2.16, 3.6-3.8), strike-slip fault systems (Figures 2.14, 2.18, 2.20) mantled gneiss domes (Figs. 2.12, 2.16, 3.13, 3.21), and metamorphic core complexes (Figure 3.15).

3.4 Contraction as an Explanation of Earth Dynamics

3.4.1 CONTRACTION ACTING ALONE

Despite the attraction of a cooling Earth, both Scheidegger (1963) and Bott (1971) concluded that contraction acting alone is inadequate to produce the crustal shortening measured in the Earth's many tectonic belts. For both geological and seismological reasons, this conclusion appears to be well founded. They gave several reasons; three of which and one of our own are crucial.

1. The total amount of shortening measured across the Earth's foldbelts far exceeds what can be inferred on theoretical grounds, whether one uses the contraction model of MacDonald (1963) or of Jeffreys (1970). Even if one accepts Bucher's (1955, p. 357-360; 1956, p. 1306) outstanding demonstration that apparent (measured) shortening can be and generally is four to five times true shortening, the contraction hypothesis cannot explain all true shortening in foldbelts. (Lyttleton's [1982] theoretical estimate of 2,000 km of shortening adequately explained the measured shortening, but his hypothesis requires cataclysmic geological events that need to be sought in the field.)

2. Contraction alone is unable to explain the origins of all types of tectonic belts---- compressional foldbelts, tensional rift zones (including midocean ridges), and strike-slip zones.

3. Ritsema (1957, 1960) and Scheidegger (1963) observed that earthquake first-motion studies show that strike-slip motions are most common in Benioff zones, not just in strike-slip and rift zones. Contraction alone cannot explain the ubiquitous strike-slip component.

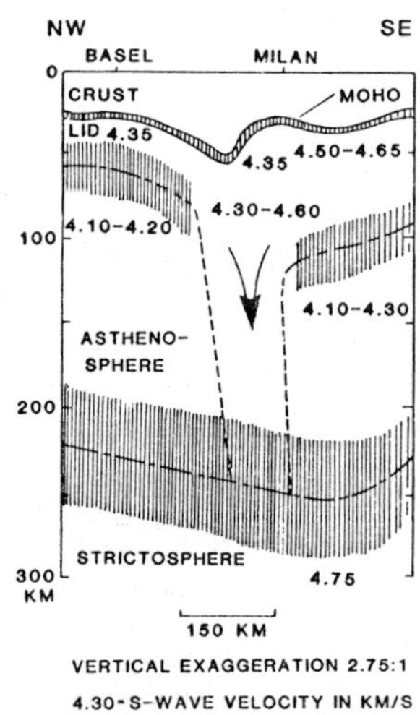

Fig. 3.16. Northwest-southeast refraction-seismic cross section of crust and upper mantle beneath the western Alps along a line passing through Basel, Switzerland and Milan, Italy. (1) The near-vertical slab marked with a black arrow constitutes a large *Verschluckungszone* of the lithosphere that extends to the strictosphere, penetrating the entire asthenosphere. (2) Numbers are S-wave velocities, the 4.20-km/s value corresponding to the 7.7-7.9-km/s P-wave velocity. (3) The lid is the hard mantle in the lithosphere above the asthenosphere. (4) Finally, the hatched areas indicate the range of uncertainty of the locations of the bottoms of (a) the crust, (b) the lithosphere, and (c) the asthenosphere. From Mueller (1983).

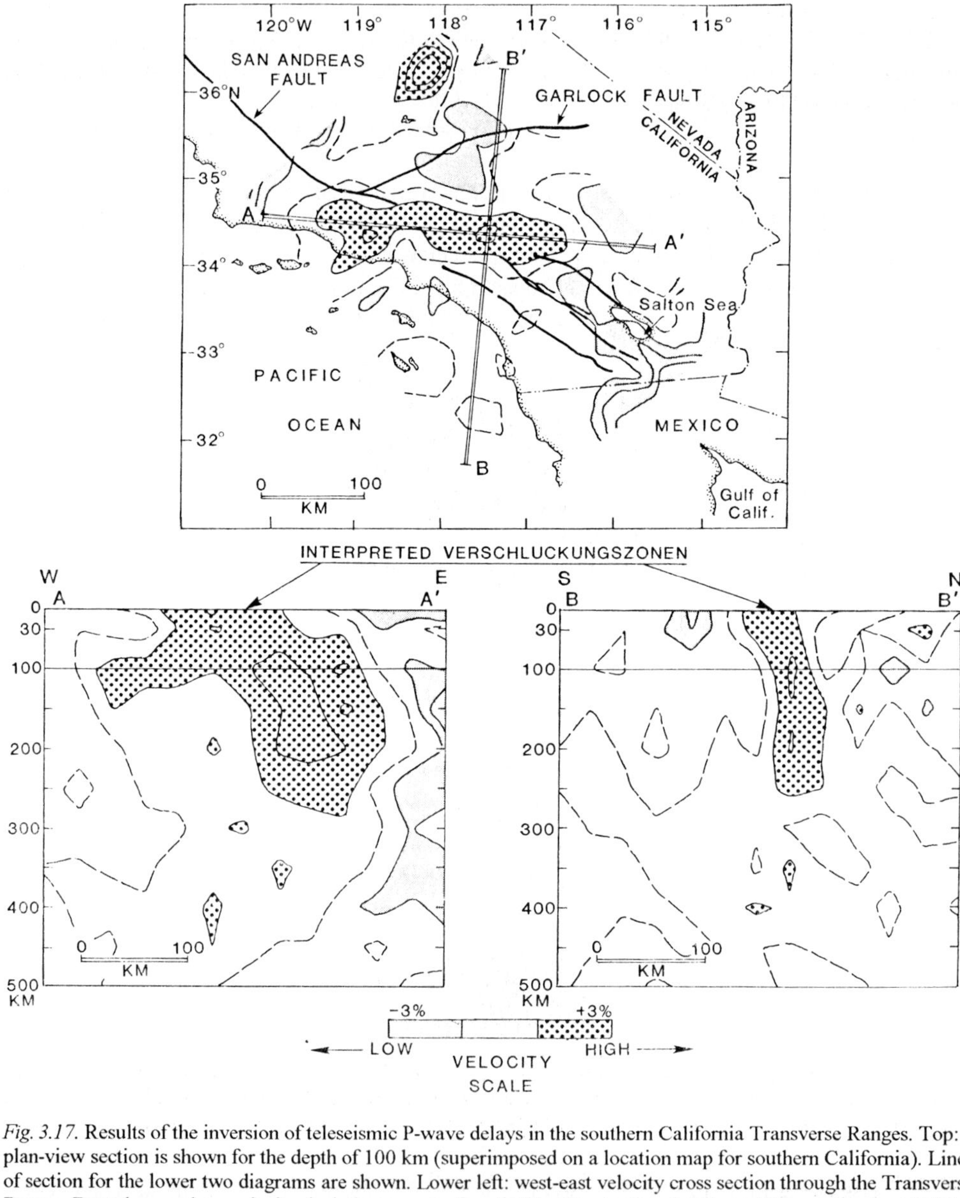

Fig. 3.17. Results of the inversion of teleseismic P-wave delays in the southern California Transverse Ranges. Top: a plan-view section is shown for the depth of 100 km (superimposed on a location map for southern California). Lines of section for the lower two diagrams are shown. Lower left: west-east velocity cross section through the Transverse Ranges. Dotted areas show velocity deviations greater than 1.5%. (Greater than 1.5% = cold areas; less than 1.5% = hot areas.) 100-km depth line is shown. This section, parallel with the Transverse Ranges, shows a *Verschluckungszone* that reaches ca. 150-km depth in the west and 250-km depth in the east. Lower right: south-north velocity cross section across the Transverse Ranges. Symbols are the same as for section A-A'; lower left. The high-velocity anomaly shows as a slab-like body dipping 80° to 90° toward the north that extends to approximately 250 km. From Humphreys et al. (1984).

4. Contraction theory requires that foldbelts are concentrated in and adjacent to oceanic trenches. This is not observed. More than 50% of the Earth's foldbelts lie at great distance from the surface trace of a Benioff zone, and all Jurassic-Cenozoic foldbelts lie within the high heat-flow bands illustrated on Figure 2.26. This cannot be explained by any Earth-dynamics hypothesis yet proposed. However, if contraction could lead to tectogenesis of large parts of the lithosphere far removed from Benioff zones, the preceding objections to the contraction hypothesis would be irrelevant.

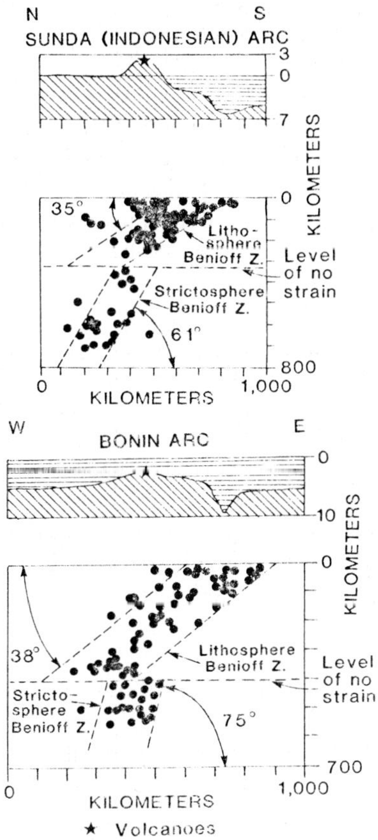

Fig. 3.18. Two cross sections of western Pacific volcanic arcs and the distribution of the hypocenters beneath them are shown. As Benioff (1954) observed, these two sections demonstrate that the lithosphere Benioff zone need not be continuous with the strictosphere Benioff zone, a fact that is explained by contraction theory but not by plate tectonics.

3.4.2 Contraction Acting as the Trigger For Tectogenesis

Figure 2.26, as we have discussed, portrays the bands of high heat flow that crisscross the Earth's surface. We believe that this reticulate network of high heat-flow bands is underlain by a network of interconnected magma chambers. In our surge-tectonics hypothesis, these magma chambers are the mantle diapirs discussed in preceding sections and summarized in part in

Table 2.1. Figure 2.26 indicates that these mantle diapirs, or magma chambers, are interconnected. The interconnected channels comprise the surge channels of surge tectonics. If the hot material in these channels is sufficiently mobile, lateral flow through them should be possible provided a pressure gradient is present. The compression already present in the lithosphere would provide the force needed to initiate and maintain flow. We emphasize that such flow would be temporally discontinuous (i.e., episodic) and, when it did occur, would be extremely slow. Flow velocities are discussed later.

We have pointed out that in a cooling Earth, the lithosphere by definition is everywhere and at all times in a state of compression (Fig. 3.1; Davison, 1887; Scheidegger and Wilson, 1950; Bucher, 1956; MacDonald, 1959, 1963, 1965a; Jeffreys, 1970; Meyerhoff et al., 1972; Stacey, 1981; Lyttleton, 1982). The compression is concentrated in planes tangent to the Earth's surface, and is equal in all directions. James C. Meyerhoff (pers. comm., 1988) called this equiplanar tangential compression, and this compression in the lithosphere is what Bucher (1956) referred to (incorrectly) as all-sided compression.

The only elements in the lithosphere that disturb this approximately equiplanar tangential stress state are the surge channels. Flow can take place in these channels wherever a pressure gradient develops. For example, the escape of lava from a channel lowers the pressure at that point, and equiplanar compression, acting at right angles to the surge-channel walls, mobilizes the fluid elements inside the channel until pressure equilibrium is restored. The presence above active channels of channel-parallel fault, fracture, and fissure systems indicates that (1) flow takes place along the full length of each tectonic belt and (2) the channels are in communication with the Earth's surface through the fault-fracture-fissure system.

Surge channels and their fault-fracture-fissure systems constitute zones of weakness in the lithosphere. Because (1) the channels are the only bodies in the lithosphere that, owing to their potential to contain mobile fluids, they have the capacity to upset the state of equiplanar tangential compression, and because (2) they are constantly losing their contents to the surface, lithosphere compression ultimately destroys them. Their deformation and ultimate destruction are the essence of tectogenesis. Thus the cooling process in the Earth's strictosphere effectively guarantees the presence within the lithosphere of a powerful mechanism for tectogenesis.

3.5 Review of Surge and Related Concepts in Earth-Dynamic Theory

Several workers proposed, on both theoretical and geological-geophysical grounds, the presence of bodies similar to surge channels in the lithosphere. Others developed concepts much like those that are the basis for surge tectonics.

3.5.1 Surge Channels

Our study of this topic was by no means exhaustive; we may have missed important references that deal with the concept of surge-channel-like edifices in the lithosphere and uppermost mantle. A particularly good example of a surge-channel-like feature was proposed by Vogt (1974) for the Iceland region, including the Kolbeinsey, Reykjanes, and Faeroe-Greenland ridges. He wrote (p. 116, 118), "In the model I assume there is a pipe-like region below the spreading axis, extending subhorizontally away from a plume such as Iceland . . .. This mid-oceanic pipe extends from the base of the axial lithosphere, about 5 or 10 km deep, down to maximum depths (30 to 50 km?) from which basalt melts segregate and rise. Tholeiitic fluids would be released from the entire pipe; origin depths of 23 km for the Mid-Atlantic Ridge and 16 km for the East Pacific Rise . . . would approximate depth to the center of the pipe. The ultrabasic mush in this pipe is assumed to be flowing away from the hot spot at a rate determined by pipe diameter, viscosity, and horizontal pressure gradient . . .." Vogt (1974) estimated that flow ranged laterally outward from Iceland from 500 to 600 km. A similar study by Gorshkov and Lukashevich (1989) suggested that channels are present beneath the full length of the midocean ridge system. According to them, flow beneath the Mid-Atlantic Ridge would be from hot spots located beneath Antarctica in the south and Iceland in the north, with the two flows converging near the equator.

Other midocean ridges where some type of axis-parallel flow and/or rift propagation have been postulated include the Juan de Fuca Ridge (Shih and Molnar, 1975) and the Galapagos Rift (Hey and Vogt, 1977). After the discovery of systematic segmentation along the midocean ridges beginning with the East Pacific Rise (Lonsdale, 1982), Lonsdale, MacDonald, Fox, and others commenced a series of investigations that led to a general postulate of ridge-parallel flow in the midocean ridges (e.g., Macdonald et al., 1988; Winterer et al., 1989). Macdonald et al. (1988) proposed that mantle diapirs rise beneath the centers of each ridge segment, thereby accounting for the greater elevations of the central parts of such segments. From the crest, ridge-parallel lateral flow commences. Such flow halts in the depressed areas between adjacent segments because of mutual impingement.

Sonographs of the midocean ridges show that the Macdonald et al. (1988) hypothesis is not tenable. Where the diapirs rise in the centers of the ridge segments, radial and/or annular structures should be present. Where the flows from adjacent segments impinge, compressional structures should be present. Neither structural form is observed. Instead, linear fractures and faults extend for hundreds of kilometers along strike, with interruptions only at transform faults, devals, and overlapping spreading centers. In several of the transform fault zones and devals, we traced individual fault traces through the fault zone from one ridge segment to the next, which negates the Macdonald et al. (1988) hypothesis.

Surge-channel, or interconnected mantle diaper systems have been reported from small oceanic basins and continental areas. A well-known example of the former is the Tyrrhenian Sea west of Italy, where geophysical techniques and very high heat flow show the presence of a very large diapir-like body at shallow depths (Locardi, 1988).

Among continental examples, the Fergana Valley in Soviet Central Asia is the best documented. Kuchay and Yeryemin (1990) discovered a very large pipelike body, or channel, below this large east-west-striking late Cenozoic structural depression nestled among the western ranges of the Tian Shan. In their summary (p. 45), Kuchay and Yeryemin concluded: "Interpretation and interpolation of geophysical data from the [Fergana Valley] lead to the conclusion that the base of the 'granitic' layer undergoes partial melting, as a consequence of which there forms a layer of lowered viscosity that has been identified seismically as a zone of reduced velocity above the Conrad discontinuity. It is possible to consider this zone as a 'granitic' asthenochannel, a subhorizontal layer that has a high strain rate and in which relative lateral displacement takes place between the contents of the asthenochannel and the surrounding rock layers. The absence of a gravity anomaly, which is an indication of variations in thickness and other properties within the zone of reduced velocity, suggests that the 'granitic' layer and the 'granitic' asthenochannel have the same density."

3.5.2 Use of the Surge Concept in Tectonics

We have found three parallel and presumably independent derivations of the surge-tectonics concept. The most recent of these originated with Hollister and Crawford (1986) who used "tectonic surge" to describe rapid vertical uplift in the structural core of a bilaterally deformed foldbelt (i.e., in the center of what we call a kobergen). Hollister and Crawford (1986, p. 560) opined that "Weakening of the crust [by] anatexis and accompanying development of melt-lubricated shear zones . . ." is essential to rapid vertical uplift. Paterson et al. (1989, p. 116) referred to this type of tectonics as surge tectonics. Tobisch and Paterson (1990) used this term in two or three areas of southeastern Australia. Their usage is close to our own.

A second, but earlier employment of the term "surge" in tectonics is from Coward (1982). He noted during studies of the Moine thrust system of northwestern Scotland that the central parts of thrust sheets commonly are thrust farther than the ends of the thrust sheets, and likened the extended central part of the sheet to a surge (as in a breaking wave, the central part sweeps--surges---forward). This usage subsequently was retained and applied to the British and Irish variscides by Murphy (1985, 1990), Coward and Smallwood (1984), and Cooper et al. (1986).

The third and, as far as we can determine, oldest use of the term in tectonics in recent literature was by Meyerhoff and Meyerhoff (1977). They proposed that asthenosphere surges (1) from beneath the Asian continent, (2) between North and South America, and (3) between South America and Antarctica produced the eastward-facing island arcs in the three regions. The idea was used subsequently to explain the complexities of Caribbean tectonics (Morris et al., 1990). Morris et al. (1990) used the term surge tectonics (coined by Bruce D. Martin). The paper was written during 1987-1988 and was submitted to and accepted by the Geological Society of America in 1988. Regardless, the Paterson et al. (1989) use of the term in print precedes by five months that by Meyerhoff et al. (1989). The term surge tectonics is used in this book in the same sense that it was employed by Taner and Meyerhoff (1990).

3.6 Geotectonic Cycle of Surge Tectonics

The asthenosphere alternately expands (during times of tectonic quiescence) and contracts (during tectogenesis). Thus, when the asthenosphere is expanding, the surge channels above it, which are supplied from the asthenosphere, also are expanding; and when tectogenesis takes place, the magma in the surge channels is expelled. Tectogenesis is triggered by collapse of the lithosphere into the asthenosphere along the 30°-dipping lithosphere Benioff zones. The following is Meyerhoff et al.'s (1992b) interpretation of the approximate sequence of events during a geotectonic cycle (Fig. 3.19).

1. The strictosphere is always contracting, presumably at a steady rate, because the Earth is cooling.

2. The overlying lithosphere, because it is already cool, does not contract, but adjusts its basal circumference to the upper surface of the shrinking strictosphere by (1) large-scale thrusting along the lithosphere Benioff zones, and (2) normal-type faulting along the strictosphere Benioff zones. These two types of deformation, one compressive and the other tensile, are complementary and together constitute an example of the Navier-Coulomb maximum shear stress theory (Jaeger, 1962; Jaeger and Cook, 1979).

3. The large-scale thrusting of the lithosphere is not a continuous process, but occurs only when the lithosphere's underlying dynamic support fails. That support is provided mainly by the softer asthenosphere and frictional resistance along the Benioff fractures. When the weight of the lithosphere overcomes the combined resistance offered by the asthenosphere and Benioff-zone friction, lithosphere collapse ensues. Because this process cannot be perfectly cyclic, it must be episodic; hence tectogenesis is episodic.

4. During the anorogenic intervals between lithosphere collapses, the asthenosphere volume increases slowly as the strictosphere radius decreases (Fig. 3.19). The increase in asthenosphere volume is accompanied by decompression in the asthenosphere.

5. Decompression is accompanied by rising temperature, increased magma generation, and lowered viscosity in the asthenosphere, which gradually weakens during the time intervals between collapses.

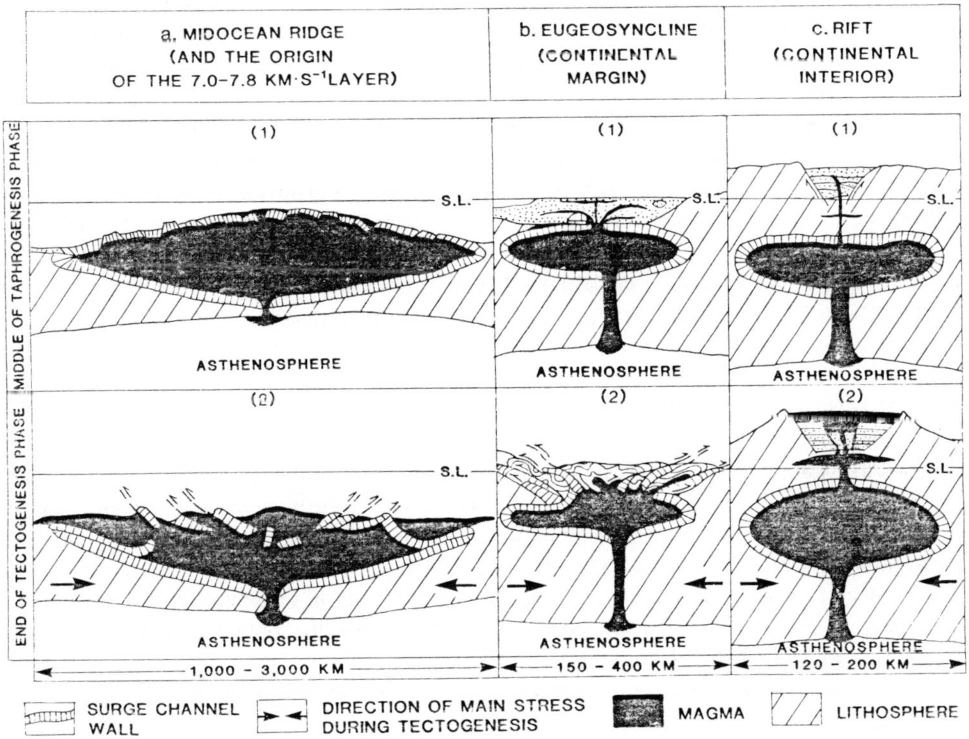

Fig. 3.19. The geotectonic cycle in the ocean basins (midocean ridge, for example), the continent-ocean basin transition (eugeosyncline), and the continental interior (rift basins). *Midocean ridge*: during the taphrogenic phase of the geotectonic cycle (**a-1**), the ridge fills with magma. During the tectogenesis phase (**a-2**), the ridge bursts and massive basalt floods cover the ocean floor. *Eugeosyncline*: during the taphrogenic phase, the eugeosyncline subsides, accumulating volcanic and sedimentary rocks (**b-1**). During the tectogenesis phase (**b-2**), alpinotype folding and thrusting occur. *Continental rift*: during taphrogenesis (**c-1**), continental rifting and volcanism occur. During tectogenesis (**c-2**), rifting is accelerated and volcanism increases as the surge channel fills to near the bursting point. If the lithosphere above the surge channel is somewhat thinner, a germanotype foldbelt may develop, as illustrated on Figure 4.19. From Meyerhoff et al. (1992b).

6. Flow in the asthenosphere is predominantly eastward as a consequence of the Earth's rotation (Newton's Third Law of Motion; Meyerhoff and Meyerhoff, 1977). Magma flow in the surge channels above the asthenosphere also tends to be eastward, although local barriers may divert flow in other directions for short distances. Coriolis force also must exert an important influence on asthenosphere and surge-channel flow, which by its nature is Poiseuille flow. Therefore, the flow at the channel walls is laminar and is accompanied by viscous, or backward drag. The viscous drag produces the swaths of faults, fractures, and fissures (streamlines) that are visible at the surface above all active tectonic belts. These bands or swaths are examples of Stokes's Law (one expression of Newton's Second Law of Motion).

7. During lithosphere collapse into the asthenosphere, the continentward (hanging wall) sides of the lithosphere Benioff zones override (obduct) the ocean floor (Ampferer, 1906; Ampferer and Hammer, 1911). The entire lithosphere buckles, fractures, and founders. Enormous compressive stresses are created in the lithosphere.

8. Both the lithosphere and the strictosphere fracture along great circles at the depth of the strictosphere's upper surface, as predicted by theory (Jeffreys, 1959, 1970, 1976) and demonstrated in the laboratory (Bucher, 1956). Only two partial great circle fracture zones survive on the Earth today. These include the fairly extensive, highly active Circum-Pacific great circle and the almost defunct Tethys-Mediterranean great circle.

9. When the lithosphere collapses into the asthenosphere, the asthenosphere-derived magma in the surge channels begins to surge intensely. Wherever the volume of the magma in the channels exceeds their volumetric capacity, and when compression in the lithosphere exceeds the strength of the lithosphere that directly overlies the surge channels, the surge-channel roofs rupture along the cracks that comprise the fault-fracture-fissure system generated in the surge channel by Poiseuille flow before the rupture. Rupture is bivergent, whether it forms continental rifts, foldbelts, strike-slip zones, or midocean rifts. The foldbelts develop into kobergens, some of them alpinotype and some of them germanotype. The *tectonic style of a tectonic belt depends mainly on the thickness and strength of the lithosphere overlying it* (Fig. 3.19).

10. Tectogenesis generally affects an entire tectonic belt and, in fact, may be worldwide; the worldwide early to late Eocene tectogenesis is an example (Meyerhoff et al., 1992b). This indicates that the lithosphere collapse that generates tectogenesis transmits stresses everywhere in a given belt at the same time. Thus Pascal's Law is at the core of tectogenesis. Sudden rupture and deformation of surge channels may therefore be likened to what happens when someone stamps a foot on a full tube of toothpaste. The speed or rapidity of tectogenesis, then, is related to the number of fractures participating in the event, as well as to the thickness of lithosphere involved, the size of the surge channel or surge-channel system, the volume and types of magma involved, and related factors.

11. Once tectogenesis is completed, another geotectonic cycle or subcycle sets in, commonly within the same tectonic belt.

3.7 Pascal's Law---the Core of Tectogenesis

Pascal's Law (or theorem) states that pressure applied to a confined liquid at any point is transmitted undiminished through the fluid in all directions and acts upon every part of the confining vessel at right angles to its interior surfaces and equally upon equal areas. This law applies in part to all fluids, but wholly to Newtonian fluids; it is the principle behind every hydraulic machine, notably the hydraulic press. A most important condition of Pascal's Law is that the pressure (force per unit area) acts equally upon equal areas. This condition lies at the very core of tectogenesis.

The Earth, according to our surge-tectonics hypothesis, is a very large hydraulic press. Such a press consists of three essential parts: a closed vessel, the liquid in the vessel, and a ram or piston. The collapse of the lithosphere into the asthenosphere is the activating ram or piston of tectogenesis. The asthenosphere and its overlying lithospheric surge channels---which are everywhere connected with the asthenosphere by vertical conduits---are the vessels that enclose the fluid. The fluid is magma generated in the asthenosphere. The magma fills the lithosphere channels. When the piston (lithosphere collapse) suddenly compresses the channels and the underlying asthenosphere, the pressure is transmitted rapidly and essentially simultaneously through the worldwide interconnected surge-channel network; the surge channels burst and the tectogenesis is in full swing. The compression everywhere of the asthenosphere compensates for the fact that the basaltic magma of the surge channels is non-Newtonian.

A possible objection to this simple picture of tectogenesis is that the sudden application of pressure against the surge channels would consolidate the magma in the channels, and thereby prevent the bursting of the channel roof. This would be true if the channels had no communication with the surface at the onset of tectogenesis, but this is not the case. As Meyerhoff et al. (1992b) noted, the channels are connected to the surface by swaths of belt-parallel faults, fractures, and fissures.

A second possible objection is that the magma in surge channels is non-Newtonian; i.e., it is too viscous to transmit the added stress to all of the interconnected parts of the surge-channel system. This objection would be valid for a tectonic model in which the added stress is applied only at a single point in the system. In a contracting Earth, however, compression in the lithosphere is omnipresent. Hence, the added stress is applied everywhere along the interconnected lithosphere channels so that the viscosity argument is invalid; the added stress is being applied at an infinite number of points in the system. As shown in Figure 3.19, the thickness of the lithosphere overlying each channel is extremely important, because the thickness determines the resulting tectonic style of the channel during tectogenesis---rift valley, germanotype foldbelt, alpinotype foldbelt, midocean rift, and so forth.

Although Pascal's Law applies to all tectonic settings, it is especially important in midocean ridge systems. The law states that pressure applied to a confined liquid acts equally on equal areas of the walls of the confining vessel. The surge channels beneath midocean ridges can be thousands of kilometers wide. Hence, they are much larger than their continental-margin and continental counterparts. Moreover, the lithosphere above midocean ridges is only 10 to 15 km thick, less than half of the thickness found in a continental-margin/continental setting. This means that the total force acting on the walls of a midocean-ridge surge channel is vastly greater than in any other setting. Thus, during tectogenesis, midocean ridges presumably rupture throughout their lengths and across widths far greater than those of continental surge channels, thereby producing veritable magma floods on the ocean floors. If one keeps in mind the fact that the most massive Phanerozoic continental flood volcanism took place from Late Permian though Middle Jurassic time (Yoder, 1988; Meyerhoff et al., 1992b), such magma flooding in the oceans during the same time interval would account for the fact that the oldest basalts thus far penetrated by deep sea drilling beneath the abyssal plains are Middle Jurassic. On the midocean ridges themselves, however, basalts older than Middle Jurassic are common (Meyerhoff et al., 1992a).

3.8 Evidence for the Existence of Surge Channels

3.8.1 Seismic-Reflection Data

As noted above, reflection-seismic techniques (Oliver et al., 1976; Nelson, 1988) have shown that the continental crust of the upper lithosphere is divisible in a very general way into an upper moderately reflective zone and a lower highly reflective zone (for a summary, see Thompson and McCarthy, 1990). Closer scrutiny of the newly-acquired data soon revealed the presence in the lower crust of numerous cross-cutting and dipping events. When many of these cross-cutting events were perceived to be parts of lens-like bodies, various names sprang up: rift pillows, shear pods, low-strain lozenges, anastomosing shear networks, and so forth. Strictly nongenetic names include lenses, lenticles, lozenges, and pods (Klemperer, 1987, 1988; Reston, 1988; Finlayson et al., 1989; Thompson and McCarthy, 1990). Finlayson et al. (1989) found that the lenses have P-wave velocities of 7.0-7.8 km/s, commonly with a low-velocity zone in their middle. Thus we equate the lenses with the pods of "anomalous lower crust" and "anomalous upper mantle" that we discussed in a preceding section. Klemperer (1987) noted that many of the lenses are belts of high heat flow. Hyndman and Klemperer (1989) observed that the lenses generally have very high electrical conductivity.

Meyerhoff et al. (1992b) discovered that there are two types of undeformed reflective lenses, and that many of these lenses have been severely tectonized. The first type of lens is transparent in the middle (Fig. 3.20); the second type is reflective throughout (Fig. 2.11). Tectonized lenses also may have transparent interiors, or parts of interiors; many, however, are reflective throughout (Fig. 3.21). Where transparent zones are present (Fig. 3.20), bands of high heat flow, bands of microearthquakes, belts of high conductivity, and bands of faults, fractures, and fissures are present. Where a transparent layer is not present, high heat flow and microearthquakes are absent; bands of faults, fractures, and fissures, and belts of high electrical conductivity, however, are commonly still present. Meyerhoff et al. (1992b) also found that lenses with transparent interiors are younger than those without transparent interiors; moreover, there is a complete spectrum of lenses from those with wholly transparent interiors to those without.

The best explanations of these observations are that (1) the lenses with transparent interiors are active surge channels with a low-velocity zone sandwiched between two levels of 7.0 to 7.8 km/s material; (2) the lenses with reflective interiors are former surge channels now cooled and consisting wholly of 7.0 to 7.8 km/s material; and (3) the tectonized lenses are either active or former surge channels since converted into kobergens by tectogenesis.

3.8.2 Seismic-Refraction Data

After Revelle (1958) discovered the presence of a body of 7.0-7.8 km/s material on midocean ridges (the East Pacific Rise), a similar body was discovered on the Mid-Atlantic Ridge, and the general lens shape was reported for the first time (Ewing and Ewing, 1959). Subsequently Talwani et al. (1965) combined gravity and seismic data, and detailed the lens shape of the

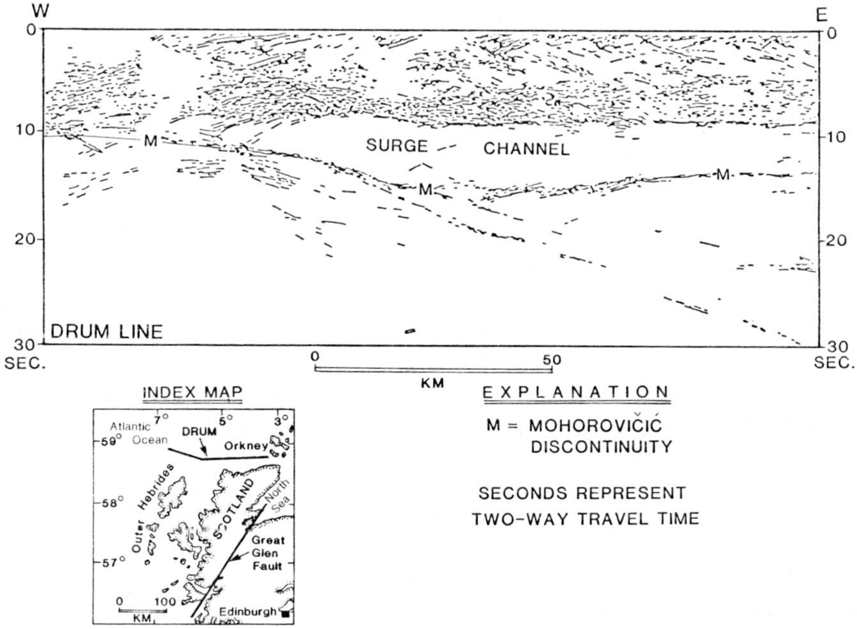

Fig. 3.20. West-east reflection seismic line (tracing) from offshore northwestern Scotland. Note the lens-shaped surge channel with an acoustically transparent low-velocity zone that probably contains some melt. Acoustically transparent surge channels, in our interpretation, are active, associated with high heat flow, microearthquakes, and related phenomena. From Klemperer and Hobbs (1991).

surge channel across the entire Mid-Atlantic Ridge (Fig. 2.27). Fuchs and Landisman (1966) found a similar but much smaller lens beneath the Upper Rhine graben with a P-wave velocity of 7.6 km/s. Now such a 7.0-7.8-km/s lens is known to underlie every well-studied tectonic belt, regardless of tectonic origin. Figure 3.6 shows a 7.0-7.8-km/s lens under Lake Baykal (Krylov et al., 1979; Sychev, 1985); Figure 3.13 shows a similar lens under the northern Appalachians (Taylor et al., 1980). Meyerhoff et al. (1992a, 1992b) summarized the worldwide evidence for the presence of such lenses under every type of tectonic belt. These same authors showed that the lenses under older tectonic belts contain no low-velocity zone, but that lenses in younger tectonic belts contain low-velocity zones.

3.8.3 Seismotomographic Data

Seismotomographic data, wherever detailed studies have been made, indicate that the lenses seen in seismic-refraction and seismic-reflection studies form an interconnected, reticulate network in the lithosphere. Although only one highly detailed seismotomographic study has been made on a continental scale---this in China, it leaves no room for doubt that the 7.0-7.8-km/s lenses with transparent interiors and the seismotomographically detected low-velocity channels in the lithosphere are one and the same (Liu Futian et al., 1989; Liu Jianhua et al., 1989; Sun Ruomei et al., 1991, 1993). Figure 2.31 shows the active surge channels at a depth of about 50 km in southwestern China. Figure 3.9 is a seismotomographic cross section to a depth of 300 km across the Yunnan surge channel shown in Figure 2.31. Figure 3.14 is a more

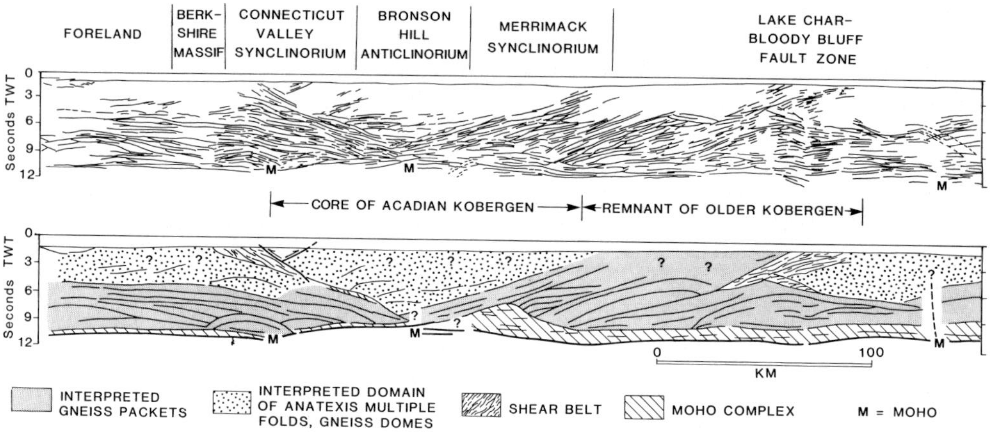

Fig. 3.21. Top: Tracing of a northwest-southeast reflection-seismic line across the northern Appalachian Mountains of the northeastern United States. The right side of the line is the shoreline of the Atlantic Ocean. Line of section is close to those of Figures 2.12 and 3.13. Note the "winged" structure over the Bronson Hill anticlinorium (Fig. 2.16I), which was the root zone of the Appalachian kobergen; the "wings" are the western and eastern thrust zones of the Appalachian kobergen respectively. From Phinney (1986). *Bottom*: Taylor's (1989a) interpretation of the reflection-seismic line (above).

detailed cross section (above a depth of 65 km) of the central part of Figure 3.9. Velocity data from refraction surveys have been added. The 7.6-7.9 km/s layer below 50 km is the Yunnan surge channel shown in Figure 2.31; the low-velocity layer between 22 and 44 km (5.4-6.0 km/s) is a part of the Yunnan surge-channel complex (Liu Jianhua et al., 1989). Using seismotomographic techniques, it will be possible to map active surge channels over the world with comparative ease. The reader should note that Figure 3.22 shows a seismotomographic image of the active kobergen of the Hengduan Shan-Shaluli Shan that overlies the Yunnan surge channel, a further demonstration of the validity of our tectonic interpretation.

3.8.4 Surface-Geological Data

Direct evidence for the existence of surge channels comes from tectonic belts themselves, and from one type of magma flood province. The latter includes rift igneous rocks that crop out nearly continuously for their full lengths. Examples include the rhyodacitic Sierra Madre Occidental-Sierra Madre del Sur extrusive and intrusive belt of Mexico and Guatemala, some 2,400 km long; the 1,600-km-long Sierra Nevada-Baja California batholith belt; the 4,000-km+ batholith and andesite belt of the Andes south of the equator; the 4,000-km-long Okhotsk-Chukotka silicic volcanic belt; the 5,800-km-long Wrangellia linear basaltic province extending from eastern Alaska to Oregon, which erupted in less than 5 Ma; and many other similar continental magma belts. The ocean basins are equally replete with them, ranging from the 60,000-km-long midocean ridge system through the 5,800-km-long Hawaiian-Emperor island and seamount chain to many similar belts of shorter lengths. Geochemical studies also show that most of these belts are interconnected. Another linear flood-basalt belt, which has been studied only relatively recently, is the subsurface Mid-Continent province that extends 2,400 km from Kansas through the Great Lakes to Ohio (Figs. 3.23, 3.24).

3.8.5 OTHER DATA

Other data mentioned in the preceding sections corroborate the interconnection of active surge channels. One of these is the coincidence of the 7.0-7.8-km/s lenses of the active surge channels (Figs. 2.9, 2.31, 3.6, 3.9, 3.14, 3.20) with the belts of high heat flow (Fig. 2.26) and with belts of microseismicity. Both the presence of high heat flow and microseismicity indicate that magma is moving within active surge channels.

However, an even more dramatic example is the June 28, 1992, Landers, California, earthquake-related activity shown on Figure 3.25. This figure shows that the 7.5-magnitude earthquake was strong enough to affect areas up to 1,250 km from the epicenter (Wallace, 1992; Kerr, 1993; Hill et al., 1993) and provides an example of Pascal's Law in action. Given the importance of Pascal's Law in surge-channel systems, the fact should be noted that the viscosity of the magma in the surge channels affected by the Landers event is sufficiently low that, when the stress was applied at a single hypocentral point (Landers), the effects could still be transmitted for 1,250 km!

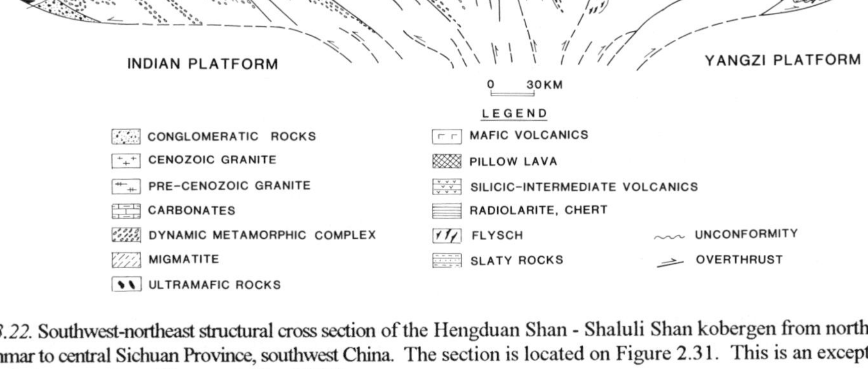

Fig. 3.22. Southwest-northeast structural cross section of the Hengduan Shan - Shaluli Shan kobergen from northeastern Myanmar to central Sichuan Province, southwest China. The section is located on Figure 2.31. This is an exceptionally active kobergen. From Wang and Chu (1988).

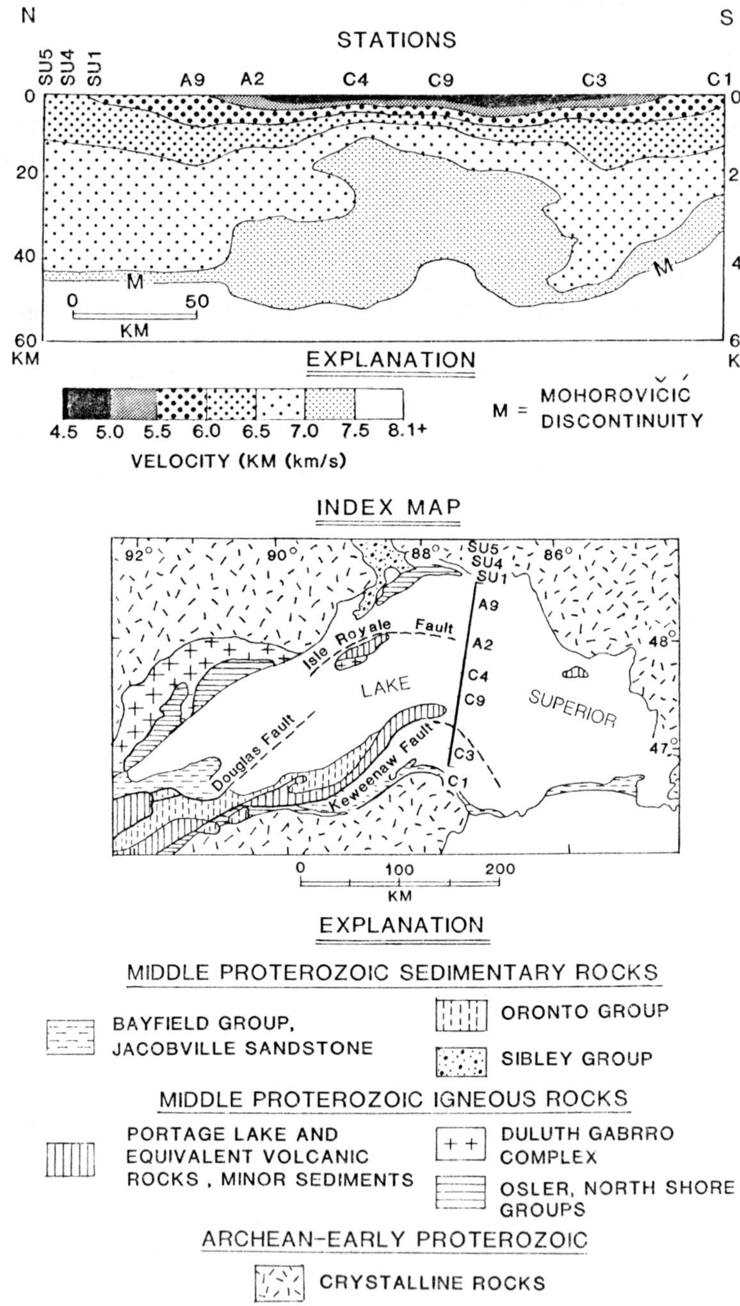

Fig. 3.23. North-south seismotomographic cross section, Lake Superior region, United States and Canada. The P-wave velocity structure is shown. Note the two-tiered structure of the 7.0-7.5 km/s material, a structure that indicates that it is an image of a deformed kobergen. From Hamilton and Mereu (1993).

3.9 Geometry of Surge Channels

3.9.1 Surge-Channel Cross Section

In cross sections, surge channels have a variety of shapes, and are of many sizes and depths within the lithosphere (Meyerhoff et al., 1992b). Two models proposed for sill-and-laccolith complexes may be ideal representations of surge-channel complexes because, despite the differences in scale, the same physical principles apply. Corry (1988) published the "Christmas Tree" model shown in Figure 2.8; Bridgwater et al. (1974) published the more complex model shown in Figure 3.26. Either of these could be cross sections of surge channels. Both are multitiered with one or more magma chambers above the main chamber. Both are formed on the basis of Newton's Law of Gravity or, more specifically, the Peach-Köhler climb force (Gilbert, 1877; Weertman and Weertman, 1964; Gretener, 1969; Weertman, 1971; Corry, 1988; Huppert et al., 1986; Walker, 1989).

Seismotomographic images are available from hundreds of surge channels in different parts of the world. They are complex structures as Figures 3.9 and 3.27 demonstrate. Figure 3.27 is a tracing of a seismotomographic image of the multitiered Yunnan surge channel (Fig. 2.31; Liu Futian et al., 1989; Liu Jianhua et al., 1989; Sun Ruomei et al., 1991, 1993). As we have noted, this channel is highly active (Figs. 2.31, 3.9, 3.14, 3.22). It is also quite large for a continental channel.

Figure 3.23 shows a partly deformed ("kobergenized") inactive Proterozoic surge channel that underlies the Midcontinent Rift of central North America (Hamilton and Mereu, 1993). A surge-tectonic structural interpretation is shown on Figure 3.24. The figures show that the channel before deformation consisted of a large lower chamber at the Mohorovičić discontinuity, and a smaller, higher chamber at about 20 km. Kobergenic structure developed during tectogenesis (Fig. 3.24). The P-wave velocity in the channel is 7.0 to 7.2 km/s, and the top of the channel extends to within 10 km of the surface. The Midcontinent Rift was a major flood-basalt province of 1,100 Ma.

3.9.2 Surge-Channel Surface Expression

Study of Figures 2.8, 2.9, 2.11, 2.31, 3.6, 3.9, 3.13, 3.14, 3.20, 3.23 and 3.24 might lead one to believe that surge channels are everywhere fairly simple structures expressed at the surface by a single belt of earthquake foci, high heat flow, bands of faults-fractures-fissures (streamlines), and related phenomena which, during tectogenesis, deform into a single kobergen. Although this simple picture is true of many kobergens, it is not true of all. Study of Figures 3.26 and 3.27 suggests that, during tectogenesis of the surge-channel complexes shown on these figures, two or more parallel kobergens may exist at the surface. Such a complex surface expression is in fact quite common. Well-documented examples are found in the Western Cordillera of North America, the Mediterranean-Tethys orogenic belt (including the Qinghai-Tibet Plateau), and the Andes, *inter alia*. Within the Western Cordillera, the Qinghai-Tibet Plateau, and the Andes, we have found four or more parallel kobergens side by side at the surface as documented and illustrated by Meyerhoff et al. (1992b).

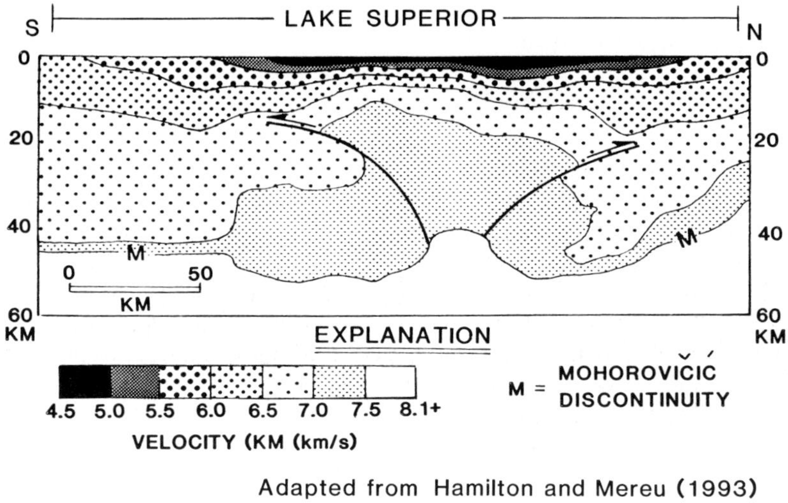

Fig. 3.24. Our structural interpretation of the fossil surge channel, or kobergen, illustrated in Figure 3.23.

3.9.3 Role of the Mohorovičić Discontinuity

The principal forces acting on the lithosphere are compression, rotation, and gravity. We have discussed the first two briefly and now endeavor to describe gravity's role in lithosphere development.

Gravity controls the depths of all magma chambers which, because the Earth cools at the asthenosphere level, have to be in the upper asthenosphere and lithosphere. The mantle above the asthenosphere is believed to be peridotite from which basalt has already been extracted (Ringwood, 1979). It has a P-wave velocity just below the Mohorovičić discontinuity greater than 7.9-8.0 km/s. The oceanic crust above the Mohorovičić discontinuity has been assumed until very recently to have a P-wave velocity of 6.8 km/s and less; the continental crust above the discontinuity has been assumed to have a P-wave velocity of about 6.5 km/s and less.

The liquid generated by the removal of basalt from the mantle cannot be ordinary basalt, but may be akin to tholeiitic picrite (Green et al., 1979). In any case, it is less dense than the peridotite above the asthenosphere and more dense than the basalt found in deep-sea drillholes. Therefore, while in the asthenosphere, it is gravitationally unstable. Consequently Newton's Law of Gravity works to bring the magma upward to a level where it is gravitationally stable. The mechanism for this is the Peach-Köhler climb force (Weertman and Weertman, 1964; Weertman, 1971). This force compels the magma to rise through available conduits to its level of neutral buoyancy, i.e., the level where the lithosphere density and the magma density are the same (Gilbert, 1877; Weertman and Weertman, 1964; Gretener, 1969; Weertman, 1971; Huppert et al., 1986; Ryan, 1987, 1993; Corry, 1988; Walker, 1989). At this level the magma can only move horizontally.

As we have stated repeatedly, the P-wave velocity of the walls of active surge channels is in the 7.0-7.8-km/s range; that of inactive channels is 7.0-7.8 km/s throughout. In the crustal models generally accepted until the late 1980s, there was no layer having a velocity in the 7.0-7.8-km/s range. In recent years, however, a lower crustal layer has been reported in large areas of North America and elsewhere with velocities in the 7.0-7.8-km/s range (Mooney and Braile,

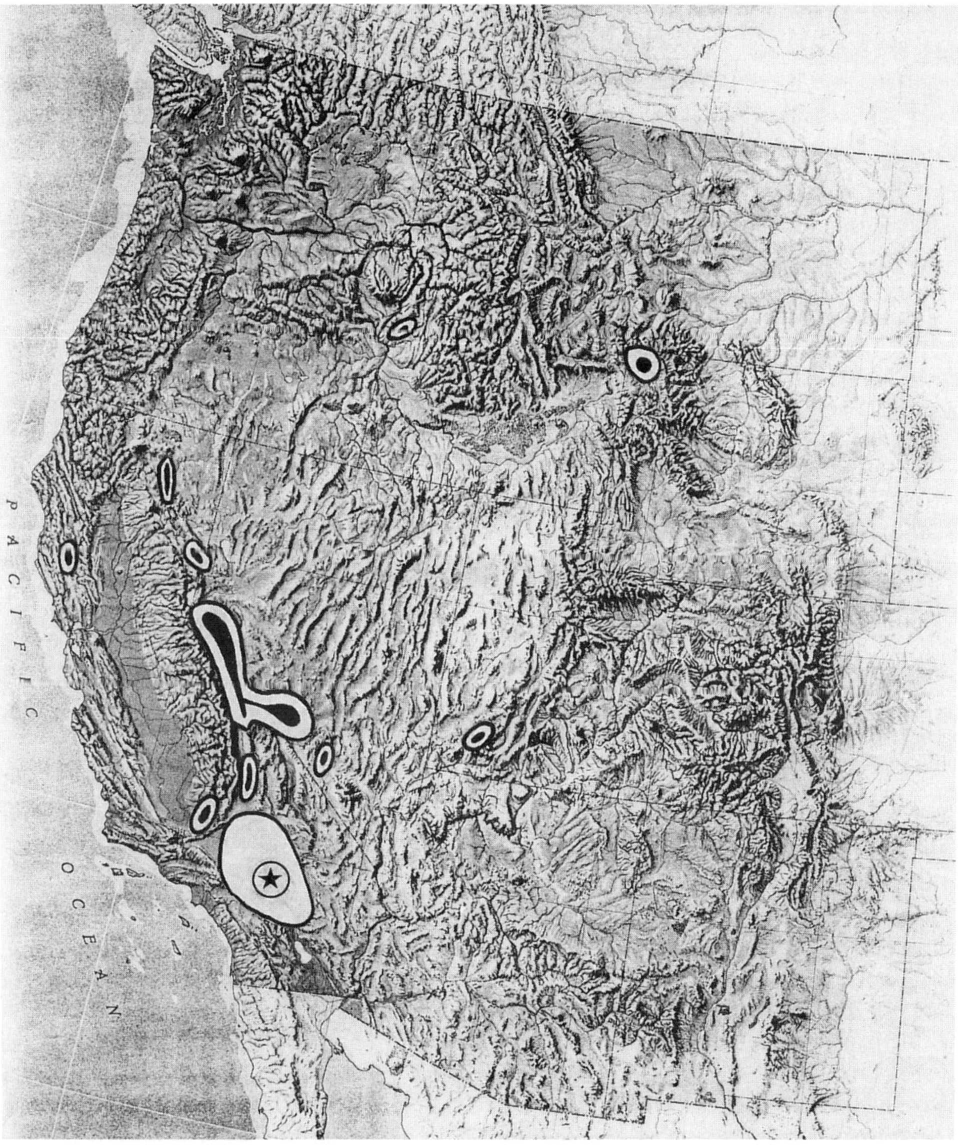

Fig. 3.25. Sketch map showing the location of the Landers, California, earthquake of June 28, 1992, and the various regions that recorded seismic and geothermal effects from the quake. From Kerr (1993) and Hill et al. (1993). Effects were felt up to 1,250 km from the Landers epicenter, thus showing the effects of Pascal's Law.

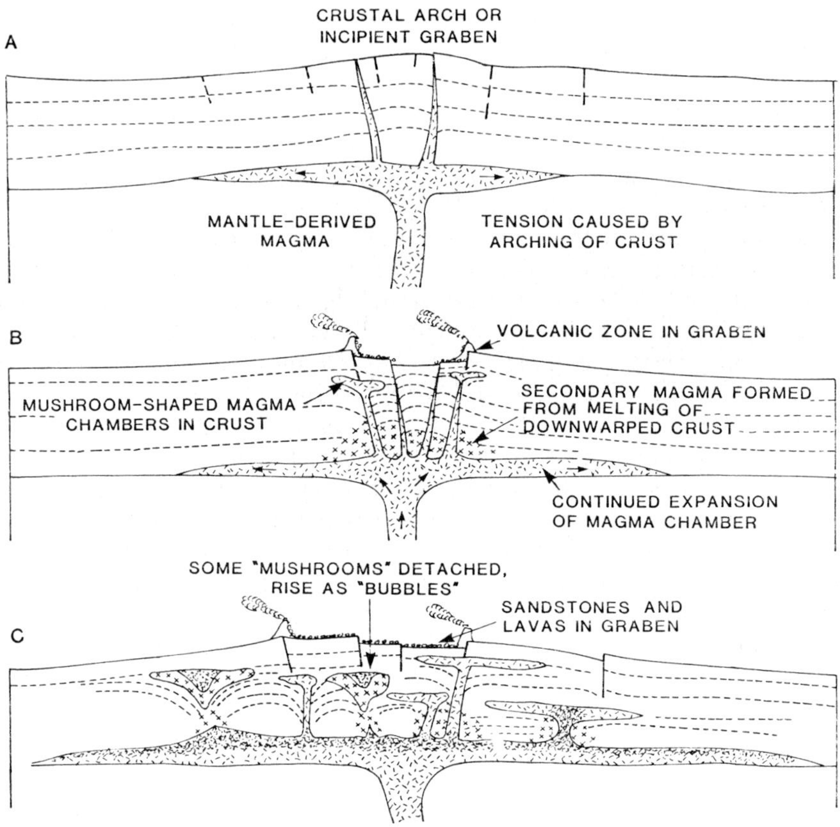

Fig. 3.26. Bridgwater et al. (1974) presented this figure to show one possible model for a sill and laccolith complex that uparches the lithosphere, thereby causing graben formation. The difference between this model and that of Corry (1988; Figure 2.8) is that the positions of the sills and laccoliths here are more random. Bridgewater et al. (1974) referred to this laccolith-sill complex as a "nested series of mushrooms."

1989). Mooney and Meissner (1992) in fact state that the layer is omnipresent, is ca. 3.5 km thick, and is a transition zone between the mantle and crust.

Thus, when the postulated tholeiitic picrite magma reaches the Mohorovičić discontinuity (i.e., the zone between 8.0-km/s mantle below and 6.6-km/s above), it has reached its level of neutral buoyancy and spreads laterally. Under the proper conditions---abundant magma supply and favorable crustal structure---a surge channel can form. We suggest the possibility that the entire 7.0-7.8-km/s layer may have formed in this way. In support of this suggestion, we note that the main channel of every surge channel studied, from the Archean to the Cenozoic, is located precisely at the surface of the Mohorovičić discontinuity. This indicates that the discontinuity is very ancient, perhaps as old as the Earth itself. This fact and the great difference in P-wave velocities above and below the Mohorovičić surface suggest in turn that the discontinuity originated during the initial cooling of the Earth. Hence, Mooney and Meissner's (1992) "transition zone" was the level of neutral buoyancy at the time the 7.0-7.8-km/s material was emplaced.

The suboceanic Mohorovičić discontinuity is some 10-15 km below sea level; its subcontinental counterpart is deeper, at about 40 km below sea level. Hence the neutral buoyancy levels of the two regimes are separated vertically by 25 to 30 km. This fact probably is closely related to the fact that surge channels underlie all continental shelf/slope breaks. Thus these breaks are transition zones between continental and oceanic regimes. MacDonald (1963, 1964) long ago stated that this transition zone must be the site of vertical faulting. Writing of the differences in the vertical distribution of radioactive heat sources under continents and ocean basins, MacDonald (1963, p. 589) stated that these differences ensure "...that the boundaries between them [continents and ocean basins] become discontinuities in subsurface vertical motion. The boundary regions are thus narrow regions where faults are formed and volcanic activity is concentrated." In summary, the ocean-continent boundary is a weakness zone due to differing crustal thicknesses and compositions on either side of that boundary, and the accompanying changes in temperature regimes, specific gravities, and velocities. This weakness zone is therefore a focus for faulting and ascending magma.

It is a fact that volcanism does take place along continental margins, not just in the "active" margins, but in the "passive" ones as well, as for example along the eastern seaboard of North America (Pe-Piper and Jansa, 1987) and the Atlantic margin of Africa (Martin, 1975; Lehner and de Ruiter, 1977; Martin, 1982). In fact, a recent study of North America's eastern seaboard by Holbrook and Kelemen (1993) shows the presence beneath the outer shelf of a linear pod of volcanic material at the Mohorovičić discontinuity. The pod has a P-wave velocity of 7.2-7.5 km/s and is at least 3,300 km long. It should be added that the above explanation does not mean that all surge channels form at ocean-continent boundaries; rather, surge channels are found in every tectonic setting, one of which is the ocean-continent boundary.

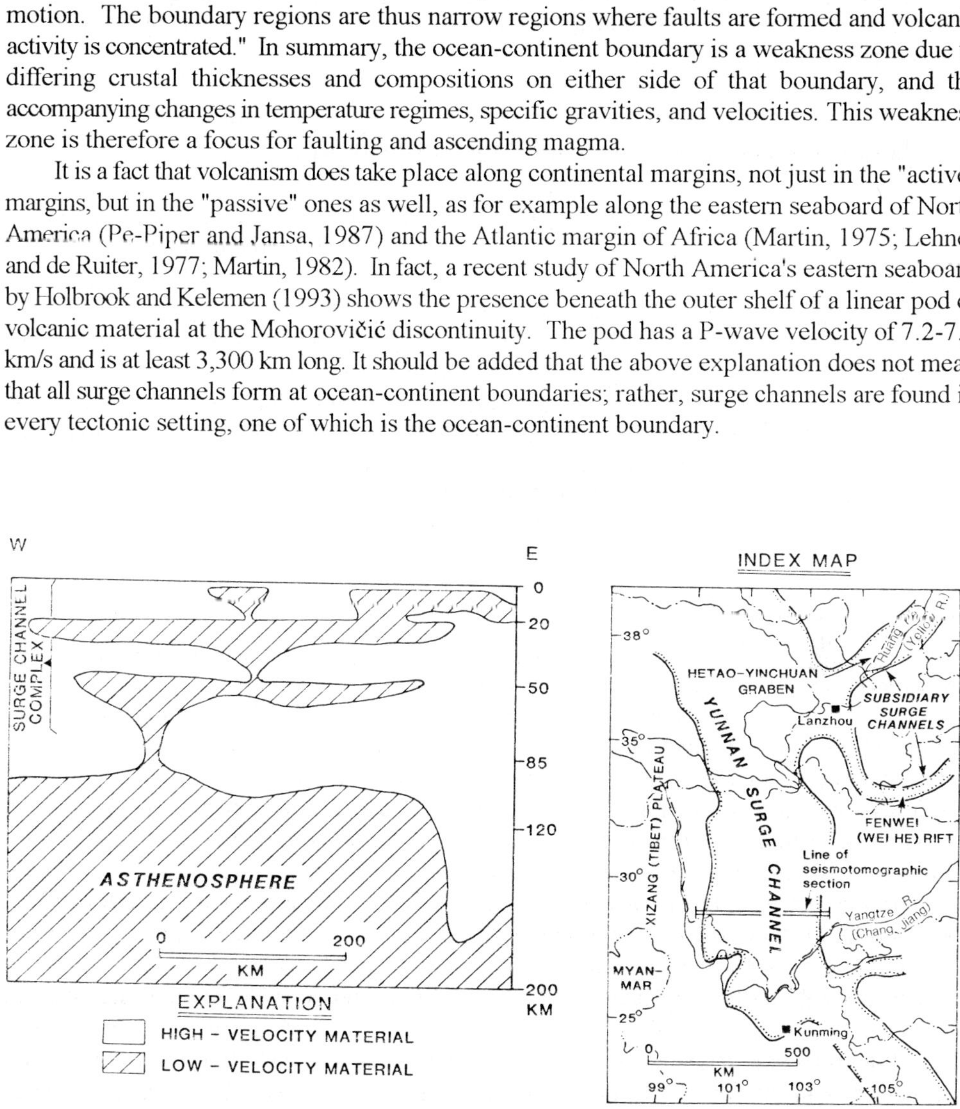

Fig. 3.27. West-east seismotomographic cross section of Yunnan surge channel, from Sun Ruomei et al. (1991). Note the similarity of this "real-life" model to the Bridgwater et al. (1974) model shown in Figure 3.26.

3.9.4 Formation of Multitiered Surge Channels

The formation of the Christmas-tree-like structures (Figs. 2.8, 3.26) at the Mohorovičić discontinuity is simply an extension of the larger scale process of magma transfer from the asthenosphere to the discontinuity. Once surge channels are established at the discontinuity, the same processes take over that brought the magma to the discontinuity in the first place, specifically, magma differentiation in the channels and the Peach-Köhler climb force (Weertman and Weertman, 1964; Weertman, 1971). After lighter magmas have formed by differentiation and related processes, they rise to their own neutral buoyancy levels, forming channels above the main surge channel (Figs. 3.23, 3.27).

3.10 Demonstration of Tangential Flow in Surge Channels

Several workers have postulated flow tangential to the Earth's surface in magma chambers. Vogt (1974) thought he had found evidence for outward flow from the Iceland "hot spot" for distances up to 500-600 km. Gorshkov and Lukashevich (1989) postulated that tangential flow took place along the full length of the Mid-Atlantic Ridge, partly toward the north from a "hot spot" in Antarctica and partly toward the south from an Iceland "hot spot." Shih and Molnar (1975) postulated ridge-parallel propagation along the Juan de Fuca Ridge. Hey and Vogt (1977) suggested that ridge-parallel propagation took place along the Galapagos Rift. Macdonald et al. (1988) and Winterer et al. (1989) proposed that diapirs underlie each major segment of the East Pacific Rise and that, as magma enters each ridge magma chamber, it parts into two oppositely flowing streams parallel with the ridge. All of these suggestions are unlikely and even impossible for reasons outlined by Meyerhoff et al. (1992b). However, ridge-parallel flow *has* occurred.

More recently, suggestions that tectonic belt-parallel flow takes place in continental foldbelts have attracted some proponents. Nicolas (1987) proposed a type of lateral horizontal flow for the Oman ophiolite similar to that suggested by Macdonald et al. (1988) for the East Pacific Rise. Archanjo et al. (1992) proposed flow parallel with large strike-slip faults in the Proterozoic of Brazil. Kuchay and Yeryemin (1990) proposed foldbelt-parallel flow in the western Tian Shan ranges of Asia, and called the magma chambers *asthenochannels*. Makeyeva et al. (1992) proposed a similar flow model for the entire Tian Shan. Except in a very few examples (e.g., Nicolas, 1987), no solid evidence for such flow has been presented.

However, positive evidence for flow does exist. Data acquired from field mapping, satellite imagery, reflection seismology, sonography, and seismotomography reveal two simple geometric proofs of linear tangential flow in surge channels which, according to seismotomography, generally lie above 85 km (Liu Futian et al., 1989; Liu Jianhua et al., 1989; Sun Ruomei et al., 1991, 1993). Both are examples of Newton's Second Law of Motion, one of them utilizing Stokes's Law and the other Bernoulli's Theorem. We review each.

3.10.1 Linear Tangential Flow: Belts of Parallel Faults, Fractures, and Fissures

Sonographs of oceanic ridges, field mapping on land, and satellite images of subaerial tectonic belts, as well as recent studies of the inner planets and the moons of all planets, reveal the presence in *all* tectonic belts of hundreds, even thousands, of linear, belt-parallel faults, fractures, and fissures. These are streamlines in fluid mechanics. Examples are shown on Figure 3.28. We stress the fact that identical fault-fracture-fissure systems characterize *all active tectonic belts, regardless of type*. Such belts are linear or curvilinear and may extend along strike for hundreds, even thousands of kilometers. The individual faults, fractures, and fissures do not have the jagged, zig-zag geometry of tensile fault systems (Fig. 3.29).

The close similarity between these fault-fracture-fissure belts and the streamlines generated by viscous drag on bedrock at the base of mountain glaciers and ice streams is evident (Fig. 3.30). The streamlines so generated in glaciers are examples of Stokes's Law (e.g., Blatt, 1983, p. 210-211), and their very presence shows that flow parallel with them generated them.

We infer that the linear faults-fractures-fissures of tectonic belts are also streamlines. Their presence is proof that somewhere directly beneath them belt-parallel flow is occurring (examples of Stokes's Law). The fact that the faults-fractures-fissures are visible at the surface indicates that they, like the streamlines of glaciers, are also produced by viscous drag, because it is self-evident that the brittle rocks in which they occur are not actively flowing. Moreover, as is well known, ample evidence exists deep in tectonic belts to show that at one time at great depths flow did take place (e.g., Cloos and Cloos, 1927; Balk, 1937; Ramberg, 1952; Cloos and Dietz, 1954). Proof that flow activity still occurs in many tectonic belts is the fact that microearthquakes occur repeatedly, accompanied by high heat flow and small-scale vertical displacements (Stegena, 1990; Tongue et al., 1992).

3.10.2 Linear Tangential Flow: Elongate Segmentation Parallel with Tectonic Belts

The phenomenon of segmentation (Fig. 2.16) is the second manifestation of Newton's Second Law of Motion in tectonic belts, and is related to surface tension. All scientists are familiar with surface tension. For example, at the interface between a liquid and a gas, forces develop on the liquid's surface which cause it to segment into discrete drops (Munson et al., 1990, p. 28). If the liquid is water lying motionless on a flat waxy surface, the water breaks up into round droplets. In contrast, water flowing across the same waxy surface is segmented into elongate interconnected segments parallel with flow (Fig. 2.16) rather than discrete round droplets. Segmentation in both cases occurs because of the presence of unbalanced cohesive forces acting on the molecules at the water's surface. The molecules in the interior of the water body are surrounded by molecules that are attracted to one another equally, whereas the molecules exposed at the surface are subjected to a net force pulling toward the interior. The intensity of the molecular attraction per unit of length along any line on the water's surface is the *surface tension* (Munson et al., 1990, p. 28-29). This manifestation of the law of gravity is repeated in larger scale features such as ocean currents, clouds, lava flows, lava tunnels, and lava tubes (Sears et al., 1974, p. 67; Mandelbrot, 1983, p. 99). Thus all fluids under the proper set of physical conditions segment and form self-similar fractal sets (Mandelbrot, 1983, p. 4; Moon, 1987, p. 23; Crovelli, 1992, p. 19).

A flowing column of water with a free upper surface is held together by surface tension. As the column flows along, it becomes increasingly unstable as small disturbances in it are amplified. Thus segmentation of the water column increases, and sinuosity is amplified, ultimately breaking the column into discrete bodies (Mandelbrot, 1983, p. 99; Tritton, 1988, p. 243-244).

The mechanism that destabilizes the water column may be understood in terms of Figure

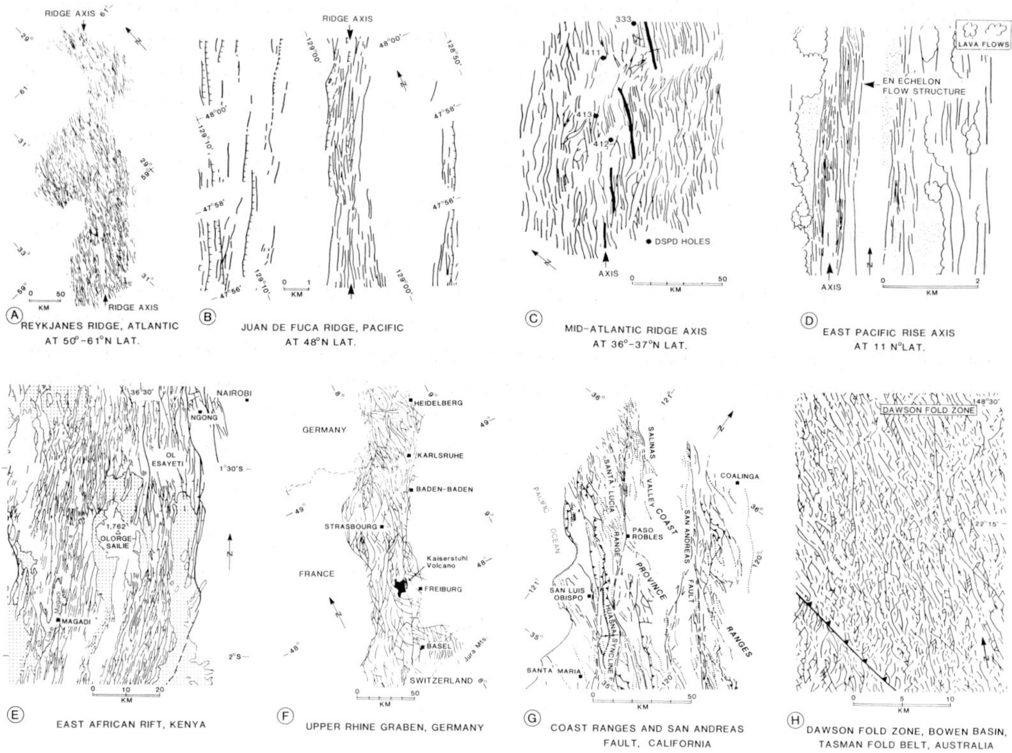

Fig. 3.28. Fault-fracture-fissure systems on different types of tectonic belts. These are good examples of laminar flow; i.e., each of the eight drawings is an illustration of Stokes's Law (one expression of Newton's Second Law of Motion), specifically, that the fault-fracture-fissure directions show the direction of the latest motions directly beneath the surface on which the fault-fracture-fissure pattern appears. Note the fractal-type scale distribution (i.e., the great diversity in scales). Figure 3.28D shows *en échelon* structure, which is further proof that subsurface flow produced these patterns. Figure 3.28F shows anastomosing (braided) and horsetail (splay) patterns, as well as segmentation, further evidence of their flow origin. Figure 3.28H shows evidence of two flow patterns, one north-northwest and the other north-south. From Meyerhoff, in press.

3.31, which shows the type of disturbance that grows. The water column's diameter changes along its length; i.e., its along-strike surface curvature changes. These changes produce pressure changes inside the water column. The pressure changes are needed to balance the action of the surface tension. Provided that the along-strike (longitudinal) wavelength of the column is long enough, the across-strike (azimuthal, or cross-sectional) curvature dominates the longitudinal curvature. Hence the pressure gradient inside the water column is maximal where the radius is at a minimum and minimal where the radius is at a maximum (i.e., the Bernoulli Theorem conditions are approximated except that a water column has both free and flexible surfaces).

In the case of ocean currents such as the Gulf Stream, segmentation is increased further by the bottom topography over which the currents move. The presence of rugged bottom topography heightens the pinch-and-swell segmentation pattern. Again the Bernoulli Theorem conditions are approximated except that ocean currents also have a free surface. Figure 2.16A shows one example of ocean-stream segmentation from Kelly and Gille (1990).

The same principles operate in a viscous fluid (viscosity may be described as the internal friction of a fluid). Lava flows, for example, exhibit marked segmentation (Fig. 2.16B; Greeley, 1982). In a column of flowing lava the pressure gradient at right angles to the flow direction works more strongly against the viscous forces, thereby enhancing the appearance of segmentation. Lava becomes increasingly viscous as it cools, with the result that the surface

Fig. 3.29. This is a photograph of a tension crack in pavement. Compare with Figure 3.28. Note the jagged fracture trace. Tension plainly was at right angles to the crack, and is the stress system shown on Figures A.1e and A.1f. The stress system that produced the crack in this figure clearly is at right angles to the stress system that produced the fault-fracture-fissure patterns of the midocean ridges illustrated on Figures 3.28A through 3.28D. Hence, movements beneath midocean ridges are *parallel with the ridges*, not orthogonal to them. From Meyerhoff et al. (1992b).

Fig. 3.30. Upper Lambert Glacier (ice stream), Antarctica. Flow is from bottom to top. Streamlines show that the flow is laminar. This is another example of Stokes's Law (see Blatt, 1983). This satellite photograph was obtained through the courtesy of Baerbel K. Lucchitta and the late Harold Masursky, U. S. Geological Survey, Flagstaff, Arizona. Compare with Figure 3.28.

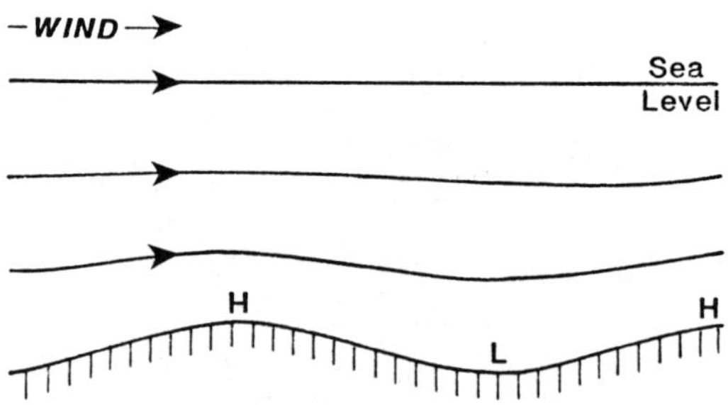

Fig. 3.31. Mechanism producing segmentation in water bodies such as the Gulf Stream, adapted from Gill (1982). The illustration shows a water body, such as the Gulf Stream, flowing from left to right. Topographic irregularities over which the water body flows (the hachured line is the water-body floor) cause the flowing current to alternately constrict and expand. This in turn causes segments of slower moving water to alternate with segments of faster moving water, which is what produces the segmentation observed in the Gulf Stream (Figure 2.16A). This is an example of Bernoulli's Theorem.

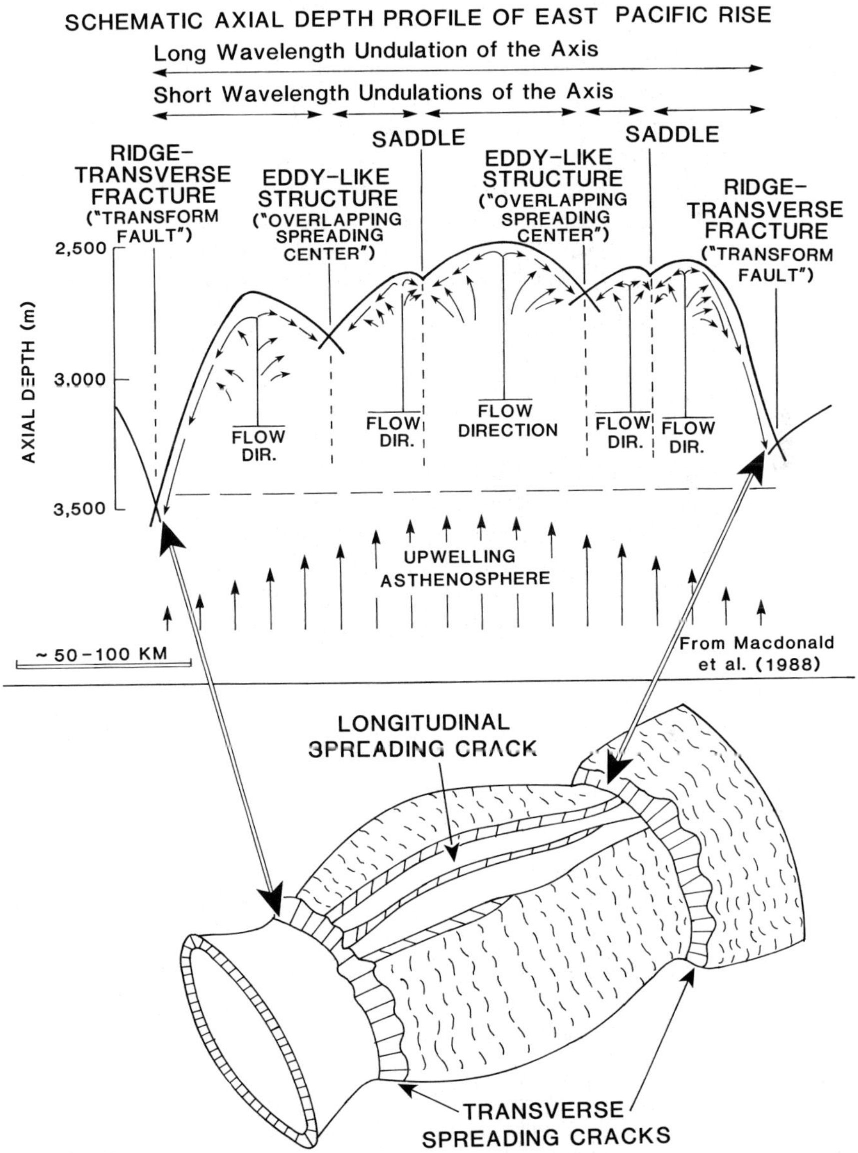

Fig. 3.32. This is another example of Bernoulli's Theorem. The figure compares (schematically) a segmented midocean ridge (East Pacific Rise) with a lava tube, or tunnel. Note that the "transform faults" occur in the bathymetric lows of the midocean ridge; hence, such faults cannot be true "transform faults" but are examples of the effects of the increased velocity of magma movements in constricted parts of magma chambers (surge channels). The "transform faults" (which Meyerhoff et al. [1992b] renamed *ridge-transverse fractures*) thus are a consequence of motions predicted by Bernoulli's Theorem, or equations. Modified in part from Yamagishi (1985).

tension increases sharply. Ultimately the azimuthal pressure approaches in magnitude the combined forces of friction (resistance to forward flow) and viscosity so that a strongly segmented lava tube of the type illustrated in Figure 3.32 is constructed. Flow in the lava column is a type of shear flow and therefore is subject to the types of instabilities that shear flows undergo. When a flow or column of lava cools sufficiently to form a lava tube or tunnel, 100% Poiseuille flow---and therefore the Bernoulli Theorem conditions---takes over and instability stresses develop. The commonest of these is the Kelvin-Helmholz instability in which the shear is produced by lava flowing more rapidly next to the highly convex (upward) parts of the lava tube, thereby accentuating the segmentation of the tube (Drazin and Reid, 1981, p. 1-22; Tritton, 1988, p. 267-270).

In our opinion, lithosphere surge channels are direct analogues of lava tubes and lava tunnels, the principal difference being that flow is not at the Earth's surface. Figure 2.16 shows how we have come to this conclusion. This observation, however, does not change the primary cause of the segmentation, which is differential gravitational attraction (analogous to surface tension).

Segmentation identical to that of the Gulf Stream and lava flows is clearly visible at the Earth's surface beneath every conceivable type of tectonic belt (Fig. 2.16C-2.16H; Meyerhoff et al., 1992b). Figure 2.16, in addition to showing segmentation in the Gulf Stream and a lava flow (which flowed at the Earth's surface), shows identical segmentation in several belts in which flow is deep beneath the Earth's surface, midocean ridges (Macdonald et al., 1987; White, 1989), linear island and seamount chains (Cheng et al., 1987), submarine "aseismic" ridges (GEBCO, 1975-1982), continental rifts (Ebinger, 1989), continental and island volcanic arcs (Marsh and Carmichael, 1974; Marsh, 1979; Ramberg, 1981; Blakely and Jachens, 1990), alpinotype foldbelts (Williams, 1978; Nicolas, 1989), germanotype foldbelts (Hamilton, 1988), strike-slip and normal fault zones (Zandt and Furlong, 1982; Arabasz et al., 1992), belts of gneiss domes and metamorphic core complexes (Macgregor, 1951; Thompson et al., 1968), and continental shelf edges (Moore et al., 1992). In addition, nearly identical segmentation is a characteristic of diapir fields (Trusheim, 1960; Ramberg, 1967, 1981), batholithic belts (Tikoff and Teyssier, 1992), lava fields (Camp and Roobol, 1992), and lava tubes and tunnels (Fig. 3.32; Yamagishi, 1985). The great range of sizes of these different segments demonstrates their self-similar fractal nature (Mandelbrot, 1983). Numerous additional examples of tectonic-belt segmentation were published by Meyerhoff et al. (1992b). Figure 3.32 shows how Meyerhoff et al. (1992b) related lava-tube segmentation to midocean-ridge segmentation.

The development of several linear island and seamount chains in the western Pacific Basin is directly analogous with the development of lava tubes and lava tunnels. Two examples are the Hawaii-Emperor chain (Fig. 3.33) and the Louisville chain (Figs. 2.29, 2.16). The surge channel under the Hawaii-Emperor chain (5,800 km long) broke out of the Asian continent at the cusp between the Kuril-Kamchatka and Aleutian arcs in Campanian-Maastrichtian time (ca. 80 Ma; Decker, 1989) and moved steadily southeastward to its present position beneath Hawaii (Fig. 2.29). Similarly the 4,800-km-long Louisville chain may have broken out of the Australasian region at the cusp between the Tonga and Kermadec arcs in Maastrichtian time (ca. 70 Ma; Cheng et al., 1987) and subsequently moved southeastward along what is now the Eltanin fracture zone to the crest of the East Pacific Rise at 0.5 Ma. Smoot (1994a, b), however, has proposed that this surge of asthenosphere flow instead moved north to the East Mariana Ridge (Smoot, 1994a, b).

Examples of identical segmentation abound in the tectonic belts of the inner planets and

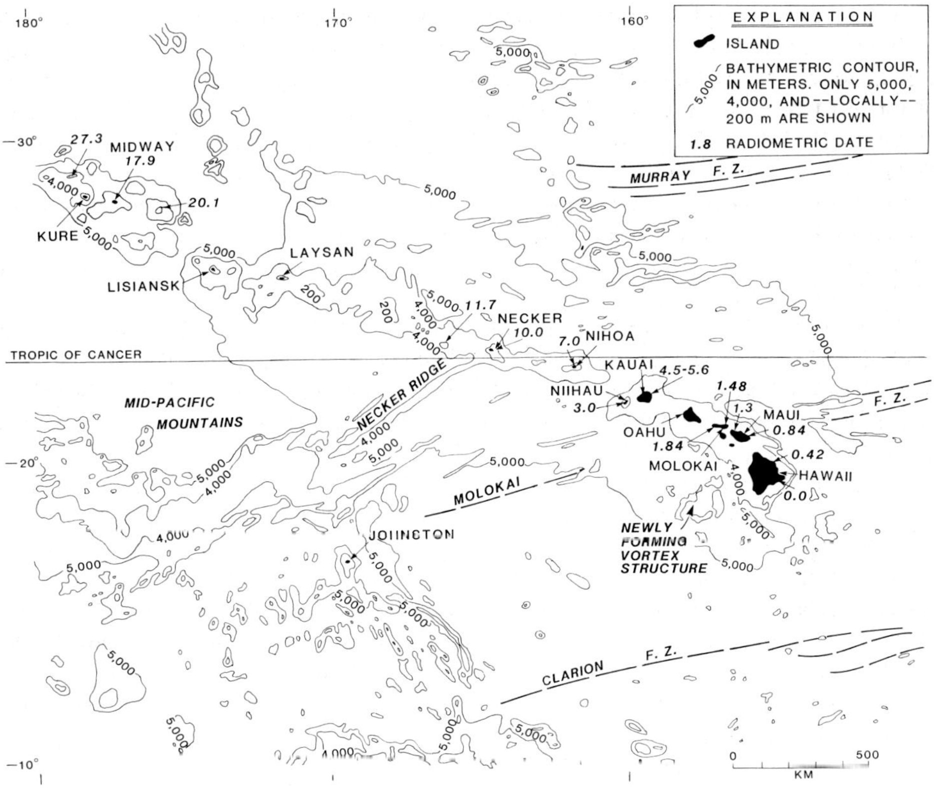

Fig. 3.33. The eastern part of the Hawaiian-Emperor linear island and seamount chain. This is an outstanding example of an eastward-migrating eastern breakout channel (Figures 2.29, 3.39). The radiometric dates provide an accurate measure of the rate of movement of fluids in these surge channels, if one keeps in mind that breakout channels are *newly forming* channels, not well-established channels that have been in existence for more than 100 Ma. (Motion in well-established channels could easily be several centimeters per year faster than motion in breakout channels.) Note that segmentation in the figure is fractal (at several scales). Also note that, from Oahu to Hawaii in the American state of Hawaii, a new vortex seems to be forming. From Meyerhoff, in press.

many moons, both in the inner and outer planets. Fine examples have been illustrated on Mars (Blasius et al., 1977), Venus (Bills and Fischer, 1992), and Io (a moon of Jupiter; Yamaji, 1991). The same segmentation can be observed in any set of linear gravity anomalies (Hildenbrand et al., 1982) and magnetic anomalies (Cady, 1991). Segmentation is also a characteristic of linear epicenter arrays (Savostin et al., 1983; Costain and Bollinger, 1991; Arabasz et al., 1992) and high heat-flow belts (Tuezov, 1988). Most recently, linear low-velocity channels (surge channels) in the lithosphere have been found to show the same pattern of segmentation (Liu Futian et al., 1989; Liu Jianhua et al., 1989; Sun Ruomei et al., 1991, 1993).

To this point we have discussed only the type of segmentation produced during horizontal flow. Segmentation also can be caused by vertical motion, specifically by the Rayleigh-Taylor instability, which is a special case of Kelvin-Holmholz instability (Rayleigh, 1883; Taylor, 1950; Drazin and Reid, 1981). In the Rayleigh-Taylor instability, the hard lower lithosphere (high density) is underlain by the "soft" asthenosphere (lower density). Gravitational potential energy is released spontaneously by the upward penetration of the low-density material in the form of diapirs, which develop a more or less regular spacing pattern (Ramberg, 1967, 1981; Marsh and Carmichael, 1974; Marsh, 1979; Drazin and Reid, 1981; Poliakov and Podladchikov, 1992; Zaleski and Julien, 1992). However, diapirs rising in this manner from the asthenosphere would produce a spacing pattern like that of a salt-dome province which has multiple parallel rows of domes and salt ridges (Trusheim, 1960; Ramberg, 1967, 1981; Poliakov and Podladchikov, 1992). Instead, mainly single rows are observed beneath tectonic belts, not multiple parallel rows (Ramberg, 1967, 1981; Marsh and Carmichael, 1974; Marsh, 1979). This observation means that, if diapirs are rising beneath tectonic belts, they come from

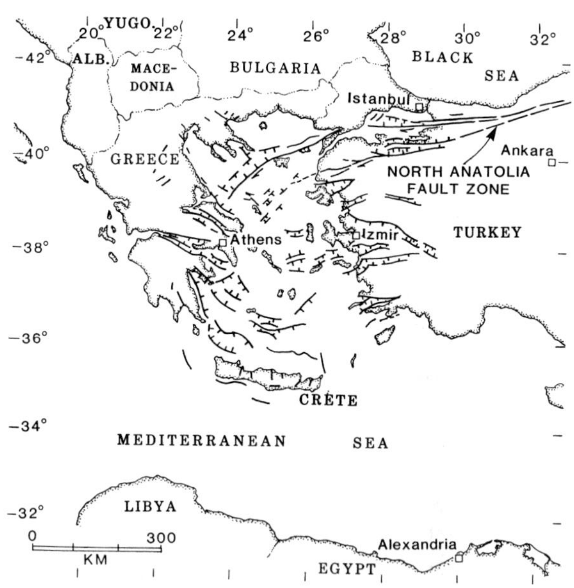

Fig. 3.34. Aegean vortex structure: distribution of Cenozoic grabens and normal faults. Note the circular pattern. From Şengör (1979) and Zanchi and Angelier (1993).

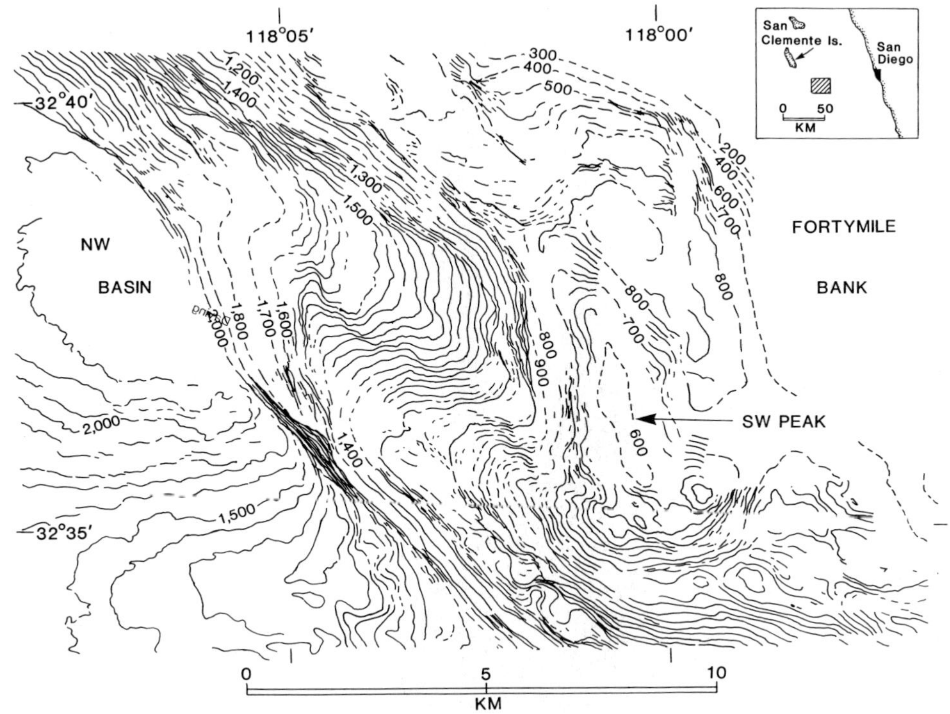

Fig. 3.35. San Clemente fault zone, southern California borderland west of San Diego. This is an outstanding example of a small vortex structure which, at this stage in its development, is what we call an *incipient vortex*, closely akin to the "overlapping spreading center" of plate tectonics. See Figure 3.36. From Legg et al. (1989). (In plate tectonics, such structures commonly are called "pull-apart" structures.)

linear magma chambers and not directly from a blanket-like body such as the asthenosphere (Meyerhoff et al., 1992b). This also clearly implies that linear, not vertical, segmentation mechanisms dominate in tectonic belts, as shown by the linear shapes of the segments in them and the parallel associated belts of faults, fractures, and fissures. However, there is vertical communication among the magma chambers of surge-channel complexes so that the Rayleigh-Taylor instability may have some effect on segment spacing in the surge channels (Fig. 3.27).

3.10.3 Vortical Tangential Flow

The existence of vortical geological structures up to several hundred kilometers across has been known for nearly 70 years. Li Siguang (J. S. Lee) first identified large-scale vortices in eastern Asia in the 1920s (Li Siguang, 1929, 1939, 1973). Wellman (1966) subsequently identified the huge Dasht-i-Lut vortex in eastern Iran (Fig. 2.19). Since 1966, large vortices have been reported by Huang Qinghua (1980), Neev and Hall (1984), Neev et al., (1985), Rousseau (1987, 1992), Westaway (1990), Zeng Zuoxun (1990), and Meyerhoff et al. (1992b). Vortices have been alternatively interpreted as median massifs (Kober, 1928), orogen collapse structures (Dewey, 1988a), overlapping spreading centers (Macdonald et al., 1988), pull-apart basins

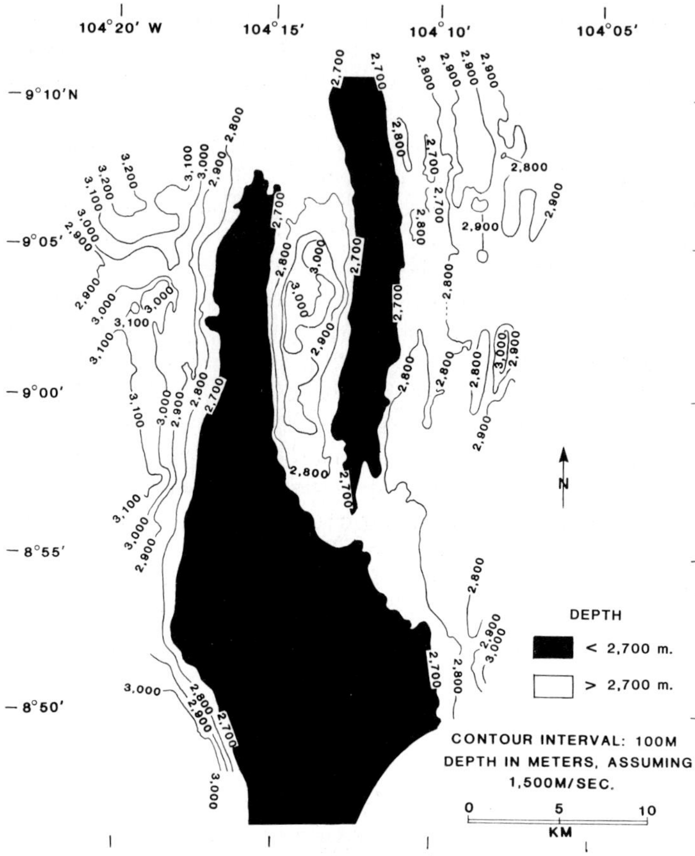

Fig. 3.36. An incipient vortex from the East Pacific Rise at 9° N latitude. According to Macdonald et al. (1987, p. 995), this structure is formed by flow parallel with the East Pacific Rise axis; it is what Macdonald et al. refer to as an "overlapping spreading center." A complete gradation exists between the ridge-transverse fractures of Figure 3.36 and the full-scale vortices of Figures 2.19, 3.34, 3.37, and 3.38. The *en échelon*-like structures of Figures 3.28D, 3.35, and 3.36 are intermediate. From Macdonald et al. (1987).

(Legg et al., 1989), and microplates (Searle et al., 1989), to name a few. Examples we have studied and identified include the Galicia Bank, the Tyrrhenian Sea, the Pannonian Basin, the Aegean Sea-western Turkey, Lake Victoria (East Africa), several in southwestern China, the Banda Sea, the North Fiji Basin, the Galapagos Rift, the Easter Island and Juan Fernández "microplates," the island of Hawaii, the Columbia River Plateau, San Francisco Bay, the Mojave block, and the Colombia Basin in the Caribbean Sea. Figures 3.34 and 2.19 show respectively the Aegean Sea and Dasht-i-lut vortices. For comparison, Figure 3.35 shows a small vortex in the Southern California Borderland west of San Clemente (Legg et al., 1989), Figure 3.36 shows an "overlapping spreading center" on the East Pacific Rise for comparison (Macdonald et al., 1984), Figure 3.37 shows the Lake Victoria vortex between the Eastern and Western Rifts of East Africa, and Figure 3.38 is the Easter Island vortex (Rusby, 1992).

Macdonald et al. (1988) discovered a complete gradation in structural form and style between the "transform" faults and the "overlapping spreading centers" (*ridge-transverse fracture zones* and *incipient vortices* respectively in this paper's terminology). Meyerhoff et al., (1992a) found a complete gradation in form and style between the "overlapping spreading centers" (incipient vortices) and the fully developed vortices (Figs. 2.19, 3.34-3.35, 3.37-3.38). If Macdonald et al.'s and Meyerhoff et al.'s conclusions are correct, then the ridge-transverse fracture zones, incipient vortices, and full-scale vortices have a close genetic relationship.

Macdonald et al. (1988) pointed out that ridge-transverse fracture zones occupy bathymetric deeps, the lowest parts of the midocean ridges that separate major ridge segments (Fig. 3.32). Bathymetric deeps also are associated with the incipient and full vortices. This is verified by the GEOSAT data, which show vortex structures in regions of negative gravity anomalies (Smoot and Leybourne, in review). In the case of the incipient vortices, the deep occupies the space between the two overlapping ridge segments (Fig. 3.36), which is the center of the incipient vortical motions that produce the deep. In the case of the full-scale vortices, the deep lies in the vortex center, as would be expected (Figs. 2.19, 3.34-3.35, 3.37-3.38).

Bernoulli's Theorem provides an explanation for the ridge-transverse faults ("transform faults"). Magma flowing through a surge channel will move more rapidly in the constricted (bathymetric deep) zones (point H of Fig. 3.31). As we noted above, the pressure gradient at right angles to the flow direction is greatest in the most constricted regions. We suggest that the increased pressure and velocity together produce the ridge-transverse fractures. The incipient vortices ("overlapping spreading centers" of plate-tectonic terminology) are *en échelon* structures, simple eddies created by the horizontal flow of magma in the surge channels, just as simple eddies form in ocean currents where the current encounters friction.

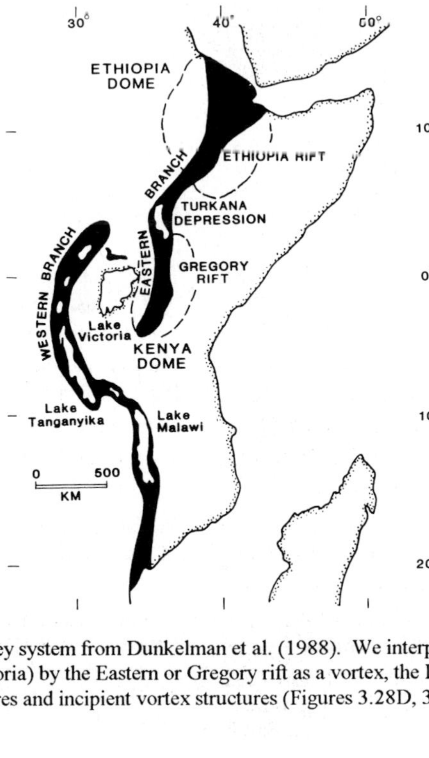

Fig. 3.37. East African rift-valley system from Dunkelman et al. (1988). We interpret the large-scale overlap of the Western Rift (west of Lake Victoria) by the Eastern or Gregory rift as a vortex, the Lake Victoria vortex. Note close similarity to *en échelon* structures and incipient vortex structures (Figures 3.28D, 3.35, 3.36).

Meyerhoff et al. (1992b) suggested that the large-scale vortex structures form wherever the contents of two surge channels "collide." They pointed out that, where two surge channels intersect, the combined volume of the lava from the two channels is too large to be held in the surge-channel space available at the point of "collision." The two flows, each from a different direction, thus form a large-scale vortex structure. This mechanism can be demonstrated as the most likely cause of several flood-basalt provinces such as the Columbia River province, the Emeishan Basalt province, the Panjal Trap province, and many others (Meyerhoff et al. 1992b; see Chapter 6 for details).

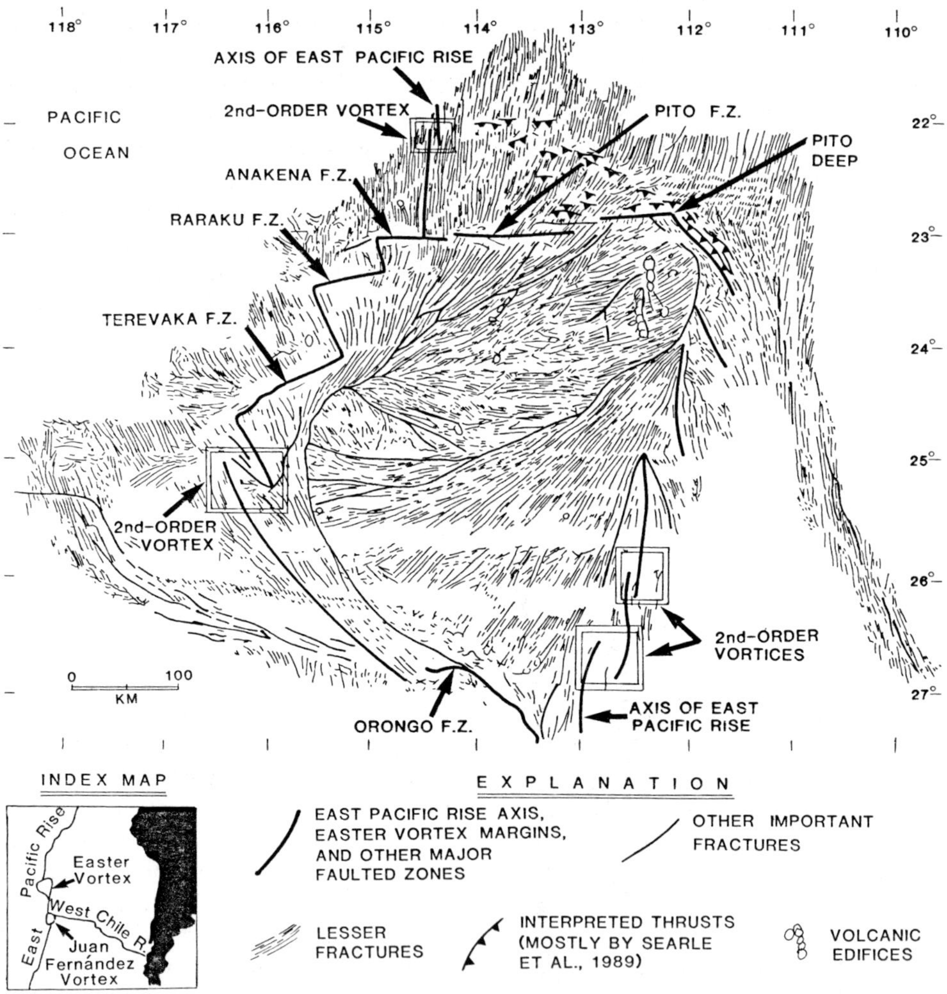

Fig. 3.38. Easter Island vortex from Rusby (1992). This is drawn from side-scanning sonar images. Note: the East Pacific Rise axis forms a giant "overlap" structure (compare with Figures 3.35-3.37) that encloses most of the vortex; the position of the thrust faults in the northeast quadrant of the vortex indicates that motion was formed by counterclockwise motions, not by clockwise motions as Rusby (1992) claims; and small incipient vortices (second-order vortices) are developed in several places along the East Pacific Rise crest.

3.11 Mechanism for Eastward Surge

The ultimate cause of all motions in the asthenosphere and the surge channels is the Earth's rotation. We have already noted that differential lag between the lithosphere and strictosphere causes eastward surging of the asthenosphere and is responsible therefore for the eastward movements that we have demonstrated (using Chinese examples) in the asthenosphere and surge channels (an example of Newton's Third Law of Motion; Meyerhoff and Meyerhoff, 1977). We believe that Coriolis force is very important in producing that eastward motion because any flow in the asthenosphere (and the surge channels) must be very large in scale.

Large-scale geophysical motions take place not only in the asthenosphere but also in the oceans and atmosphere. Gill (1982, p. 75) wrote that, "For large-scale motions in the ocean and atmosphere, the dominant terms by far in the equation of motion...are the gravitational acceleration **g** and the vertical component of the pressure gradient, which approximately balances it. In other words, none of the other acceleration terms in [the equation] approaches the gravitational acceleration. In the atmosphere, for instance, winds are of the order of 10 m s^{-1}, so the Coriolis acceleration is about 10^{-3}m s^{-1}, i.e., less than the gravitational acceleration by a factor of 10,000!" However, as Pedlovsky (1979, p. 21) noted, "The Coriolis acceleration [eventually] becomes dominant for large-scale flows, not because the Coriolis acceleration becomes large, but because the relative [acceleration of the velocity] become[s] so feeble when the scale of the motion is large. . .. [We] conclude that when the time scale of the *motion* exceeds the rotation period, the Coriolis acceleration exceeds the relative acceleration in importance." This will be even truer of the slow-moving asthenosphere and surge-channel fluids.

3.12 Classification of Surge Channels

3.12.1 INTRODUCTION

Figure 3.39 and Table 3.2 show a more recent surge-channel classification modified from Meyerhoff et al. (1992b). First, however, we present some general observations related to surge-channel classification.

A purely descriptive classification is desirable. Geology's and geophysics's greatest weaknesses of the past 30 years are that so much has been classified in plate-tectonic terms. As examples, a midocean ridge is no longer a midocean ridge; it is a "spreading center." A Benioff zone is no longer a Benioff zone; it is a "convergence zone." A wrench or strike-slip fault is no longer a wrench or strike-slip fault; it is a "transform fault" or a "zone of oblique convergence." The removal of objectivity and use of genetic terms in earth science does great harm to the progress of the science. Herein we try to avoid any genetic connotations with the classification used here, while admitting that the use of the term "surge channels" also has its dangers.

Surge channels---or, to use the more objective term, interconnected magma chambers---and tectogenesis are the principal Earth-cooling mechanisms, as pointed out by the Basalt Volcanism Study Project (1981). They are, therefore, very important in the Earth's development and, whether they are called surge channels or magma chambers, they must be regarded as a vital mechanism for the transfer and dissipation of heat from the Earth's interior.

We do not pretend to know all of the factors that control the locations of surge channels.

The positions of continental margins exert a major control on the locations of midocean ridge channels and continental-margin channels; in fact, the possibility must be raised that surge channels determine the positions of continental margins, and not the reverse. Meyerhoff et al. (1992b) showed that the position of the lithosphere Benioff zone governs the position of active-margin surge channels. Lithosphere thickness and the presence and type of lithosphere weakness zones are important. The presence of continental shields and deep continental roots does not seem to affect surge-channel development adversely. For example, at the present time the whole Baltic Shield is being invaded by a network of surge channels. Similarly, much of the Canadian Shield, large areas of the African cratons, and parts of Southeast Asia are the sites of active surge-channel development (Meyerhoff et al., 1992b). Of all the factors that control surge-channel location and development, the Earth's rotation is probably the most important. Figure 3.39 and the following discussion show this clearly.

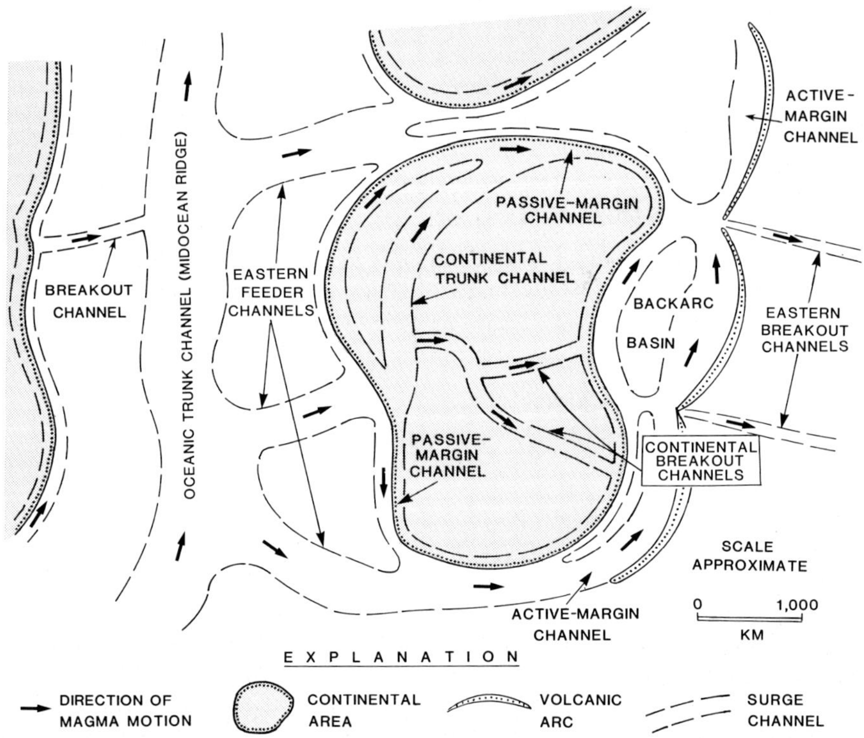

Fig. 3.39. Surge-channel classification. See text and Table 3.2 for discussion of each type. From Meyerhoff et al. (1992b).

3.12.2 OCEAN-BASIN SURGE CHANNELS

Three types of surge channels are present in ocean basins. These are *oceanic trunk channels*, *eastern feeder channels*, and *eastern breakout channels*. The first two are of oceanic origin; the last is of continental origin. Therefore we discuss only the first two in this section.

Oceanic trunk channels are the most important and the largest of all types of surge channels (Fig. 3.39). These are the midocean ridges. They comprise a system approximately

Table 3.2. Tentative classification of surge channels (see Fig. 3.39). From Meyerhoff et al. (1992b; figure numbers have been updated to this publication).

Category	Comments
A. Ocean-basin surge channels	
1. Trunk channels	These are the midocean ridges and their principal branches. They are thousands of kilometers long and can be 1,000 to >3,000 km wide.
2. Feeder channels	These are large branches of trunk channels. They are thousands of kilometers long and are typically 200 to 600 km wide. They become the continental trunk channgels. They carry magma from midocean ridges to the adjacent continent *east* of a midocean ridge. They invariably branch from the *eastern sides* of midocean ridges (Fig. 3.6).
B. Surge channels of continental margins	
1. Passive margine channels	All continental margins appear to be underlain by active surge channels. In fact, surge channels are one of main *raisons d'étre* of continental margins.
2. Active margin channels	These are very large channels that underlie backarc basins. They are quite high and are dammed by Benioff zones (Figs. 4.5, 2.12).
3. Breakout channels	These are long linear channels, 20 to 150 km wide that typically underlie "linear island and seamount chains." They grow eastward (Figs. 3.5, 4.13), commonly showing eastward age progression. They usually are short-lived. Their western terminus typically is at the cusp where two volcanic arcs meet.
C. Continental surge channels	
1. Trunk channels	See ocean-basin feeder channels.
2. Branch channels	These are small channels 20-150 km wide, several hundred kilometers long, branching from the *eastern* sides of trunk channels.

60,000 km long that crosses every ocean basin. Their widths range from slightly less than 1,000 km to more than 3,000 km.

Eastern feeder channels are large branches of trunk channels (Fig. 3.39). Because of the lag effect produced by the Earth's rotation, they invariably branch from the *eastern* sides of the midocean ridges. They carry magma from the midocean ridges to the adjacent continents *east* of the midocean ridges and thereby feed magma to the continental trunk channels; in fact, they are the western segments of continental trunk channels. Typically they are 1,000+ km long and 150 to 800 km wide. They are analogous to (1) eastern and (2) continental breakout channels which we discuss in the next two sections.

3.12.3 Continental-Margin Surge Channels

All continental margins are underlain by active surge channels. We recognize *active-margin channels* and *passive-margin channels*. All channels branching from active- and passive-margin surge channels break out from the *eastern* sides of the marginal channels, again reflecting the influence of the Earth's rotation. Such channels we call *eastern breakout channels* (Fig. 3.39). We have discussed many examples from the western Pacific Basin where a breakout channel typically originates at the cusp between adjacent volcanic arcs or at the junction between two arc segments, that is, at a weak point in the Benioff zone (Figs. 2.29, 3.33; see Ranneft, 1979). They are analogous to eastern feeder channels and eastern branch channels, and in some respects are indistinguishable from them. Typically they are 20 to 100 km wide. Some are several thousands of kilometers long, such as the 5,800-km-long Hawaiian-Emperor island/seamount chain (Figs. 2.29, 3.33, 3.39). Breakout channels characteristically are short-lived in contrast to long-lived trunk and feeder channels (<100 Ma vs. >500 Ma) and some show age progressions eastward (e.g. Louisville, Hawaiian-Emperor, and other island and seamount chains; Cheng et al., 1987; Decker, 1989). Breakout channels tend to become inactive as they progress eastward through the lithosphere (Meyerhoff et al., 1992b).

3.12.4 Continental Surge Channels

These channels lie within continental massifs and are of two principal types, *continental trunk channels* and *continental breakout channels*. The continental trunk channels (Fig. 3.39) are continuations of the eastern feeder channels of the ocean basins, and perhaps should not be given a different name. They are 150 to 800 km wide and commonly are thousands of kilometers long. A major example is the feeder-channel-continental trunk-channel system that branches eastward from the Mid-Atlantic Ridge to underlie the whole Tethys-Mediterranean tectonic belt from Gibraltar to the Solomon Islands and beyond. Locally the system anastomoses extensively, breaking up into subparallel channels (e.g., Xizang [Tibet]). Part of this surge-channel system in southwestern China is displayed on Figures 2.31, 3.9, 3.14, 3.22, 3.27.

Continental breakout channels abound. Analogous to eastern breakout and eastern feeder channels, the continental breakout channels originate only on the eastern sides of trunk channels (Fig. 3.39). Figure 2.31 shows the beginnings of three such branch channels in southwestern China, including the Fenwei (Shanxi) and Hetao-Yinchuan Cenozoic rifts.

Further research has found another type of continental surge channel which may also be present in ocean basins. We tentatively call these *cross channels*. They are channels that are coeval with trunk or branch channels in the same region, but which cross them at a high angle, commonly at right angles (Fig. 3.40). Figure 3.40 shows the northwest-striking Late Cretaceous---middle Eocene Laramide surge channels of the Rocky Mountains. Cutting across the northwest-trending Laramide channels is the Colorado Mineral Belt, which formed in a channel crossing the other Laramide channels at 90°. The Colorado Mineral Belt cross channel formed at a different depth than the other channels and, therefore, was able to "cross" the coeval northwest-southeast channels and to grow independently toward the northwest.

3.13 K Structures

Eastern feeder channels, eastern breakout channels, and *continental breakout channels* commonly are characterized by a distinctive structure which we call *K structures* (Fig. 3.41), following an hypotheses for the formation of Halmahera Island, Indonesia, by Hall and Nichols (1990). Where a surge channel breaks outward and eastward from a north-south-striking channel, it forms a V-shaped feature that opens *eastward*. This "V" plus the north-south-striking surge channel from which the eastward-opening channel emerges forms a K-shaped structure. Figure 3.41 shows three of these---the Galapagos K structure in the eastern Pacific Ocean, probably of late Proterozoic age (Meyerhoff et al., 1992b), the Lashio K structure of east-central Burma, also of probable Proterozoic age (Bender, 1983), and the Halmahera K structure of probable Mesozoic age. The larger, nearby Sulawesi K structure, probably is of late Paleozoic age. Thus K structures are the characteristic structures formed by surge-channel breakouts.

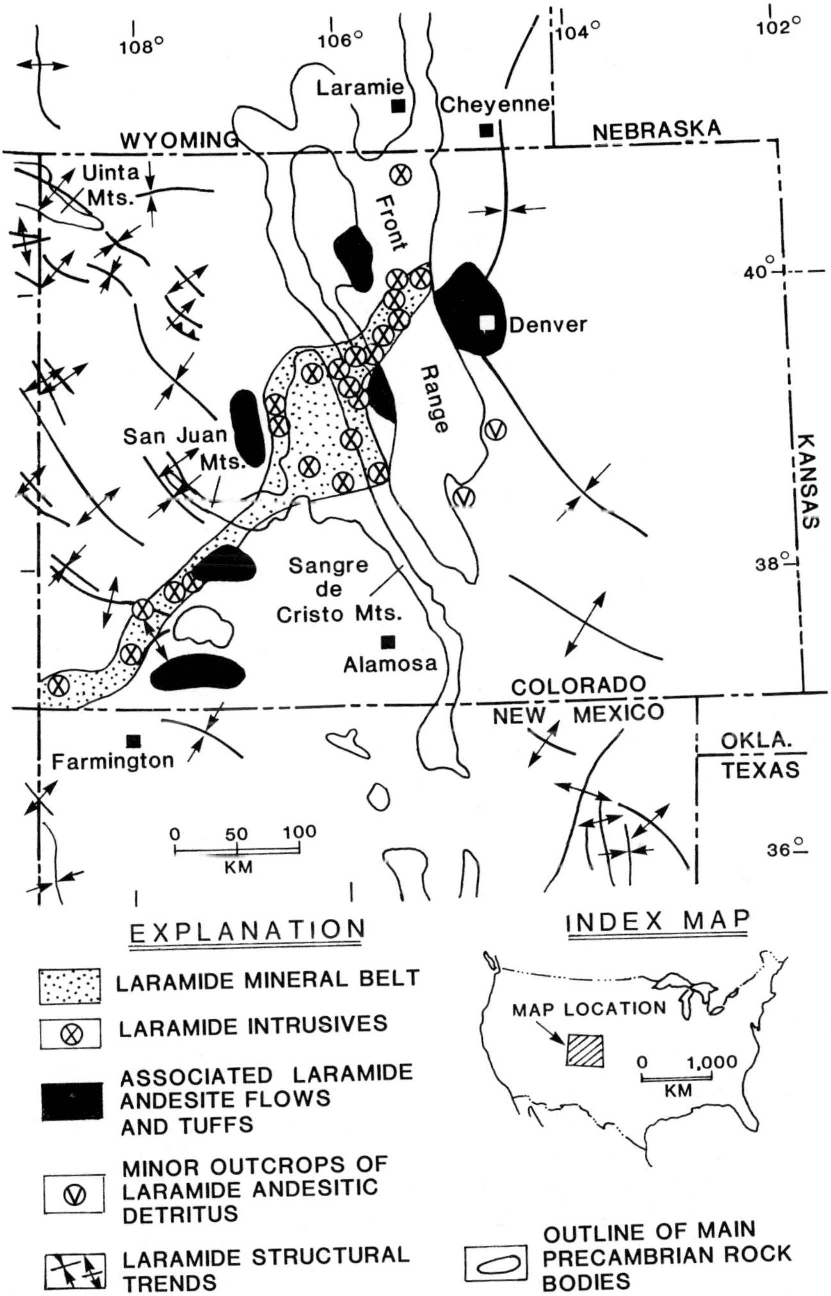

Fig. 3.40. Colorado Mineral belt cross channel, western United States. The map is based in part on Tweto (1975). The northwest-southeast structural trends are Laramide and are of the same age as the northeast-southwest Colorado Mineral belt cross channel. Our explanation for this phenomenon is that the "normal" surge channels in the area are the northwest-southeast features, with flow toward the southeast. In this model, the Colorado Mineral belt cross channel is a continental breakout channel at a different depth than the northwest-southeast channels.

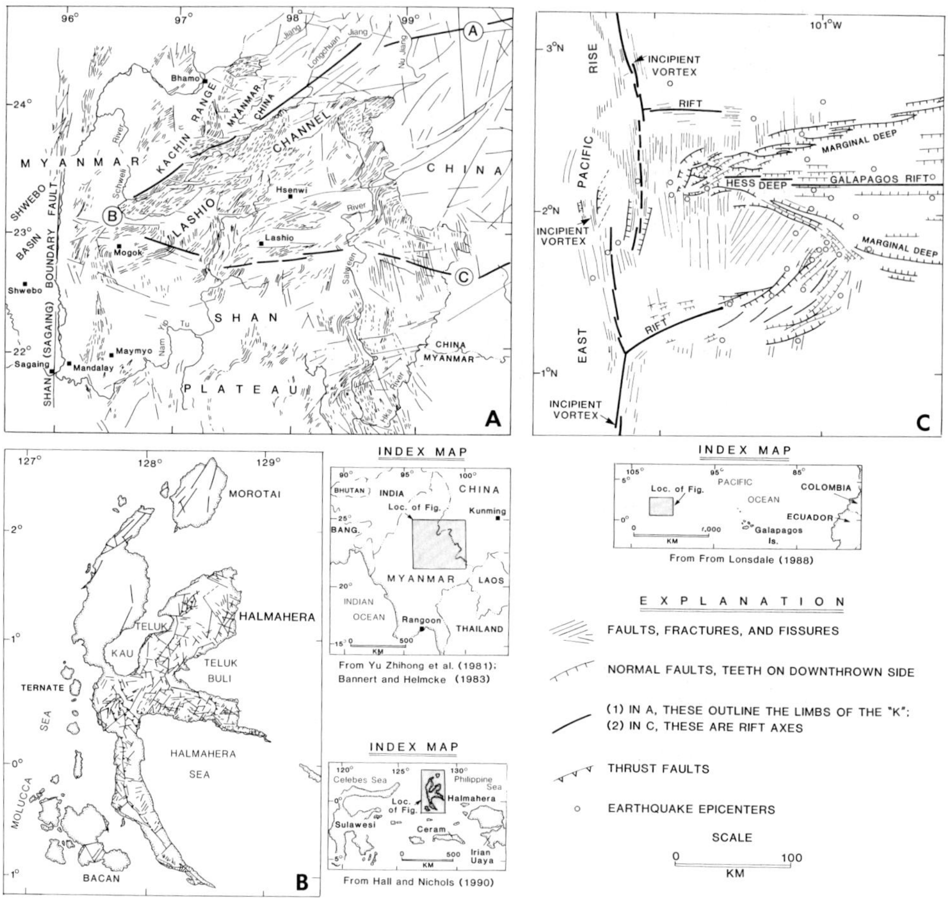

Fig. 3.41. Three K structures--(A) Lashio (Myanmar), (B) Halmahera (Indonesia), and (C) Galápagos (East Pacific Rise). These are (respectively) eastern feeder channels, eastern breakout channels, and continental breakout channels. See the text for discussion. Compiled from Yu Zhihong et al. (1981), Bannert and Helmcke (1983), Lonsdale (1988), and Hall and Nichols (1990).

Table 3.3. Characteristics of surge channels. Explanation of symbols: x = Yes, applicable; - = No, not applicable; ? = unknown, general for lack of data; * = linear valleys and basins generally form as a surge channel loses its contents through tectogenesis and other causes; + = + and - negative anomalies over tectonic belts have some significance, but we have not studied these sufficiently. From Meyerhoff et al. 1992b;figure numbers have been updated to this publication). Additional examples are listed in Meyerhoff et al. (1992b).

Tectonic feature generated by channel	Channel status		Tectonic regime			Figure number
	Active	Inactive	Oceanic	Transi-tional	Conti-nental	
Linear to curvilinear lithosphere breaks						
Long linear zones of faults, fractures, and fissures	x	x	x	x	x	3.28
Horst-and-graben complexes, rifts	x	x	x	x	x	3.6
streamline (strike-slip) fault, suture, *Verschluckungzone*, tectonic line	x	x	?	x	x	2.23
Thrust faults	x	x	x	x	x	4.14
Horsetail structures	x	x	?	x	x	2.4
Eddies and vortex structures	x	x	x	x	x	5.1
Structures related directly to lithosphere breaks						
Magmatic arcs, eugeosynclines, alpino-and germanotype foldbelts	x	x	x	x	x	2.16
Tectonostratigraphic terranes	x	x	-	x	?	4.11
Fissure eruptions and volcanic fields	x	x	x	x	x	3.9A
Aligned plutons	x	x	x	x	x	4.14
Lines of thermal springs	x	-	x	x	x	2.25
Kimberlite dikes, diatremes, ring complexes	-	x	?	x	x	----
Dike swarms	x	x	x	x	x	----
Ophiolite belts	x	x	x	x	x	----
Mélange belts	x	x	x	x	x	----
Metamorphic belts (mantled gneiss domes, core complexes, inverted gradients)	x	x	?	x	x	2.23
Submarine ridges (aseismic ridges)	x	x	x	-	-	4.1
Structure indirectly related to lithosphere breaks						
Fold trends	x	x	x	x	x	4.15
Stretching lineations	x	x	?	x	x	----
	x	x	x	x	x	----
Morphotectonic features						
Linear river courses	*	x	-	?	x	2.4
Linear topographic divides	x	x	x	x	x	2.5
Basins, linear evaporite basins	*	x	?	x	x	2.20
Plateaus	x	?	x	x	x	----
Midocean ridges	x	x	x	-	-	3.1
Feeder channels (oceanic rises)	x	x	x	-	-	4.1
Breakout channels (linear island and seamount chains)	x	x	x	-	-	2.27
Geophysical characteristics						
+ Bouguer gravity anomaly	-	x	x	x	x	3.8
- Bouguer gravity anomaly	x	-	x	x	x	2.9
Midocean ridge magnetic anomalies[+]	x	?	x	-	-	3.3

Table 3.3. Continued.

Tectonic feature generated by channel	Channel status: Active	Channel status: Inactive	Tectonic regime: Oceanic	Tectonic regime: Transi-tional	Tectonic regime: Conti nental	Figure number
High-conductivity, low-resistivity magnetotelluric anomaly	x	?	x	x	x	4.10
Bands of microearthquakes	x	-	x	x	x	4.13
High heat flow bands (>55 mWm^{-2})	x	-	x	x	x	4.14
Bands of "anomalous upper mantle" (P-wave velocity 7.0-7.8 km/s) These bands form lenses on:	x	x	x	x	x	3.2
(a) refraction-seismic lines	x	x	x	x	x	2.12
(b) reflection-seismic lines	x	x	x	x	x	2.11
Interior of reflection-seismic lens is transparent	x	-	x	x	x	2.10
Interior of reflection-seismic lens is filled with reflectors	-	x	x	x	x	2.11

3.14 Criteria for the Identification of Surge Channels

The principal criteria for surge-channel identification have been discussed on the preceding pages in several contexts which are summarized here and on Table 3.3. For a more detailed summary, the reader is referred to Meyerhoff et al. (1992b).

Surge channels, active and inactive.

1. A linear or curvilinear tectonic belt is present. Such belts include all rifts, foldbelts, metamorphic belts, and strike-slip zones.
2. The belts have linear surface swaths of faults, fractures, and fissures, commonly exhibiting splays, or horsetail structure. Strike-slip zones and "tectonic lines" are common.
3. Belt-parallel stretching lineations are present.
4. A rough bisymmetry characterizes each belt, whether rift or foldbelt. In the latter case, flower structures, "popups," and larger kobergens are developed.
5. Aligned plutons, orogenic and anorogenic, are a characteristic with or without an associated foldbelt. Aligned ophiolites, belts of mélange, volcanic centers and associated volcanic fields, kimberlite dikes, diatremes, ring structures, and mineral belts are characteristic. Zoned metamorphic belts, including inverted metamorphic zones, are characteristic.
6. Linear river valleys are present in some areas.
7. Linear high-conductivity magnetotelluric and electrical anomalies are present.
8. Vortex structures may or may not be present.
9. Flood basalt provinces commonly are associated.
10. A lens (which may or may not be tectonized) of 7.8-7.0 km/s material underlies the belt.

<u>Inactive surge channels.</u>

1. A linear Bouguer positive gravity anomaly is present.
2. A linear positive magnetic anomaly is present.
3. The linear lens-shaped pod of 7.8-7.0 km/s material is reflective throughout, whether tectonized or not.

<u>Active surge channels.</u>

1. Moderate to high elevations are present in all active belts, whether germanotype or alpinotype foldbelts, continental rifts, midocean ridges, linear island and seamount chains, or other submarine "aseismic" ridges. However, under continental shelf edges, only a buried structural arch is present, although a rugged, elevated coastal topography commonly is present (e.g., India's west coast, the southeastern coasts of China and Vietnam, and the eastern Korean coast, among many, as noted by Cox [1989]).
2. A high-heat flow belt is associated.
3. A belt of microseismicity is present.
4. Lines of thermal springs are evident.
5. A small negative Bouguer gravity anomaly commonly is present.
6. The 7.8-7.0-km/s lens is present with transparent zones in the lens center; in some cases, the lens is wholly transparent.

CHAPTER 4

EXAMPLES OF SURGE CHANNELS

In this chapter, we present actual examples of the major types of surge channels, following the outline of Table 3.2 and Figure 3.39. We also discuss the criteria for recognizing the different types of surge channels, listed on Table 3.3.

4.1 Ocean-Basin Surge Channels

4.1.1 Mid-Atlantic Ridge

The Mid-Atlantic Ridge is the best known of the ocean's trunk channels. Shortly after Revelle's (1958) discovery of 7.0-7.8-km/s anomalous upper mantle in the southern part of the East Pacific Rise, Ewing and Ewing (1959) showed that such a layer is also present on the Mid-Atlantic Ridge and that it pinches out on either flank (Fig. 2.9). Where the anomalous upper mantle is absent, crust with a P-wave velocity of 6.4 to 6.8 km/s directly overlies normal mantle (8.0-8.5 km/s). The overall shape of the anomalous upper mantle bodies under the Mid-Atlantic Ridge is that of a thin, up to 30-km-thick lens in the axial zone of the ridge, up to 1,000 km wide, and thousands of kilometers long.

Figure 2.9, from Talwani et al. (1965), shows three interpretations of Mid-Atlantic Ridge structure based on refraction-seismic and gravity data. These sections are based on data from 27° to 47° N latitude. The models shown here are greatly simplified, as Pavlenkova's (1989) later work has shown. However, they represent some truly pioneering work that has stood the test of time.

Figure 3.2 from White (1989) is a seismic-refraction section parallel with the Mid-Atlantic Ridge axis between 37° and 40° N latitude, the same area as Figure 2.9, but at right angles to the latter. As Meyerhoff et al. (1992a) noted, Figure 3.2 shows the pinch-and-swell geometry that characterizes all midocean ridges. Yamagishi (1985) pointed out that the same pinch-and-swell geometry typifies all lava flows and tunnels (Fig. 3.32), and is produced by the surging of lava through the tube, or tunnel. The pinch-and-swell structure exhibited in Figure 3.2 is further evidence that ridge-parallel flow produced the midocean ridges.

A low-velocity layer almost certainly is present in the interior of the 7.0-7.8-km/s lens. This is supported in part by a band of microearthquakes that coincides very closely with the high heat-flow anomaly at the ridge axis. Bouguer gravity maps show a negative value at the ridge crest. The negative values are greatest between transform faults, and least at the fault zones themselves (Lin et al., 1990), thereby providing strong support for our interpretation.

Farther south along the Mid-Atlantic Ridge, at 11° S latitude, Pavlenkova (1989) and several colleagues collected detailed gravity, magnetic, heat-flow, and seismic-refraction data. Figure 3.3 shows a more complex internal ridge structure than Figure 2.9. Here, at 11° S, the ridge is 1,000 km wide and the lens of 7.7-7.8-km/s material is nearly as wide as the ridge itself at depths of 26-36 km. Above this zone are two more low-velocity zones, 450-600 km wide. A distinct heat-flow anomaly is present, and the Bouguer anomaly is nearly identical with that on Figure 2.9.

The presence of the 7.0-7.8 km/s lens shown on Figure 2.9 does not easily fit into the plate-tectonic model, a problem recognized by Drake and Nafe (1968) and Vogt et al. (1969). Theoretically this material, if it is generated at the ridge crest, should be present everywhere in the ocean basins. Vogt et al. (1969, p. 595-596) discussed the significance of the lateral pinch-out, writing that if this anomalous material is produced during sea-floor spreading, "We would then require a physical process whereby the low-velocity [anomalous] mantle somehow becomes segregated into oceanic layer and normal mantle as it cools and withdraws from the axis." Drake and Nafe (1968, p. 185) expressed the opinion that the 7.0-7.8-km/s lenses were transient and their disappearance away from the ridge crests ". . . may be related to the changes in elevation associated with tectonic activity." The problem remains unsolved by plate tectonics, but is an unavoidable consequence of surge tectonics.

4.1.2 East Pacific Rise

The velocity structure of the East Pacific Rise is like that of the Mid-Atlantic Ridge, except the East Pacific Rise channel system is much broader than that of the Mid-Atlantic Ridge (Menard, 1960; Talwani et al., 1965). The greatest difference between the two ridges is the presence in the East Pacific Rise of very shallow low-velocity lenses with high electrical conductivity, possibly containing melt (Reid et al., 1977; Filloux, 1982). Figure 4.2 shows a recent interpretation of the geometry and velocity structure of one shallow lens (Harding et al., 1989). Unlike the Mid-Atlantic Ridge----at least in exploration to date----the shallowest low-velocity lens of the East Pacific Rise may extend to within 0.8 km of the sea floor (Caress et al., 1989). The presence of such shallow activity implies that magma in the principal surge channel beneath the East Pacific Rise crest is differentiating continuously. Such differentiation would explain the occurrences of extremely young andesite, dacite, rhyodacite, and silica-rich volcanic products along the rise (Thompson et al., 1989).

4.1.3 Midocean- Ridge Magnetic Anomalies

Our discussion of the midocean ridges would not be complete without at least a few comments concerning the origin of midocean ridge magnetic anomalies, particularly considering the major role that these features play in the plate tectonic hypothesis. Agocs et al. (1992) demonstrated that magnetic anomalies of midocean ridges can be explained better by magnetic-susceptibility contrasts than by magnetic field reversals, as proposed in the Morley-Vine-Matthews hypothesis (Morley and Larochelle, 1964; Vine and Matthews, 1963). Meyerhoff et al. (1992a) proposed that ridge-parallel fault, fracture, and fissure patterns of midocean ridges (Fig. 3.28) are the underlying cause of the linearity of magnetic anomalies. When new magnetic anomaly sets were found in the western Pacific----illustrated on Figure 4.3----sets that conflicted with others in the Pacific basin, an explanation was proposed to conform with plate-tectonics concepts (Fig. 4.3). Larson and Chase (1972, p. 3641) concluded that the different anomaly patterns of Figure 4.3 ". . . must have been generated by a system of five spreading centers joined at two triple points." They stated that all correlations among the magnetic profiles were established "by eye" (p. 3631). No transformations to the pole were attempted, although the sets come from greatly different latitudes (52° N, 40° N, 5° N, Agocs et al., 1992). (That is, no adjustments were made to reduce the magnetic profiles to a common latitude so meaningful correlations might be attempted.)

A complex plate-tectonics history has been proposed for the Pacific basins. In contrast, if

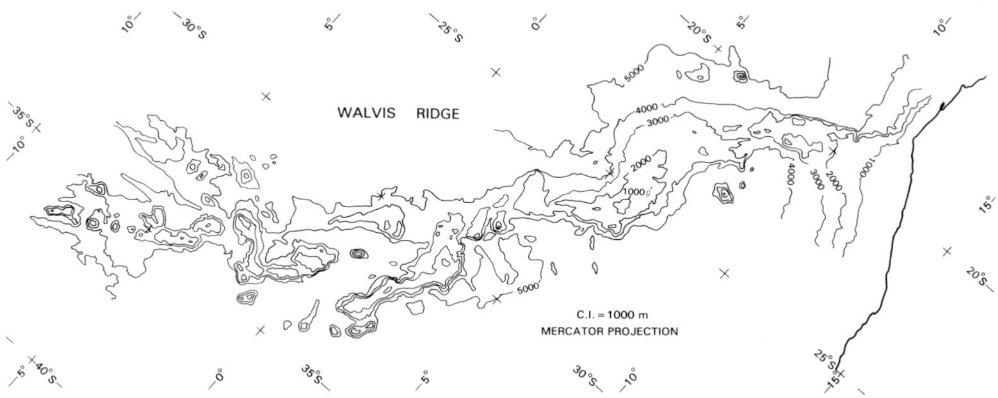

Fig. 4.1. The Walvis Ridge feeder channel, South Atlantic Ocean. Note the sinuous, anastomosing, flow-like pattern (see Figure 2.16C). Eddylike features are prominent south of 30°S latitude. Compare this with Figures 2.4, 2.19, 2.27, 3.35, 3.37, and 4.15.

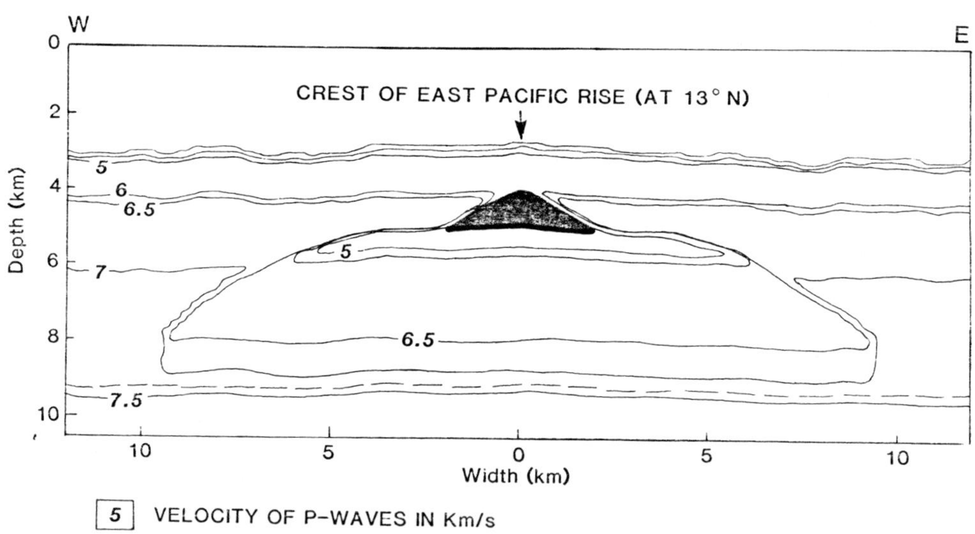

Fig. 4.2. West-east velocity profile across a shallow magma chamber on the crest of the East Pacific Rise at 13°N latitude. This section is based on a series of expanding-spread seismic profiles and common-depth-point reflection lines (CDP). The solid black may be a zone of large melt percentage as determined from CDP lines by Harding et al. (1989). This figure illustrates a fundamental difference between the East Pacific Rise and other midocean ridges, namely, the presence of very shallow magma chambers. Another difference is that the East Pacific Rise is buoyed by an exceptionally large surge channel; it may be buoyed partly by upwelling asthenosphere as well (Figure 4.5). We believe that this swell is produced by relatively unimpeded eastward flow across the broad Pacific basin. Dammed against the two American cratons, flow is forced between them, or farther poleward (Figure 4.5). In contrast, in the Atlantic basin, asthenosphere swelling can take place only where a flow enters from the Pacific, mainly between 10° and 60°N latitude, but to some extent south of 30°S latitude. From Meyerhoff et al. (1992b).

the magnetic anomaly sets of the western Pacific were produced by once-active surge channels, a coherent and internally consistent pattern of flow (as shown by the magnetic anomalies) should emerge. To test this hypothesis, we chose the Phoenix anomalies (Fig. 4.3). We then updated the Weissel (1981) map of magnetic lineations of the western Pacific, and added the anomaly patterns discovered since 1980. The result is Figure 4.4, to which we added black arrows indicating our interpretation of flow directions. A coherent flow pattern is apparent. The pattern indicates that flow from beneath Asia and Australia converges between the Yap and Tonga trenches and continues east-northeastward beneath the Phoenix anomalies. A carefully planned magnetic survey across the region would be a relatively inexpensive means of testing further this surge-tectonics model.

Fig. 4.3. This figure, from Nakanishi et al. (1992), shows various magnetic-anomaly lineation sets in the northwestern Pacific basin. Previous workers (e.g., Larson and Chase, 1972) have proposed a complex plate-tectonics models, which is, as Figure 4.4 shows, unnecessary if viewed from the surge-tectonics hypothesis.

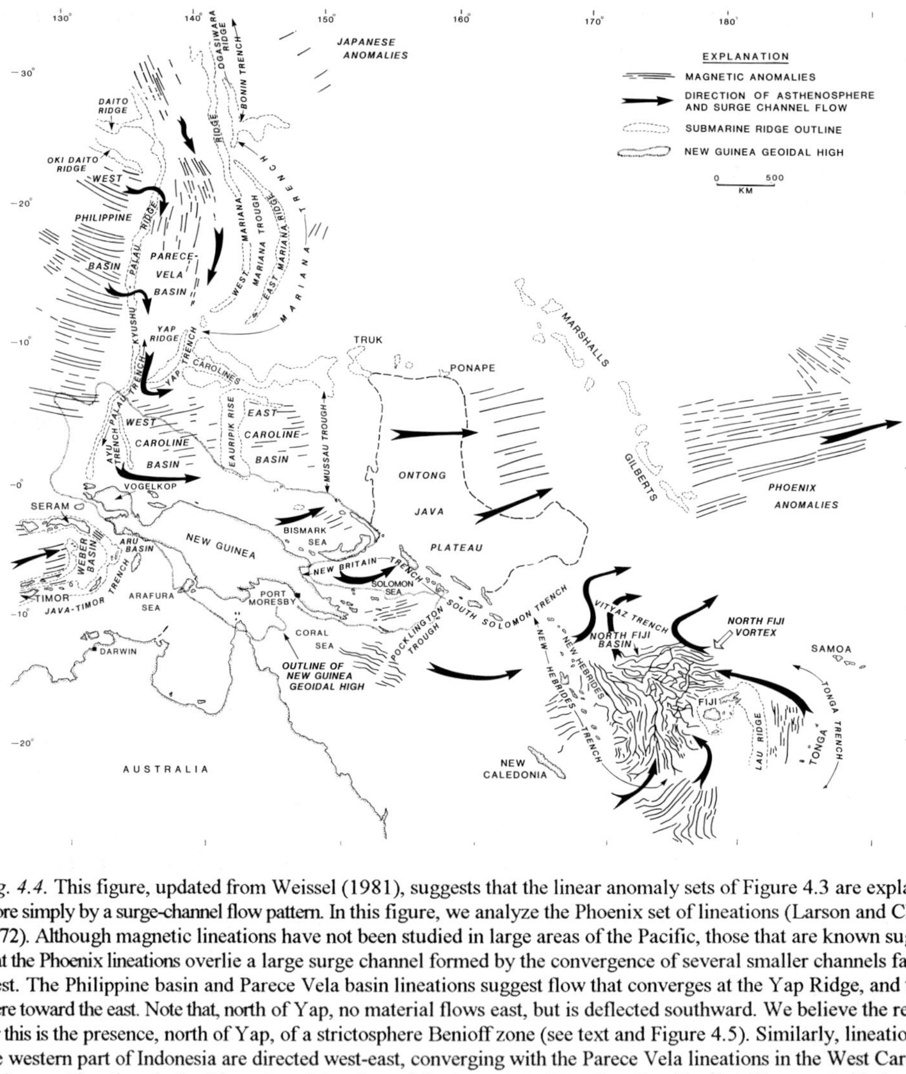

Fig. 4.4. This figure, updated from Weissel (1981), suggests that the linear anomaly sets of Figure 4.3 are explained more simply by a surge-channel flow pattern. In this figure, we analyze the Phoenix set of lineations (Larson and Chase, 1972). Although magnetic lineations have not been studied in large areas of the Pacific, those that are known suggest that the Phoenix lineations overlie a large surge channel formed by the convergence of several smaller channels farther west. The Philippine basin and Parece Vela basin lineations suggest flow that converges at the Yap Ridge, and turns there toward the east. Note that, north of Yap, no material flows east, but is deflected southward. We believe the reason for this is the presence, north of Yap, of a strictosphere Benioff zone (see text and Figure 4.5). Similarly, lineations in the western part of Indonesia are directed west-east, converging with the Parece Vela lineations in the West Caroline basin area. Lineations in the Bismark and Coral seas join the eastward-trending Caroline lineations, probably beneath and just east of the Ontong Java Plateau. East of Australia, north-to-south directed lineations collide in the North Fiji basin vortex structure, whence they turn eastward and northeastward in the Phoenix direction. We think that the Japanese and Hawaiian lineation sets (Figure 4.3) have similar explanations. To sum up: the various magnetic lineations sets in the Pacific basin originated above different surge channels active at different times in the past.

Additional data support the surge-tectonics model in this region. Furumoto et al. (1976) discovered a 20-42-km-thick 7.0-7.7-km/s layer beneath the Caroline basins, Ontong Java Plateau, and New Guinea geoidal high (Fig. 4.4; Table 3.3). This anomalous upper-mantle lens falls in a band of high heat flow (Fig. 2.26; Gramberg and Smyslov, 1986), which also coincides with a well-defined band of microearthquakes (Hegarty and Weissel, 1988). In a study of the Ontong Java Plateau, Sandwell and MacKenzie (1989) found that the geoid-topography ratio of the Ontong Java Plateau region is too high to be explained by an isostatic compensation model. They concluded that it and the New Guinea geoidal high (Fig. 4.4) are sustained partly by thermal-buoyancy forces in the lower half of the lithosphere. This conclusion supports the surge-tectonics model.

4.1.4 Feeder channels

Feeder channels are much smaller than midocean ridges (Figs. 3.39, 4.1), typically 400-700 km wide and 2,000-4,500 km long (Fig. 4.1). Some may have been branches or parts of midocean ridges. Pacific examples include the Nazca, Carnegie, Cocos, Tehuantepec, and Mendocino ridges. The East Pacific Rise itself enters North America via the Gulf of California and the Great Basin, and serves as a feeder channel. Atlantic examples are the Agulhas, Walvis (Fig. 4.1), Cameroon, Cape Verde, Azores-Gorringe, and Iceland-Faeroe ridges, rises, and plateaus. Indian Ocean examples include Broken Ridge and the Chagos-Laccadive Ridge.

Their morphology is quite complex (Fig. 4.1); many of them strike northeastward (Walvis, Cameroon, Nazca, Cocos, Tehuantepec). Most of them show a sinuous to anastomosing pattern. Figure 4.1 shows clear evidence in its sinuosity and eddylike patterns for an origin by flow (see also Fig. 2.16).

Feeder channels typically connect midocean ridges with the continent(s) that generally lie east of them. Feeder channels are not present between midocean ridges and the continents west of them. These ridges are little studied, and information concerning them is scarce. Plate tectonics includes them in a broad category that embraces several types of ridges, such as the aseismic ridges.

We selected three feeder channels from which data are abundant: the Canary Islands, Cape Verde Rise, and Walvis Ridge (Fig. 4.1). Typically, just like the surge channels beneath the continents, they have a 7.0-7.8-km/s anomalous lens present (Bosshard and Macfarlane, 1970; Goslin and Sibuet, 1975). They exhibit high microseismicity (Sykes, 1978) and high heat flow (Gramberg and Smyslov, 1986; V. Čermák, pers. comm., 1989). Volcanic activity, another surge-channel characteristic, continues to the present on the Cape Verde and Canary islands (Dillon and Sougy, 1974), but has not taken place along the Walvis trend since latest Cretaceous or earliest Tertiary time (Moore et al., 1984). The Walvis trend continues northeastward onshore for at least 1,500 km as a line of ring-shaped volcanic structures; the Lucapa graben and its kimberlite pipes also are on this trend (Sykes, 1978; Beloussov, 1980). We summarized these characteristics on Table 4.1, which should be compared with Tables 3.6, 4.2, and 4.3.

4.2 Surge Channels of Continental Margins

Breakout channels are located mainly in ocean basins. We classify them with continental-margin channels because they owe their origin to processes generated at the continental margins, not to processes generated in the ocean basins. Active margin channels are channels that are dammed behind Benioff zones (Figs. 3.10, 4.5). Undoubtedly these channel systems are structurally complex. We know the least about passive margin channels, the existence of which we did not suspect until recently. Our preliminary studies indicate that all passive continental margins may be sustained by small, deep channels. If true, surge channels are basic not only to lithosphere structure, and to continental structure in particular, but also to continental configuration (Fig. 3.39).

4.2.1 Breakout Channels

Breakout channels originate along the eastern sides of continents and in magmatic arcs adjacent to them (Fig. 3.39; Table 3.2). They are included among continental-margin channels because of their clear genesis from continental-margin phenomena. In the literature, these have been called linear island and seamount chains, linear volcanic chains, and hot-spot trails (Jackson, 1976; Shaw et al., 1980; Okal and Batiza, 1987). They form another group of submarine ridges that, like feeder channels, have been lumped under the term aseismic ridges. They have several unique characteristics.

1. Their western termini, at least in the Pacific basin, are in the cusps between adjacent magmatic arcs (Fig. 2.29). Examples include the Obruchev Rise-Hawaiian-Emperor chain (Fig. 2.29), Louisville Ridge (Fig. 2.27), Caroline Ridge (Fig. 2.29), Christmas Island Ridge, Cook-Austral chain (Fig. 2.28), the Gilberts (Kiribati, Fig. 2.29), Marshalls (Fig. 2.29), Mid-Pacific Mountains (Fig. 2.29), Samoa chain (Fig. 2.29), Society Islands, the Tuamotus, and several others. Western Atlantic examples include the Newfoundland-Milne seamounts, New England-Corner seamounts, and the Rio Grande Rise.

Cusps are where breakout channels originate, presumably because they are the weakest links in the Benioff zones of the western Pacific. We suggest that lithosphere compression, combined with eastward asthenospheric movement produced by Earth rotation, builds up stresses west of the Benioff zones (Fig. 4.5), stresses that are relieved from time to time by a breakout of excess magma or magmatic mush.

2. The eastern termini of these channels commonly are in the middle of nowhere (e.g., the Hawaiian Ridge), although some of them extend to the western flanks of a midocean ridge.

3. Almost all such channels are in the western Pacific. We have not found such channels in the eastern Pacific.

Table 4.1. Some common characteristics of feeder channels. From Meyerhoff et al. (1992b).

Feeder channel	High heat flow	Hight seismicity	7.0-7.8 km/s layer	Linear fracture pattern	Volcanic rocks
Canary Islands	x	x	x	x	x
Cape Verde Rise	x	x	x	x	x
Walvis Ridge	x	x	?	x	pre-Tertiary

Sources: Fúster and others (1968a, 1968b); Bosshard and Mcfarlane (1970); Dillon and Sougy (1974); Goslin and Sibuet (1975); Sykes (1978); van der Linden (1980); Courtney and White (1986); Gramberg and Smyslov (1986); Newhall and Dzurisin (1988); Vladimir Čermák, written communications (1989).

Note: These ridges have been studied very little. As a consequence, the fine velocity structure, conductivity, and many other attributes are unknown.

4. The channels are small----80-300 km across, but generally 100-200 km----and long, commonly 1,000 to 5,500 km. They are sinuous (Fig. 2.27), exhibiting eddy-like features (Fig. 2.27). They are studded with volcanoes in the form of seamounts and guyots. The Louisville Ridge of the southwestern Pacific has at least 60 volcanoes (Lonsdale, 1988).

5. They show age progressions in many areas (e.g., Louisville Ridge; Fig. 2.27; Hawaiian Ridge). Where age progressions are present, the youngest age is usually at the eastern terminus and the oldest age at the western (Jackson, 1976; Shaw et al., 1980; Cheng et al., 1987). Some of these west-to-east age progressions were used as proof of Morgan's (1971) hot-spot concept, but most of the island chains show no age progression at all (Fig. 2.28), and a few show a reverse progression (Jackson, 1976; Turner and Jarrard, 1982).

6. All geophysical data available to us indicate a surge-channel origin, in conformity with the criteria listed in Table 3.3. A 7.1-7.8-km/s layer has been found under the Hawaiian-Emperor chain and the Line Islands (Brune, 1969; Hill, 1969; Sutton et al., 1971; Ellsworth and Koyanagi, 1977). Ellsworth and Koyanagi (1977) found low-velocity zones beneath the Hawaiian Ridge. Von Herzen et al. (1989) found a band of elevated heat flow under the Hawaiian Ridge to Midway Island (Fig. 3.35). Gramberg and Smyslov (1986), on the basis of few control points, found elevated heat-flow bands beneath the Tuamotus, Marquesas, Carolines, and Mid-Pacific Mountains (Figs. 2.29, 3.35). Active volcanism and microearthquake activity are known from the Hawaiian Ridge, Tuamotus, Society Islands, and the Cook-Austral volcanic chain (Talandier and Kuster, 19976; Ellsworth and Koyanagi, 1977; Talandier, 1989). Talandier (1989) pointed out that studies of active volcanism, microseismicity, and heat flow have barely begun in large areas of the Pacific basin.

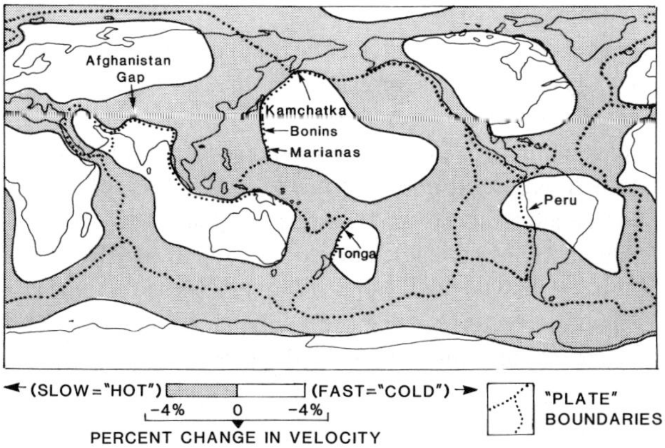

Fig. 4.5. Generalized seismotomographic map of the 50-150-km depth range, upper mantle. Low velocity, high-temperature (heat flux >55 mW/m^2) areas are shaded; high-velocity, low-temperature areas are white. The figure illustrates several important points: (1) the Benioff zone in the western Pacific basin, from the western tip of the Aleutian arc to the southern tip of the Marianas arc, is a barrier to eastward flow of "hot" material from Asia (thus explaining the northeast and south-to-southeast splays of surge channels on Figures 5.3, 5.6, 5.8-5.12, 5.17-5.18); (2) most ancient cratonal areas are in the white or "cold" areas, showing that they (a) probably are rooted in the mantle (see Figure 1.1) and (b) also are barriers to surge channels; (3) the Afghanistan Gap is active today; (4) both "hot" and "cold" areas cross continent-ocean boundaries; and (5) true convection cells do not exist, as demonstrated by the facts that (a) hot upper mantle is associated with the midocean-ridge system of the eastern Pacific and (b) with the island (magmatic) arcs of the western Pacific. All data from Woodhouse and Dziewonski (1984) and Dziewonski and Woodhouse (1987).

7. Breakout channels are short-lived, with lifespans of 20 to 115 Ma, in contrast to 400-1,000+ Ma for other types of channels. Unlike trunk or feeder channels, they die out in the rear as their leading, forward edges progress eastward.

4.2.2 Active Margin Channels: Western Pacific Basin

Because of the Earth's rotation, subcontinental asthenosphere west of the Pacific basin flows outward, mainly eastward from the Australasian landmasses. This relative eastward movement is believed to produce the backarc basins that dominate continental-margin tectonics in the western Pacific (Fig. 4.5). Figure 4.6, from Meyerhoff and Meyerhoff (1977), illustrates this concept. Figure 4.6 also suggests that the shallow-dipping lithosphere Benioff zone can deflect the mobile hot material upward beneath the backarc basin, thereby explaining the high heat flows observed across large areas behind magmatic arcs.

Figure 3.10 is Rodnikov's (1988) more modern conceptualization of backarc-basin structure. It shows the velocity, density, and thermal structure of a typical western Pacific magmatic arc-backarc basin complex, and it is based on refraction-seismic, gravity, and heat-flow data. We have added two shallow-level surge channels to illustrate our concept of the further complexity beneath backarc basins.

Table 4.2 shows some of the key characteristics of surge channels (Table 3.3) in relation to ten magmatic arcs, nine in the western Pacific and one in the Caribbean. Comparison with Tables 3.6, 4.1, and 4.3 illustrates the point that many seemingly different megastructures have similar geophysical and geological characteristics.

Figure 3.11 is a southwest-to-northeast seismic-refraction section across the northeastern corner of the Lesser Antilles arc, as interpreted by Officer et al. (1959). The well-developed 7.0-7.6-km/s lens is evident. Edgar et al. (1971) have since mapped this layer westward into the Colombian and Venezuelan basins.

Figure 2.17 shows four west-to-east seismic-reflection profiles across the Taiwan and Philippine island arcs. Their locations appear on Figure 2.29. We modified Létouzey et al.'s (1988) geology slightly on the northernmost line through Taiwan (Fig. 2.17a), preferring instead the work of Suppe (1981, his fig. 9). The kobergenic structure on all four lines needs no elaboration. In the southern profiles (Fig. 2.17d), the kobergen has split into two branches. The western branch passes through Mindoro and Palawan to Sabah and Borneo. The eastern branch continues to the Molucca Sea between Sulawesi (Celebes) and Halmahera in northeastern Indonesia. Both exhibit the distinctive kobergenic style. Surge tectonics explains in a natural and unforced way the coexistence of two or more foldbelts in the same region with parallel and coeval histories of volcanism, sedimentation, and tectogenesis. Our model eliminates the needs for postulating the presence of two or more coexisting Benioff zones, numerous microplates, and strongly contorted slab configurations.

Figure 4.7 shows three cross sections along the eastern Luzon kobergen of Figure 17d, except that these sections are farther south in the Molucca Sea between Sulawesi and Halmahera (Fig. 2.29). Figure 4.7b is a structural interpretation based on reflection and refraction seismology, and field work (Silver and Moore, 1978). These authors stated that their data provide ". . . clear documentation of a two-sided symmetrical collision zone" (p. 1689). Subsequent gravity modeling by McCaffrey et al. (1980) produced Figure 4.7c. The black prong approaching the surface on both Figures 4.7b and 4.7c appears to be a sliver of faulted mafic to ultramafic surge-channel roof. Regardless, the kobergenic style is unmistakable. The

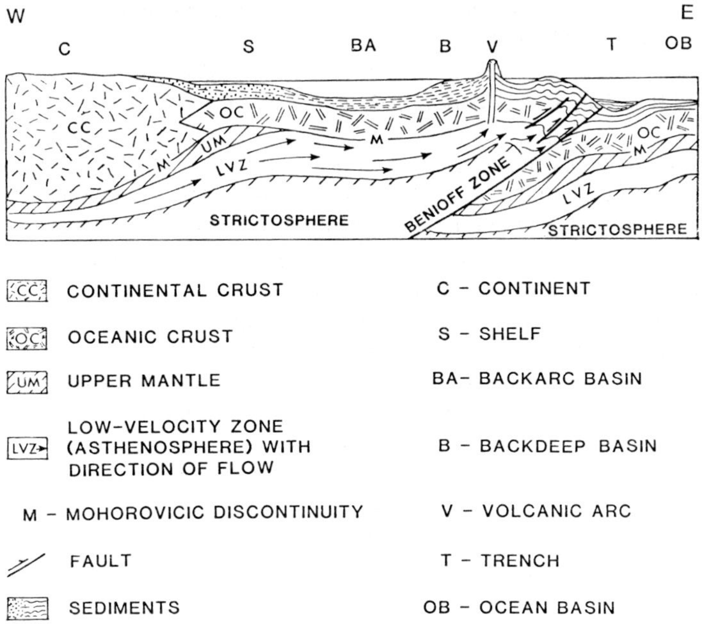

Fig. 4.6. Schematic structural cross section of a volcanic arc showing the emplacement of asthenosphere between the lithosphere and the stereosphere. Eastward-directed movement of the asthenosphere, caused by the Earth's rotation, made it possible for the backarc basins of the western Pacific, the Caribbean, and the Scotia Sea to form, and--in our opinion--produced the transitional-type crust that characterizes backarc basins. From Meyerhoff and Meyerhoff (1977).

up-faulted prong (black) produces a sharp gravity high within a regional Bouguer gravity low. The close similarity with gravity profiles across the Alps (Fig. 2.13) and the Appalachians (Fig. 3.13) is apparent.

Figure 4.8 shows an east-west kobergen in the Banda arc (Breen et al., 1989). The north-vergent thrust faults north of Sumbawa, Flores, and Wetar are documented by seismic-reflection studies and by direct observations using SeaMARC II sonographs (Silver et al., 1986; Breen et al., 1989).

4.2.3 Active Margin Channels: North American Western Cordillera

In the eastern Pacific, backarc basins are absent for reasons discussed. Benioff zones are present in parts of the eastern Pacific rim, but not in others, presumably because, being fixed in time and space, they were overrun by lateral continental accretion. One area of the Western Cordillera that has a remnant lithosphere Benioff zone is the Cascade Range extending from northern California to southernmost British Columbia.

Figure 3.4 is a north-south seismic-refraction line across the Great Basin part of the Western Cordillera. The line extends southward from southern Idaho to Lake Mead on the Arizona-Nevada border, crossing the Great Basin surge-channel complex at an oblique angle. This channel complex underlies the entire Great Basin, is 500 to 600 km wide between the Sierra Nevada and the Wasatch Range, has a P-wave velocity of 7.3-7.8-km/s, and is 25 to 33 km deep (Landisman et al., 1971; Shurbet and Cebull, 1971; Fuis et al., 1987; Mooney and Braile, 1989). Shurbet and Cebull (1971) calculated its thickness to be not less than 10 to 12 km.

Figure 3.4 also shows another prominent feature of the Great Basin, a 2-6-km thick low-velocity zone (5.0-5.8 km/s) at a depth that ranges from 8 to 10 km (Prodehl, 1970, 1979; Landisman et al., 1971; Fuis et al., 1987). This low-velocity zone and the one below are nearly identical to two similar zones that underlie the Hengduan Shan (Yunnan Himalaya) channel of southwestern China (Figs. 2.31, 3.9, 3.14). The principal difference is that the Hengduan Shan channel underlies an active foldbelt (Fig. 3.22), whereas the Great Basin channel system underlies an active extensional system superimposed on a formerly active foldbelt. This seemingly contradictory situation is explained naturally in surge tectonics.

Another feature portrayed on Figure 3.4 is a 180-km-wide lens of 7.0-km/s material centered below Mountain City, Nevada. Although we have not completed our study of the Western Cordillera, our preliminary mapping indicates that this lens is an inactive remnant of a Paleozoic surge channel--a biconvex lens--that extends from the Pacific Ocean basin through the Mendocino Escarpment to a south-to-north orientation below the Western Canada basin.

Table 4.2. Some common characteristics of selected arcs.[1,2] From Meyerhoff et al. (1992b).

Volcanic arc	Rift	Wrench system	Foldbelt	Kobergen is present	High heat flow	High seismicity	7.0-7.8 km/s layer	Midcrustal low-veolocity zone	Negative gravity anomaly	Linear fracture pattern	Volcanic rocks
Bonin	•				X	X	X		X	X	X
Indonesia	•	•	•	•	X	X	X	X	X	X	X
Japan			•	•	X	X	X	X	X	X	X
Kuril-Kamchatka	•		•	•	X	X	X			X	X
Marianas	•				X	X	X	X		X	X
Phillipines	•	•	•	•	X	X	X			X	X
Ryukyu	•				X	X	X			X	X
Taiwan	•	•	•	•	X	X	X		X	X	X
Tonga-Kermadec	•				X	X	X	X	X	X	X
Lesser Antilles			•	•	X	X	X		X		X

[1]Blank spaces indicate that we found no literature on these topics. This does not mean that there is no literature. It may also mean that the topic has not yet been studied.

[2]Data sources by region are listed below: *Bonin arc:* Murauchi et al. (1968), Watanabe et al. (1977), Sychev and Sharaskin (1984), Honza and Tamaki (1985), Gramberg and Smyslov (1986). *Indonesia:* Curray et al. (1977), Milsom (1977), Silver and Moore (1978), Hamilton (1979), Bowin et al. (1980, 1982), McCaffrey et al. (1980), Silver et al. (1986), Eva et al. (1988), Mount (1988), Newhall and Dzurisin (1988), and Breen et al. (1989). *Japan:* Murauchi et al. (1968), Utsu (1971), Watanabe et al. (1977), Yoshii (1983), Sychev and Sharaskin (1984), Grambert and Smyslov (1986), Oike and Huzita (1988), Shimazu (1988), Aihara (1989), and Hirahara et al. (1989). *Kuril-Kamchatka:* Vlasov and Belova (1964), Utsu (1971), Udintsev (1972), Fedotov (1973), Gorshkov (1973), Sychev and Sharaskin (1984), Gramberg and Smyslov (1986), and Newhall and Dzurisin (1988). *Marianas:* Pushcharovskiy (1972), Segawa and Tomoda (1976), Watanabe et al. (1977), Karig et al. (1878), Bibee et al. (1980), LaTraille and Hussong (1980), Hussong and Sinton (1983), Sinton and Hussong (1983), Eguchi (1984), Sychev and Sharaskin (1984), Gramberg and Smyslov (1986, and Bloomer et al. (1989). *Philippines:* Silver and Moore (1978), Cardwell et al. (1980), Divis (1980), McCaffrey et al. (1980), Weissel (1981), Gramberg and Smyslov (1986), Stéphan et al. (1986), Létouzey et al. (1988), Mount (1988), and Sarewitz and Lewis (1988). *Ryukyus:* Murauchi et al. (1968), Udintsev (1972), Watanabe et al. (1977), Sychev and Sharaskin (1984), Kobayashi (1985), Gramberg and Smyslov (1986), Kizaki (1986), Létouzey et al. (1988), and Ouchi et al. (1989). *Taiwan:* Ho (1982), Biq et al. (1985), Grambertg and Smyslov (1986), Richard et al. (1986), Tsai (1986), Létouzey et al. (1988), and Liu and Yu (1989). *Tonga-Kermadec:* Raitt (1956), Menard (1964), Karig (1970), Sclater et al. (1972), Watanabe et al. (1977), Gramberg and Smyslov (1986), Newhall and Dzurisin (1988), and Pelletier and Louat (1989). *Lesser Antilles:* Officer et al. (1959), Bowin (1976), Westercamp (1979), Westercamp and Tomblin (1979), and Gramberg and Smyslove (1986).

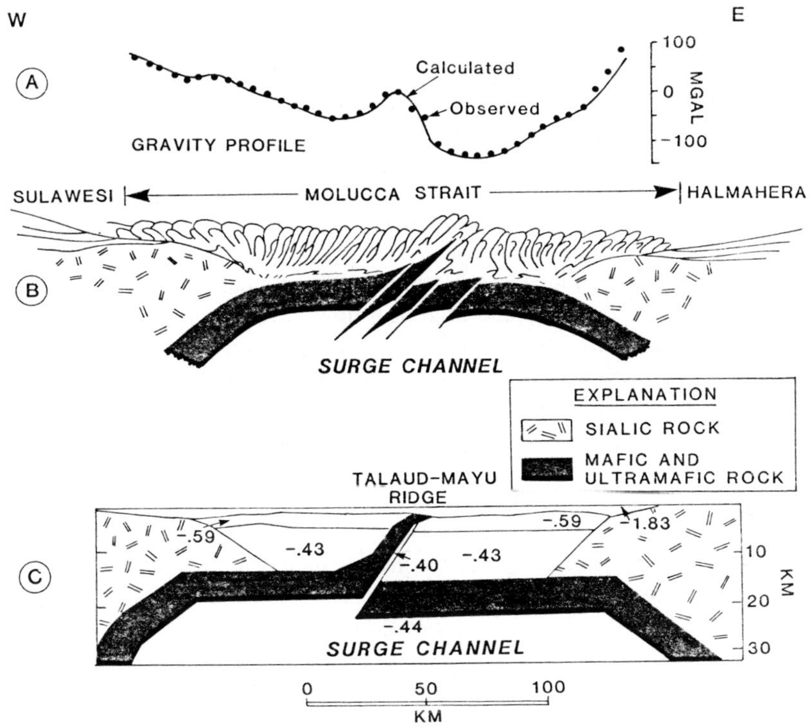

Fig. 4.7. Three west-east cross sections across the Molucca Sea between Sulawesi and Halmahera, Indonesia (from Silver and Moore [1978] and McCaffrey et al. [1980]). The location of the cross sections is shown on Figure 2.29. **A**--is a gravity profile across the line of section. **B**--is a structural interpretation--schematic--based on seismic and geologic data. **C**--is the gravity model that satisfies both A and B. The kobergen shown here is the same kobergen that underlies Luzon (Figures 2.17a-2.17c). The kobergen of Figure 4.8 is part of the same system.

Figure 4.9 presents seven east-west profiles across the Great Basin at 38° N latitude. Eaton et al. (1978) compiled these profiles and found a striking overall east-to-west symmetry of several features in the Great Basin along a north-south symmetry axis located between 115° and 116° W longitude. The symmetrical elements shown on Figure 4.9 include heat flow (which is high), earthquake (seismicity) distribution, the low-velocity zone at 8 to 10 km depth, magnetism, gravity, topography, temperature distribution (including the location of the Curie isotherm), and depth to the Mohorovičić discontinuity. The Great Basin as a whole occupies a huge regional Bouguer gravity low (Eaton et al., 1978). The cause of the symmetrical features does not lie at the surface, because symmetry of the surface geology from one side of the basin to the other is lacking. The only geological surface phenomenon that exhibits symmetry is the distribution of Quaternary volcanic rocks, which are concentrated along the eastern and western margins of the Great Basin (Eaton et al., 1978). Thus, the cause of the symmetrical features lies at depth.

Additional phenomena characteristic of the Great Basin include a pronounced (at all scales) north-south fault, fracture, and fissure pattern (Fig. 2.1; Stewart, 1978); a strong low-resistivity (high conductivity) magnetotelluric anomaly (Fig. 4.10; Gough, 1984); metamorphic core complexes (analogous to mantled gneiss domes of the Appalachians; Coney 1980; Wust, 1986; Davis and Lister, 1988); gravity detachment fault complexes (Wust, 1986; Davis and Lister, 1988); and kobergen structures (bivergent foldbelts). The linear fault and fracture pattern of Figure 2.1, as on midocean ridges, suggests (following Stoke's Law) north-south mantle flow under the Great Basin. Comparison of Figure 2.1 with figures 9, 10, and 11 of Meyerhoff et al. (1992a) illustrates the striking similarity between the linear structures of midocean ridges and that of the Great Basin. Moreover, as Gough (1984, p. 430) observed, ". . . in the western US the close association of electrical conductivity with the seismic low-velocity layer, and with high heat flow, makes it very probable that the conductive layer is, in fact, a layer of the uppermost mantle containing a molten mineral fraction."

Heezen (1960) considered the Great Basin to be a continuation of the East Pacific Rise-Gulf of California belt of seismicity and high heat flow. Plate tectonics, however, suggests extensive offset along the San Andreas fault zone, which separates the Great Basin from the Gulf of California. As a consequence, Heezen's (1960) concept was discarded. We believe that Heezen was correct, and the Gulf of California belt of high seismicity and heat flow is continuous into the Great Basin. If the belt is not a continuous one, it is necessary to attribute to coincidence the juxtaposition of the Gulf of California province with the Great Basin. To illustrate the similarity, some of the Gulf of California's salient tectonic features are discussed.

Figure 3.17 (compare with Great Basin data on Figs. 3.4, 4.9) shows the great similarity of geological and geophysical features. These include high heat flow (Lawver and Williams, 1979), elevated seismicity (Ortlieb et al., 1989), high electrical conductivity (Fig. 3.12, Gough, 1984, 1989), a lens of 7.0-7.8-km/s material (Phillips, 1964; Fuis and Kohler, 1984), a lithosphere low-velocity zone (York and Helmberger, 1973), a system of north-south-striking parallel-subparallel linear faults, fractures, and fissures (Ortlieb et al., 1989), and a Late Jurassic through middle Eocene kobergen (Fig. 2.34; Rangin, 1984). Both belts, after undergoing

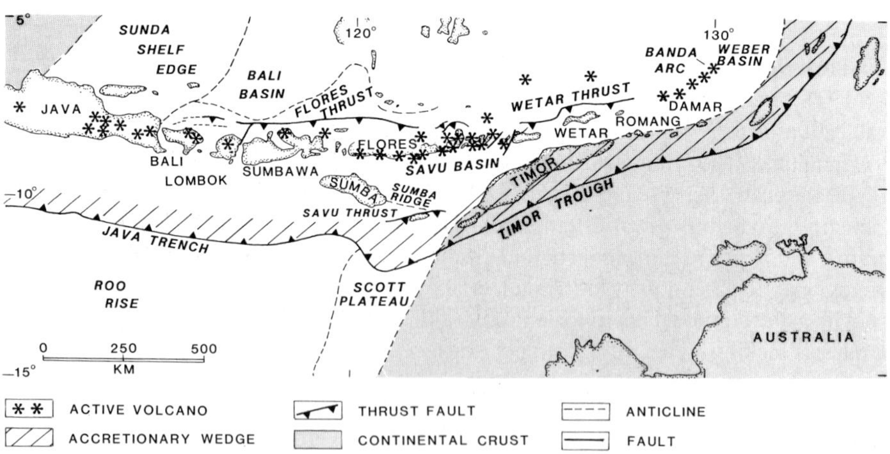

Fig. 4.8. East-west surge channel and kobergen of the Banda arc (from Breen et al., 1989). Together, Figures 2.17, 4.7 and 4.8 illustrate a surge-channel system through a distance of 6,500 km, which is perhaps 60% of the length of this active margin surge-channel system. Location of figure is shown on Figure 2.29.

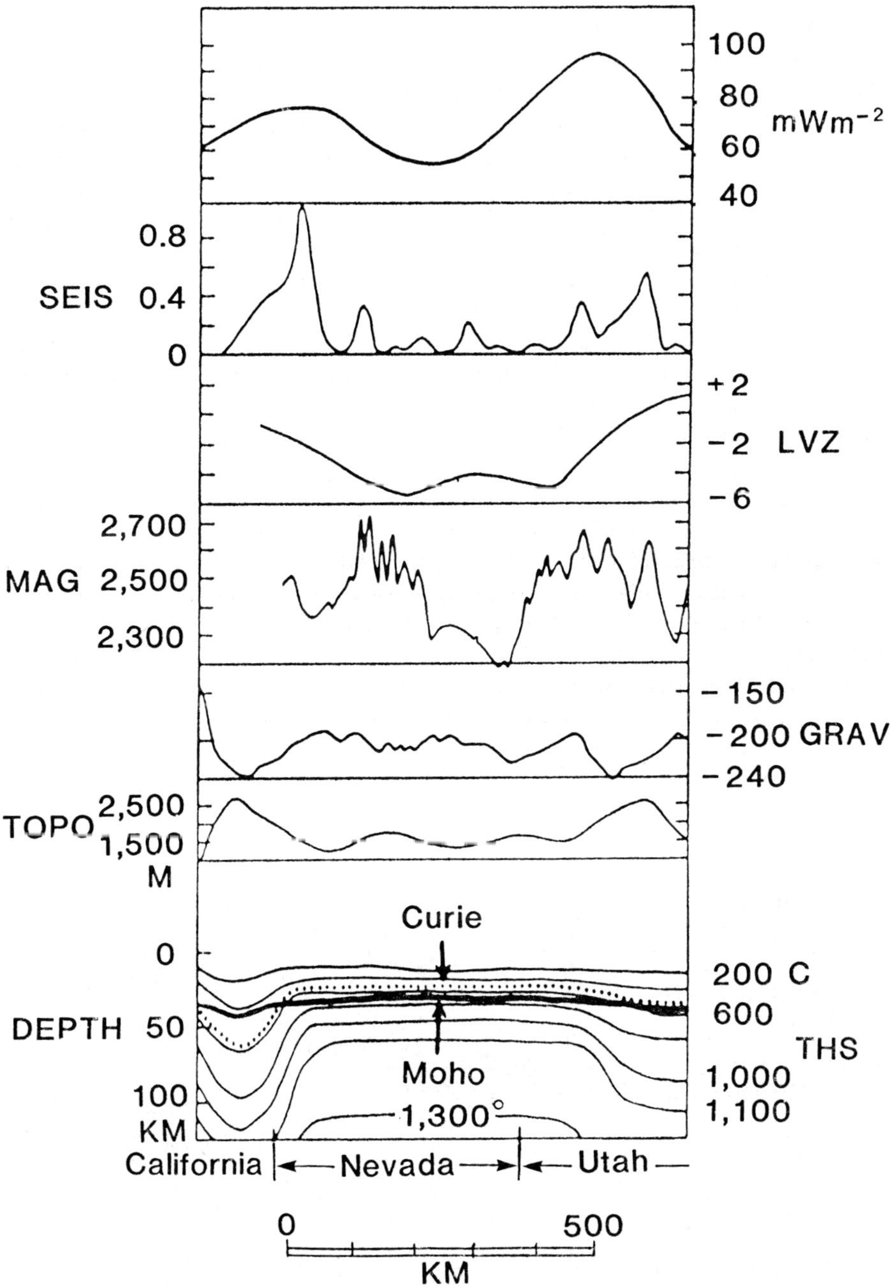

Fig. 4.9. Seven west-to-east profiles across the Great Basin of Utah and Nevada. The location of the profiles is shown on Figures 4.10 and 4.11. Note the overall bilateral symmetry of each--heat flow, intensity of seismicity, low-velocity zone arrival-time shifts, magnetic signature, gravity expression, topography, depth to Curie point, and thermal structure. From Eaton et al. (1978).

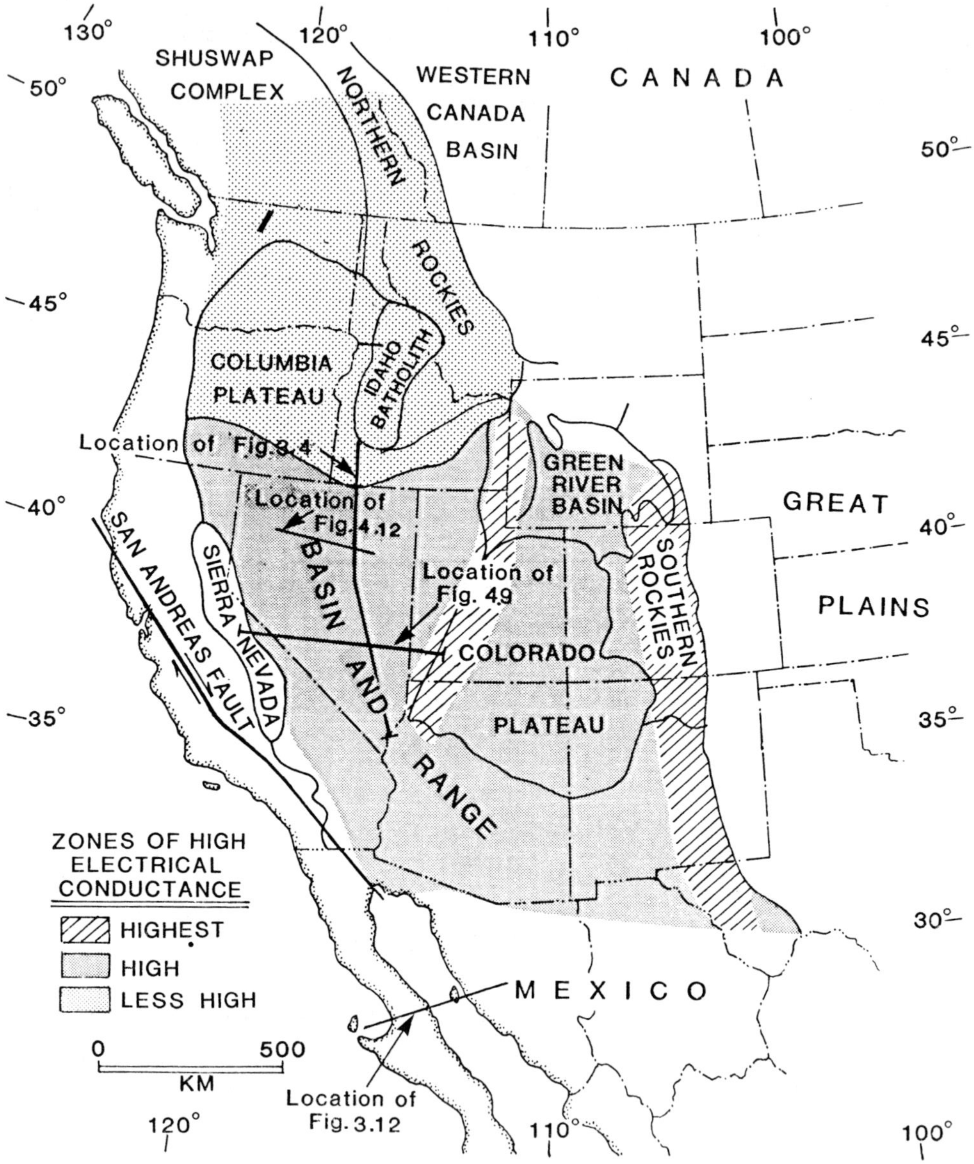

Fig. 4.10. Zones of high electrical conductivity, western Cordillera of the United States (from Gough, 1989). Two north-south, very high-conductivity, low-resistivity zones are present, one beneath the eastern edge of the Southern and Central Rocky Mountains, the other beneath the eastern margin of the Great Basin. Following Gough (1989), we believe that the high conductivity is related directly to the percentage of mafic melt in the surge-channel complex underlying the region. In our interpretation, the melt is in surge-channel complexes. This figure suggests that an interconnected surge-channel system underlies the whole Western Cordillera of the United States. The locations of Figures 3.4, 3.12, 4.9, and 4.12 are shown.

a long succession of Late Jurassic through middle Eocene compressive events, became sites of tensile stress beginning in late Eocene time, a stress regime that continues in both areas today.

The kobergen tectonic styles of the Gulf of California (Fig. 3.12) also characterize the Great Basin, except that in the Great Basin several kobergens have developed (Fig. 4.11). Much of the Great Basin, however, is not adequately mapped, so that detailed reconstructions of the several kobergens there are not yet possible.

Geologists working in the Western Cordillera, despite the clear evidence in many ranges for extensive westward thrusting, have been loath to treat the Western Cordillera as anything but a monovergent, west-to-east-directed, tectonic belt. Hershey (1903, 1906) was the first to find westward-verging thrusts, these in the Klamath Mountains. Suess (1909) and Kober (1921,

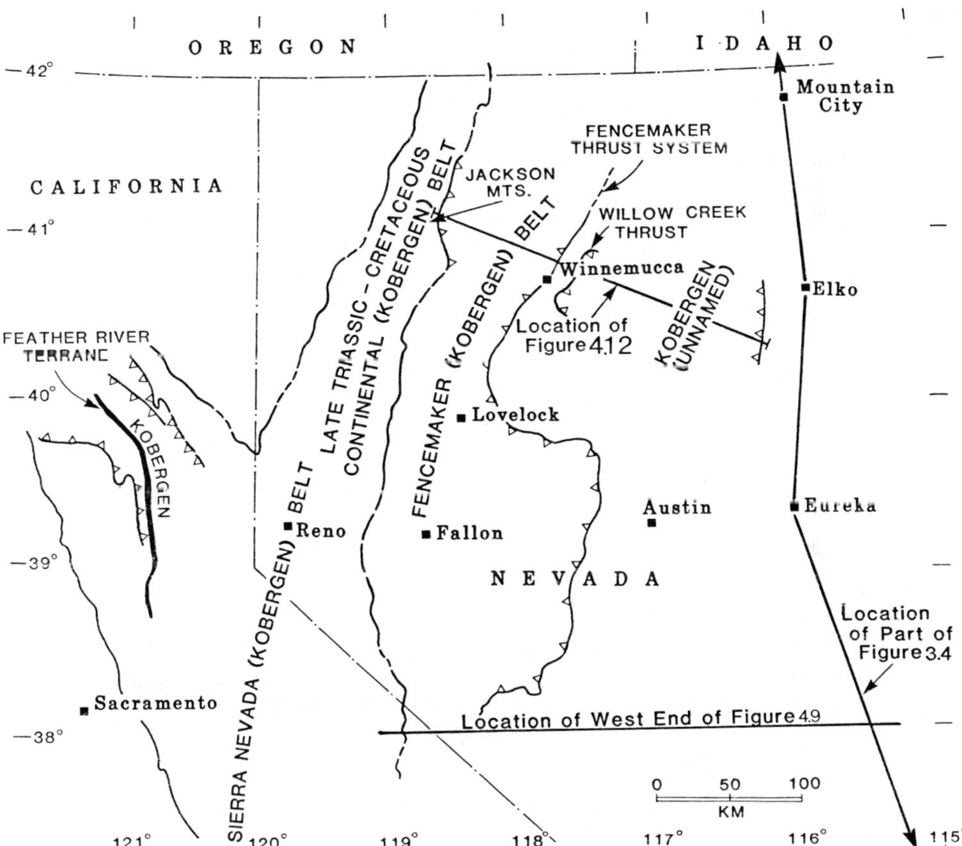

Fig. 4.11. Index map of part of the Great Basin. This figure shows the locations of identified parallel to subparallel kobergens in this region. From west to east, these are: (1) Sierra Nevada kobergen, which bifurcates in the north into a Paleozoic-middle Cretaceous ocean-margin kobergen (Feather River belt) and a Late Triassic-middle Cretaceous kobergen in northwestern Nevada; (2) the Late Triassic-middle Cretaceous Fencemaker kobergen; and (3) a coeval unnamed kobergen. A fourth kobergen probably lies east of this kobergen. Published structural data from the region still are inadequate to outline accurately the extent of the identified kobergens. Locations of parts of Figures 3.4, 4.9 and 4.12 are shown From Meyerhoff et al. (1992b).

1925, 1928) interpreted Hershey's (1903) discovery to mean that the Western Cordillera, like the Alpide system (Fig. 2.15), was a bivergent system. Similar concepts were published by Burchfiel and Davis (1968) and by Yeats (1968), but as plate tectonics increasingly dominated geological thought, all aberrations from the monovergent model were attributed to backthrusting during the alleged underthrusting of North America by eastward-moving plates of oceanic lithosphere (e.g., Burchfiel and Davis, 1975). The same explanation has been used for the Western Cordillera of Mexico (Rangin, 1984, 1986) and Canada (Brown and Tippett, 1978). Proponents of bivergent thrusting are few; a notable example is Krebs (1975).

As detailed mapping continues, a very different tectonic pattern emerges (Figs. 4.11, 4.12), specifically one of linear, parallel to subparallel, bivergent foldbelts from the Sierra Nevada to the Central and Southern Rocky Mountains. For example, in western and central Nevada, Roure and Sosson (1986), and Speed et al. (1988) postulated the presence of various allochthonous terranes that have been transported eastward onto the North American block. Speed et al. (1988) suggested that some of these terranes may have rafted thousands of kilometers from some unknown part of the Pacific basin, and became attached to North America during late Paleozoic, Triassic, and possibly later times. In fact, the proposed allochthons, although tectonically disrupted and partly allochthonous, are not far from their original positions and are in their original positions with respect to each other. We interpret them to be the facies belts of the Alps literature (Bertrand, 1897; Trümpy, 1960). Each group of facies belts has been deformed bivergently, as both Roure and Sosson (1986) and Speed et al. (1988) have shown (Figs. 4.11, 4.12). When mapping of the area shown on Figure 4.11 is completed in detail, at least five Mesozoic kobergens will occupy the area of Figure 4.11. These kobergens have continued to be the sites of Cenozoic surge-channel activity (Fig. 3.4, 4.9).

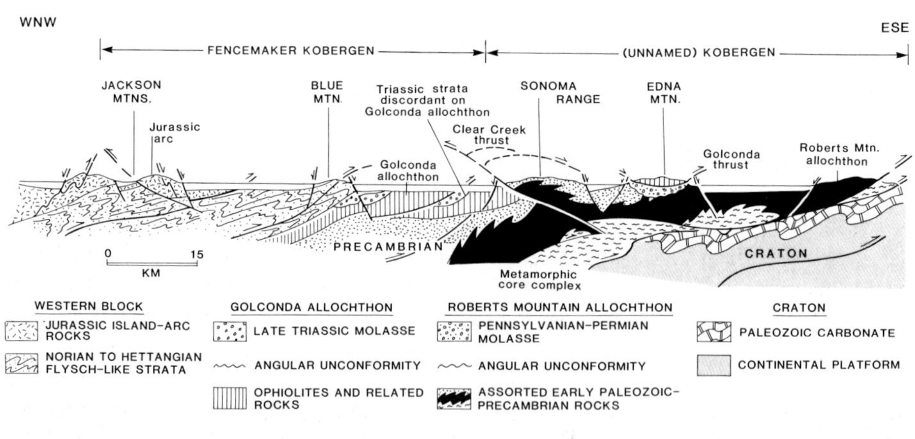

Fig. 4.12. Northwest-southeast structural cross section across Late Triassic-middle Cretaceous kobergens, northwestern Nevada. The location is on Figures 4.10 and 4.11. Modified from Roure and Sosson (1986). In our interpretation, the Fencemaker kobergen underlies a deep-marine backarc basin during Late Triassic time; the unnamed kobergen farther east underlies the continental crust east of the backarc basin. The Sierra Nevada kobergen underlies the Sierra Nevada volcanic arc west of the Fencemaker kobergen (see Figure 4.11).

Farther east in the Central and Southern Rocky Mountains, surge channels have been active only since Late Jurassic time (providing yet another example of eastward migration resulting from the Earth's rotation). Because the lithosphere beneath the Central and Southern Rocky Mountains is thicker than that beneath the Great Basin, the tectonic style is quite different, with germanotype structure predominating.

Recently, Eaton (1986, 1987) published two epic studies of the geological evolution of the Southern Rocky Mountains, which extend 1,200 km north-south, cresting in a band 150-200 km wide, from the Medicine Bow Range in south-central Wyoming to the Franklin Mountains at El Paso, Texas (Fig. 2.5). He called attention to several characteristics of the Southern Rockies, to which we added a few more.

1. The Southern Rocky Mountains form a north-south unit that cuts across the northwest-southeast-trending Laramide structures (Fig. 2.5). They are a post-middle Eocene uplift, largely Oligocene through Quarternary (Fig. 2.5). The Laramide northwest-southeast trends are the trends of the Late Jurassic-early Eocene surge channels of the region.

2. The north-south, linear, Neogene to present, normal faults indicate a tensile stress perpendicular to the range (Figs. 2.1, 2.5; Stewart, 1978; Easton, 1986). These faults occupy a belt 150 to 200 km wide, a typical width of many intracontinental surge channels (Figs. 2.13, 2.20, 3.5, 3.6, 3.12, 4.12-4.14).

3. The Rio Grande Rift occupies the southern two-thirds of the crest of the range. The system of faults that bound the rift extends into Wyoming (Fig. 4.15; Easton, 1986, 1987).

4. An asthenosphere bulge underlies the entire Southern Rocky Mountains.

5. A band of high seismicity extends the full length of the range (Smith, 1978).

6. A belt of high heat flow is present (Lachenbruch and Sass, 1977).

7. A lens of 7.1-7.35-km/s material is well established (Prodehl, 1970, 1979; Prodehl and Pakiser, 1980). This lens lies at a depth of 33 to 55 km (Mitchell and Landisman, 1971)

8. A crustal low-velocity zone is present about 18 km above the 7.1-7.35-km/s lens (Mitchell and Landisman, 1971).

9. A low-resistivity, high-conductivity magnetotelluric anomaly parallels the range (Fig. 4.10, Gough, 1984, 1989).

10. A negative Bouguer anomaly is present.

11. The 550° C isotherm is sharply elevated beneath the range.

12. The range contains Cenozoic volcanic rocks in several places along the Rio Grande rift and north of it.

If our surge-tectonics concepts are valid, these data show that an active surge channel underlies the entire range, a conclusion that finds support in a discovery by Eaton (1986). He drew five topographic profiles across the Southern Rocky Mountains at five different latitudes; he did the same for the Mid-Atlantic Ridge. He then paired them, as shown on Figure 4.16. He noted the great similarity between Southern Rocky Mountains and Mid-Atlantic Ridge profiles, writing (p. 175) that they ". . . demonstrate a remarkable, but coincidental, similarity in first-order morphology between a continental ridge and a slowly spreading oceanic ridge. . .. The cross-sectional dimensions of width and height of the crestal ranges, themselves, are the same, as are the topographic relief and slopes . . .. One can see that . . . the concave rises of the paired profiles track one another faithfully for hundreds of kilometers." Although Eaton (1986) considered the "remarkable similarity" to be wholly coincidental, we do not. According to our surge-tectonics concepts, both the Southern Rocky Mountains and the Mid-Atlantic Ridge have

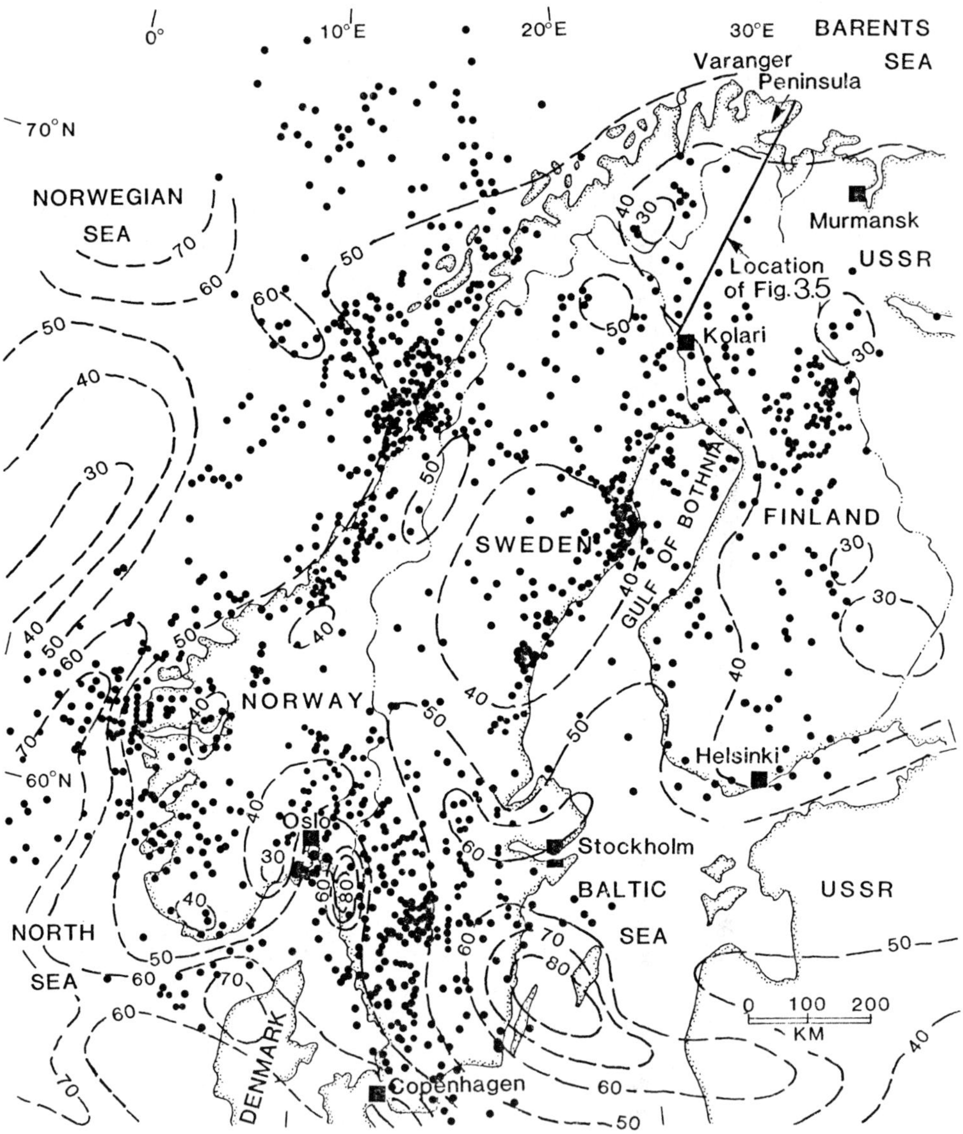

Fig. 4.13. Epicenters and heat flow of the Scandinavian Caledonides and Baltic Shield. This illustration shows several features: (1) A middle Paleozoic and older foldbelt (Caledonides) beneath Norway is underlain by an active surge channel; (2) A stable craton has been penetrated by an extensive surge-channel system; (3) stable cratons and inactive foldbelts can later become the sites of surge-channel activity; (4) Many--probably all--major linear uplifts (e.g., Caledonides) and linear depressions (e.g., Gulf of Bothnia, Gulf of Finland, central Baltic Sea) are underlain by active surge channels. The location of Figure 3.5 is shown. The figure was compiled from Çermak and Rybach (1979), Luosto et al. (1989), Mueller and Ansorge (1989), and Talbot and Slunga (1989).

to be underlain by active surge channels. The conclusions reached here for the Southern Rocky Mountains apply equally to the Central Rockies.

Additional supporting evidence is the fact that most Central and Southern Rocky Mountain ranges have kobergenic structure. This is seen on Figure 4.15. Figure 4.15 also shows another feature of surge-channel flow that we mentioned in a preceding section. This is the segmentation of the ranges, a continental pinch-and-swell structure, texture that characterizes linear flow phenomena everywhere (Figs. 2.16, 2.27, 3.2, 3.32, 4.1 and many others).

Figure 4.17 is a structural cross section across the Colombian Andes, which we introduce here for comparison. The section extends from the Pacific basin to the continental craton. Several kobergens, parallel and partly coeval, exist side by side, just as in the North American Western Cordillera. Alpinotype structure is developed in kobergen 2 near the Pacific coast. Germanotype structure dominates in the east.

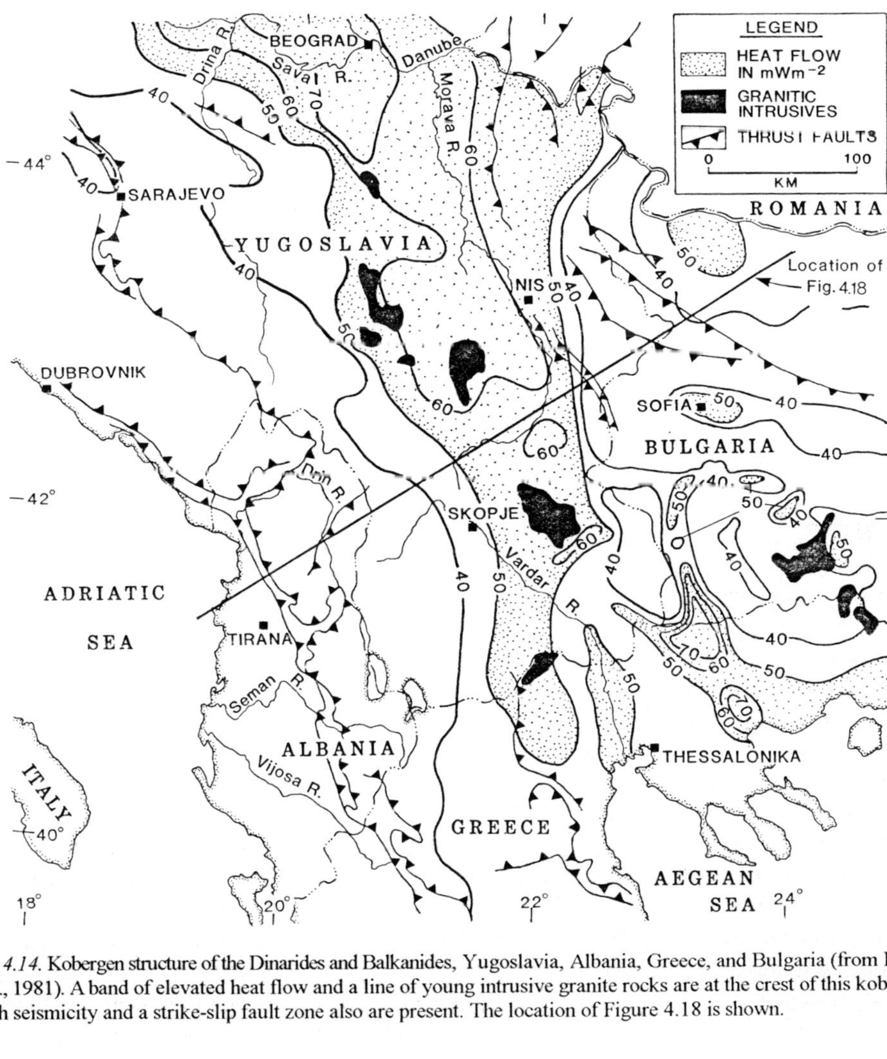

Fig. 4.14. Kobergen structure of the Dinarides and Balkanides, Yugoslavia, Albania, Greece, and Bulgaria (from Hurtig et al., 1981). A band of elevated heat flow and a line of young intrusive granite rocks are at the crest of this kobergen. High seismicity and a strike-slip fault zone also are present. The location of Figure 4.18 is shown.

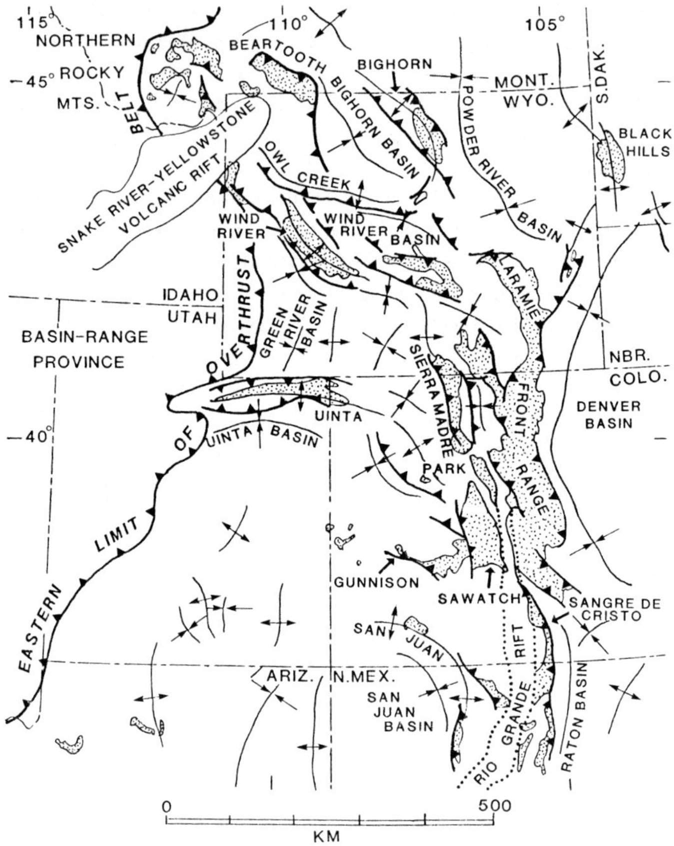

Fig. 4.15. Central Rocky Mountains germanotype structures: uplifts (stippled), thrust faults (low and high angle), anticlines, and synclines are shown. This figure shows that each uplift, or combination of uplifts, in the Central Rocky Mountains has kobergen structure and therefore is sustained by a surge channel. This conclusion is supported not only by the geological data but also by the geophysical data--heat flow, seismicity, and so forth. Note the segmentation of the ranges and see Figure 2.16F. Compiled mainly from Gries (1983) and Hamilton (1988).

4.2.4 Passive Margin Channels: Caledonides and Appalachians

Figure 4.13 shows Scandinavia, Finland, the Baltic Sea, and adjacent watermasses west of northern Europe. Epicenters (Talbot and Slunga, 1989) also are shown, together with the heat-flow contours (Čermák and Rybach, 1979). The figure shows that the entire region, mainly the Scandinavian Caledonides and ancient Baltic Shield, is seismogenic. High concentrations of epicenters in the Skagerrate-Kateggat region (Oslo-Copenhagen), the Gulf of Bothnia, and the Norwegian coast indicate that Precambrian and Paleozoic tectonic trends influence the configurations of some of the surge channels. Our conclusion regarding the basis of epicenter distribution is that one or more laterally connected low-velocity zones underlies the entire region north and west of the Baltic Sea.

Supporting data are copious. At the Gulf of Bothnia, Luosto et al. (1989) and Mueller and Ansorge (1989) found a shallow low-velocity zone (6.15 km/s) 3 to 10 km thick at a depth of

7 to 10 km (Fig. 3.5). About 23 to 31 km deeper is a 15-km thick lens of 7.0-7.3-km/s material that deepens from 30 km in the north (Fig. 3.5; Luosto et al., 1989) to 42 km near the mount of the Gulf of Bothnia (Mueller and Ansorge, 1989). A bouguer gravity minimum (Simonen and Mikkola, 1980) and a high-conductivity anomaly (Ádám, 1983) are present beneath the Gulf of Bothnia channel. Farther west in the mountains close to the Norwegian coast, Muir-Wood (1989) discovered a swarm of linear, seismically active, high-angle, reverse faults in zones of high heat flow and strong microseismicity.

Quite a different situation exists in the northern Appalachians. Here, high heat-flow readings are scarce and microearthquake activity is minimal. Figure 3.13 shows a geological cross section from Rodger (1987). The presence of a kobergen is at once evident, with its center or root zone at the Bronson Hill anticlinorium, an extensive north-south zone of mantled gneiss domes (analogous to metamorphic core complexes) in New England (Thompson et al., 1968). Seismic-refraction studies reveal the presence below the Bronson Hill anticlinorium of a lens of anomalous upper mantle with a P-wave velocity of 7.0 km/s (Fig. 2.12; Taylor et al., 1980). A narrow, sharp, positive magnetic anomaly overlies the root zone (Zietz, 1982), as does a positive Bouguer gravity anomaly (Hildenbrand et al., 1982) and a heat-flow anomaly (Birch et al., 1968; McCartan and Architzel, 1988). The highest grade of metamorphism is at the Bronson Hill anticlinorium, with decreasing grades toward both the northwest and southeast. Reverse metamorphic zonation (inverted isogradic surfaces), another kobergen characteristic, is common (Thompson et al., 1968). The winged, or kobergenic structure shows up clearly on a seismic-reflection line shot here (Fig. 3.21; Taylor, 1989a).

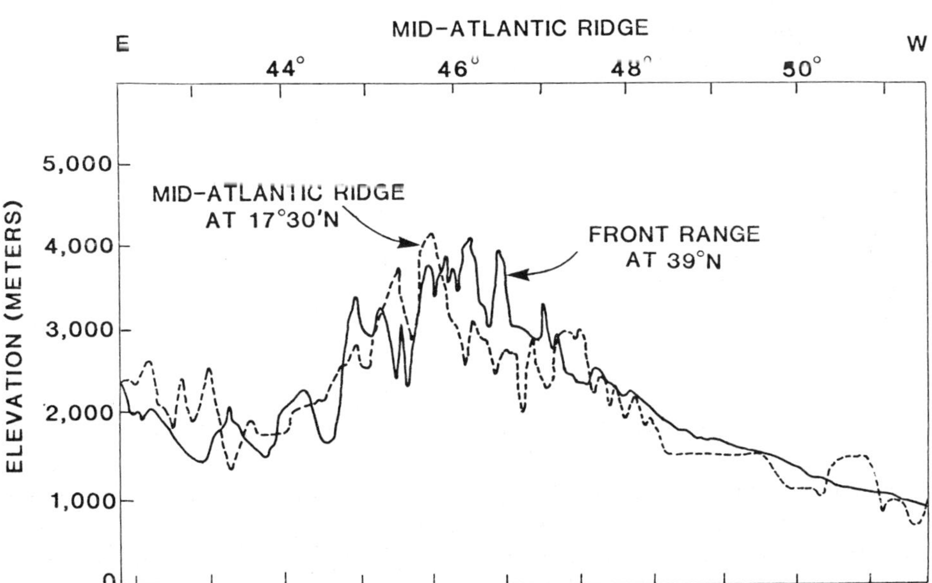

Fig. 4.16. Superimposed topographic profiles of the Mid-Atlantic Ridge and the Southern Rocky Mountains (from Eaton, 1986). Eaton (1986) commented on the"...remarkable, but coincidental similarity.." of the two features. The similarity, in our opinion, is not coincidental, as the two features in our interpretation were produced by a surge channel.

We interpret the lens of 7.0-km/s material as the upper part of a surge-channel complex that was active at least from late Proterozoic through Middle Devonian time, when Acadian tectogenesis deformed the region. This surge-channel system is dead, or nearly so. We say this because the positive Bouguer gravity anomaly indicates that little or no low-density material remains in the lens of 7.0-km/s material, and the level of microearthquake activity is very low. The positive heat-flow anomaly might indicate that some hot material remains in the channel, but Jaupart et al. (1982) related the high heat-flow readings to shallow radioactivity associated with silicic margins of the nearly Jurassic-Cretaceous White Mountains Magma series, which formed from a post-Paleozoic surge channel that traverses the whole Paleozoic foldbelt from northwest to southeast.

Figure 2.16 shows the segmented Bronson Hill anticlinorium side by side with other equally segmented tectonic belts. The remarkable similarity of these supposedly diverse features, ranging from the root zone of an alpinotype foldbelt (Bronson Hill anticlinorium) to young rifts in a tensile regime, directly confirms our contention that all tectonic belts----alpinotype or rift----originate by flow parallel with the tectonic belt.

4.3 Continental Surge Channels

We tentatively recognize two categories of continental surge channels (Table 3.2, Fig. 3.39). The first is the trunk channel, or continental trunk channel, which is a continuation of oceanic feeder channels. We also mentioned the East-Pacific-Rise to Great-Basin continuum, in which an oceanic trunk channel becomes a continental trunk channel. The smaller channels that branch from continental trunk channels are branch channels. Like the feeder channels in the ocean basins, they branch only from the eastern sides of their parent trunk channels, thereby providing one more demonstration of the important role played by Earth rotation. Thus the

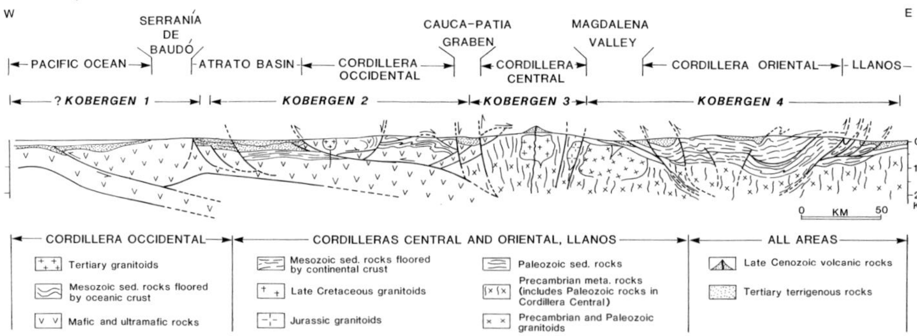

Fig. 4.17. Andes of Ecuador and Colombia: the development side by side of alpinotype (left) and germanotype (right) kobergens (from Mégard, 1987). On the far west, just offshore and adjacent to the coastal plain, a Cenozoic kobergen may be in the process of formation (kobergen 1?). Adjacent to and east of kobergen 1? is the Late Cretaceous-Tertiary kobergen 2. Kobergen 3 also is a young feature, the site of an active volcanic arc. Kobergen 4, adjacent to the Precambrian craton, is a germanotype structure.

branch channels originate in a manner similar to that of breakout channels, except that a Benioff zone is not present.

4.3.1 Lithosphere Thickness and Tectonic Style

The style of tectogenesis in any given tectonic belt, or part of one, reflects directly the lithosphere thickness at the location of that belt. Consequently, at this stage in our research we see no obvious way to subdivide surge channels in continental environments. In general, tectonic belts closest to the continental margin, or located in thin lithosphere, deform in alpinotype style. In thicker lithosphere, germanotype style is common and, deep in the continental interiors, rift valleys of the East African type are more likely to form than any other type of tectonic belt.

In this section, we illustrate some Eurasian and North African foldbelts, followed by consideration of some rift zones around the world. The alpinotype examples probably are indistinguishable from the active margin channels of the last category of surge channels, surge channels of continental margins. In Table 4.3 we combined some continental-margin tectonic belts, as well as belts from other regimes, with those of the continental interiors. This table, based on the criteria presented in Table 3.3, should be studied together with Tables 3.6, 4.1, and 4.2. The similarities among all the features are evident.

Table 4.3. Some common characteristics of selected tectonic belts. An "X" means that the feature is present, and "No" means it is not. A blank space means that feature may well be present, but we found no literature on the topic. From Meyerhoff et al. (1992b).

Tectonic belt	Rift	Wrench system	Foldbelt	Kobergen is present	High heat flow	High conductivity anomaly	High seismicity	7.0-7.8 km/s layer	Midcrustal low-velocity zone	Negative gravity anomaly	Linear fracture pattern	Volcanic rocks
Alps			•	•	X	X	X	X	X	X	X	X
(North) Anatolia Fault		•	•	•	X		X				X	X
Appalachians			•	•	X			X		No	X	X
High-middle Atlas			•	•	X	X	X	X	X	X	X	X
Balkanides-Dinarides			•	•	X	X	X	X	X	X	X	X
Baltic shield					X	X	X	X	X	X	X	No
Lake Baykal	•				X	X	X	X	X	X	X	X
East Pacific Rise	•				X	X	X	X	X	X	X	X
Fenwei graben	•				X	X	X	X		X	X	X
Great Basin	•		•	•	X	X	X	X	X	X	X	X
Gulf of California	•		•	•	X	X	X	X	X	Locally in north	X	X
Hetao-Yinchuan	•				X	X	X	X		X	X	
Himalayas (Yun-Nan; Hengduan)			•	•	X	X	X	X	X		X	X
Mid-Atlantic Ridge	•				X	X	X	X	X	X	X	X
San Andreas F.Z.		•	•	•	X	X	X	X	X	X	X	X
Canadian Cordillera			•	•	X	X	X	X		X	X	X

4.3.2 YUNNAN HIMALAYA (HENGDUAN SHAN)

Figures 2.31, 3.9, and 3.14 are from the seismotomographic imagery study conducted in China by Liu Futian et al. (1989) and Liu Jianhua et al. (1989). Through this study, Liu Futian and his Chinese colleagues inadvertantly discovered a network of shallow interconnected lens-shaped bodies in the lithosphere of China. The great importance of this discovery cannot be overstated. Liu Futian et al. (1989) had no explanation for their discovery, and noted only that a close relationship seemed to exist between the seismotomographic images and China's gross tectonic subdivision.

Figure 2.31, from Liu Jianhua et al. (1989), depicts a 1.4-million-km^2 area in southeastern Asia, centered in Yunnan Province. The figure portrays a large north-northwest-trending trunk channel in the west from which at least three much smaller branch channels branch eastward. Figure 3.9 shows the changes in depth that take place through short distances. Causes of the depth changes are unknown but likely include changes in lithosphere composition and density. If so, depth variations are at least partly a response to gravity requirements (Peach-Köhler climb force; Weertman and Weertman, 1964; Weertman, 1971; Ryan, 1987; Corry, 1988; Walker, 1989). Figure 3.14, also from Liu Jianhua et al. (1989), shows the velocity structure beneath Yunnan Province as interpreted from seismic-refraction data. The low-velocity zone at 25-42 km, just as in the East Pacific Rise, is well above the underlying anomalous upper mantle with P-wave velocities of 7.6 to 7.9 km/s, and contains lighter magma products that presumably were differentiated from the 7.6-7.9-km/s layer. The overall similarity between the velocity structure of the Yunnan compressional belt of southwestern China and that of the Great Basin (Fig. 3.4) should be noted.

Figures 2.31, 3.9, and 3.14 also illustrate differences in surge-channel sizes. The large trunk channel in the west is about 500 km wide and thousands of kilometers long, with a height of 17 to 20 km. It is associated with high heat flow (Ma Xingyuan, 1986a, 1986b), elevation of the Curie point (Chen Zongji, 1987), lines of hot springs (Ma Xingyuan, 1986a; Wang Jun et al., 1986; Zhu Meixiang, 1986), intense seismicity (Ma Xingyuan, 1986a, 1986b; Chen Zongji, 1987; Yan Zhide, 1987), a high-conductivity (low resistivity) magnetotelluric anomaly (Liu Guodong, 1987a, 1987b), a lens of 7.5-7.9-km/s material (Ma Xinguan, 1986a, 1986b), and numerous parallel strike-slip faults that are 1,200 km long or more (Ma Xingyuan, 1986a; Chen Zongji, 1987; Yan Zhide, 1987). A nearly symmetrical Tertiary kobergen overlies the 7.5-7.9-km/s lens (Wang and Chu, 1988), and is illustrated on Figure 3.22.

In the northeastern part of the map area of Figure 2.31, two small surge channels branch eastward to northeastward from the large channel beneath western Yunnan Province. The northern of these (approximately 38° N latitude) underlies the active Hetao-Yinchuan rift. The surge channel is approximately 180 km wide; its associated graben is 50 to 80 km wide and at least 540 km long. Formation began in middle to late Eocene time in the southwest and worked northward and eastward. The Hetao-Yinchuan graben system is associated with high heat flow (Zhang Buchun et al., 1985; Wang Jun et al., 1986; Ye Hong et al., 1987), intense seismicity (Zhang Buchun et al., 1985; Ma Xingyuan, 1986a), a gravity low (Zhang Buchun et al., 1985), a high-conductivity (low resistivity) magnetotelluric anomaly (Zhang Buchun et al., 1985), a linear upward bulge of the Mohorovičić discontinuity and the asthenosphere (Zhang Buchun et al.,1985; Ma Xingyuan, 1986a), volcanics, and strike-slip faults hundreds of kilometers long (Zhang Buchun et al., 1985; Ma Xingyuan, 1986a).

The surge channel just south of the Hetao-Yinchuan channel (approximately 35° N latitude) is overlain directly by another active graben system, the Fenwei rift (Wang Jingming, 1987). This surge channel is 30 to 100 km wide. Its associated graben is 20 to 90 km wide and 1,000 km long. The graben began to form in the southwest during middle Eocene time, and gradually extended northeastward to the present time (Zhang Buchun et al., 1985; Wang Jingming, 1987). The Fenwei rift system is associated with high heat flow (Wang Jun et al., 1986; Chen Guoda, 1989), numerous Pliocene and Pleistocene volcanoes (Ma Xingyuan, 1986a), intense seismicity (Zhang Buchun et al., 1985; Ma Xingyuan, 1986a; Ye Hong et al., 1987), a negative gravity anomaly (Zhang Buchun et al., 1985), a high-conductivity (low resistivity) magnetotelluric anomaly (Zhang Buchun et al., 1985; Liu Guodong, 1987a), a linear upward bulge of the Mohorovičić discontinuity and the asthenosphere (Zhang Buchun et al., 1985; Ma Xingyuan, 1986a), anomalous upper mantle with a P-wave velocity of 7.8 to 7.9 km/s (Ma Xingyuan, 1986a), and numerous straight faults with a continuity of up to several hundred kilometers (Zhang Buchun et al., 1985; Ma Xingyuan, 1986a; Wang Jingming, 1987).

From the preceding, the close genetic relations among the three structures are apparent---- the Yunnan kobergen, the Hetao-Yinchuan rift, and the Fenwei rift (Figs. 2.31, 3.9, 3.14, 3.22). Figure 2.31 demonstrates that the three are interconnected by the same surge-channel system. This is another example demonstrating the close genetic link between tensile and compressive tectonic belts.

4.3.3 Dinarides-Balkanides

The foldbelts of the Balkans are associated with an unusually symmetrical kobergen (Fig. 4.14, 4.18). Figure 4.14, from Hurtig et al. (1981), shows westward-verging nappes in the west, and eastward-verging nappes in the east. A belt of high heat flow (up to 75 mWm^{-2}) is symmetrical with respect to the kobergen, being concentrated along its core, from which nappes extend outward up to 200 km on either side (Aubouin, 1977). A high-conductivity magnetotelluric anomaly has been mapped from Hungary into the northern part of the kobergen (Gough, 1989). The core of the kobergen has been intruded by numerous young plutons (Fig. 3.10; Hurtig et al.,

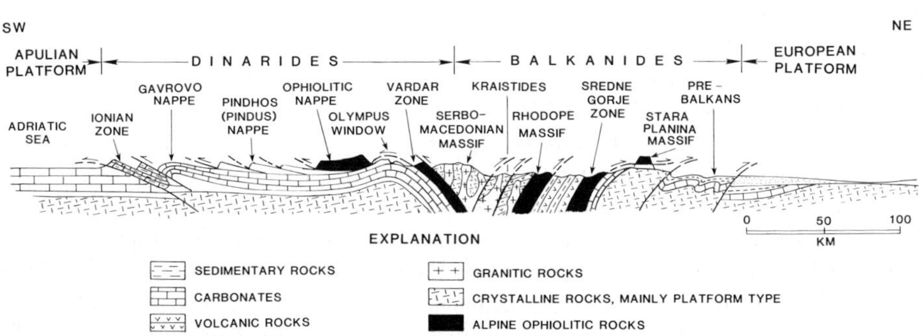

Fig. 4.18. Southwest-northeast structural cross section of the Dinarides-Balkanides kobergen. This is another spectacularly developed symmetrical kobergen (see Figure 3.22). The location of the cross section is on Figure 4.14. From Aubouin (1977).

1981) and, in the Vardar zone (Fig. 4.18), Neogene volcanic rocks crop out (Dimitrijević, 1974). The kobergen also has high seismicity, both at the crest and in still-active thrust faults on both flanks (Dragašević, 1974). At depth is a crustal low-velocity zone above a lens of 7.0-km/s material (Dragašević, 1974). The sparse gravity control in our possession (we have one line across the full kobergen) shows a negative Bouguer anomaly. We call attention to the great similarity between this kobergen and that of the Yunnan Himalaya, or Hengduan Shan (Figs. 3.22, 4.18).

4.3.4 ALPS

Figure 2.13 is a structural cross section through the western Alps. It is based on classical field geology, to which we added data from various geophysical surveys. The kobergen structural style is evident (Fig. 2.15; Kober, 1925; Frei et al., 1989). The Ivrea body probably was part of the original surge channel. Its P-wave velocity is 7.2 to 7.6 km/s (Miller et al., 1982; Mueller, 1983; Frei et al., 1989). This body, now displaced considerably from its original position, is about 20-25 km thick, 40 km wide, and several hundred kilometers long in an east-west direction. It is very close to the root zone of the western Alps (Rutten, 1969). Its mafic to ultramafic rocks are associated closely with the Caledonian granulites (Laubscher and Bernoulli, 1983). The Ivrea body, however, cannot have been in the surge-channel roof because its rocks indicate that temperatures probably did not exceed 800° C (Hunziker and Zingg, 1980). The body is closely associated with the Insubric Line (Gansser, 1968), part of an extensive wrenchlike fault system that extended from the western Alps to the Carpathians (Ádám, 1987). The Ivrea zone has been termed seismically dead (Gansser, 1983b) and lies between two east-west zones of high heat flow (Čermák and Ryback, 1979); Deichmann and Rybach, 1989) associated with bands of high seismicity (Deichmann and Rybach, 1979; Deichmann and Rybach, 1989) and very rapid uplift (Miller et al., 1982; Gansser, 1983b; Mueller, 1983). In addition, at least two low-velocity zones are present (Angeheister et al., 1972; Laubscher, 1978; Miller et al., 1982; Mueller, 1983).

Parallel to and closely associated with the Ivrea zone-Insubric Line and its eastern counterparts is a row of late Alpine intrusions (late Eocene-early Miocene) that formed in an extensional belt. This same extensional belt contains linear to curvilinear wrench faults (Royden et al., 1983; Laubscher, 1988; Ratschbacher et al., 1989). The Ivrea zone-Insubric Line and its eastern continuations are underlain by a high-conductivity anomaly (Brodie and Rutter, 1987) and are known from magnetotelluric studies to continue into the Pannonian basin via the Balaton Line (Ádám et al., 1986; Ádám, 1987). These magnetotelluric, low-resistivity, high-conductivity anomalies originate in bodies 6 to 17 km deep, just below the hypocenters of the associated seismic belt (Ádám, 1987). In all likelihood, the anomaly sources are the fluid contents of one or more shallow low-velocity zones (that is, part of the surge-channel complex).

The metamorphic grades of the Alps are related closely to the position of the Ivrea zone-Insubric Line (Fig. 2.13). The highest metamorphic grades are closest to the Ivrea zone and its eastern equivalents, and decrease away from that zone (Mueller, 1983; Frei et al., 1989). Stretching lineations generally are parallel to subparallel with the structural strike (Ellis and Watkinson, 1987; Selverstone, 1988; Ratschbacher et al., 1989).

Figure 2.13 also shows, whereas a positive Bouguer gravity anomaly overlies the Ivrea zone, two bands of high seismicity and heat flow respectively north and south of the zone are associated with negative Bouguer gravity anomalies (Čermák and Rybach, 1979; Mueller, 1983; Bayer et al., 1989).

Figure 2.13 shows another important phenomenon. The 5.4-km/s low-velocity zone north of the Alps bends upward beneath the Helvetic zone, reaching the Earth's surface in the Aar massif (Miller et al., 1982; Mueller, 1983). This suggested relationship cannot be proven, but Miller et al. (1982) and Mueller (1983) highlighted the likelihood that many former low-velocity zones in the lithosphere now crop out at the surface. The rapid uplift of the entire Alps chain from the Helvetic to the southern Alps, including the Ivrea zone (Fig. 2.13), suggests to us that a single surge channel once underlay this part of the Alps (Ivrea zone). Subsequent tectogenesis broke up the surge channel, dividing it into at least two parts, one north of and one south of the Ivrea zone. Seismicity data suggest the two channels are linked laterally beneath the Ivrea zone, if only by a very thin and limited low-velocity zone (channel). The importance of the Alps example is apparent. It provides an example of a surge channel that through tectogenesis was deformed and segmented; yet the surge-channel system is still very active.

4.3.5 Middle and High Atlas of Morocco

The Middle and High Atlas of Morocco illustrate the germanotype style. In surge tectonics, germanotype foldbelts develop above surge channels well removed (approximately 200-300 km) from oceanic lithosphere and where continental lithosphere is moderately thick. Continental crust appears to have thickened appreciably during a succession of Proterozoic and Paleozoic events that are recorded in outcrops north of, south of, and within the High and Middle Atlas (Wallbrecher, 1988; Piqué and Michard, 1989). Figure 4.19, from Froitzheim et

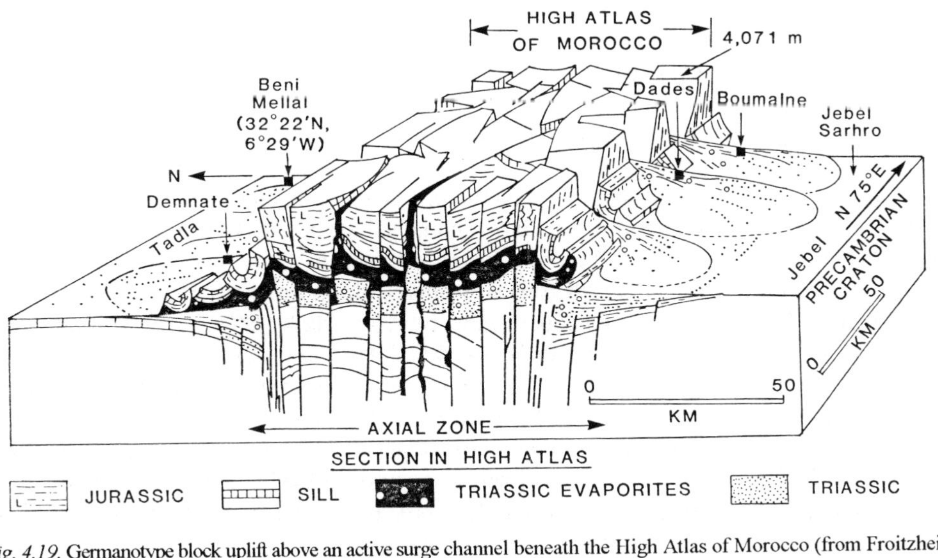

Fig. 4.19. Germanotype block uplift above an active surge channel beneath the High Atlas of Morocco (from Froitzheim et al., 1989). This style of kobergen structure develops wherever the overlying lithosphere is between approximately 20 and 35 km thick. Note the thick development of Late Triassic evaporites (in black). These evaporites consist largely of halite. Distribution of the Late Triassic evaporites suggests that they were precipitated directly above a surge channel. If so, the channel has remained active from at least Late Triassic time to the present.

al. (1988), shows the block uplift of a part of the western High Atlas. Outcrops of Proterozoic and Paleozoic rocks show that they accumulated on a sialic basement. This basement during Jurassic-Cenozoic time broke along near-vertical faults except at the margins of block uplifts. There, along the northwestern and southeastern margins of the range, small reverse faults and some low-angle thrusts developed, with thrusting northeastward in the northwest and southeastward in the southeast (Fig. 4.19; Froitzheim et al., 1988; Jacobshagen et al., 1988; Piqué and Michard, 1989).

Phenomena associated with the High and Middle Atlas and confirming our interpretation were described by Schwarz and Wigger (1988). They include high heat flow; thermal springs; a high-conductivity, low-resistivity, magnetotelluric anomaly; a belt of high seismicity down the crest of the western High Atlas and the Middle Atlas; a negative Bouguer gravity anomaly; Quarternary volcanic rocks; and bundles of linear faults and fractures parallel with the strikes of the ranges. A broad lens of 7.2-7.8-km/s material underlies both ranges at a depth of 29 to 35 km; it is overlain near the top of the lower crust, or the base of the middle crust, by a 6.2-km/s low-velocity layer between 17 and 29 km (Schwarz and Wigger, 1988).

4.3.6 Baykal Rift System

Figure 4.20 shows the 2,500-km long Lake Baykal rift system of south-central Siberia. Mapping of the rift system indicated that it began to form about middle Eocene time in the southern part of modern Lake Baykal, and then propagated both to the east-northeast for some 1,500 km and toward the west-southwest for 500 km.

Figure 3.6 is a seismic-refraction profile across the rift zone (Krylov et al., 1979; Sychev, 1985). The surge channel beneath this half-graben is clearly visible at a depth of approximately 30 to 50 km. The P-wave velocity of much of the channel is 7.7 km/s (Krylov et al., 1979; Sychev, 1985) with a center low-velocity zone with velocities in the 5.4-6.0-km/s range (Krylov et al., 1979). The conduit joining this channel with the asthenosphere is at the northern margin of the depression, directly beneath the main fault zone. The lens of 7.7-km/s material is 20 km thick and 170 km wide. These dimensions are typical for large numbers of surge channels, although there is a wide range of widths (compare Figs. 2.9, 2.17, 2.27, 2.31, 3.3-3.4, 3.10, 4.2-4.3, 4.7-4.8).

The Baykal rift zone overlies a band of high heat flow (Zorin and Osokina, 1984). A low-resistivity, high-conductivity, magnetotelluric anomaly develops at 10 to 20 km, and probably is a shallow low-velocity zone in which the contents are partly molten (Popov, 1987). Cenozoic volcanics are scattered along the rift zone (Genshaft and Saltykovskiy, 1989). The Bouguer gravity anomaly is strongly negative (Artemjev and Artyushkov, 1971). Lake Baykal and its associated rifts lie in a belt of intense seismicity (Golonetskiy and Misharina, 1978), with hypocenters between 0 and 30.5 km, just above the top of the surge channel shown on Figure 3.6 (Vertlib, 1978).

A striking feature of the Baykal surge channel is its apparent extreme asymmetry with respect to the feeder conduit. Most surge channels we have studied thus far, with the exception of deformed channels (Fig. 2.13), are fairly symmetrical with respect to feeder conduit.

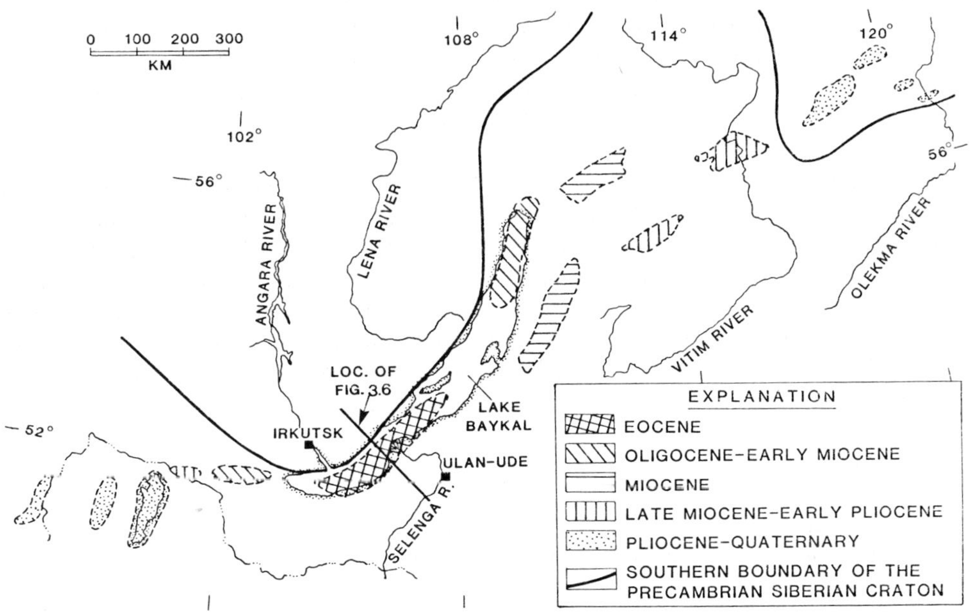

Fig. 4.20. Lake Baykal and vicinity, south-central Siberia. The figure shows the probable chronology of development of the fault depressions in the Baykal rift system. From Logatchev and Florensov (1978). Note that, whereas the principal growth was northeastward from the vicinity of Irkutsk, some growth was westward.

4.3.7 East African Rift System

Like the Baykal rift, the East African rift zone has a 7.3-7.5-km/s lens at 20 to 30 km (Mooney et al., 1983), a long history of volcanism and thermal spring activity, a shallow low-velocity zone (based on the presence of a high-conductivity layer above the anomalous lens; Wood, 1983), a negative Bouguer anomaly (Browne and Fairhead, 1983), and high heat flow (Crane and O'Connell, 1983). The high microseismicity of the rift has long been known (Tobin et al., 1969).

Figure 2.3 shows another characteristic, specifically, the presence of a rift-parallel fault, fracture, and fissure system (Mohr, 1987). This rift-parallel system is offset repeatedly along strike, much as the ridge-parallel fractures of the midocean ridges are offset at transform fault zones. We interpret these fracture offsets as further evidence for the presence of a surge channel system beneath the rift.

4.3.8 Mississippi Embayment

This region is known best in modern tectonic studies for generating some of North America's more spectacular earthquakes in 1811 and 1812 (Hamilton and Johnston, 1990). This earthquake zone----the New Madrid zone----lies on a band of epicenters that extends from the Ouachita Mountains to the Gulf of St. Lawrence (Sykes, 1978, his fig. 14). Part of the epicenter system appears on Figure 2.6. The associated fault, fracture, and fissure system appears on

Figure 2.7 (Fletcher et al., 1978; Hildenbrand, 1982; Hinze and Braile, 1988). McCartan and Architzel's (1988) heat-flow map of the eastern United States, despite the few data points, showed elevated heat flow in the New Madrid region (55-65 mWm^{-2}).

Ervin and McGinnis (1975), Fletcher et al. (1978), Hildenbrand (1982), and Mooney et al. (1983) reported the presence of a 450-km wide lens, 26 to 40 km below the surface, of anomalous low-density mantle material at the base of the crust with a P-wave velocity of 7.2 to 7.5 km/s (Fig. 3.7). This lens lies not only beneath the Mississippi embayment but also beneath a late Proterozoic graben system called the Reelfoot rift (Ervin and McGinnis, 1975). This rift contains an unusually thick section of late Proterozoic-Early Cambrian sedimentary rocks (Hinze and Braile, 1988). A positive Bouguer gravity anomaly characterizes the rift (Fig. 3.7). We believe the presence of a positive anomaly indicates that the surge channel present between about 36° and 38° N latitude has reached its waning stages. It has been a long-lived channel, because it is associated with numerous intrusions of different ages (Phipps, 1988). These include Early to Middle Devonian (399-383 Ma) alnöite and lamprophyre diatremes with associated dikes; Early Permian (287-275 Ma) kimberlite and lamprophyre dikes and sills; and Late Cretaceous (93-80 Ma) sills, dikes, and stocks of nepheline syenite and related rocks (Phipps, 1988). The intrusions are related closely to the fault systems flanking each side of the Reelfoot rift, and some are quite large (Hildenbrand, 1982).

4.3.9 Rhine Graben

The Rhine graben has all of the characteristics of the Baykal, East African, and other continental rifts (e.g., Mooney et al., 1983). This graben has an unusually well-developed fault, fracture, and fissure pattern (Fig. 2.4; Illies and Greiner, 1979). Throughout, the fracture pattern contains horsetail structures that demonstrate flow toward the north-northeast.

4.4 Surge Channels in Zones of Transtension-Transpression

4.4.1 San Andreas Fault

Figures 2.2, 2.20-2.22 illustrate some features of the San Andreas fault zone critical to its structural interpretation. In surge tectonics, the major strike-slip zones of the Earth's lithosphere are generated at or near the crests of surge channels, and create a tectonic style that may be regarded as intermediate between that of a tensional rift zone and that of the alpinotype foldbelt.

We first concluded that strike-slip systems owe their origin to surge-channel activity after reading Lachenbruch and Sass's (1980) study of heat flow along the San Andreas fault. They found that heat flow under most of the California Coast Ranges average 80 mWm^{-2} and is not concentrated along the fault zone. Therefore, they reasoned, friction along the fault zone does not generate the observed heat flux. If the heat is not the result of shear, they deduced further, then the observed heat flow can be explained best by "extreme mantle upwelling" (p. 6185), with asthenosphere conditions reaching a depth of only about 20 km, especially in the northern Coast Ranges. Such a zone was found by Zandt (1981) and Zandt and Furlong (1982), and is illustrated on Figure 2.20. We interpret the upwelled zone as evidence that the upwelling is produced by a surge channel. Note that, north of the Transverse Ranges of southern California, the surge channel becomes shallower, reaching to less than 20 km about 25 km southeast of Cape Mendocino, which is in effect the 20 km deduced by Lachenbruch and Sass (1980).

The surge channel (as we interpret it) on Figure 2.20 exhibits three characteristics. These are (1) rather constant width, (2) constantly changing depth (from 20 to 70 km), and (3) a position that shifts from one side of the San Andreas fault to the other, for example, just south of Hollister where the Calaveras fault splits off (Fig. 2.20). The depth changes are similar to those observed elsewhere, such as western Yunnan (Figs. 2.31, 3.9).

Smith (1978) showed that most intense seismicity in central California is concentrated along the Calaveras fault east of San Francisco Bay, and on the San Andreas fault from San Francisco south. North of San Francisco, seismic activity on the San Andreas fault decreases markedly with the major activity along the Hayward-Calaveras system and its northward continuations.

The transfer of the zone of maximum seismicity from the San Andreas to the Hayward-Calaveras system----based on Smith (1978)----is shown schematically on the inset map on Figure 2.20. The overlap pattern produced by zones of maximum seismicity is analogous, we believe, to the overlapping spreading centers of midocean ridges (Macdonald and Fox, 1983).

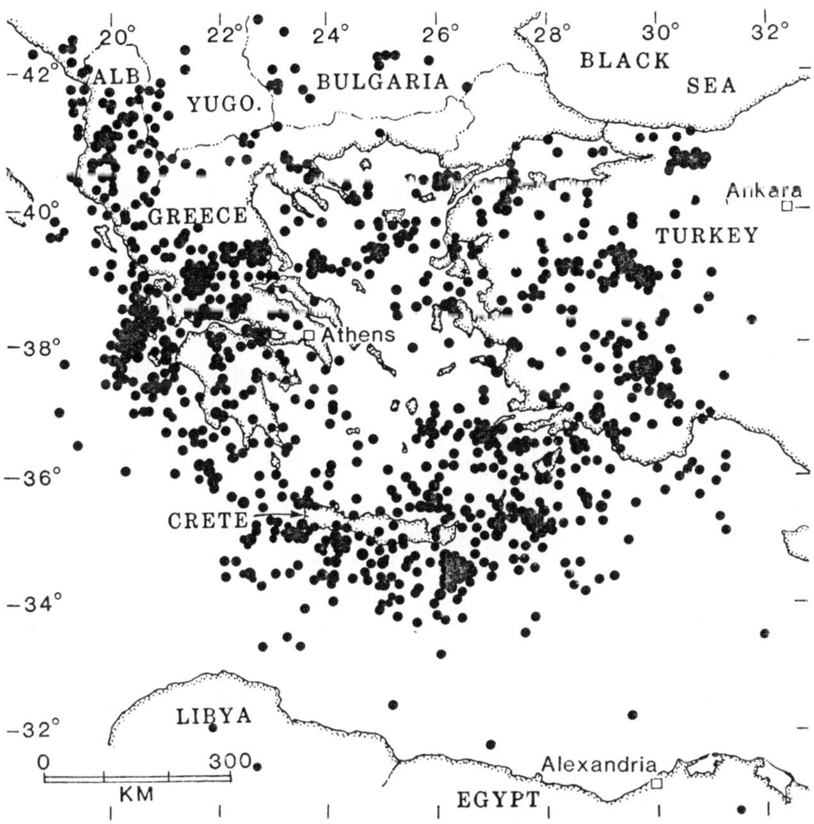

Fig. 4.21. Aegean vortex structure: earthquake epicenter pattern from Papazachos and Comninahis (1977). Note the concentration of earthquake epicenters around the rim of the structure.

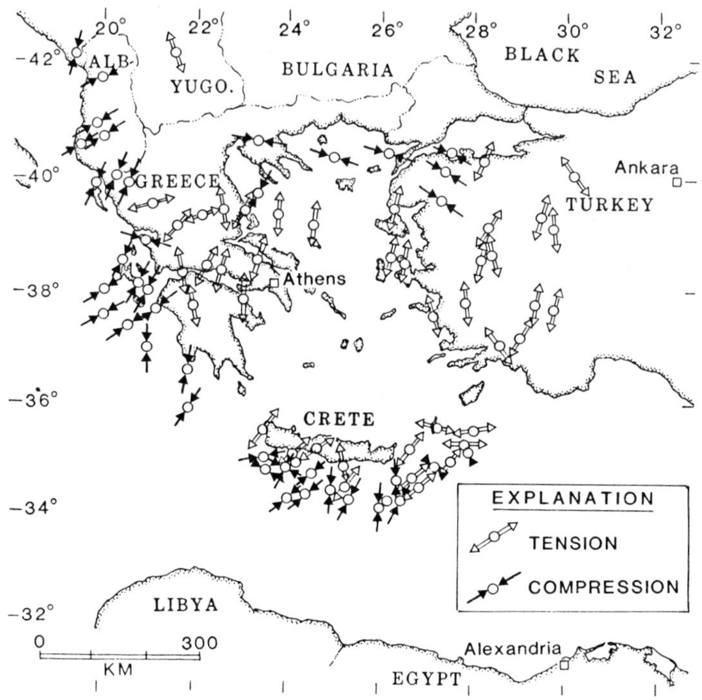

Fig. 4.22. Aegean vortex structure: stress distributions. Tensile stress is in the center; compressive stresses are concentrated around the rim of the structure. From Papazachos and Comninahis (1977).

As Meyerhoff et al. (1992a) explained, such features likely are caused by eddylike motions in the upper asthenosphere. The shape and direction of this eddylike feature demonstrate that motions below the San Andreas fault zone are parallel with that zone. The fact that San Francisco Bay directly overlies one of these features is not, in our opinion, coincidental, but is related directly to the presence of this eddylike feature. In fact, depressions are everywhere found where eddy structures are present (Figs. 3.34, 3.35, 3.37, 4.21-4.23).

Figure 2.2 strongly supports our interpretation of flow parallel with the San Andreas fault system. The map portrays the known faults, fractures, and fold axes of part of the San Andreas fault system (Johnson and Page, 1976). Just as the dominant pattern of faults, fractures, and fissures of midocean ridges suggests ridge-parallel flow, these faults, fractures, and fold axes in the California Coast Ranges indicate flow parallel with them. Figure 2.2, like Figures 2.1 and 3.28, is an illustration of Stoke's Law (Sears et al., 1974; Blatt, 1983; Appendix).

Additional facts support our contention that a surge channel underlies the San Andreas fault zone. We discussed the high heat flow (80 mWm^{-2}; Lachenbruch and Sass, 1980), intense seismicity (Smith, 1978), and linear fault, fracture, and fold-axis pattern (Fig. 2.2; Johnson and Page, 1976). Of equal importance is the presence of a 7.3-7.6-km/s lens (Blümling et al., 1985) beneath the surface east (Blümling and Prodehl, 1983) and west (Walter and Sharpless, 1987) of the San Andreas fault. Figure 2.21 shows the layer beneath a 200-km segment of the Diablo Range, just east of the fault (Blümling and Prodehl, 1983). A low-velocity zone of 5.2-5.7-km/s material occupies the middle crust above the 7.3-7.6-km/s lens (Fig. 2.21), just as in midocean ridges (Fig. 3.3), the Himalayan foldbelt of western Yunnan (Fig. 3.14), the Alps (Fig. 2.13), Lake Baykal (Fig. 3.6), the Baltic Shield (Fig. 3.5), the Great Basin (Fig. 3.4), and elsewhere. A gravity low is associated with the fault (Robbins, 1968), but this may reflect that a vertical low-velocity zone--caused by the lower density of crushed rock in the fault zone--characterizes

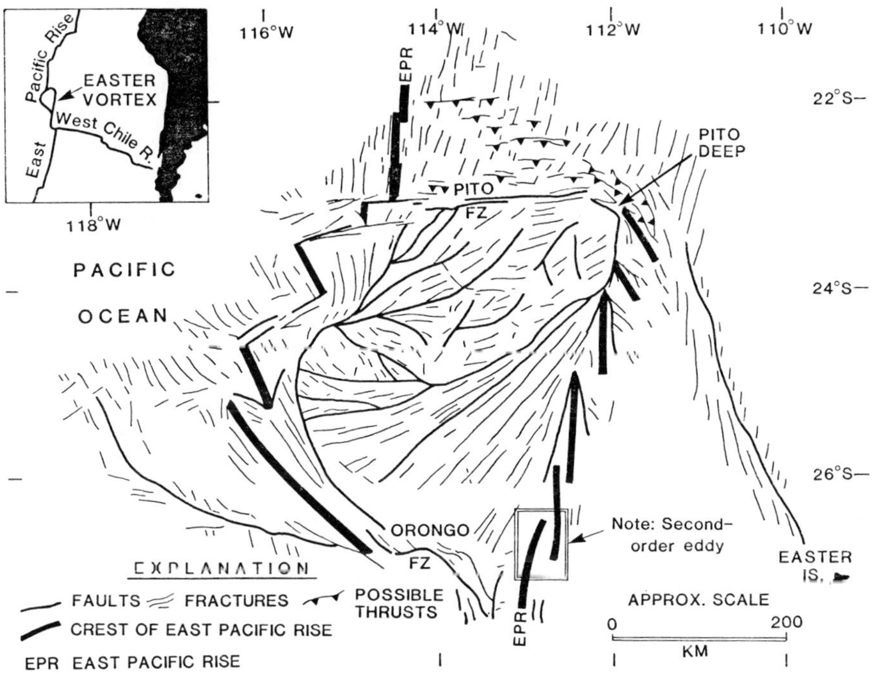

Fig. 4.23. Easter Island vortex structure. This is a simplified tectonic map based on GLORIA and SeaMARC II sonographs. Note: (1) the East Pacific Rise axis forms a gigantic overlapping spreading center that encloses most of the vortex. (2) The orientations of the various fault sets and the presence of northward-verging thrusts north of the Pito fracture zone show that the vortex formed by counterclockwise motions and not by clockwise motions, as Searle et al. (1989) suggest. (3) Small second-order overlapping spreading centers are developed in places along the East Pacific Rise crest. From Searle et al. (1989).

the fault (Feng and McEvilly, 1983). A high-conductivity magnetotelluric anomaly is associated with the San Andreas fault at both the surge-channel depth (7.3-7.6 km/s, Lienert et al., 1979) and at the shallow low-velocity zone depth (Phillips and Kuckes, 1983).

The San Andreas fault lies precisely at the crest of a developing kobergen (Fig. 2.22), that is, along the plane of maximum tension at the crest of the surge channel beneath it. This is illustrated clearly on the cross sections of Figure 2.22 (Zoback et al., 1987; Namson and Davis, 1988). The evidence of this figure has major implications for the origin of strike-slip faults. The structures east of the San Andreas fault are east-vergent; those west of it are west-vergent. This is true not only of the fault between approximately 32° and 36° N latitude where these sections are located (see Figs. 2.2, 2.21), but also of the San Andreas fault in northern and central California (Wentworth et al., 1984; Wentworth, 1987; Wentworth and Zoback, 1989).

We are aware that our interpretation of the San Andreas fault zone is unorthodox, as is our concept that lateral movements along strike-slip zones are small. Some of our reasons are presented in a paper by Martin (1992).

4.4.2 North Anatolia Fault

Our interpretation of the San Andreas fault zone applies to strike-slip faults elsewhere, as shown in Figure 2.18, a structural cross section through the North Anatolia fault zone in northern Turkey (Létouzey et al., 1977). The presence of a kobergen centered on the North Anatolia fault is evident. Structures verge northward north of the fault, and southward south of it. In addition, this fault is associated with high heat flow (Čermák and Rybach, 1979); numerous thermal springs (Tezcan, 1979); intense, destructive, and shallow earthquakes (Şengör, 1979); a lens of anomalous upper mantle (Fig. 2.18; Létouzey et al., 1977); scores of linear faults, fractures, and fissures close to and parallel with the fault (Dewey et al., 1986); and Quaternary volcanism (Pasquarè et al., 1988).

4.4.3 Other Major Strike-Slip Zones

Other examples of major strike-slip zones located in the geometric center of a kobergen include the North Pyrenean fault of the Pyrenees Mountains of Spain and France (Fig. 2.14; Choukroune, 1989), the Insubric Line of the Alps (Fig. 2.13; Gansser, 1968), the Alpine fault of New Zealand (Fig. 2.24, Kingma, 1974; Wood, 1978), the Longitudinal Valley of Taiwan (Fig. 2.17a; Létouzey et al., 1988), and the Philippine fault (Figs. 2.17c, 2.17d; Létouzey et al., 1988). In each example, structural vergence on opposite sides of the fault is away from the fault, that is, the structural style is that of a kobergen. Opposing structural vergence centered on postulated major strike-slip zones is not addressed in any Earth-dynamics hypothesis.

CHAPTER 5

THE TECTONIC EVOLUTION OF SOUTHEAST ASIA--A REGIONAL APPLICATION OF THE SURGE-TECTONICS HYPOTHESIS

5.1 Surge Tectonic Framework

In this chapter, we give a surge-tectonic interpretation of the geological evolution of southeastern Asia. This part of the book should be treated as interpretative and not as an attempt at a definitive statement on the tectonic history of southeastern Asia (Fig. 5.1). We also present a series of paleotectonic surge-channel maps to illustrate our interpretation. These maps (Figs. 5.2, 5.3, 5.6, 5.8-5.12 and 5.17-5.18) portray the major stable areas (massifs and platforms) and paleogeothermal phenomena (surge channels) of southeastern Asia from the beginning of Sinian time (850 Ma) through the present. Insofar as possible, we identified both former and still active surge channels on the basis of the criteria for the identification of surge channels that were listed in Chapter 3. The literature sources used for these maps are listed in the bibliography: an asterisk appears before each source used.

Except for Figure 5.3, which is a map of the geothermal phenomena still active in the region, the maps show mainly the locations of formerly active (and some still active) volcanic and plutonic belts. Where igneous rocks of a particular time interval are absent, we located the surge channels by using Stokes's Law and Bernoulli Theorem phenomena and, as a guide, by using also the positions of the stable areas from Figure 5.2 (i.e., we mapped coeval linear tectonic features such as faults, fractures, fissures, horsts, grabens, anticlinal and synclinal axes, and the like). In a few places it was possible to map linear evaporite basins to trace the location of a specific surge channel, or surge-channel complex. Time constraints led us to concentrate on the mapping of paleogeothermal phenomena. Because of this, we have undoubtedly overlooked some surge channels or surge-channel complexes within each time interval mapped. Regardless of these probable omissions, we believe that the overall tectonic patterns shown by the igneous belts will not change appreciably with the addition of new or overlooked data and, therefore, that the addition of such data will not invalidate the overall interpretation.

5.1.1 The North-South Zone

Unlike the other paleotectonic maps, Figure 5.3 is a map of modern geothermal data---areas of current high heat flow, hot springs, active volcanism, and active strike-slip faults. Inspection of Figure 5.3 reveals at once that, in nearly the entire region east of 103° E longitude, higher-than-normal heat flux is characteristic; and, in the region west of 103° E, with the exception of the Qinghai-Tibet Plateau, surge channels and surge-channel complexes are more clearly defined. This observation suggests that almost the entire region east of 103° E is underlain by a single huge channel complex with multiple magma chambers (i.e., low-velocity layers) containing partial melts.

The 103° E longitude "line" marks what Chinese workers term "The North-South-Trending Belt" or "North-South Zone" (Figs. 5.2, 5.3-5.5; Yin Xiuhua et al., 1980; Wang Hongzhen, 1985; Ma Xingyuan, 1986a, p. 14-16). This nearly north-south-striking belt extends

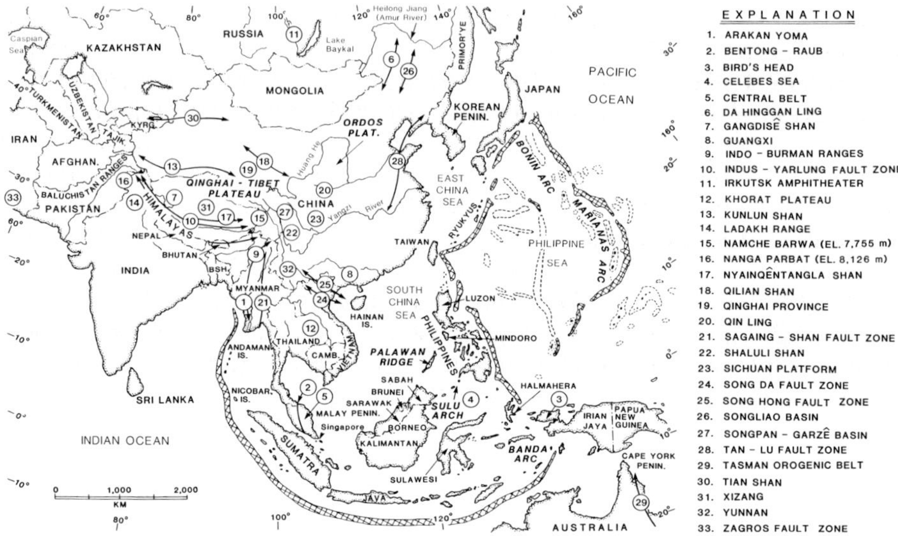

Fig. 5.1. Index map to place names in southeastern Asia. From Meyerhoff, in press.

from the Song Hong (Red River) fracture zone on the China-Vietnam border to the apex of the Irkutsk Amphitheater in Asiatic Russia (Figs. 5.1-5.3). Major structural features along it, in addition to the apex of the Irkutsk Amphitheater, include (1) the western margin of the Ordos Platform and associated features---the eastern flank of the faulted Helan Shan, the Hetao-Yinchuan graben, and the western north-south segment of the Huang He (Yellow River; Figs. 5.1, 5.2); (2) the eastern terminus of the Qilian Shan (Fig. 5.1); (3) the western margin of the Sichuan Platform (Fig. 5.1); and (4) the eastern flank of the great foldbelts emanating and turning southward from the east-west structures of the Qinghai-Tibet Plateau, thereby forming the Eastern (Namche Barwa) syntaxis of the Himalayas (Fig. 5.2). The Sichuan Platform was consolidated before ca. 1,850 Ma and the Ordos Platform before ca. 2,600 Ma (Wang Hongzhen, 1985; Tang Hanzhang, 1986; Ma Xingyuan, 1989). Therefore, the original structures along which the North-South Zone is developed are very early Proterozoic or---more likely---Archean.

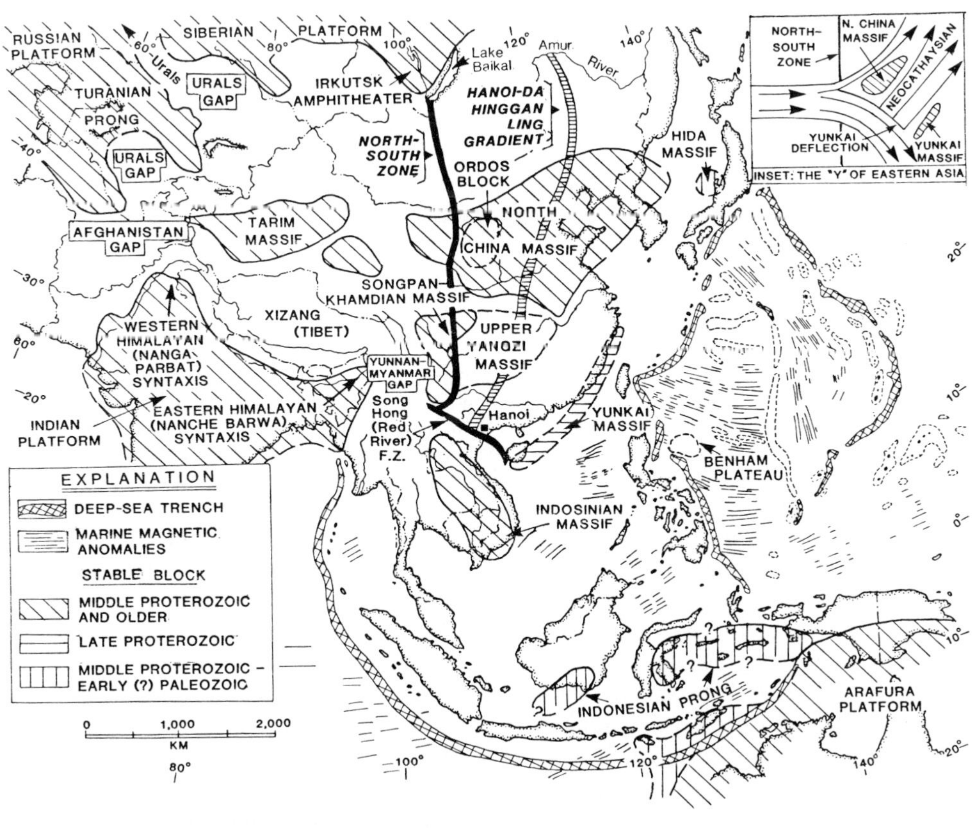

Fig. 5.2. Map of southeastern Asia showing platforms, massifs, principal gaps, North-South Zone, Hanoi-Da Hinggan Ling gradient, and marine magnetic anomalies. The inset map shows our concept of Asia's fundamental "Y" structure, splaying toward the east. From Meyerhoff, in press.

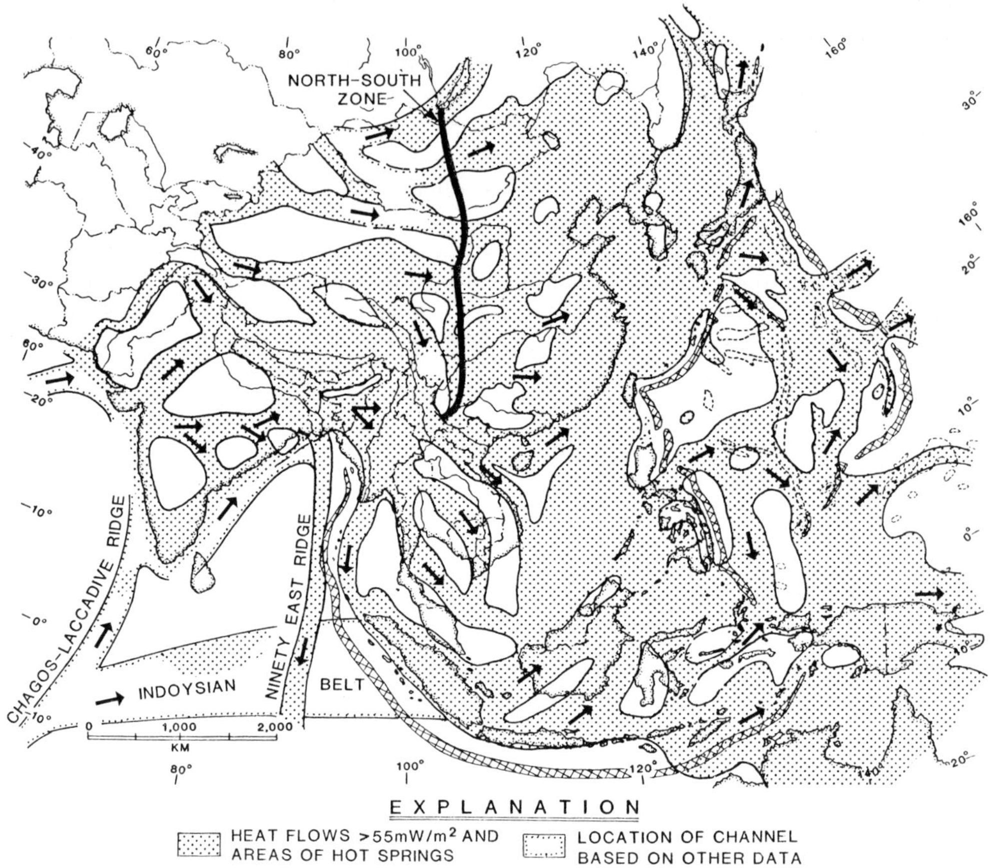

Fig. 5.3. Map of southeastern Asia showing (stippled pattern) all areas of high heat flow (>55 mW/m^2) and hot springs. Note anastomosing (braided) channel pattern west of the North-South Zone; note that high heat flow covers more than 90% of the continental and marginal-sea area east of the North-South Zone. (The North-South Zone is explained in the text.) Note: Channels shown here and on Figure 2.26 do not everywhere coincide because we had fewer data when we drew Figure 2.26 several years ago and had to generalize much more. From Meyerhoff, in press.

We have mentioned that the average heat flow in the region is highest east of the North-South Zone. Ma Xingyuan (1986a,b, p. 14-16) lists additional features of the zone. Two are especially noteworthy. Structural trends west of it are generally east-west, but have a north-northeast strike east of it (Neocathaysian strike of Li Siguang, 1939). Earthquake epicenters are concentrated along the zone from Mongolia to Vietnam (Ma Xingyuan, 1989, pl. 25).

Other differences between the western and eastern sides of the North-South Zone occur only in the region east of the Qinghai-Tibet Plateau and are not obvious farther north. These differences involve, for example, a sharp change in crustal thickness, with thickness values ranging between 60 and 70 km beneath the plateau, but only 42-46 km just east of it (Fig. 5.4). Bouguer gravity values between the Plateau and the region east of it differ by 300 to 400 mgal (Fig. 5.5). A major high-heat-flow band, which we interpret to be a continental trunk surge channel (Fig. 3.39), underlies the eastern margin of the plateau (Figs. 2.31, 3.9, 3.14, 3.27). This band is associated with 7.6-7.9 km/s material (Fig. 3.14), a high-conductivity magnetotelluric anomaly (Ma Xingyuan, 1986a, b, p. 15), and shallow magma chambers above the main magma chamber, or trunk channel (Fig. 3.27; Liu Futian et al., 1989; Liu Jianhua et al., 1989; Sun Ruomei et al., 1991, 1993). Moreover, the entire margin of the plateau is interpreted as an active kobergen (Fig. 3.22; Wang and Chu, 1988). As the band rims the plateau, the origin of the plateau is closely linked to surge-channel development.

5.1.2 Neocathaysian (North-Northeast) Strikes

Figure 5.6, a paleotectonic map showing some of the Sinian surge channels of southeastern Asia, demonstrates that Neocathaysian strikes predominated in eastern China at least since the beginning of Sinian time (ca. 850 Ma). A prominent feature of the Mohorovičić discontinuity (Fig. 5.4) and Bouguer gravity (Fig. 5.5) maps is the steepened gradient that strikes north-northeast across the full width of China, 3,800 km, from Hanoi on the Song Hong (Red River) in Vietnam to the Amur River (Heilong Jiang) north of the Da Hinggan Ling Gradient (Figs. 5.1, 5.2). Ma Xingyuan (1986a,b, p. 16) observed that this steepened gradient is associated with a 100-km-wide band of high electrical conductivity and a lens of high-density rock, which suggest that the zone of steepened gradient is the site of an ancient surge channel.

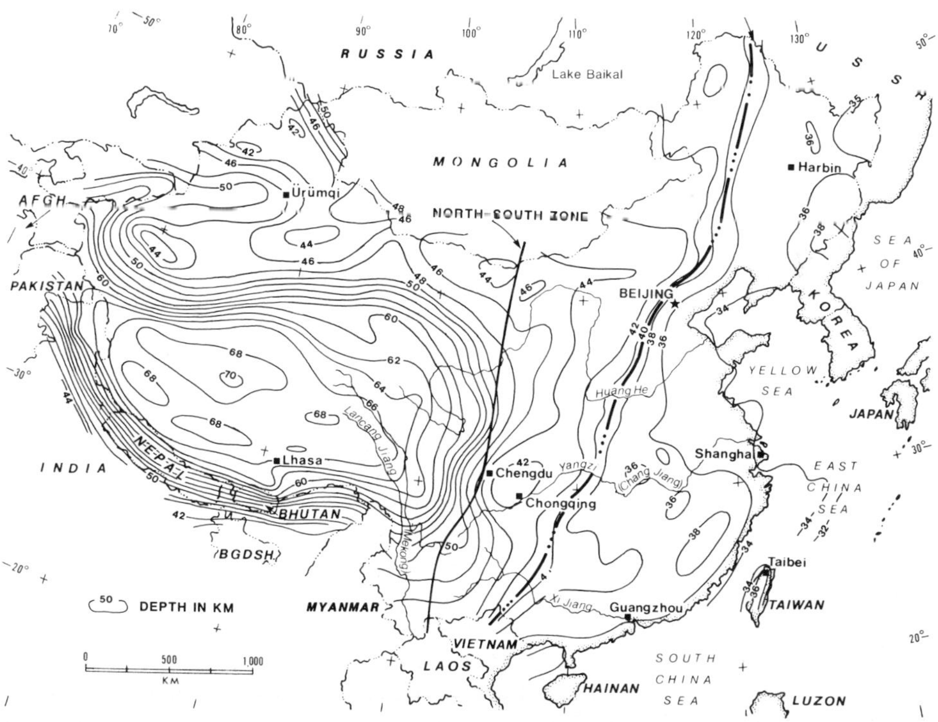

Fig. 5.4. Depth to Mohorovičić discontinuity in China from Wang Shangwen (1983). North-South Zone and Hanoi-Da Hinggan Ling gradient have been added.

5.1.3 Role and Position of Benioff Zone

As mentioned previously, the average heat flow in China east of the North-South zone is substantially higher than it is west of that zone. This elevated heat flow (considerably greater than 55 mW/m^2 in large areas) is associated with large clusters of hot springs and high ground-water temperatures, higher seismic activity, larger numbers of active faults, and some volcanic activity (the only other region of recent volcanism in China is the Qinghai-Tibet Plateau). The increase in geothermal activity in eastern China is not of geologically recent origin, but is easily recognizable from Triassic time to the present (Figs. 5.3, 5.12, 5.17-5.18); it also is evident during Sinian through Ordovician times (Figs. 5.6, 5.8).

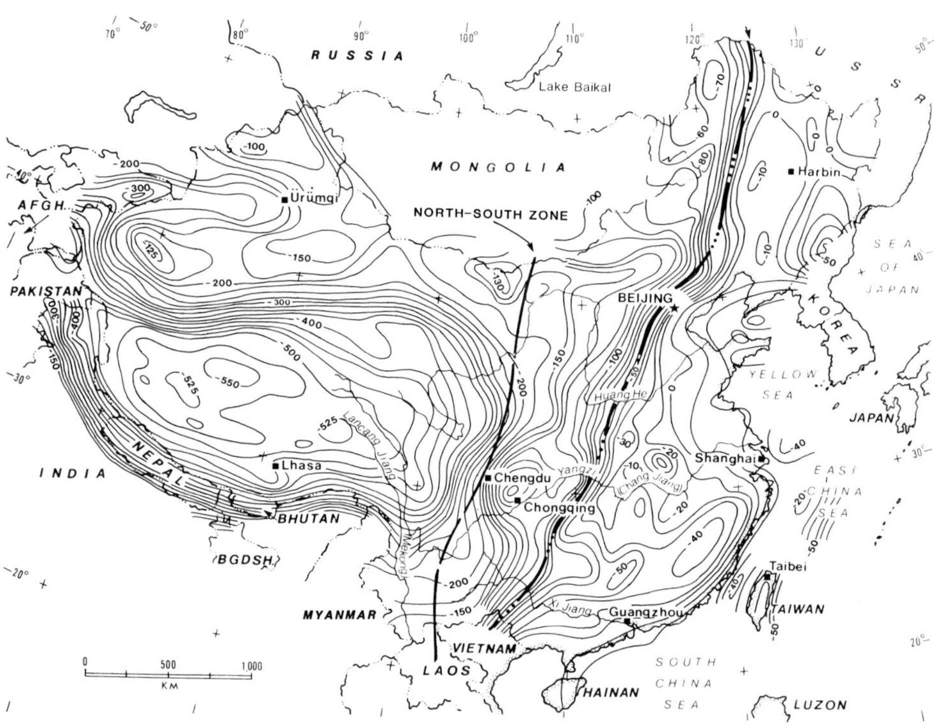

Fig. 5.5. Bouguer gravity map of China from Wang Shangwen (1983). North-South Zone and Hanoi-Da Hinggan Ling gradient have been added.

Meyerhoff et al. (1992b, p. 327-329) explained the concentration of heat flow in eastern Asia as a damming effect caused by the presence of Benioff zones in the western Pacific basin. As pointed out earlier by Meyerhoff and Meyerhoff (1977), island-arc genesis probably is caused by eastward asthenosphere surge resulting from the Earth's rotation. In the western Pacific, damming of asthenosphere flow is observed from the western tip of the Aleutian arc to

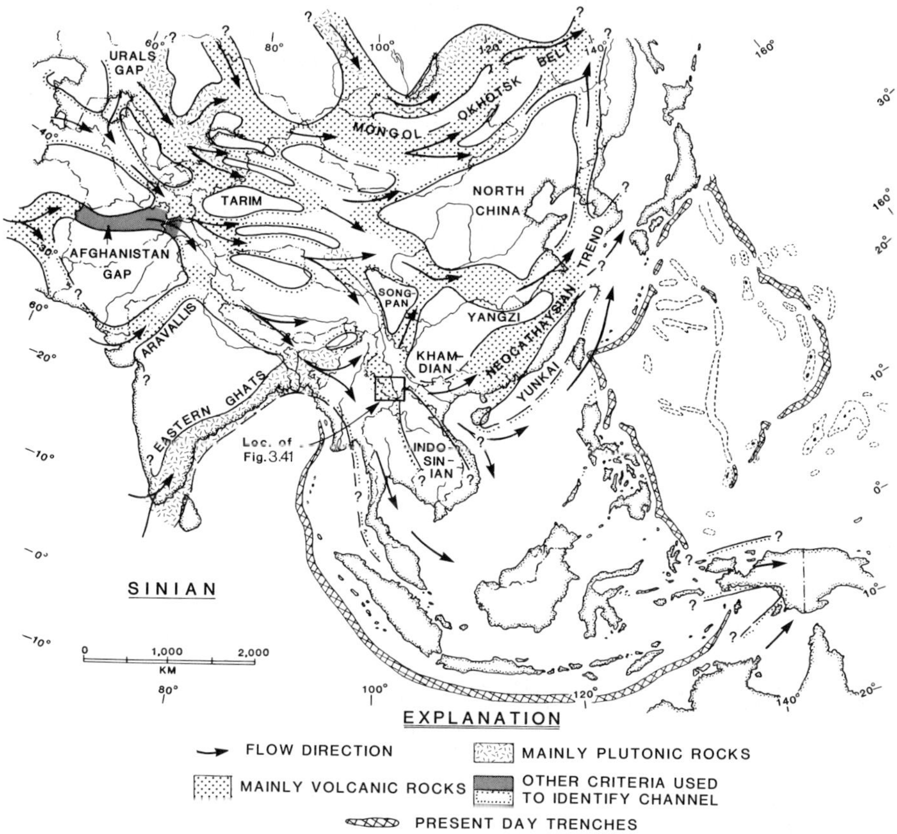

Fig. 5.6. Paleotectonic (surge channel) map for the Sinian, southeastern Asia. Note anastomosing, braided, splay pattern that indicates west-to-east flow. From Meyerhoff, in press.

the southern tip of the Marianas arc, especially along the Kuril-Kamchatka-Marianas trench in the north, and along the Tonga-Kermadec trench in the south. One possible explanation for this effect is demonstrated on deep seismicity cross sections of the region, which suggest that damming is related either to compressional forces in the western Pacific in relation to Benioff zones and/or due to the presence of deeply rooted blocks.

On the eastern side of the Pacific basin, the damming effect of the Benioff zone is readily explained by examining the relative depths of the crust and asthenosphere---i.e., on the eastern side of the Benioff zone, the continents are rooted to depths of up to 400 km or more, whereas the asthenosphere extends to a depth of 200-250 km. Thus, the continental blocks themselves prevent continued eastward flow of the asthenosphere.

West of the Pacific basin, eastward flow in the asthenosphere and in the overlying surge channels is effectively blocked so that high heat flux is concentrated under eastern Asia. Figure

4.5, from Woodhouse and Dziewonski (1984), demonstrates how effective the Benioff zones are in their role as dams in the northwestern Pacific basin. We conclude, therefore---on the basis of the paleotectonic maps presented here (Figs. 5.6, 5.8-5.12 and 5.17-5.18)---that the position of the Benioff zone in the western Pacific basin has remained constant since the beginning of Triassic time, and more likely, since pre-Sinian time. Such an interpretation is a necessary consequence of a contracting Earth, because the Benioff zones are the Earth's cooling cracks.

Strongly supporting our contention that the Benioff zone forms a dam to the eastward flow of melts in the asthenosphere and surge channels of eastern Asia is the geometry of the surge channels themselves. The channels splay northeastward in northeastern Asia, and southward or southeastward in southeastern Asia. These observations also demonstrate that a barrier to eastward flow was present along the western margin of the Pacific Basin.

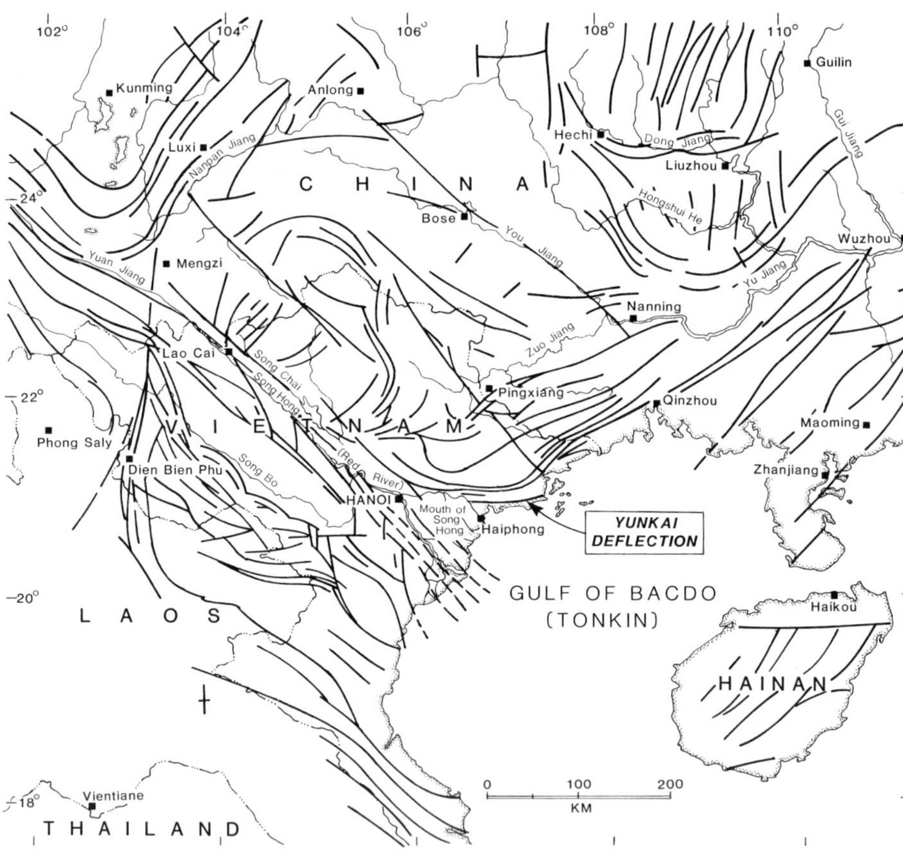

Fig. 5.7. Structural trends in southern China and northern Vietnam showing the Yunkai deflection, which has forced some of the southeastward flow from the Yunnan-Myanmar Gap (Figure 5.2) to be deflected north-northeastward. This deflection, caused by the Yunkai massif beneath the modern Hainan Island (Figure 5.2), produced the Neocathaysian structural trend of eastern Asia. The Yunkai massif, according to radiometric data from Tang Hanzhang (1986), is at least 850-1,000 Ma. Map compiled from ECAFE-UNESCO (1971), Markovskiy (1975), Chinese Academy of Geological Sciences (1979), and Chikov and Nguyen (1992). From Meyerhoff, in press.

5.1.4 Geometric Patterns Formed by Platforms, Massifs, and Surge Channels

The pattern shown by the *platforms* and *massifs* (essentially the same thing; other more-or-less synonyms are *blocks* and *cratons*) is illustrated on Figure 5.2, which was constructed from the maps published by Spizharskiy (1966), Markovskiy (1975), Choubert and Faure-Muret (1976-1985), Hamilton (1979), Wang Hongzhen (1985), and Hutchinson (1989a). Figure 5.2 shows four platforms flanking the eastern Asiatic foldbelts, the Russian and Siberian platforms on the north (Suess's [1901, p. 26] Angara continent) and the Indian and Arafura platforms on the south (Suess's Gondwana continent). Between the northern and southern platforms are Asia's foldbelts and several smaller platform areas that we call *massifs*, all of which played important roles in the tectonic evolution of southeastern Asia. The most important massifs are the Tarim, North China, Songpan-Khamdian (the western part of the larger Yangzi massif), Hida, Yunkai, and Indosinian (Fig. 5.2).

The massif extending from west to east through the Indonesian Archipelago is largely hypothetical, and is based on the identification of "continental fragments" by several workers in the region, most notably Bemmelen (1949), Visser and Hermes (1962), Audretsch et al. (1966), Hermes (1968), Haile (1969, 1974), Markovskiy (1975), Hamilton (1979), Ray (1982), Pieters et al. (1983), and Hartono and Tjokrosapoetro (1986). We have called it the "Indonesian prong." Its shape and east-west strike suggest that it very likely is a fragmented Proterozoic or older surge-channel complex that included the Indosinian massif (Fig. 5.2).

For the southeastern extension of the Russian (East European) platform, we used the term "Turanian prong" rather than the Russian terms "Turanian platform" or "Turan platform" (Spizharskiy, 1966; Zonenshain et al., 1990). Their usage of the word "platform" implies that it is separate from the Russian platform, which it is not. Although the history of the Turanian prong has differed somewhat from that of the Russian platform, much of the difference is related to the fact that the Alpide foldbelt borders the southern margin of the Russian platform and, therefore, younger tectonic events have affected it long after the Variscan-Hercynian tectogeneses which produced the Uralian and Kazakhstan foldbelts farther north.

5.1.5 Overall Eastward-Directed Flow Pattern of Asia

Figures 5.6, 5.8-5.12 and 5.17-5.18 show the surge-channel patterns of southeastern Asia. The channels form *an overall anastomosing or braided flow-like pattern* in the western half of the region, threading their way in and around the various massifs that lie between the northern ("Angaran") and southern ("Gondwanan") platform regions. In the eastern half of the map area, the channels fan outward to form prominent eastward-facing splays or horsetail patterns---southward, southeastward, and eastward in southeastern Asia, and eastward to northeastward in northeastern Asia (see especially Figs. 5.9-5.11, 5.17-5.18). *The whole pattern strongly resembles one that would be produced by a fluid moving from left to right (west to east) through a narrow opening (e.g., the mouth of a bottle, or a river gorge, or Central Asia) into a large container (the bottom of the bottle, or an open river valley, or eastern Asia).* This statement describes the essence of our interpretation of the surge-tectonic evolution of central and eastern Asia during Phanerozoic time.

5.1.6 Fundamental Elements of the Surge-Tectonic Evolution of Southeastern Asia

Since at least the beginning of late Proterozoic time, the tectonic belts of eastern Asia have been underlain by eastward-moving magma in the upper asthenosphere and its overlying surge channels. These eastward movements are sustained by the rotation of the Earth, as proposed by Meyerhoff and Meyerhoff (1977) and Meyerhoff et al. (1992b). Most of the eastward-moving belts passed through two gaps ("choke points") between Precambrian platforms in the western part of the map area. These are the Urals and Afghanistan Gaps shown on Figure 5.3. The Urals Gap was active only before Late Triassic time and was destroyed by late Paleozoic Uralian tectogenesis. The Afghanistan Gap is still active, and shows clearly on the seismotomographic map of Woodhouse and Dziewonski (1984; see Fig. 4.5).

Following the conditions of the Bernoulli Theorem, fluid moving through such gaps would move at a greater velocity than fluid in the wider areas east and west of the gaps. The more rapid motion in the Afghanistan Gap offers an explanation for the more severe deformation and the greater elevations present in this region of the Western Himalayan (Nanga Parbat) syntaxis (Fig. 5.1). In fact, the existence of such gaps or "choke points" provides a general explanation for the origin of syntaxes, or "structural knots," as they are called by some.

The same explanation accounts for the presence, west of the Western Himalayan (Nanga Parbat) syntaxis, of the huge Dasht-i-Lut vortex shown on Figure 2.19. Eastward-moving fluid would be partly dammed west of the Afghanistan Gap, and the damming could easily initiate vortical motion in the surge-channel complex west of the Gap.

Eastward movement of fluids from the Afghanistan Gap would be controlled primarily by the platforms north and south of the region (Suess's [1901, p. 26] *Angara Festland* and *Gondwana Continent*, respectively). The smaller massifs between the two platform areas would exert only secondary control over the eastward motion of asthenosphere and surge-channel fluids, because these massifs---their positions and shapes---are themselves products of the eastward fluid motions. This last statement is shown clearly by the anastomosing patterns of the channels through and around them (e.g., Figs. 5.8-5.12, 5.17-5.18; see Tritton and Davies [1985], Tritton [1988], and Munson et al. [1990] for a physical explanation).

Nevertheless, the gaps between these smaller massifs would behave mechanically in just the same way as the Afghanistan Gap. For example, the Yunnan-Myanmar Gap (Fig. 5.2) between the Indian Platform on the west and Yangzi massif (Songpan-Khamdian massif) on the east has structures (and a major syntaxis) just like those between the Indian Platform and the Turanian prong. Thus, both the Western Himalayan (Nanga Parbat) and Eastern Himalayan (Namche Barwa) syntaxes lie adjacent to important gaps, or "choke points," in the region (Fig. 5.2).

With the preceding hypothesis in mind, we now present a model of the surge-channel and flow patterns from Sinian time to the present.

5.2 Surge-Tectonic History

5.2.1 General

With the exception of parts of China and Mongolia, data on pre-Paleozoic rocks are too sparse to permit the construction of reliable Precambrian paleotectonic maps. Consequently, we made no attempt to study times older than the Sinian (i.e., before ca. 850 Ma).

During the 850-Ma interval that we did study, *the same basic tectonic patterns, or surge-channel geometries persisted.* The positions of the channels and channel complexes did shift somewhat, but very little. At the local level (e.g., the Indian Platform), small surge channels came and went. Yet, despite these minor shifts and changes, the gross patterns remained unchanged (Figs. 5.2, 5.3, 5.6, 5.8-5.12, 5.17-5.18). This observation alone contradicts the plate-tectonic paradigms for the geological evolution of southeastern Asia. Even when thoroughly documented contrary tectonic views of southeastern Asia to those held by the plate-tectonics proponents were published, they were not considered (e.g., Gansser, 1964, 1977, 1981, 1983a, 1983b, 1991; Meyerhoff, 1970; Petrushevsky, 1971; Saxena, 1971a, 1971b, 1978; Meyerhoff and Meyerhoff, 1977; Haller, 1979; Auden, 1981; Chatterjee and Hotton, 1986; Saxena et al., 1986).

The basic tectonic pattern of southeastern Asia resembles a giant letter "Y" turned on its side (see inset map on Fig. 5.3). The stem of the "Y" lies west of the North-South Zone (Fig. 5.2), where the overall strike of the surge channels is east-west. Individual channels and channel complexes do change trend along strike, but only in an anastomosing (braided) manner and pattern whose overall direction is approximately constant.

The northern branch of the "Y" turns northeastward through Mongolia and northeastern China. This branch is north of the east-west-striking Tarim and North China massifs, is mainly of late Proterozoic-Paleozoic age, and is known as the Mongol-Okhotsk geosyncline, or foldbelt (Figs. 5.6, 5.8; Zonenshain et al., 1990). Its surge channels are aligned mainly with the Afghanistan Gap, and only partly with the Urals Gap (Fig. 5.2). East of the North-South Zone, the northern surge channels, now deformed into kobergens, gradually swing into a northeast strike. North and east of the Ordos block (Fig. 5.2), they become wholly confluent with the northeast- to north-northeast-striking Neocathaysian trend of eastern China.

The southern branch of the "Y", in contrast, is more complex than the northern. It lies south of the Tarim and North China massifs (Fig. 5.2), and is aligned entirely with the Afghanistan Gap. It includes many foldbelts (Huang Jiqing, 1978, 1984). In the east, it turns southward, passing through the Yunnan-Myanmar Gap between the Indian Platform on the west and the Songpan-Khamdian massif on the east (Fig. 5.2). The latter massif was consolidated by at least 2,500-1,800 Ma (Tang Hanzhang, 1986).

South of the Yunnan-Myanmar Gap, the southern branch splits in two parts. The western branch passes west of the Indosinian massif, then southward beneath the present Malay Peninsula, and eastward under what is now Indonesia. The eastern branch passes east of the Indosinian massif, its southwestern part extending southward and rejoining the western branch. The northeastern part of the eastern branch, however, which coincides with the Song Hong (Red River) fracture zone, terminates the southern end of the Neocathaysian trend *at right angles*. Surface and Landsat mapping shows that the northeastern part of the eastern branch is deflected into the Neocathaysian trend west of Hainan Island. This deflection is called the *Yunkai deflection*, and is illustrated on Figure 5.7 (ECAFE-UNESCO, 1971; Markovskiy, 1975; Chinese Academy of Geological Sciences, 1979; Chikov and Nguyen, 1992).

The age of the Yunkai deflection is unknown, but it must postdate the formation of the Yunkai massif (Fig. 5.2). According to Tang Hanzhang (1986), the Proterozoic basement exposed on Hainan Island was consolidated in late Proterozoic time, between 1,000 and 600 Ma. Figure 5.6 shows that the deflection already existed by Sinian time. Hence, both the deflection and the Yunkai massif are older than 850 Ma.

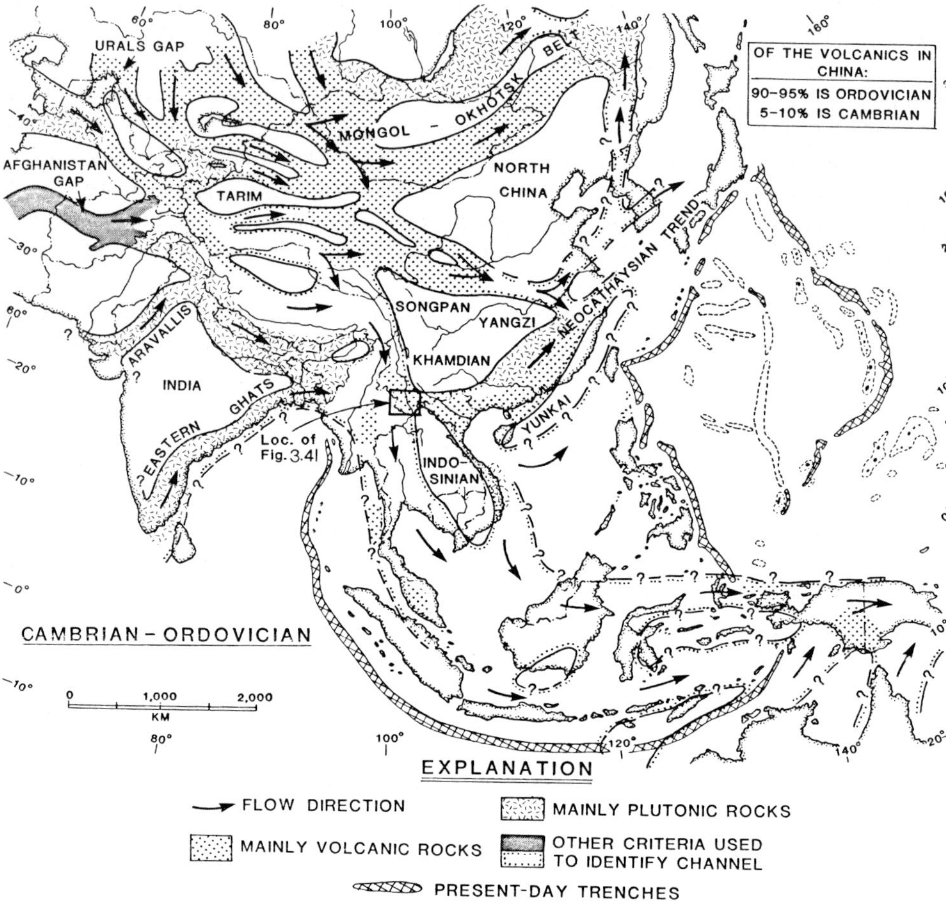

Fig. 5.8. Paleotectonic (surge channel) map for the Cambrian-Ordovician, southeastern Asia. Note anastomosing, braided, splay pattern that indicates west-to-east flow. From Meyerhoff, in press.

A nearly identical structural situation is present in Hebei, Shandong, Henan, Anhui, and Jiangsu Provinces in northeastern China. Zhu Ying (1979) portrayed an east-west trend striking toward the present coast. There, at the location of the modern Tan-Lu fault zone, the west-to-east trend is deflected abruptly north-northeastward by an ancient basement feature which controls the trend of the highly active Tan-Lu fault (Fig. 5.1). As in the case of the Yunkai deflection, the north-northeast trend described by Zhu Ying (1979) parallels the Neocathaysian trend. This structural feature is described also by Ren Jishun et al. (1987, p. 48-50).

5.2.2 Pre-Sinian

The principal continental masses of southeastern Asia seem to have acquired their present sizes and shapes by Sinian time. The only major areas whose pre-Sinian existence is in doubt are the island-arc regions where, except for Japan, pre-Paleozoic rocks do not crop out. This fact has been used to bolster the widely accepted claim that the arcs are geologically young, and that their presumed continuous formation throughout Phanerozoic time is proof that the continents are still growing by accretionary processes.

Most of the island-arc areas are unmapped in detail; many critical parts of the arcs are unmapped and unsampled. The land area within the arcs south of Japan and northwest of the Solomon Islands is just over 3,000,000 km^2, only slightly smaller than India. Of this amount, less than 5% has been mapped in detail. The shallow-water areas within these island arcs comprise an additional 7,500,000 km^2. With few exceptions, the water-covered areas range from poorly known to unknown. In fact, so much of the southeastern Asia island-arc region is unstudied that definitive statements about the geological evolution of these regions hinder rather than help geological progress (e.g., Hamilton, 1979; Hutchinson, 1989a, 1989b).

New geological discoveries are constantly being made in the island-arc areas. For example, Paleozoic fossils were first found in the Philippines in 1961 (Easton and Melendres, 1963) and now many parts of the Philippines are known to have, at the very least, a Paleozoic basement. Proterozoic granitic clasts were found in 1982 on the Vogelkop (Bird's Head; Fig. 5.1) of western New Guinea (1,250 Ma; Pieters et al., 1983). The discovery of granitic clasts in northwestern Irian Jaya suggests that a surge channel may have been present there before 1,250 Ma (Irian Jaya).

Geochemical studies of the region's igneous rocks have been underway for nearly two decades, particularly in island-arc provinces (e.g., Stanton's [1978, 1994] compilations of the geochemistry of the volcanic rocks of the Solomon Islands). A few recent isotopic studies of some of the older igneous rocks have yielded a few surprises. For example, reversely discordant patterns displayed by zircons from igneous stocks on the Malay Peninsula yield U-Pb upper intersection ages ranging from 1,700 to 1,500 Ma west of the Bentong-Raub fault zone, and ages ranging from 1,400 to 900 Ma east of the fault zone (Fig. 5.1; Liew and Page, 1985; Pitfield et al., 1987; Hutchinson, 1989a; and others). Thus large areas of the island-arc region also could be underlain by Proterozoic or older crust, and the assumption of a youthful, Phanerozoic basement is unwarranted. Until similar studies (substantial field work and extensive field sampling) are completed, much of southeastern Asia's island-arc region will remain *terra incognita*, and conclusions made without an adequate sample base, especially from basement rocks, will be very risky.

5.2.3 Sinian-Ordovician (Figs. 5.6, 5.8)

Figures 5.6 and 5.8 are paleotectonic surge-channel maps for Sinian and Cambrian-Ordovician times respectively. In this study, the time scale of Harland et al. (1990) has been used for defining ages. Figure 5.6 for the Sinian shows only half as much information as appears on Figure 5.8 for the Cambrian-Ordovician. Yet it is apparent that the surge-channel pattern must have been almost the same throughout the ca. 411 Ma represented by the two maps. (This 411 Ma is nearly half of the total 850 Ma time period reviewed for this study!) Figure 5.8 shows clearly that the southeastward-opening splay pattern of southeastern Asia was well developed by the end of Ordovician time. Therefore, Benioff zones must have been present in approximately the same positions as the existing ones.

The east-west geosynclines of central Asia were fully developed (Spizharskiy, 1977, 1979; Huang Jiqing, 1978, 1984; Wang Hongzhen, 1985; Zonenshain et al., 1990), and the

Afghanistan Gap was developing with the initiation of the Afghanistan Channel (Weippert et al., 1970; Wolfart and Wittekindt, 1980). The Yunnan-Malaya geosyncline (Burton, 1967) had formed, although it did not have all the characteristics of a true geosyncline; T.T. Khoo (<u>in</u> Ichikawa et al., 1990) called this geosyncline the "Yunnan-Malaya block." In India, the Proterozoic geosynclines of the Aravallis and Eastern Ghats manifested themselves as dying surge channels from which late orogenic granitic plutons formed (Verma, 1991, p. 11-14). A very important fact is that, wherever geosynclinal or platformal basins were present from the Arctic to the Indian Ocean, the basins and platforms were marine. Nonmarine depositional areas were limited in number and in extent.

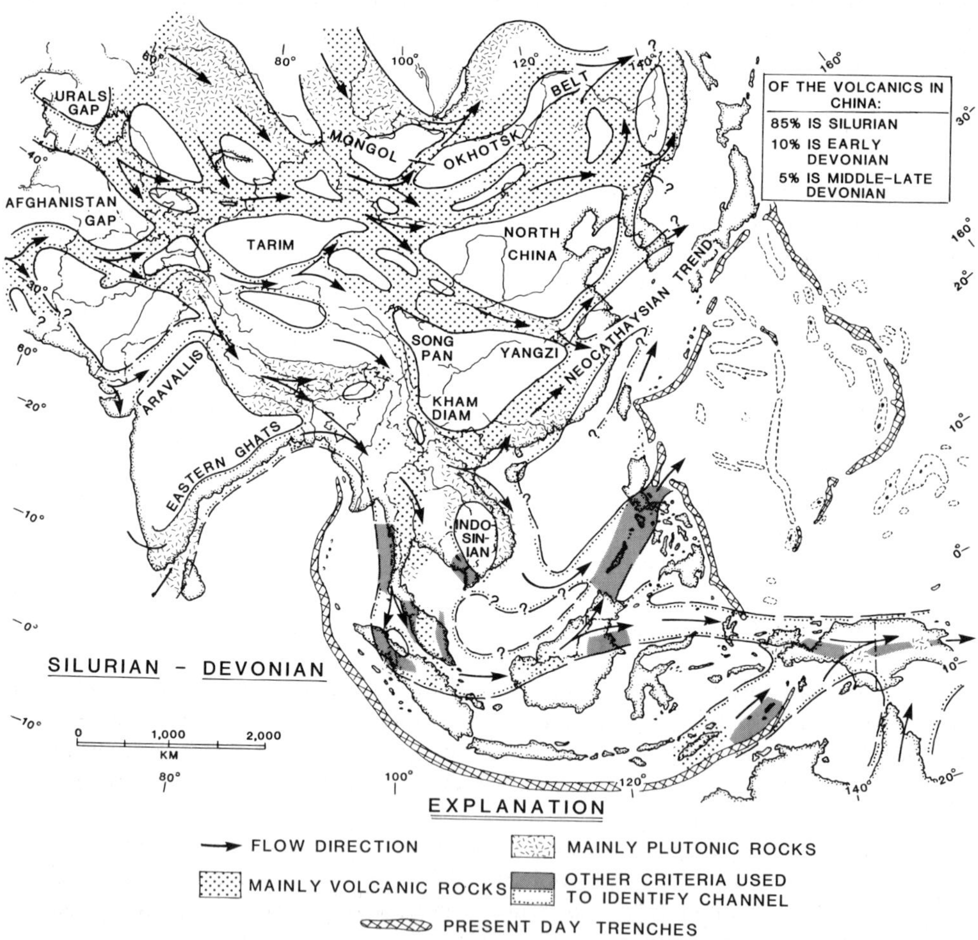

Fig. 5.9. Paleotectonic (surge channel) map for the Silurian-Devonian, southeastern Asia. Note anastomosing, braided, splay pattern that indicates west-to-east flow. From Meyerhoff, in press.

In Myanmar, the Lashio east-west structure probably appeared in Proterozoic time (Fig. 3.41A). This east-west structure, which breaks up the normal north-south structural trends of the region, has been the subject of some speculation (e.g., Searle and Ba, 1964; Goosens, 1978; Bannert and Helmcke, 1983; Bender, 1983; Hutchinson, 1989a). The east-west deflection that occurs at Mogok (22°55' N, 96°29' E; Figs. 3.41A, 5.6, 5.8) forms a K structure that opens toward the Song Hong (Red River) fault zone (Figs. 3.41A, 5.2, 5.6, 5.8). Because this structure has every appearance of a breakout channel, we have called it the Lashio surge channel, or Lashio breakout channel. It is associated with Late Cambrian (?) silicic volcanic rocks (Bawdwin Volcanic Formation; Searle and Ba, 1964). A radiometric date from the associated metamorphic and igneous rocks is 730 Ma, which may place the upper age limit of the Lashio channel in the middle Sinian (Hutchinson, 1989a, p. 45). The Lashio channel's eastward extent is uncertain, but it is on strike with the long-active Song Hong (Red River) fracture zone (Figs. 5.2, 5.6, 5.8). Ren Jishun et al. (1987, p. 145) discussed several types of geological and geochronological field evidence that suggest that the Song Hong fracture zone is at least as old as late Proterozoic, that is, approximately the same age as the Lashio channel. Hence, the two features may well have been joined at one time.

5.2.4 Silurian-Devonian (Fig. 5.9)

During Late Ordovician time, major parts of China began to emerge along a broad east-west arch that included the Tarim and North China massifs (Fig. 5.2; Wang and Liu, 1980; Wang Hongzhen, 1985; Meyerhoff et al., 1991). By Middle and Late Silurian time, the emerged area had spread southeastward to encompass the Yangzi massif. During all of Devonian time, China---and therefore much of the area discussed in the present work---was divided into separate marine regimes, one in the north and another in the south. Of the volcanic rocks extruded during this time interval, 85% are Silurian (Fig. 5.9).

North of the area considered here, the Siberian and Russian Platforms began to emerge and, by Late Devonian time, were fully emergent (Spizharskiy, 1979). In contrast, Southeast Asia remained a shallow-water platform region that included a few deep-water troughs (Hutchinson, 1989a). The site of the modern Indonesian archipelago, on the other hand, seems to have consisted largely of deep-water troughs that surrounded small, shallow-water continental platforms, much as today (Rutten, 1940; Visser and Hermes, 1962; Pieters et al., 1983; Tate, 1991; Tate and Hon, 1991). Despite numerous model-dependent interpretations to the contrary (e.g., Hamilton, 1979), there is no direct evidence that this region was wholly oceanic. Much of it may have been underlain by sizable areas of continental crust.

The Indian subcontinent remained emergent and the northeast-striking Aravalli and Eastern Ghat foldbelts became inactive (Verma, 1991). Volcanism in the similarly northeast-striking Neocathaysian belt of southeastern China ceased (Wang Hongzhen, 1985) as "Caledonian" tectogenesis effectively crushed and destroyed the surge channels of that region (compare Figs. 5.8 and 5.9).

Tectogenesis during the Silurian and Devonian was limited to the truly mobile geosynclinal zones (Suess, 1901; Argand, 1924; Gregory, 1929; Li Siguang 1939, 1973; Huang Jiqing, 1945, 1978, 1984; Bemmelen, 1949; Visser and Hermes, 1962; Hermes, 1968; Wang and Liu, 1980; Bender, 1983; Wang Hongzhen, 1985; Yang Zunyi et al., 1986; Ren Jishun et al., 1987; Chen Guoda, 1989; Hutchinson, 1989a, 1989b; Ren and Chen, 1989; Zonenshain et al., 1990, Meyerhoff et al., 1991). These same authors also described the widespread epeirogenic movements that took place in adjacent platform regions. In Southeast Asia, true alpinotype tectogenesis did not occur, but germanotype tectogenesis and epeirogenic movements were widespread, as shown by the presence of large hiati, redbeds, and similar phenomena (e.g., Rutten, 1940; Bemmelen, 1949; Glaessner et al., 1950; Visser and Hermes, 1962; Haile et al., 1977; Pieters et al., 1983; Khoo, 1984; Tan, 1984; Khoo and Hada, 1986; Bhat, 1987; Hutchinson, 1989a; Tate, 1991; Tate and Hon, 1991). The overall splay patterns on Figure 5.9 show that, in addition to the western Pacific oceanic trenches, the trench system extending from the Andaman-Nicobar Islands of the Indian Ocean to New Guinea was also in place.

The most important observations about eastern Asia during Silurian and Devonian times are that (1) northern Asia became strongly emergent, including a large percentage of China (especially along and north of the east-west Tarim-North China massifs axis); (2) India remained stable; and (3) Asia southeast of China remained largely submerged. Thus, as northern Asia rose, the southeastern part either stabilized or sank.

5.2.5 Carboniferous-Late Permian (Figs. 5.10-5.11)

Figure 5.10 is a paleotectonic (surge-channel) map for the Carboniferous and Early Permian; Figure 5.11 is a paleotectonic (surge-channel) map for the Late Permian. Figure 5.11 was drawn separately from Figure 5.10 in order to stress the great sizes of the Punjal, Abor, Emeishan, and other coeval flood basalts, which are such prominent features of southeastern Asian geology. Most (50%) of the volcanic and plutonic rocks on Figure 5.10 are Late Carboniferous; of the remainder, 20% are Early Carboniferous and 30% are Early Permian.

The anastomosing, braided channel patterns of Figures 5.10 and 5.11 are nearly the same as those on Figures 5.6, 5.8, and 5.9, with very few changes in positions. Neocathaysian (eastern China) volcanic rocks are nonexistent, but some sedimentary troughs with Neocathaysian strikes persist. The most important change in surge-channel patterns took place in India where the older northeast-striking Aravalli and Eastern Ghats foldbelts vanished (Verma, 1991). A new system of surge channels appeared, possibly in Late Carboniferous time but most likely in Early Permian (Asselian) time. The principal surge channel of this new system is the Narmada-Son channel (Narmada-Son rift; Verma, 1991, p. 22), which crosses the Indian Peninsula (Fig. 5.10). Two prominent southeast-striking branches are the Godavari and Mahanadi rifts. West of these, beneath the Deccan Traps, is another set of southeast-striking rifts, the Kurduvadi and Koyna, determined from seismic studies (Naqvi and Rogers, 1987, p. 203). Together, the entire system (Narmada-Son, Kurduvadi, Koyna, Godavari, Mahanadi) forms a horsetail-shaped, eastward-facing splay, showing that (1) India was approximately in its present position when the rift system formed in Early Permian time and (2) flow in the surge channels was from west to east. These channels are, in fact, still active (Gupta, 1993).

Although fossiliferous Silurian strata are known from Irian Jaya and the Asian mainland, the oldest known sedimentary rocks in the modern Indonesian Archipelago are Devonian, as we noted in the preceding section. They are platform-type, shallow-water carbonates. Their occurrences, however, are extremely scarce. The first appreciable numbers of Paleozoic localities are Carboniferous, and are known from Sumatra, Sarawak, Kalimantan, the Philippines, Timor, and several localities between Papua New Guinea and Sulawesi (Figs. 5.1, 5.10). The Carboniferous and Permian rocks exhibit a wide diversity of depositional and volcanic environments, almost as diversified as those found on the mainland of Asia. On the basis of these occurrences, we can state that fragments of most of the major Indonesian-

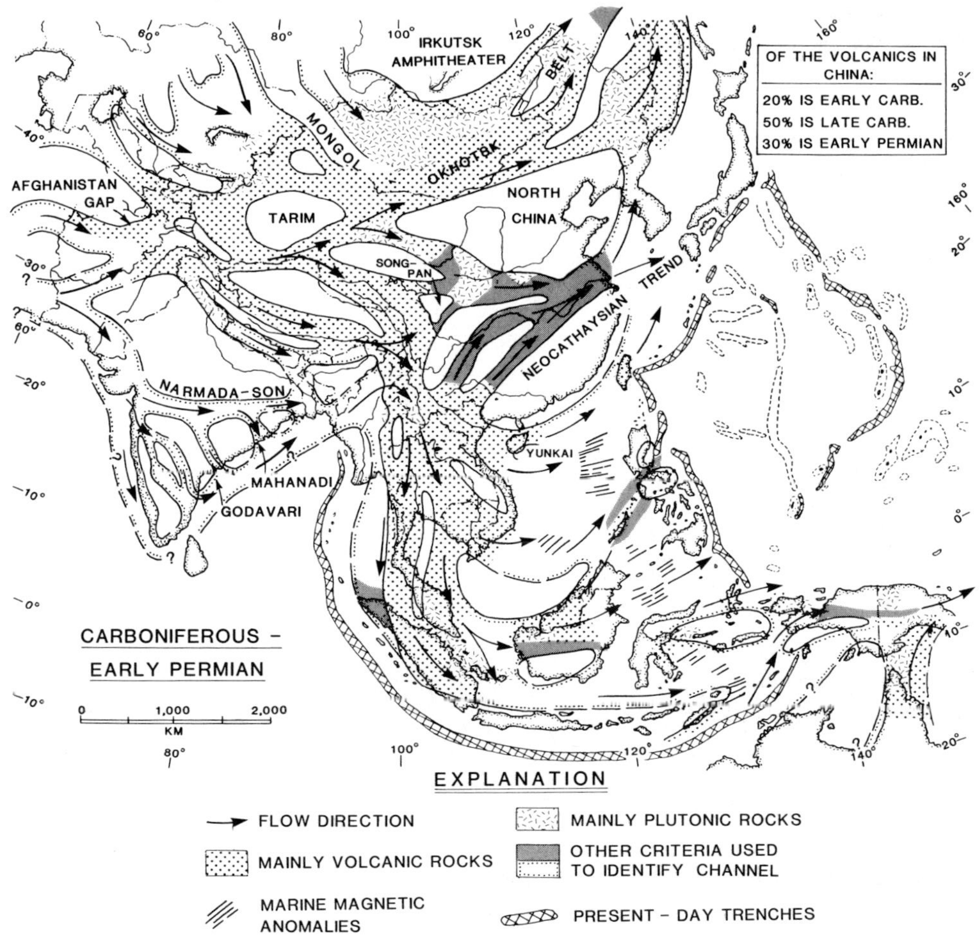

Fig. 5.10. Paleotectonic (surge channel) map for the Carboniferous-Early Permian, southeastern Asia. Note anastomosing, braided, splay pattern that indicates west-to-east flow. From Meyerhoff, in press.

Archipelago islands were present by Carboniferous time. So many of these islands include rock fragments of continental origin that we suggest that, at the very least, the continental occurrences east of Sulawesi were once part of the Australian continent. However, we do not regard them as having been rifted from Australia, but interpret them to be emergent remnants of peninsula-shaped prongs that once extended unbroken from Australia but which have since been broken up and tectonized by surge-channel invasions into the upper lithosphere north of Australia.

Figures 5.2 and 5.10 show that marine magnetic anomalies parallel the Sulu Ridge and the Palawan Ridge west of the Philippines in both the Sulu and Celebes Seas (Figs. 5.1, 5.2, 5.10). Because marine magnetic anomalies are produced by anomaly parallel flow (Agocs et al., 1992; Meyerhoff et al., 1992a), and not by sea-floor spreading, these anomalies must have been produced at the time(s) when the Palawan Ridge and Sulu Arch were formed. Because the

Philippines are at least as old as Carboniferous---and probably much older---the anomalies are at least of Early Carboniferous age.

Figures 5.2 and 5.10 also show northeast-trending marine magnetic anomalies in the South China Sea. Those in the northern part of the South China Sea just west of Luzon Island (Philippines) are on strike with the volcanic belts of the Song Hong (Red River) fracture zone and therefore could be of any age from late Proterozoic to Cenozoic. Because the most intense surge-channel activity in the Song Hong channel was between late Proterozoic and Late Triassic time, we suggest that the age of the anomalies falls within this time span.

The age of the anomalies of the South China Sea just east of southern Vietnam may be the same as those of the anomalies farther north, but these southern anomalies have a slightly different strike, more like those in the Sulu and Celebes Seas. Hence, we believe that these anomalies are Early Carboniferous, or older.

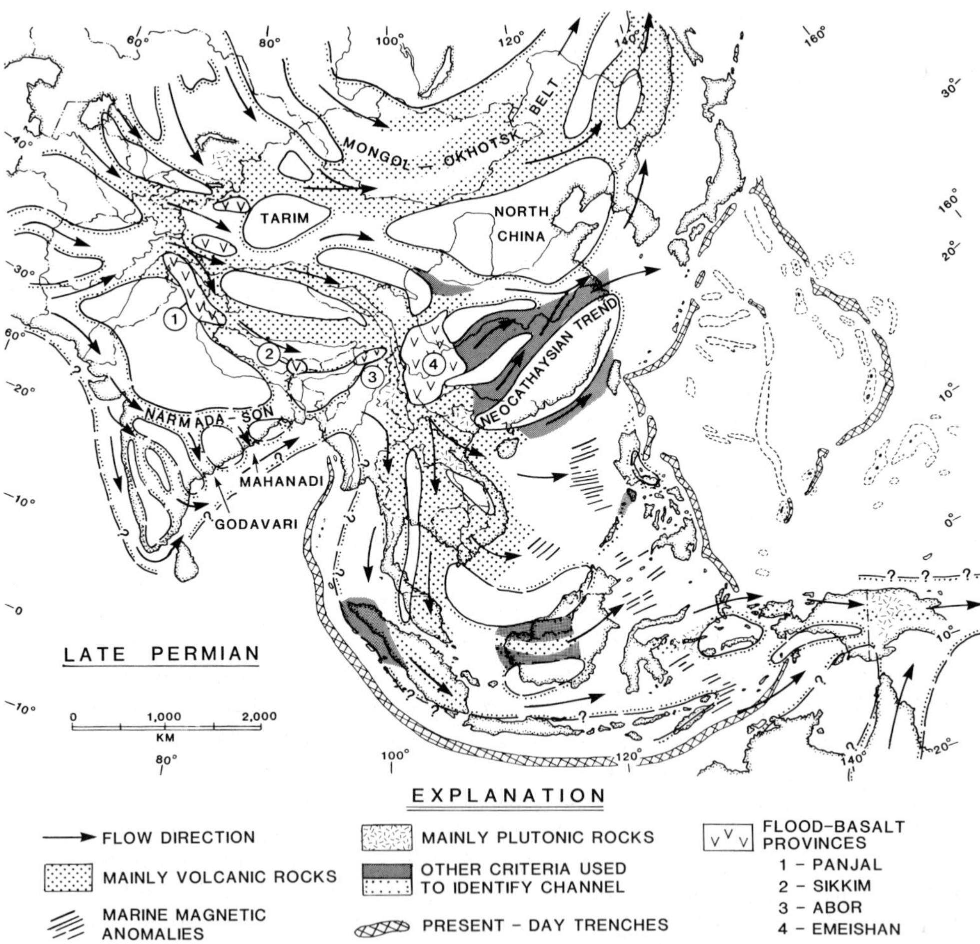

Fig. 5.11. Paleotectonic (surge channel) map for the Late Permian, southeastern Asia. Note anastomosing, braided, splay pattern that indicates west-to-east flow. From Meyerhoff, in press.

Much farther east in southwestern Papua New Guinea, O'Brien et al. (1961) found a north-northeast-striking band of magnetic anomalies, fractures, and Late Carboniferous through Triassic granitic plutons extending from the Cape York Peninsula of Australia into southwestern and central Papua New Guinea (Fig. 5.1). We interpret these linear features to be orogen-parallel structures related to northward and north-northeastward flow in the Tasman orogenic belt of eastern Australia. In Papua New Guinea, linear features gradually turn eastward to a nearly east-west direction. Hence the surge-channel flow of the Tasman orogenic belt at one time, at least, may have joined the principal eastward flow from southeastern Asia into the Pacific Basin (Meyerhoff et al., 1992b, Fig. 4.4).

Continental conditions predominated in northern China, Mongolia, and Asiatic Russia during Carboniferous-Permian times. Marine seaways lay mainly in central and southern China, Southeast Asia, and the Indonesian Archipelago. In the Mongol-Okhotsk geosynclinal belt of southern Asiatic Russia, however, a very narrow seaway persisted (Zonenshain et al., 1990, p. 102-104). West and east of the Siberian Platform in the West Siberian basin and Vilyuy syneclise-Verkhoyansk trough respectively, shallow marine seas persisted until early Late Carboniferous time (Zapivalov et al., 1978; Meyerhoff, 1982; Danilov and Budinkov, 1989; Tesakov, 1990; Zonenshain et al., 1990), to be replaced eventually by nonmarine Permo-Triassic fluviatile sediments (Kontorovich et al., 1975). Thus, by the end of Permian time, nonmarine conditions existed in all but the deepest basins north of central and southern China (Zonenshain et al., 1990). We note that, even though marine geosynclinal and platform regimes gave way to continental regimes, including the development of continental rifts (e.g., the West Siberian basin during Late Permian, Triassic, and Early to Middle Jurassic time; Meyerhoff, 1982, p. 470-480; Şengör, 1984, p. 34), *the surge-channel patterns did not change*. Hence the basic tectonic mechanism throughout this vast area remained unaltered, regardless of tectonic regime---"collisional" or extensional! Such a mechanism is not provided by plate tectonics.

As northern Eurasia rose and continental conditions took over in the north, major flood volcanism began from Western Europe to the eastern margin of the Siberian Platform. Carboniferous basaltic rocks were intruded and extruded in northwestern Europe (Glennie, 1986; Francis, 1988). Locally flood volcanism lasted into Permian time (e.g., Oslo graben). In Asiatic Russia, more than 2,000,000 km^2 of basaltic flows, sills, and dikes---up to 2 km thick---intruded and covered parts of the West Siberian basin (Zhuravlev, 1986) and the Siberian Platform (Kuznetsov and Naumov, 1975; Cherkasov, 1979; Beloussov, 1981). The Siberian trap volcanism began in Late Carboniferous time and lasted into Cenomanian time, with the bulk of the volcanic activity taking place during Late Permian and Triassic times. Thermal activity lasted well into Tertiary time. Thus the total documented time span for the volcanism alone is 213 m.y. (from 283 until 70 Ma; Kuznetsov and Naumov, 1975; Cherkasov, 1979; Zhuravlev, 1986). Several thousand radiometric dates from approximately 1,000 localities spread across 6,000,000 km^2 corroborate the statements made here.

Assertions such as those by Renne and Basu (1991) that all Siberian flood volcanism took place in one brief episode centered around 248-249 Ma do not consider the facts that (1) their samples came from a single section, and (2) the Siberian Platform flood volcanics have been dated from more than 1,000 localities spread over a 6,000,000 km^2 area! Renne and Basu (1991) focused their research on this 1 million year period in their search for an explanation for extinctions at the Permian - Triassic boundary. Although most of the flood volcanism is Late

Permian and Triassic, the older and younger traps associated with the Late Permian-Triassic rocks cannot be disregarded.

We suggest that the epeirogenic uplift that caused most late Paleozoic seas to withdraw from northern Asia resulted from the elevated heat flux produced by the invasion of the Siberian trap magmas into the northern Asian lithosphere. The trap magmas invaded a huge area---several million square kilometers---so that a very large region inevitably was heated and, therefore, involved in uplift.

Far to the south in southern China and in the northern part of the Indian subcontinent, smaller-scale, partly coeval flood volcanism also took place. The principal floods were the Panjal (and Phe) volcanics (>90,000 km^2), the Abor volcanics (>6,000 km^2), and the Emeishan Basalt (>300,000 km^2). They and a few other approximately correlative flood basalts are shown on Figure 5.11. The Panjal, Abor, and Emeishan are approximately the same age, mid-Permian (Bhat, 1984; Radhakrishna, 1987; Zhang Yunxiang et al., 1990), and hence are partly equivalent to the Siberian basalts.

The locations of at least three of the flood-basalt areas are noteworthy. The Panjal volcanics are associated with the Afghanistan Gap and the Western Himalayan syntaxis; the Abor and Emeishan volcanics are associated with the Yunnan-Myanmar Gap and the Eastern Himalayan syntaxis. This close spatial relationship implies that the increased velocity of the surge-channel magmas in the gap areas (following Bernoulli's Theorem), and, therefore, the increased pressure against the surge-channel walls, in part caused the flooding at each of these "choke points." The formation of the Emeishan flood province is of special significance, because it led directly to the creation of the Yangzi Valley surge channel along which important ore deposits were emplaced in Mesozoic time.

A possibly related feature---in terms of increased pressure within surge-channel walls---is the island of Sulawesi in the Indonesian archipelago. Like Halmahera just northeast of it (Figs. 3.41, 5.2), Sulawesi is a K structure with a breakout channel developed between its eastward-pointing arms. The presence on these arms of dated Late Carboniferous, Permian, and Triassic igneous rocks suggests that the Sulawesi breakout---a response to increases in pressure within surge channels---began during Carboniferous, possibly as late as Permian, time.

We propose that the principal cause of flood volcanism is very widespread---at times, worldwide---tectogenesis during which surge channels would be collapsed and destroyed to form bilateral foldbelts (kobergens). Tectogenesis thus would create great increases in the internal pressures inside of surge channels, and thereby force excess magma to form new channels very rapidly, or---more likely--- to escape to the surface. The timing, where known, of most flood volcanism coincides almost without exception with major tectogenesis. For example, the Midcontinent rift volcanism of central North America coincides with the late Proterozoic Ottawa tectogenesis. The Deccan flood volcanism coincides with one of the world's best-known worldwide tectogeneses, the "Laramide" orogeny and its equivalents (Campanian-middle Eocene). The Columbia River basalts of the northwestern United States were erupted beginning at about 17 Ma, or just after the commencement of the Himalayan tectogenesis whose effects in Asia need no elaboration. In the Western Cordillera of North America, a major tectogenesis was in progress at precisely the same time (middle Miocene). Thus it seems logical to attribute the Siberian, Panjal, Abor, and Emeishan basaltic floods to the great pressures built up by a worldwide series of Late Carboniferous-Permian-Early Triassic tectogeneses, the "Variscan" and "Hercynian" orogenies.

5.2.6 TRIASSIC-MIDDLE JURASSIC (FIG. 5.12)

Continental conditions spread during Triassic time in northern China (north of 37° N latitude), whereas marine conditions were widespread in the south. A major east-west geosynclinal complex occupied all of what is now the Qinghai-Tibet Plateau. At its eastern end the geosynclinal trend turns due south through the Yunnan-Myanmar Gap into Southeast Asia. The thickest and most diversified marine Triassic sedimentary section known on Earth was deposited in this geosynclinal complex. This section, up to 20 km thick (Huang and Chen, 1987) and much of it preserved in the 220,000-km^2 Songpan-Garzê basin, has been claimed to be a part of the proposed Paleotethys ocean (Şengör and Hsü, 1984). Actually it is a flyschlike deposit on continental crust (Huang and Chen, 1987, p. 29-37). Farther south, the Indus-Yarlung Zangbo eugeosynclinal belt began to form in Late Triassic time (Huang and Chen, 1987) above an east-west surge channel. Farther east, a shallow-marine Triassic sea on the Yangzi massif accumulated platform deposits. Locally within the massif, deep-water troughs formed.

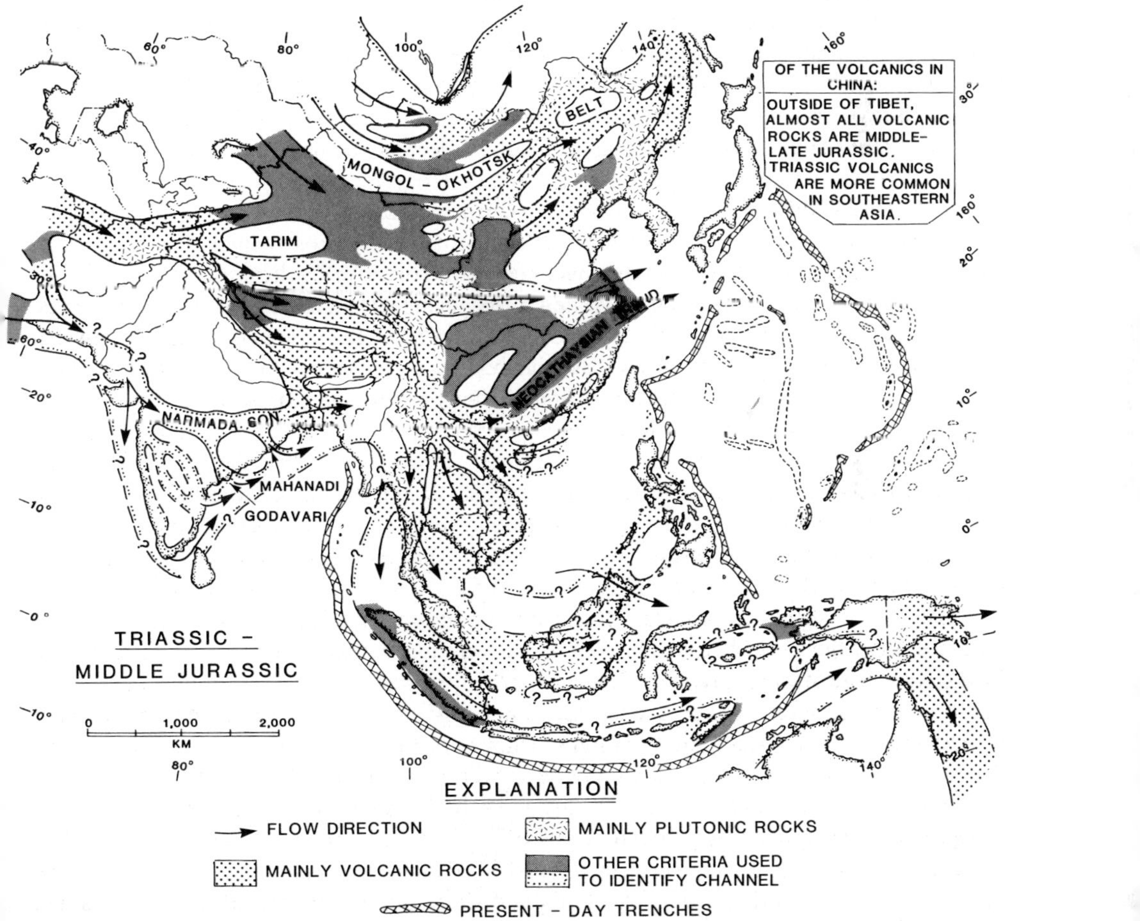

Fig. 5.12. Paleotectonic (surge channel) map for the Triassic-Middle Jurassic, southeastern Asia. Note anastomosing, braided, splay pattern that indicates west-to-east flow. From Meyerhoff, in press.

Triassic igneous rocks are scarce in northern China. Most Triassic igneous activity occurred in the geosynclinal complex beneath the present Qinghai-Tibet Plateau, in the Yunnan-Myanmar Gap, in Southeast Asia, and in the area of the present Indonesian Archipelago. In Southeast Asia, surge-channel activity pervaded the whole Indosinian massif (Figs. 5.2, 5.12). With the end of the "Hercynian" orogeny in eastern Australia, existing surge channels there were collapsed and tectonized; new channels formed. Although we have not investigated that area thoroughly, the data we do have from eastern Australia indicate that surge-channel flow reversed, from south-to-north (Australia to New Guinea) in Paleozoic time (Figs. 5.8-5.11) to north-to-south (New Guinea to Australia) after the Triassic Indosinian tectogenesis (Figs. 5.12, 5.17-5.18).

In the Philippines far to the north, the Palawan Ridge area became consolidated, and much of the southwestern Philippines appears to have become platformal (Mitchell and Leach, 1991; Amiscaray and Tan, 1992). Radiometric dates based on various methods range from 240 to 180 Ma (Middle Triassic-Early Jurassic) in this platformal region. On the Indosinian massif, the Khorat basin (plateau) may have originated in Triassic time (Cooper et al., 1989) as a breakout channel, but the basin reached its fullest development in Cretaceous time (Sattayarak, 1983). The economically important Southeast Asia tin granites were emplaced in Myanmar, Thailand, Cambodia, peninsular Malaysia, Sumatra, and southeastern China. In southeastern China, the north-northeast-trending Neocathaysian igneous-dominated trend was reactivated. Clearly, the Yunkai massif still exerted a strong deflecting effect on some of the magmas moving southeastward from the Yunnan-Myanmar Gap.

Early and Middle Jurassic volcanism was much more widespread than during Triassic time. The surge channels that had underlain the Mongol-Okhotsk geosynclinal belt in southern Siberia, Mongolia, and northern China were destroyed or severely damaged by "Hercynian" orogeny. By Early Jurassic time, however, new surge channels had formed. These grew during the remainder of Jurassic time when, in northeastern China, the newly reconstituted channel system merged with the "resuscitated" Neocathaysian surge-channel complexes that began to dominate eastern China for the first time since Ordovician-Silurian times (Figs. 5.8, 5.9).

A large breakout channel flowed easterly beginning in Triassic time from the Emeishan flood-basalt centers to the present Chinese coast. This channel, the Yangzi Valley surge channel, became the site of the Late Jurassic-Turonian-Lower Yangzi Valley plutonic-volcanic belt (Zhu Ming, 1989), an important source of metallic ores. This surge channel first manifested itself as a Triassic evaporite basin stretching from the modern Sichuan basin to the coast near Shanghai (Fig. 5.12; Huang and Chen, 1987, their Fig. 9). Neocathaysian trends became dominant once again in eastern China, presaging the intense volcanic activity with accompanying horst-and-graben formation that was to dominate eastern China from Late Jurassic time well in to the Tertiary. In fact, Wang Chengshan et al. (1993) attributed much of this Mesozoic volcanism to surge-channel activity. In India, the first mafic volcanism appeared in the Godavari rift, part of the Gondwana "splay" or "horsetail" rifts that appeared in Early Permian time (Figs. 5.10-5.12). This particular plutonic occurrence (at 170 Ma) and another in the Rajmahal depression (also 170 Ma) much farther northeast were forerunners of the younger Deccan flood volcanism of Late Cretaceous-early Tertiary time. By the end of Middle Jurassic time, almost all of southeastern Asia was emergent, with only a few exceptions. One exception was the east-west Indus-Yarlung eugeosyncline, and its north-south continuation into Burma (Ren Jishun et al., 1987, p. 90).

Indosinian tectogenesis produced many major tectonic features. We have mentioned briefly the Indus-Yarlung eugeosyncline that appeared during Late Triassic time at the crest of a large east-west surge channel just north of the Indian Platform. The Indus-Yarlung eugeosyncline, in our interpretation, formed as follows. At the same time that many other eastern Asian surge channels were filled, burst, and tectonized into kobergens (such as the kobergens of the Gangdisê Shan, Kunlun Shan, Tian Shan, Qin Ling, Malay Peninsula, and many others), the Indus-Yarlung surge channel was largely filled but insufficiently compressed to become a foldbelt. As the upper surface of the channel rose, it was stretched and a 2,400-km-long rift was formed during Late Triassic time (Huang and Chen, 1987). After sedimentary filling and later compression during the Yanshanian tectogenesis, the Indus-Yarlung eugeosyncline assumed its present form. The present straight course of the ultramafic rocks exposed along the Indus-Yarlung "suture" is a direct consequence of Poiseuille flow in the underlying active surge channel. This relatively straight fracture zone is a streamline produced by viscous drag at the roof of the Indus-Yarlung surge channel. The "ophiolites" are a mixture of ultramafic rock intruded early in the formational stage of the eugeosyncline and younger deep-marine Late Triassic through middle Cretaceous sedimentary rocks (Xigazê Group and related units; Ren and Chen, 1989). Thus the ultramafic assemblages of the Indus-Yarlung ophiolites may have originated in the same way as similar ultramafic intrusions in southwestern Asia, other parts of China, and Southeast Asia (Kashfi, 1976, 1988; Tan, 1976, 1981, 1984; Tan and Khoo, 1983, 1993; Stöcklin, 1984, 1989). Many of these "ophiolites" have been differently interpreted as sutures into which whole ocean basins were subducted (e.g., Şengör, 1984, 1989; Şengör and Hsü, 1984; Hutchinson, 1989a, 1989b). Stöcklin (1984, 1989) has shown why Şengör (1984, 1989) and Şengör and Hsü (1984) are not correct in their interpretation.

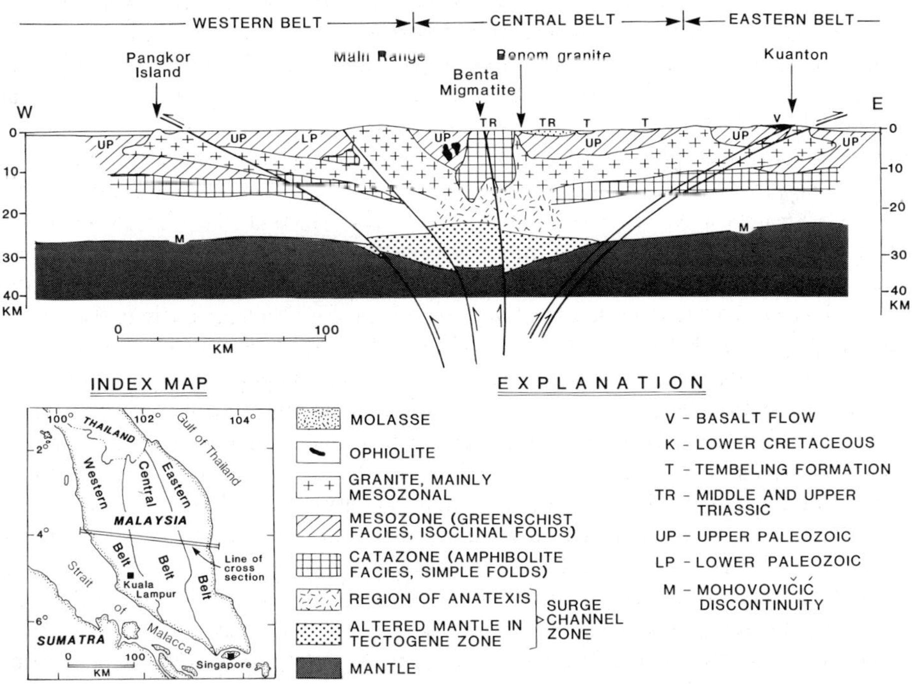

Fig. 5.13. West-east structural cross section of the Indosinian Malay Peninsula kobergen. From Hutchinson (1973).

For the most part, Indosinian tectogenesis generated kobergenic foldbelts rather than rifts such as the Indus-Yarlung eugeosyncline. Most of the foldbelts are intracontinental and, consequently, exhibit a germanotype tectonic style (horizontal and vertical movements are of nearly equal importance; steep reverse faults appear on both foldbelt flanks). Figure 5.13 is a structural cross section of the Peninsular Malaya kobergen from Hutchinson (1973). Tan (1976, 1981, 1984) correctly noted that the central part (Central Belt of southern peninsular Malaysia) is a region of extension. Figure 5.14 (from Tan, 1981, 1984; Khoo and Tan, 1983) shows the extensional character of the Central Belt and its surroundings. A comparison of Figures 5.13 and 5.14 shows that they are complementary. Figure 5.15, from Mitchell (1992), shows that the germanotype kobergen style persists northward into Thailand and Burma.

Chinese examples of germanotype Indosinian kobergens are so numerous that they would provide material for a research paper in their own right. Hence we have chosen a single example, the Tian Shan (Fig. 5.16), from Yao Huijin (1988). Yao's cross section shows, in cartoon form, the complete evolutionary cycle of the Tian Shan. The reader will note that surge tectonics---*without any change in stress system* (i.e., without reversals from extensional stress to compressional stress and *vice versa*)---explains this whole diagram within the framework of a single compressive stress field (compare Fig. 5.16 with Fig. 3.19).

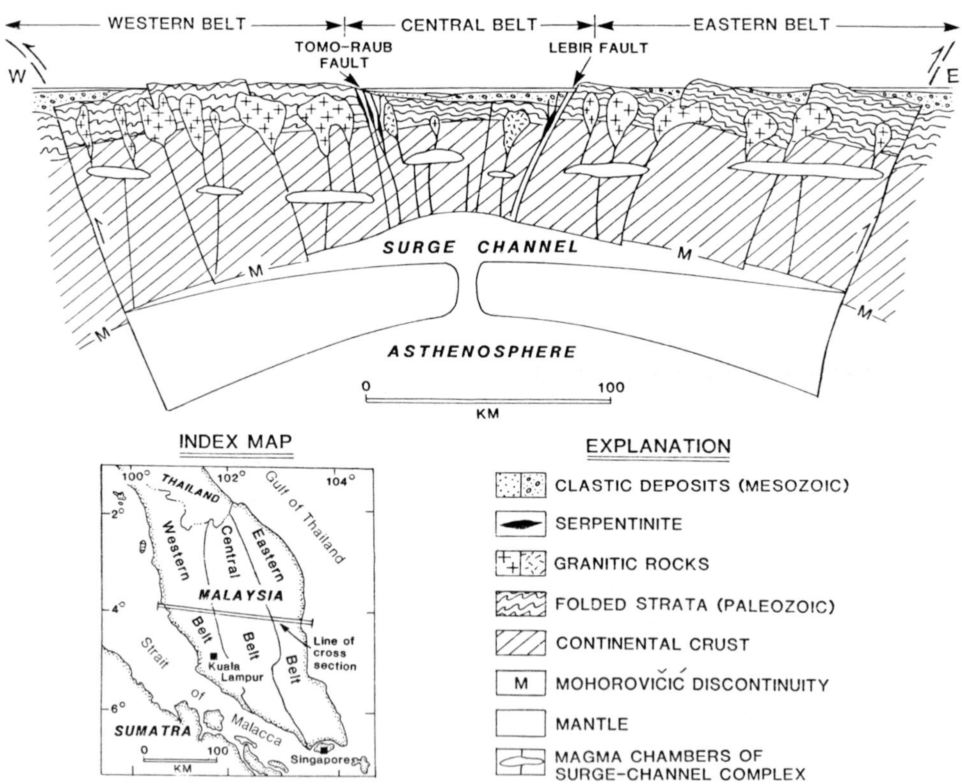

Fig. 5.14. West-east structural cross section of the crestal zone of the Malay Peninsula kobergen. Adapted from Tan (1981, 1984).

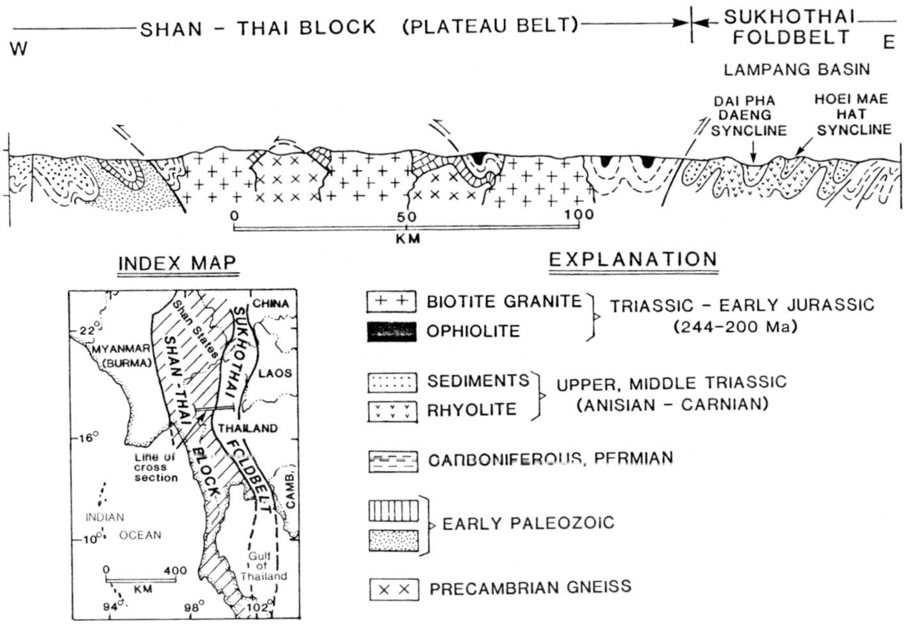

Fig. 5.15. West-east structural cross section of an Indosinian kobergen in Myanmar and Thailand. This kobergen is the northern extension of the Malay Peninsula kobergen shown on Figures 5.13-5.14. From Mitchell (1992).

5.2.7 Late Jurassic-Present (Figs. 5.17-5.18, 5.3)

The principal volcanic and other igneous activity during this time interval was in four "provinces": (1) eastern China, east of the North-South Zone, from the Song Hong fracture zone in Vietnam to the Primor'ye region of eastern Siberia (Fig. 5.1); (2) the Alpide mountain belt from Afghanistan to southern Myanmar, with branches into Vietnam and Thailand, south of the Yunnan-Myanmar Gap area; (3) the offshore magmatic or island arcs; and (4) the Deccan flood-basalt province of India.

The igneous activity in eastern China was intense, closely related to the formation of multiple northeast-striking Neocathaysian horst-and-graben systems during Late Jurassic and Early Cretaceous times. Igneous activity dropped off sharply after Cenomanian time (middle Cretaceous). The active igneous trends in eastern Mongolia also follow northeast and north-northeast trends. More than 90% (by volume) of the post-Middle Jurassic igneous rocks in eastern China are Late Jurassic and Early Cretaceous.

Marine sedimentation in eastern China was nil, with only a Late Jurassic seaway left in the northeastern part of Heilongjiang Province and a very short-lived middle Cretaceous transgression in the Songliao basin of the same province (Fig. 5.1; Chen, 1980; Wang Hongzhen, 1985).

Li and Luo (1988) described in general terms slightly more than 140 grabens and half grabens, which were Late Jurassic - Tertiary lake basins in eastern China. They presented

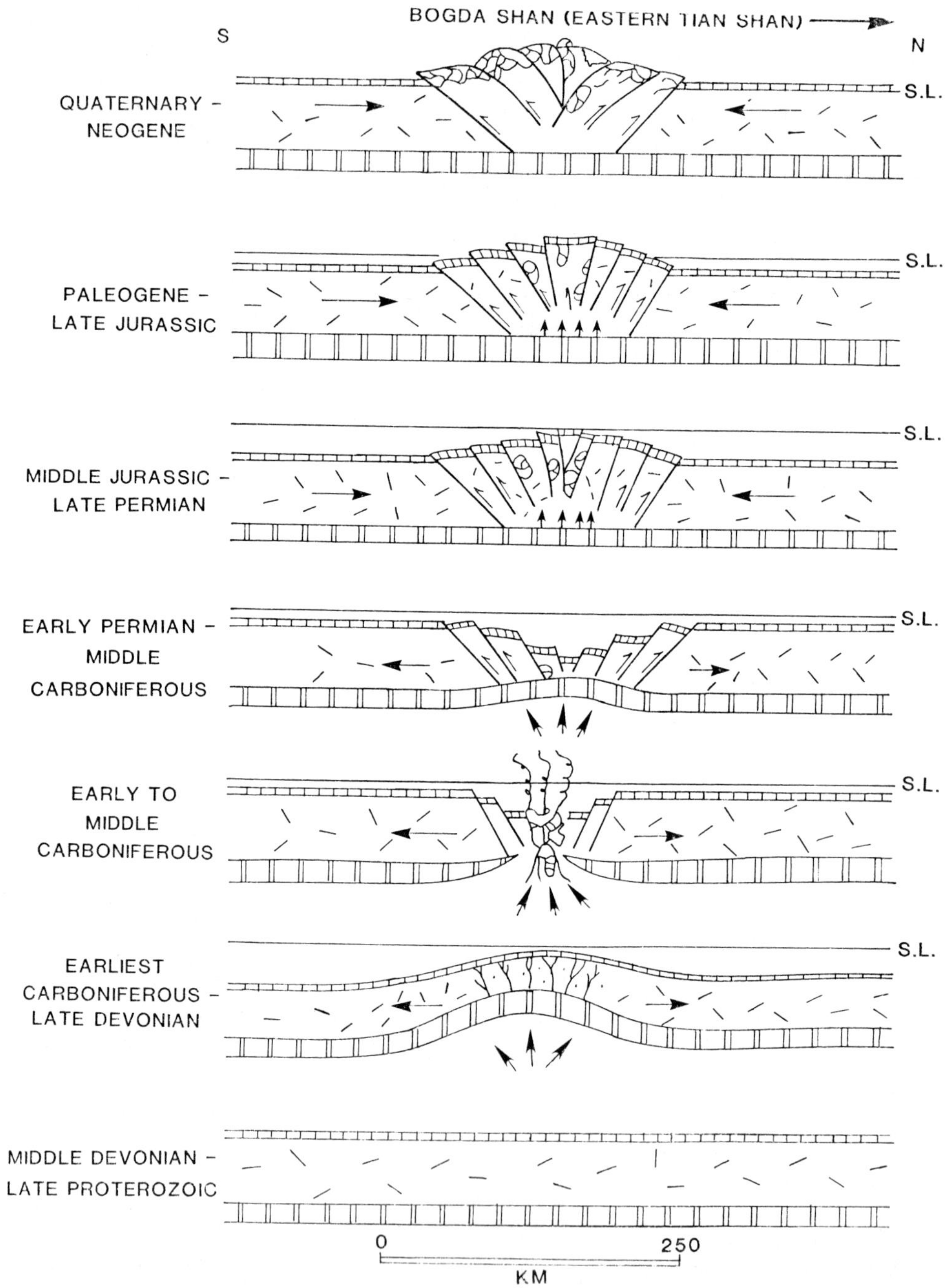

Fig. 5.16. South-north sections through the Bogda Shan (at eastern end of Tian Shan), showing the historical development into an Indosinian kobergen, and subsequently reactivated ("diwa" stage). From Yao Huijin (1988).

convincing evidence that all of eastern China was a region of very high heat flow (well over 55mW/m^2) during Late Jurassic and younger times (see Fig. 5.3).

Eastern China was not, however, a region undergoing simple tensile stress above a single surge channel complex. Multiple channels existed, as pointed out by Wang Chengshan et al. (1993) in their description of the surge-tectonics of eastern China. In many places, grabens were forming under tensile stress while nearby grabens were being mildly compressed, as is established by the many thrust systems and "flower structures" (small-scale kobergens) associated with the numerous north-northeast-striking Neocathaysian faults (e.g., Zhu Zhicheng et al., 1982; Meyerhoff et al., 1991). Major kobergens developed locally in Late Cretaceous-early Tertiary time. One very pronounced Yanshanian ("Laramide") kobergen underlies the Xuefeng Shan in western Hunan Province (Fig. 5.17; Chen Haihong et al., 1993).

Almost no volcanism took place in China west of the North-South Zone and north of the Qinghai-Tibet Plateau. What little volcanism occurred was in east-west-trending surge channels related to the anastomosing east-west surge channel complexes of pre-Late Jurassic time (Figs. 5.2, 5.6, 5.8-5.12).

In the Qinghai-Tibet region, the surge-channel complex that extended from Afghanistan to Indonesia via the Yunnan-Myanmar Gap continued to function. Most of the Qinghai-Tibet Plateau volcanism took place in association with Late Triassic to middle Eocene rifting along the Indus-Yarlung eugeosyncline. The marine seaway in this eugeosyncline withdrew in middle Eocene time as a result of tectogenesis, during and after which the largest mass of granitic rocks in eastern Asia was emplaced, the Transhimalayan batholith (Gansser, 1964, 1977, 1981, 1991). This batholith stretches east-west nearly 2,500 km through the Ladakh Range, the Gangdisê Shan, and the Nyainqêntanghla Shan. After middle Eocene time, volcanism diminished on the Qinghai-Tibet Plateau.

The uplift that created the present 4,500-m-high, 2.4-million-km^2 Qinghai-Tibet Plateau did not begin until Pliocene time (5 Ma). The rapid and steady uplift that is continuing today (4 to 5 mm/a) is faster than any other uplift rate yet measured (Gansser, 1983b). Clearly such uplift must be thermally driven. We propose the following explanation to account for the Qinghai-Tibet Plateau.

Compression during the early stages of the Himalayan tectogenesis (which began at about 24 Ma) induced the partial or entire caving in and blockage of the surge channels at the two key choke points, or syntaxes---the Western Himalayan (Nanga Parbat) syntaxis and the Eastern Himalayan (Namche Barwa) syntaxis (Fig. 5.2). Magma generated in the asthenosphere directly underlying the plateau eventually could no longer enter the surge channels beneath the plateau because the magma in them could not move eastward and southward into Southeast Asia. Because the Qinghai-Tibet surge-channels's magma could not move laterally, the magma had nowhere to go but up, which is precisely what is observed. Eventually the strength of the upper lithosphere above the Qinghai-Tibet surge channel complex will be exceeded, and collapse accompanied by severe tectogenesis will result. In fact, the large-scale thrusting that is taking place today south of the plateau may indicate that this severe tectogenesis already is in progress (Seeber et al., 1981).

In Southeast Asia south of the Yunnan-Myanmar Gap, new surge-channel splays formed, one of them through the Arakan Yoma in Myanmar, which extended farther southward into the Andaman-Nicobar Islands (Figs. 5.1, 5.17, 5.18). Foldbelt-parallel fault-fracture-fissure systems (in accordance with Stokes's Law) continued to form in active surge-channel areas. One such fault - fracture - fissure system has been recognized on the Thai - Malaysia border

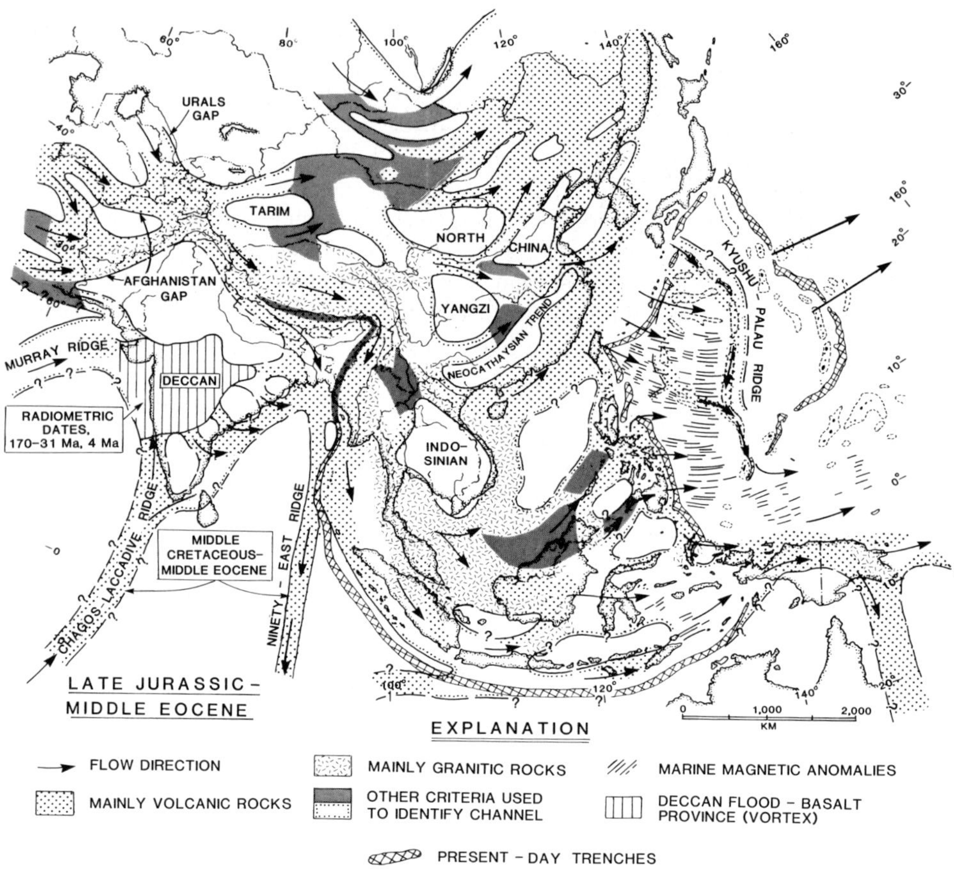

Fig. 5.17. Paleotectonic (surge channel) map for Late Jurassic-middle Eocene, southeastern Asia. Note anastomosing, braided, splay pattern that indicates west-to-east flow. From Meyerhoff, in press.

(Almashoor and Tjia, 1987). The activity of this surge-channel system underneath the Malay Peninsula is further demonstrated by the fact that volcanic breakout channels still form in this region from time to time. One such breakout possibly is the so-called "East Arm," which broke eastward from the peninsula after middle Eocene time (Fig. 5.18; T. T. Khoo, written commun., 24 March, 1993).

In the Indonesian Archipelago, the Banda Trench seems to have worked its way eastward into the Arafura Platform (Fig. 5.2), gradually severing pieces of continental lithosphere from the Australia-New Guinea landmass. Many of the surge channels of Indonesia have been active for more than 200-300 m.y. as shown by the multiple ages of the volcanic and plutonic rocks associated with them (Figs. 5.6, 5.8-5.12, 5.17-5.18). Multiple ages for the volcanic rocks associated with each channel suggest stability, not a constant shifting by means of postulated wrench faulting or motions related to "spreading centers." In fact, several mélanges in Indonesia (e.g., Java) contain sizable clasts of granitic rocks. Ketner et al. (1976) described these mélan-

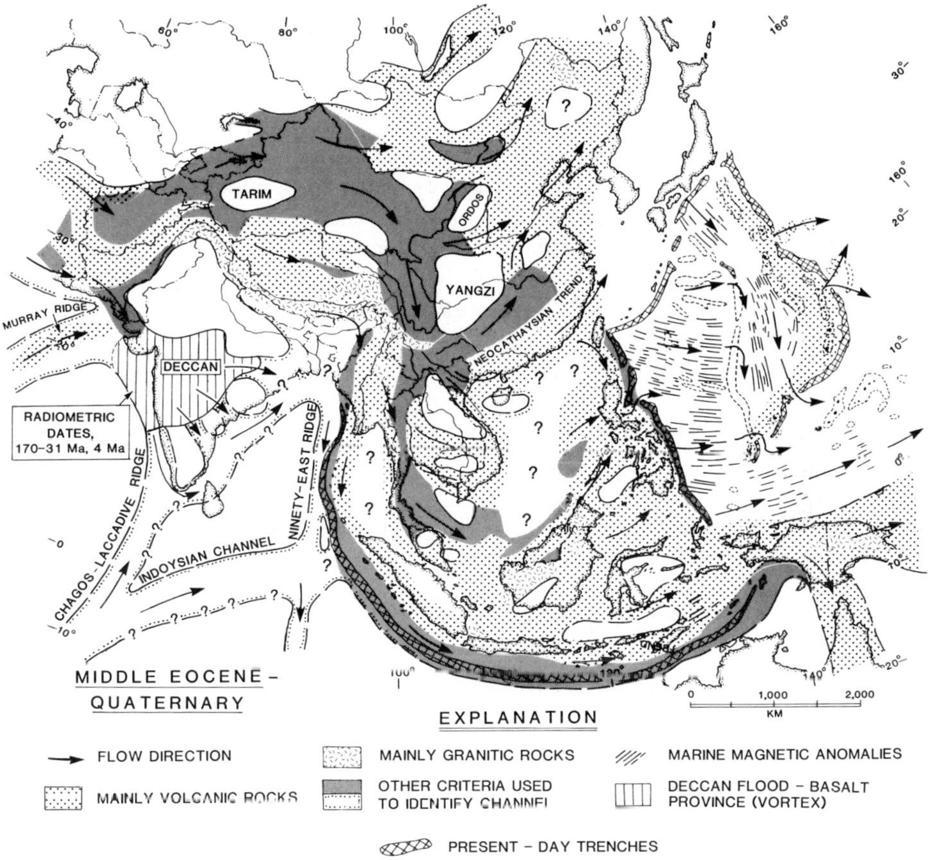

Fig. 5.18. Paleotectonic (surge channel) map for late Eocene to present, southeastern Asia. Note anastomosing, braided, splay pattern that indicates west-to-east flow. From Meyerhoff, in press.

ges and observed that the presence of granitic rocks in tectonic mélanges is inconsistent with plate-tectonic models. Elsewhere, increasing internal stresses within the channels produced a K structure in Halmahera (Fig. 3.41B; Hall and Nichols, 1990), a very good example of an eastern breakout channel. Many island arcs of this region are not true arcs but a row of straight-line segments (Ranneft, 1979).

In the western Philippine Sea, a whole "family" of east-west-striking magnetic anomalies is present (Figs. 5.2, 5.17-5.18). The anomalies underlie and parallel several submarine ridges which we interpret to be breakout channels from the Ryukyu-Taiwan-Philippine arc system on the west. Faunal and radiometric data recovered from boreholes on these east-west ridges show that they are at least as old as Early Cretaceous, and have an age range of Early to Late Cretaceous (Ozima et al., 1980). The anomalies could easily be older, because they are on strike with anomalies believed to be as old as Carboniferous (Figs. 5.2, 5.10). Like all oceanic anomalies, they show the characteristics of streamlines produced by viscous drag (Stokes's

Law), and therefore demonstrate that massive west-to-east surge-channel (and possibly upper asthenosphere) flow has taken place here from beneath the Asian continent (Meyerhoff and Meyerhoff, 1977).

Figure 4.4 (from Meyerhoff et al., 1992b) shows an interpretation of the magnetic anomaly patterns of the whole Philippine Sea and the regions south and southeast of it. The magnetic anomalies of the eastern Philippine Sea between the Kyushu-Palau Ridge on the west and the Mariana and Bonin arcs on the east strike north-south. The anomalies of eastern Indonesia strike east-northeast. Those of the Bismark and Solomon Seas also strike northeasterly. Meyerhoff et al. (1992b) suggested that the surge-channel (and, possibly the asthenosphere) flow that produced these anomalies converged on the West and East Caroline Basins, continuing beneath the Ontong Java Plateau eastward to (and beyond) the Phoenix magnetic anomalies (Fig. 4.4). The lavas and sills of the Ontong Java Plateau thus far dated range in age from Barremian to Campanian (129-83 Ma; Mahoney et al., 1993), and therefore are of the same age range as the material sampled from the east-west ridges of the western Philippine Sea (Ozima et al., 1980). Flow is still taking place in the region as evidenced by the earthquake and other activity of the surge channels of the Mariana and Bonin arcs, Banda arc (Timor region of Fig. 5.2), New Guinea, Solomon Islands, Tonga arc, and North Fiji Basin. The North Fiji Basin is interpreted as a large vortex (Meyerhoff et al., 1992b).

Far to the northwest in India, the Deccan traps cover a 500,000-km^2 flood-basalt province in northwestern India (Figs. 5.17, 5.18). The literature written on the Deccan Traps, especially the speculative papers, is enormous. Many of the papers are written to "prove" that (1) the Deccan basalts were extruded in a 1- to 4-million period of time, and (2) the gases emitted during this short time would cause the extinctions at the Cretaceous-Tertiary boundary. For example, Basu et al. (1993) claim that the Deccan Traps were extruded between 68.5 and 65 Ma to produce the worldwide extinctions. As in the case of the Siberian flood basalts, wide-

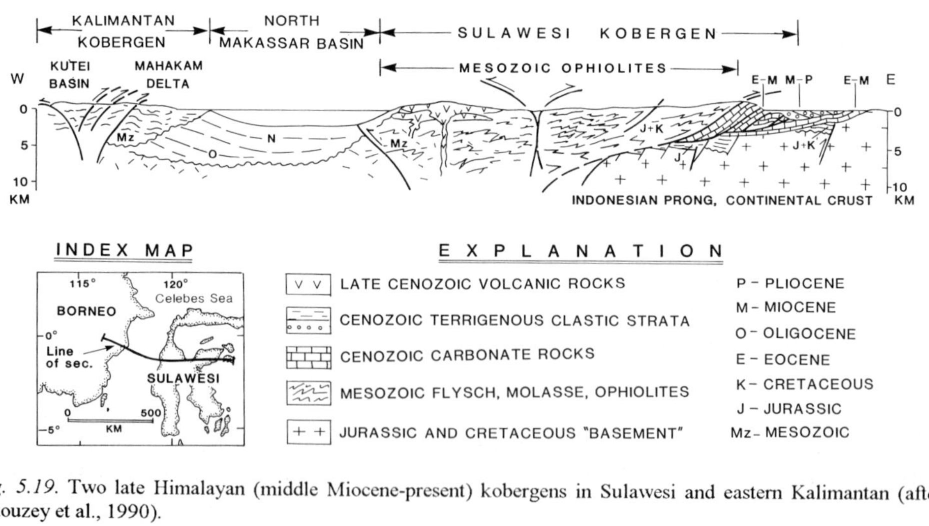

Fig. 5.19. Two late Himalayan (middle Miocene-present) kobergens in Sulawesi and eastern Kalimantan (after Létouzey et al., 1990).

spread sampling and dating of the Deccan Traps show that they were extruded over a long age span, in this case between 102 and 20 Ma, with precursor activity as old as 170 Ma and with one possible extrusion as young as 4.2 Ma (Alexander, 1981; Saxena, 1986; see the extensive bibliography on Deccan radiometric dating contained in the Saxena [1986] paper). All researchers advocating a short 1- to 4-m.y. period of extrusion (e.g., Duncan and Pyle, 1988; White and McKenzie, 1989) have not considered the literature on radiometric dating by many workers. Additionally, they have not considered the contradictory paleontological ages obtained from faunal and floral collections. If one honors all of the data, it is clear that the first Deccan Traps were extruded in Albian (late Early Cretaceous) or earlier time, that extrusion reached a maximum rate in Maastrichtian and Danian times (66.5-60.7 Ma; Jaiprakash et al., 1993), and that extrusions continued on a modest scale well into Eocene and perhaps later time (Saxena, 1986).

Field data on the Jurassic through Eocene dike patterns in India indicate that the Deccan floods were independent of the Rajmahal flood basalts of northeastern India which were extruded first between 170 and 155 Ma (Middle and Late Jurassic) and later between 112 and 70 Ma (Albian-Maastrichtian; Alexander, 1981; Saxena, 1986). The Rajmahal flood basalts occupied a channel separate from that of the Deccan Traps (Fig. 5.17).

To explain the origin of the Deccan Traps, we note the following. Dikes coeval with the Deccan Traps in northwestern India strike east-southeast through Kutch and the Kathiawar Peninsula from the active surge channel beneath the Murray Ridge (Fig. 5.17; Murthy, 1987). A second set of coeval dikes strikes north up the western coast of India from the Chagos-Laccadive Ridge (Murthy, 1987). The two dike sets (and the underlying surge channels) "collide" just north of Bombay where the two directions are inextricably mixed in a chaotic manner (Murthy, 1987). In surge tectonics, this chaotic structure is a vortex. The Deccan

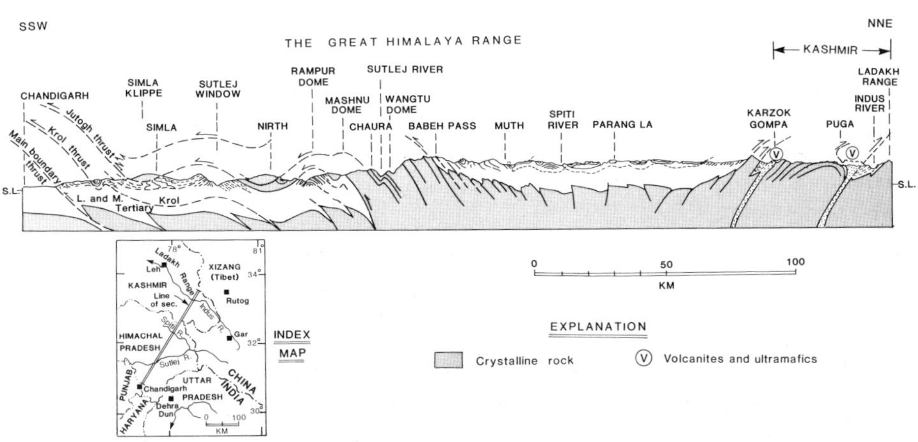

Fig. 5.20. South-north structural cross section of a Himalayan (late Eocene to the present, mainly Himalayan) kobergen in the western part of the Himalayas (from Gansser, 1964). Note that the style of tectogenesis changes from the southern flank (alpinotype) to the northern flank (germanotype). The reason for this style change is explained in the text.

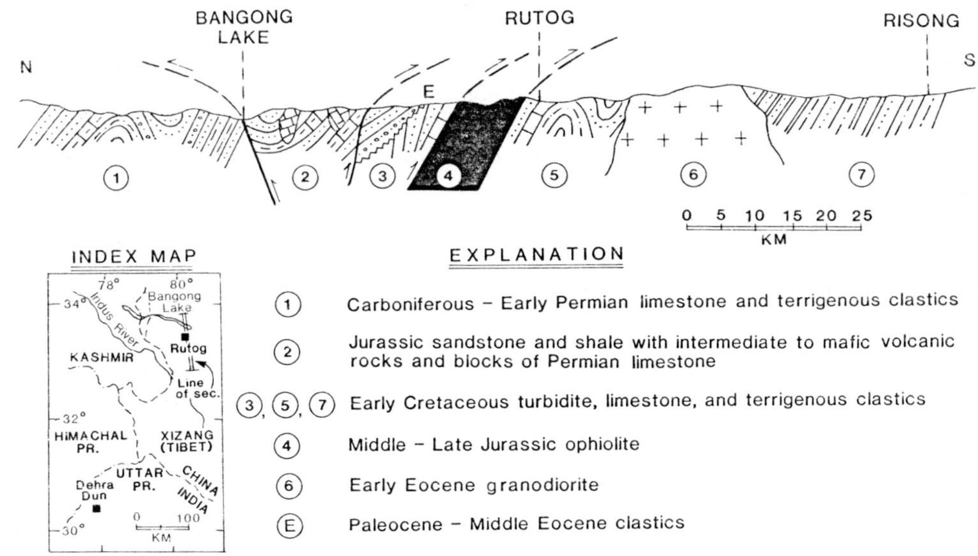

Fig. 5.21. South-north cross section on Qinghai-Tibet Plateau near Rutog. The southern end of this cross section nearly overlaps with the northern end of Figure 5.20. The Rutog kobergen is the next kobergen north of the Himalayan kobergen (whose root is in the Indus-Yarlung fault zone). Note germanotype style. From Yang Jingsui et al. (1990).

flood-basalt province, in fact, has all the criteria characteristic of a large vortex. Murthy's (1987, p. 395) mapped dike pattern shows one more interesting feature. East of the Deccan vortex, the dike swarms follow the Son-Narmada rift and join the Rajmahal channel confluently (Figs. 5.10-5.12, 5.17-5.18).

The Chagos-Laccadive Ridge has been interpreted to be a "hot-spot trace," as has the Ninety-East Ridge (Figs. 5.17, 5.18; Luyendyk, 1977; Duncan and Hargraves, 1990), and both paleontological and radiometric data have been used to support this interpretation (McDougall, 1974; Rundle et al., 1974; Luyendyk, 1977; White and McKenzie, 1989; Duncan and Hargraves, 1990). We suggest a different interpretation on the basis of (1) insufficient radiometric data, (2) the ridge's position on the east side of the Mid-Indian Ridge, and (3) the ridge's shape (curving northeastward toward India; see Figs. 5.17, 5.18). We suggest that the Chagos-Laccadive Ridge is a feeder channel of possibly Early Cretaceous age, with flow from southwest to northeast.

The Ninety-East Ridge is best known for its nearly straight north-south course for 5,500+ km from the Bengal Fan in the north to the Southeast Indian Ridge, southwest of Australia (Figs. 5.17, 5.18). It too is interpreted as a "hot-spot trace." The ridge is thought to become younger from north to south (Luyendyk, 1977). This interpretation is supported by the age of the sedimentary rocks that overlie the basalts of the channel, but is not supported by the radiometric dates obtained from the various Deep Sea Drilling Project sites (McDougall, 1974; Rundle et al., 1974). We interpret the ridge to be a north-to-south breakout channel from the Indian subcontinent. It is still an active ridge, with high heat flow (Geller et al., 1983), seismicity (Levchencko, 1989), and a 7.1 km/s layer (Detrick and Watts, 1979).

A new surge-channel complex seems to have begun to form since Miocene time in the northeastern Indian Ocean. This has been studied in considerable detail, because it forms a broad belt of high heat flow, high microseismicity, and block-fault-type topography between the Chagos-Laccadive and Ninety-East Ridges (Fig. 5.18; Neprochnov et al., 1988; Wezel, 1988; Weissel et al., 1992). In addition, east-northeast-striking ridges in this belt show steep-angled reverse faults, which are still active. Accordingly, Wezel (1988) named this broad belt the *Indoysian foldbelt* (Fig. 5.18). The Indoysian surge channel, as we call it, seems to be a breakout channel (feeder channel) from the eastern flank of the Chagos-Laccadive Ridge. Its eastern terminus is unknown; it may be the Ninety-East Ridge or the Indonesian magmatic arc (Fig. 5.18).

Also during the Late Jurassic-to-present time interval, Yanshanian and Himalayan tectogeneses took place in the mobile belts of eastern Asia. The most striking foldbelts produced during the Yanshanian and Himalayan tectogeneses are the alpinotype foldbelts that characterize the surge channels of the island-arc complexes. Germanotype kobergens, however, also were formed in the continental interiors, some of them being older Indosinian kobergens rejuvenated during the younger tectogeneses. We illustrate some typical kobergens from the Jurassic-present time interval.

Figure 2.17, from Létouzey et al. (1988), shows the Taiwan kobergen in the north, three sections across the Luzon kobergen farther south, and a section across the Palawan kobergen near Mindoro Island. These show many germanotype characteristics but numerous alpinotype characteristics are well developed (e.g., low-angle thrusts, presence of ultramafic complexes). These are late Himalayan (mainly middle Miocene-recent) features.

Figure 5.19, from Létouzey et al. (1990), shows two parallel, late Himalayan, alpinotype kobergens in Kalimantan (Borneo) and Sulawesi. Figure 5.20, from Gansser (1964), shows a classic Himalayan cross section. The kobergen is germanotype on the north, alpinotype on the south, the different styles reflecting the differences in lithosphere thickness and depths to basement on either side of the kobergen (Meyerhoff et al., 1991). Figure 5.21, from Yang Jingsui et al. (1990), shows the next kobergen north of that shown in Figure 5.20 (compare the locations of the two sections on the index maps for the two cross sections). This example, the Rutog kobergen, shows germanotype style. An example of a rejuvenated Indosinian germanotype kobergen (rejuvenated during both the Yanshanian and Himalayan tectogeneses) is seen on Figure 5.16, the Tian Shan. The rejuvenation shown is an example of Chen Guoda's (1989, 1992) "diwa" hypothesis of rejuvenated foldbelts.

CHAPTER 6

MAGMA FLOODS, FLOOD BASALTS, AND SURGE TECTONICS

6.1 Introduction

Tyrrell (1937) brought into the literature the term "flood basalts," to give substance to a concept that he felt would improve the understanding of the effusion and flow of basaltic lavas. Tyrell was dissatisfied with the term "plateau basalt," which had been current for more than a half century because it " . . . merely refers to an accidental and inessential feature resulting from levelling-up, erosion or Earth movement." Tyrrell was aware, however, that his predecessor by more than a half century, Sir Archibald Geikie, understood fully the dynamics and the framework of "plateau basalt" (Geikie, 1887), for he noted: "Sir A. Geikie actually mentions the word 'flooded' . . . ." However, Tyrrell then commented "...I hold that 'flood basalt' is a much more fitting designation for this colossal type of basalt accumulation . . . ."

The older term, "plateau basalt", had functioned with comparative efficiency and illustration, but Tyrrell's "flood basalts" gave an immediate and striking image of basalts poured out in broad areal effusion. "Plateau basalt" has continued in the literature to a considerable degree, but "flood basalts" has become by far the preferred term in mafic volcanology.

The origin of flood basalts has sparked controversy since they were first identified in the last century. The purpose of this chapter is to re-examine the critical data, including descriptions of many flood-basalt provinces, to introduce the new term "magma floods" for flood basalts--a term that we consider more appropriate and encompassing--and to propose an explanation of our own in terms of surge tectonics.

6.1.1 Significance of flood basalts

Some 63% of the ocean basins are covered with flood basalts. At least 5% of the continents are likewise covered with flood basalts. Thus 68%----a minimum figure----of the Earth's surface is covered with these basaltic rocks. Flood basalts, then, are not the oddities that many suppose them to be. In spite of this, they receive little attention among the scientific community. We examined nearly twenty geologic textbooks and reference works published since 1969, and found only two with more than three paragraphs on flood basalts. Most dealt with the topic in a single paragraph. Four textbooks did not mention the topic. Seven used only the old term "plateau basalts." Among the books that we examined were those by Verhoogen et al. (1970), Wyllie (1971), Billings (1972), Gass et al. (1972), Hills (1972), Le Pichon et al. (1973), Press and Siever (1974), Hobbs et al. (1976), Flint and Skinner (1977), Levin (1978), McElhinney (1979), Dennis (1982), Miyashiro et al. (1982), Boardman (1983), Davis (1984), Ragland and Rogers (1984), Stanley (1986), and Wicander and Monroe (1989). Such treatment----or lack of treatment----seems unusual/out of place if one considers that *flood basalts are the most important rock exposed at the Earth's surface* (Menard, 1986, p. 207).

Engel et al. (1965) long ago demonstrated that deep ocean-floor tholeiitic basalts are the oceanic equivalent of the continental flood basalts. The Basalt Volcanism Study Project (1981) differentiated between the continental flood basalts and "ocean-floor basalts," while recognizing that the principal differences were the abundance of minor and rare-earth elements. Press and Siever (1974, p. 546, 567-570) recognized the fact that the ocean-floor basalts and continental flood basalts are nearly the same, and that their differences are explained readily by contamination in the continental crustal setting. Yoder (1988), one of the world's authorities on basaltic magmas, stated essentially the same thing.

In fact, as increasing numbers of basalts are analyzed, the difference between the oceanic and continental floods blurs even further. For example, Al'mukhamedov and Zolotukhin (1991) found groups of samples from the Siberian Traps that are essentially indistinguishable from midocean ridge basalts. Fitton et al. (1991) found numerous Great Basin basalts that are chemically indistinguishable from midocean ridge basalts, and Sawlan (1991) observed a complete chemical continuum from midocean ridge basalts to the flood basalts in the Baja California, Gulf of California, and Mexican basin-and-range province.

These extremely close--in places identical--genetic relationships are well established. In a subsequent section of this chapter, we shall present geochemical data to support this statement.

6.1.2 Classification

Continental flood-basalt provinces are geometrically of two types. The first is broadly *ovate,* or even round, with the maximum diameter ranging from about 500 km (Columbia River Basalt) to more than 2,500 km (Siberian Traps). The second is distinctly *linear*, with a width of 100 to 200 km and lengths up to and even exceeding 3,000 km.

Oceanic flood-basalt provinces at first appearance are difficult to classify. However, as more seismic-refraction and bathymetric data--as well as seismotomographic data--become available, it is possible to distinguish the same two types of geometries there as well. Ovate to semi-ovate shapes characterize many oceanic submarine plateaus. The maximum diameters of these plateaus, excluding the Kerguelen Plateau, are in the order of 1,200 to 1,600 km.

Linear ridges are of two types. The larger is the midocean-ridge system with widths between 1,200 and 3,600 km; the smaller is exemplified by the various linear island and seamount chains with widths of 100-200 km and lengths of thousands of kilometers.

Ovate flood-basalt provinces include the Siberian Traps, the Deccan Traps, the Antrim Plateau Volcanics, the Ontong-Java Plateau, the Manihiki Plateau, the Columbia Plateau, the Matagalpa-Coyol (Nuclear Central America) Volcanics, the Iceland Plateau, the Maranhao-Parnaiba Basin, the Paraná Basin, the Entendeka Volcanics, the Drakensberg (Karroo) Volcanics, and the Ethiopian Plateau basalts. This list is not complete.

Linear flood-basalt provinces include the Syrian-Arabian Peninsula volcanic fields, Ninety-East Ridge, the Willouran Mafic Province, the Louisville Ridge, the Hawaii-Emperor Ridge, Wrangellia, the Midcontinent (Keweenawan) Rift, the Brito-Arctic (Thule) volcanic province, the Cape Smith Foldbelt, the Labrador Trough, the Mexican Volcanic Belt, the Patagonian lava plateau, the Kirkpatric Basalt-Ferrar Dolerite trend in the Transantarctic Mountains, and the composite of linear rifts that comprise the East African Rift system from the Dead Sea Rift to the Zambezi Delta. Again, this is not a complete listing.

Tectonism and metamorphism can severely disrupt any flood-basalt province after its formation. For example, epeirogeny has uplifted and block-faulted the whole of the Antrim Plateau Volcanics of northern Australia. As a consequence, parts of the original province have been removed by erosion. Yet study of the geological map of Australia indicates that the now separate outcrop areas of the Antrim Plateau Volcanics must once have been joined. Similarly, only very scattered, strongly flooded, and metamorphosed remains of the Willouran Mafic rocks are preserved in the Adelaide Geosyncline of South Australia, but their distribution shows that the Willouran Mafic Province is a linear flood-basalt province.

6.1.3 The Petrographic Character of Flood-Basalt Provinces

To judge from the geological literature, many earth scientists assume that flood-basalt provinces are composed mainly of basalt and little else. This characterization is justified for some provinces but is incorrect for many more. For example, the Columbia River flood-basalt province consists nearly 100% of tholeiitic basalt with small volumes of basaltic andesite and minuscule amounts of dacite and rhyolite (Swanson et al., 1979; Carlson and Hart, 1988; Hooper, 1988). In contrast the Snake River flood-basalt province on the southeastern side of the Columbia River province consists more than 50% of rhyolite and siliceous (rhyolitic) ignimbrites (Bonnichsen, 1989). A second example is the Lebombo monocline region of the Karroo flood-basalt province in southern Africa. Here are thick sequences of rhyolite (and perhaps ignimbrite) which, for most of the length of the monocline--at least 600 km--comprise 30 to 55% of the volcanic section (Dingle et al., 1983). Yet another example is the Keweenawan (Midcontinent) flood-basalt province where every region has large volumes of rhyolite associated with the basalt. Our first point is that *many flood-basalt provinces are bimodal*, and the volume of associated silicic extrusive (or intrusive) rocks can be substantial.

A second common assumption is that tholeiitic basalt and related tholeiitic rocks constitute the principal mafic rock types. Here again, the field evidence proves the assumption is incorrect. It is true that the Columbia River flood-basalt province consists 99% of tholeiitic mafic rocks. Yet the huge, 3,000-km-long Arabian flood-basalt province consists mainly of alkalic basalt. In fact, Camp and Roobol (1989) and Camp et al. (1991) refer to this example as the "Arabian continental alkalic basalt province." Thus our second point is that *many types of basalts may be present in flood-basalt provinces*. Tholeiitic basalt is just one of those types.

6.2 Descriptions of Selected Continental Flood-Basalt Provinces

We present here some brief geological descriptions of representative ovate and linear continental flood-basalt provinces in order of decreasing age. Many additional continental provinces could have been added to this list, but we believe that those selected adequately illustrate the points we wish to make. Undoubtedly, some earth scientists will not agree that all of our examples are, in fact, flood-basalt provinces. Therefore, we include data on areal extent, volume, thickness, composition, and age which led us to conclude that we were dealing with flood-basalt provinces. Data concerning the ages, areal extent and volume of these provinces and others are summarized on Table 6.1.

6.2.1 Keweenawan Flood Basalt (Proterozoic), Midcontinent United States (Linear)

The Keweenawan flood basalts are an example of a linear continental flood-basalt province of Proterozoic age. Total area and total volume of magma in these flood basalts, although clearly less than those of major continental flood basalts elsewhere in the world, enjoy a unique

Table 6.1. Ages, areal extent, and volume of major and minor magma floods.

NAME	PALEONTOLOGICAL	RADIOMETRIC	AREA(KM²)	VOLUME (KM³)	SOURCE(S)
Venters Dorp SuperGroup	Late Archean-Early Proterozoic	>2,700-2,643	300,000	450,000	Cahen et al. (1984); van der Westhuizen et al. (1991)
Cape Smith foldbelt	Early Proterozoic	1,920-1,871	13,000	52,000-65,000+	Beall et al. (1963); Easton (1983); Francis et al. (1983); Doig (1983,1987); Hoffman (1989)
Coppermine River	Middle Proterozoic	1,270-1,215	31,000-46,000	60,000-90,000	Douglas (1970); Baragar (1972); Dostal et al. (1983); Hildebrand and Baragar (1991); Sevigny et al. (1991)
Keweenawan/Mid-Continent Rift	Middle Proterozoic	1,225-1,040	120,000-200,000	400,000-1,300,000	Annells (1973); Halls (1978); Basalt Volcanism Study Project (1981); Van Schmus et al. (1987); Hutchinson and White (1990)
Sierra Ancha Diabase	Middle Proterozoic	1,140-1,150	100,000+	100,000+	Wrucke (1989); Howard (1991)
Willouran Mafic Province	Middle-Late Proterozoic	1,076-830 (>802)	210,000	100.000-420,000	Crawford and Hilyard (1990); Hilyard (1990)
Antrim Plateau	Late Proterozoic-Early Cambrian	666-654	450,000	100,000	Bultitude (1976); Basaltic Volcanism Study Project (1981); Veevers (1984)
Northwestern Europe	Latest Carboniferous-Early Permian	295-265 (peak 294-283)	130,000	Unknown >26,000;	Dixon et al. (1981); Russell and Smyth (1983); Sundvoll et al. (1990)
	Middle Jurassic-Cenomanian	175-94			
Siberian Traps — Siberian Platform — Tholeiitic + Palagonitic	Early Permian-Middle Triassic	283-240			Pol'kin (1973); Kuznetsov and Naumov (1975); Starosel'tsev (1980); Dagis and Kazakov (1984); Zhuravlev (1986)
Siberian Traps — Siberian Platform — Qtz-bearing + Normal Diorite	Middle Triassic-Early Jurassic	240-200	650,000	1,500,000	
Siberian Traps — Siberian Platform — Tholeiitic Dolerites	Early Jurassic-Maastrichtian	198-70	600,000		Kuznetsov and Naunov (1975); Cherkazov (1979)
Siberian Traps — West Siberian Basin	Late Permian-Middle Jurassic	254-175	750,000	? 300,000	Zhuravlev (1986)
Emeishan Basalts of Panxi Rift	Late Permiam-Middle Triassic	255-235 (? 219)	300,000-500,000	280,000	Wang Yunliang et al. (1989); Zhang Yunxiang et al. (1990)
Amazon	Middle Triassic-Valanginian	238-135		340,000	Mosman et al. (1986)
Newark	Middle Triassic-Albian	234-111 (211-175 for sills and flows only)	18,000-78,000	5,000-17,000	Manspeizer et al. (1978, 1989); Sutter and Smith (1979); Lanphere (1983); McHone (1984); McBride et al. (1989);

NAME	PALEONTOLOGICAL	RADIOMETRIC	AREA(KM²)	VOLUME (KM³)	SOURCE(S)
Venezuelan Andes-Cordillera Oriental	Late Triassic-Hauterivian	229-127	Several thousand	Unknown. Thicknesses up to at least 1:6 km	Irving (1975); Burkley (1976); Gonzalez de Juana et al. (1980); MacDonald and Opdyke (1984); Maze (1984); Steinitz and Maze (1984); Jaillard et al. (1990)
Wrangellia	Middle to Late Triassic (Ladinian-Carnian)	236-228	450,000	900,000-1,350,000 (Panuska says 950,000)	Jones et al. (1977); Panuska (1990)
Kirkpatric Basalt-Ferrar Dolerite	Late Triassic-Aptian (Flows are mainly Middle Jurassic)	223-114 (Flows are 182-162 Ma)	100,000-400,000	400,000-500,000	Gunn (1962); Elliot (1970, 1972);; Faure et al. (1979, 1982); Ford and Kistler (1981); Kyle et al. (1981); Dingle et al. (1983); Boyd (1991)
Morocco	Late Triassic-Middle Jurassic	219-180	Several thousand	Unknown; thickness of sills + flows ranges up to 0.5 km	Van Houton (1976, 1977); Manspeizer et al. (1978); Lorenz (1988)
Parnaiba (Maranhão) Misquito Mafic Province	Late Triassic-Barremian	±215-120	540,000 (Basalt flows = 110,000)	110,000 (Basalt flows = 11,000)	Sial (1974); Almeida (1986)
Karroo (Stormberg)	Early Jurassic-Aptian	210-115 (Main phase ca. 190-154 Ma)	1,000,000-3,000,000	200,000-420,000	Haughton (1969); Basalt Volcanism Study Project (1981); Dingle et al. (1983); Duncan (1987); Cox (1988)
Franz Josef Land	Early Jurassic-Cenomanian	203-94	Unknown	Unknown	Tarakhovskiy et al. (1983)
Tasmanian Dolerite	Middle Jurassic (Albian-Cenomanian)	174-165 (109-95)	25-30,000	12-15,000	Brown et al. (1968); Banks (1989); Ford (1989)
Rajmahal Traps	Middle Jurassic-Maastrichtian	170-69.6			Saxena (1986)
Axel Heiberg Mafic Province	Late Jurassic-Turonian	152-88 (All flows are 118-88)	12,500	Unknown	Douglas (1970); Balkwill (1978); Balkwill and Fox (1982)
Svalbard	Late Jurassic-Albian	149-110	30,000	Unknown	Harland (1973); Sokolov et al. (1973); Birkenmajer (1981)
Paraná (Serra Geral) Basalt	Late Jurassic-Albian	149-106	700,000 (Area of 1.2 mill., cited in literature is too high)	450,000 (780,000)	Amaral et al. (1966); Leinz et al. (1966); Basalt Volcanism Study Project (1981); Almeida (1986); Druecker and Gay (1987)
Labrador Shelf	Late Jurassic-Albian (Santonian ?)	145-105 (83?)			Umpleby (1979); Pe-Piper and Jansa (1987);
Entendeka (Kaokoveld) (Namiria)	Valanginian-Albian	132-108	18,000		Dingle et al. (1983); White and McKenzie (1989)

NAME	PALEONTOLOGICAL	RADIOMETRIC	AREA (KM²)	VOLUME (KM³)	SOURCE(S)
Okhotsk-Chukotka Belt	Albian-Late Oligocene	112-24	180,000-570,000	136,000-400,000	Belyy et al. (1966, 1989); Solov'yev (1966); Meyerhoff (1986); Zonenshain et al. (1990)
Sikhote-Alin Belt	Albian-Early Eocene	110-50	160,000	200,000	Belyy et al. (1966, 1989); Meyerhoff (1981b)
Deccan Traps	Albian-Pliocene	111-4.2 (Main phase, 66-60 Ma)	516,000	250,000 to 700,000 (1,500,000 ?)	Basalt Volcanism Study Project (1981); Saxena (1986); Renne and Basu (1991)
Madagascar	Turonian-Campanian	80-68			Besaire (1972); Vallier (1974)
Patagonian Plateau Basalts	Campanian-Recent	81-0.2	Up to 500,000	5,000-10,000	Baker et al. (1980); Ramos et al. (1982)
Brito-Arctic (Thulean) Province	Paleocene-Early Eocene	63-52	Up to 1,000,000		
Yakutat "Terrane"	Early to Middle Eocene	55-50			Davis and Plafker (1986)
Greater Ethiopian Volcanic Province	Early Eocene-Recent	54-5	630,000 Up to 1,400,000	150,000	Basalt Volcanism Study Project (1981); Crawford and Hilyard (1990)
Syrian-Arabian Volcanic Province	Early Eocene-Pliocene	54-5			
Nuclear Central America	Late Oligocene-Late Miocene	28-6.6			Weyl (1980)
Columbia Lavas	Early to Late Miocene	17.5-6	163,72[illegible] ± 5,000	174,300 ± 31,000	Waters (1962); Basalt Volcanism Study Project (1981); Tolan et al. (1989)
Chilcotin Plateau Basalts	Early Miocene-Pliocene	16-2	50,000	3,300	Douglas (1970); Bevier (1983)
Iceland	Early Miocene-Recent	15-0			

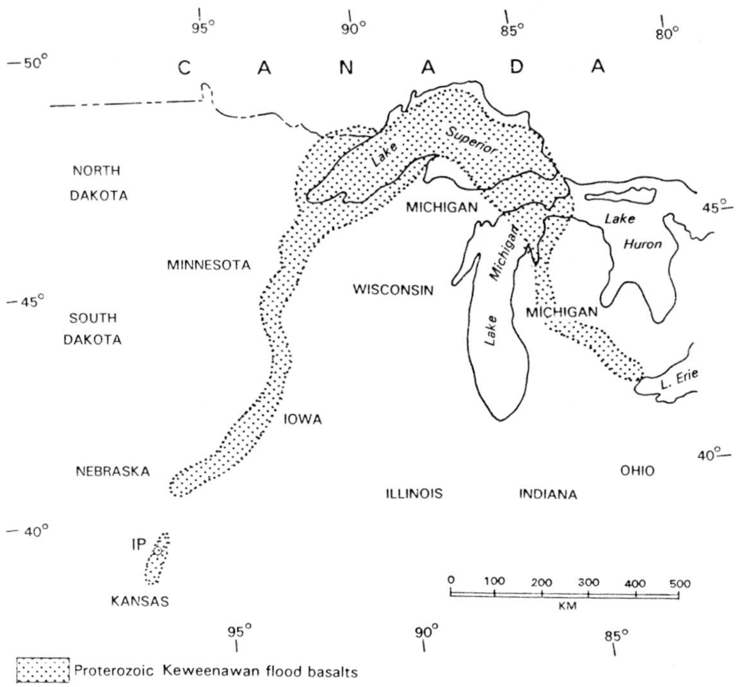

Fig. 6.1. Distribution of Proterozoic Keweenawan flood basalts (shaded areas) in the Midcontinent region of the United States: Lake Superior and south.

importance. Their existence alone is sufficient to prove that significant floods of effusive basalt occurred in the rifted epeirogenic province of the North American Midcontinent as long ago as 1 billion years. The most significant exposures are in the area of Lake Superior and its immediate surroundings. Figure 6.1, based on a diagram by "Anonymous" (1988), suggests an area of 120,000 km^2 for combined outcrop and subsurface. If we include the full 2,400 km length of the Midcontinent rift delineated by gravity data, the area is not less than 200,000 km^2 (Halls, 1978). Hutchinson and White (1990) estimated the volume to be at least 1,300,000 km^3. We estimate that the volume is not less than 1,000,000 km^3.

Green (1982) and Van Schmus and Hinze (1985) reported that exposed flows consist dominantly of rich olivine tholeiites with lesser amounts of weakly alkaline olivine basalt and tholeiitic (basaltic) andesite. These flood basalts were slightly metamorphosed, the grade of metamorphism increasing with "stratigraphic depth." Massey (1982) referred to the "normal tholeiitic suite," and to a few flows of "transitional basalt," the latter distinguishable by higher TiO_2, P_2O_5, and many minor radioactive elements. The Basalt Volcanism Study Project (1981) reported "cyclic compositional variations" in several of the local sequences of tholeiitic flood basalts.

Annells (1973) and Klewin and Berg (1991) described at some length the 5,250 m section of flood basalts that crop out at Mamainse Point on the eastern shore of Lake Superior. They reported a complex of more than 300 flows of basalt, mostly with "well-preserved wrinkled and ropy pahoehoe tops," and they classified these flows as olivine tholeiite, with an increase in olivine proportional to an increase in grain size. Annells' (1973) analyses, by weight percent, reveal the following ranges in composition:

SiO_2	49	to	50
TiO_2	1.5	to	2.3
Al_2O_3	14.8	to	17.4
Fe_2O_3	3.2	to	7.3
FeO	6.3	to	8.2
MnO		0.2	
MgO	6.4	to	6.9
CaO	9.5	to	10.2
Na_2O	2.7	to	2.8
K_2O	0.5	to	0.9
P_2O_5	0.2	to	0.3

In Figure 6.1, the major exploration well Texaco 1 Poersch appears as "1P," drilled in northeastern Kansas approximately 800 km SSW of the southwesternmost outcrop of Proterozoic Keweenawan flood basalts. Berendsen et al. (1988) recorded a drilled interval of 1,400 m of predominantly mafic rocks in lP below the Precambrian unconformity. Only 240 m consists of intercalations of nonbasaltic rock. These intercalations range from arkose and siltstone to pegmatite and gabbro. Basalt thus occupies about 1,250 m of the recorded interval. Because gabbro directly underlies the Paleozoic cover, additional mafic rocks may have been eroded during the long period that ensued before Ordovician marine waters entered this area of northeastern Kansas.

Berendsen et al. (1988) described the cuttings in 1P as microcrystalline and affected by deuteric and hydrothermal alterations whose products are hornblendic, chloritic, saussuritic, and albitic. A thin intercalation of basalt appears below the main column, at a depth only 120 m above total depth at 3,444 m.

Berendsen et al. (1988) also presented, from a study of cuttings in 1P, major-element compositions by weight percent, as follows:

SiO_2	42.4	to	46.5
TiO_2	1.05	to	2.97
Al_2O_3	14.0	to	16.1
Fe_2O_3	12.7	to	16.9
MnO	0.20	to	0.24
CaO	7.1	to	8.9
MgO	5.5	to	14.9
Na_2O	1.12	to	2.99
K_2O	1.31	to	4.99

These authors classified the basalt of 1P as mostly alkaline evidently on the basis of low percentages of SiO_2. They remarked also on the high content of magnesium and potassium. They did not provide any radiometric age for the basalts.

U-Pb, Rb-Sr, and K-Ar radiometric dating indicate this Keweenawan igneous activity began ca. 1225-1220 Ma, peaked at ca. 1110 Ma and terminated perhaps around 1010 Ma (Van Schmus et al., 1982). Thus, the Keweenawan flood basalts of Mamainse Point are stratigraphically near the boundary between middle and upper Proterozoic according to the "geological time scale" (Geological Society of America, 1984).

Two additional linear flood basalt provinces of Proterozoic age in North America include the basalts of the Coppermine River area (Baragar, 1972; Baragar and Donaldson, 1973) and the Cape Fold Belt (St. Onge et al., 1989) in north-central and northeastern Canada, respectively. The Coppermine basalts have been dated at ca. 1280 Ma, the Cape Smith basalts at ca. 1870 Ma.

6.2.2 Antrim Plateau (Proterozoic-Cambrian), Northern Australia (Ovate)

The ovate flood basalts of northern Australia hold an important position in the global regime, both by their antiquity and by their major area. Available evidence suggests that these flood basalts are Proterozoic in age at their base and Early Cambrian at their top. Figure 6.2 shows the amount of area that they cover, both by outcrops and by correlative occurrences in boreholes east of the main outcrops. This figure, compiled from two figures in Bultitude (1976), suggests to us a total of some 600,000 km^2 (shaded area). Bultitude had postulated a more conservative 300,000-400,000 km^2. In thickness, these flood basalts range from 1,500 m west of Halls Creek in Western Australia to less than 30 m near Tanami in southwestern Northern Territory. We consider, from such numbers, that the volume of flood basalts in northern Australia could amount to 60,000 km^3 or more.

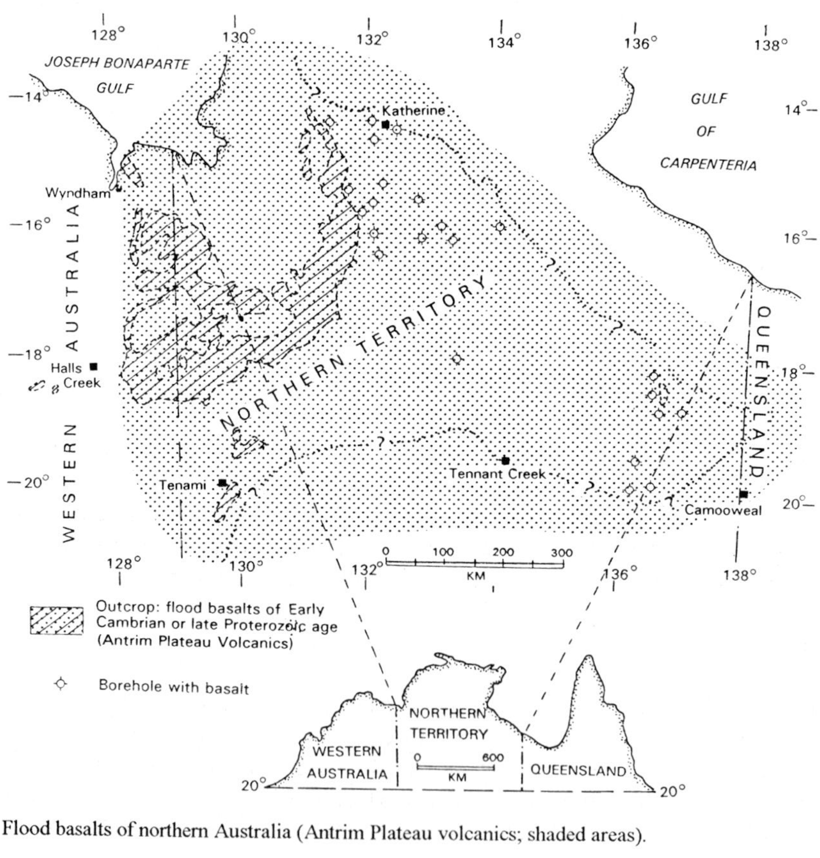

Fig. 6.2. Flood basalts of northern Australia (Antrim Plateau volcanics; shaded areas).

Bultitude referred to the outcrops of this considerable magmatic province as the Antrim Plateau Volcanics and described them as ". . . an extensive, but hitherto little-studied sequence of flood basalts in northern Australia. . . ." Petrographically, the sequence consists essentially of tholeiites. The outcrops themselves show typical layering of the lava, but much of the column consists of fine-grained massive basalt, "relatively fresh." Notable at the base of the column is a chilled zone, generally less than 1 m thick, and considerably altered. Above the main member of massive basalt, the column consists of intensely altered basalt, highly vesicular or amygdaloidal, which composes between 15 percent and 30 percent of the total column. The presence of flow banding, a scarcity of phenocrysts, and a pronounced elongation of vesicles led Bultitude (1976) to a conclusion of high fluidity in the magma at the time of effusion.

The chemical composition of the flood basalts of northern Australia is best represented by the tholeiites of the main member of massive basalt in the Antrim Plateau Volcanics. In these tholeiites Bultitude (1976) found a remarkable uniformity of composition with respect to major oxides. The most important, SiO_2, ranges from about 50 to about 56 percent by weight, but with the exception of a few samples stays between 52 and 54. Our summary of Bultitude's data follows:

SiO_2	52.0	to	54.0
TiO_2	0.60	to	2.05
Al_2O_3	12.8	to	16.1
Fe_2O_3	1.15	to	7.95
FeO	1.10	to	10.70
MnO	0.20	to	0.29
MgO	3.70	to	8.80
CaO	5.25	to	11.10
Na_2O	2.10	to	3.20
K_2O	0.11	to	4.30
P_2O_5	0.07	to	0.16

Mapping of these outcrops and examination of their correlatives in boreholes have failed so far to resolve accurately the ages of the total column of flood basalts in northern Australia. Evidence from outcrops indicates only that the top of the column lies below early Middle Cambrian marine carbonates and clastics. A disconformity or unconformity at the contact suggests that the top of the flood basalts belongs to the Lower Cambrian rather than to the Middle Cambrian (Bultitude, 1976). Sedimentary interbeds in the column of flood basalts contain stromatolites, possible indicators of organic remains, but the stromatolites make no contribution to chronostratigraphy. In Western Australia, however, the flood basalts overlie sedimentary beds whose shales have yielded radiometric dates of 666± 43 Ma and 654± 58 Ma. From these figures, an average radiometric date for the bottom of the flood basalts amounts to 660 Ma, which is chronostratigraphically equivalent to latest Proterozoic (earliest Vendian).

6.2.3 Northern Siberia (Permian-Triassic), Siberian Platform (Ovate)

The extensive flood basalts of the central Siberian Tunguska syneclise lie 2,000 km west of the Pacific Ocean and 1,200 km east of the Ural Mountains. The Tunguska syneclise itself, as drawn by the Basalt Volcanism Study Project (1981), lies half to the south of the Arctic Circle (latitude 66°32'N) and half to the north (Fig. 6.3). The longest cross-latitudinal line of the syneclise spans nearly 900 km, a length that contributes largely to a figure of 360,000 km^2 for

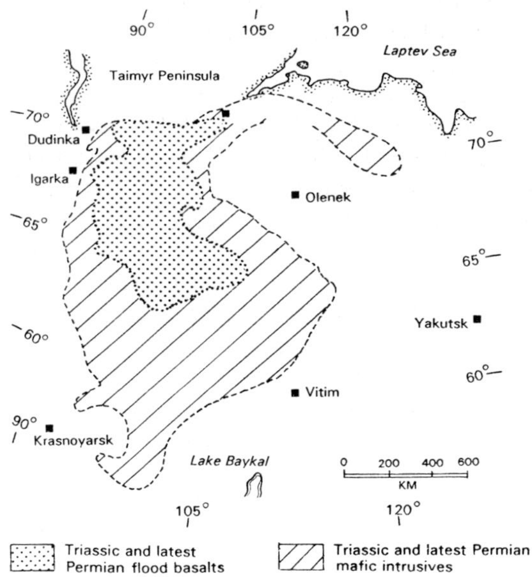

From Basaltic Volcanism Study Project (1981)

Fig. 6.3. Flood basalts (shaded areas) and mafic intrusions (hachured area) of the Tunguska syneclise, central Siberia.

total area of the visible Tunguska flood basalts. Important as such an area is, it is surrounded, and in terms of area quite eclipsed, by a singularly large spread of Triassic and latest Permian mafic intrusions. The northern edge of these mafic intrusions lies virtually coincident with the northern edge of the flood basalts at latitude N 70°, but the southern salient of the intrusions reaches as far south as the latitude of the northern shore of Lake Baykal, latitude N 55°. This large spread of mafic intrusions points to an original cover of flood basalts over an equally large area. Such a presumed original cover, probably removed by erosion in subsequent epeirogeny, would have lain on an area of 1,300,000 km^2 (Fig. 6.3).

Clues to the geochemistry of presently visible Tunguskan flood basalts appear in a study by Viswanathan and Chandrasekharam (1981), who describe the Siberian "traps" in general as high-potash olivine or quartz tholeiites with high concentrations of Rb and Sr, and in a study by Lightfoot et al. (1990). The latter group's tabulation of major element compositions by weight percent appears in summarized form as follows:

SiO_2	46.2	to	54.5
TiO_2	0.83	to	4.85
Al_sO_3	11.05	to	17.84
Fe_2O_3	9.91	to	15.56
MnO	0.13	to	0.21
MgO	2.95	to	7.53
CaO	5.09	to	12.17
Na_2O	0.64	to	3.97
K_2O	0.12	to	2.48
P_2O_5	0.07	to	1.13
LiO	0.8	to	3.6

Radiometric data from the Siberian Platform (summarized by Kuznetsov and Naumov, 1975) indicate that the radiometric dates of basalts range from 283 Ma to 70 Ma. The westward subsurface extension of the Siberian Platform is beneath the West Siberian Basin. The radiometric dates from beneath the West Siberian Basin range from 254 to 175 Ma (Zhuravlev, 1986). However, the Basalt Volcanism Study Project (1981) placed the flood basalts of the Tunguska syneclise in the supraregional framework of a single great magmatic episode of mainly Triassic age across the broad Siberian Platform. Radiometric ages of that platform reportedly range from 248 to 216 Ma. Chronostratigraphic equivalents are Permian (latest Tatarian) to Cretaceous.

6.2.4 Emeishan Flood Basalts (Permian-Triassic) of Greater Southwestern China (Ovate)

Wang Yunliang et al. (1989) in their study a small diagram showing the occurrences of Emeishan (Omeishan) basalts. North to south, these cross latitudinally from 35° N to 22° N, a distance of 1,000 km. West to east, they cross longitudinally from 98° E to 108° E, again a distance of 1,000 km. Figure 6.4, redrawn from the diagram of Wang Yunliang et al., shows by shading the area that contains all the outcrops of mapped Emeishan basalts. An area as large

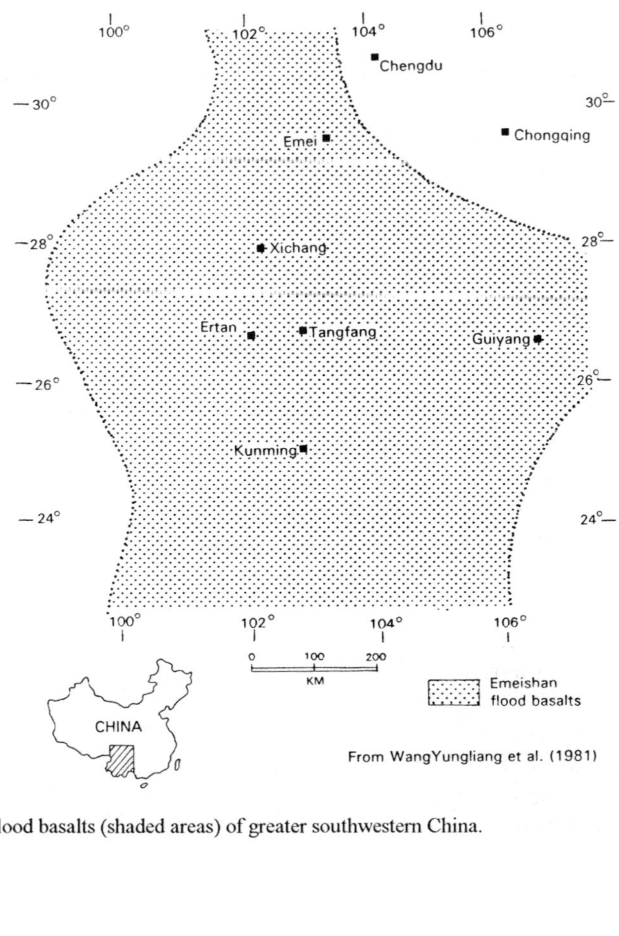

Fig. 6.4. Emeishan flood basalts (shaded areas) of greater southwestern China.

as 600,000 km^2 appears possible, considerably more than the 300,000 km^2 assigned by Wang Yunliang et al. Using the 705 m average thickness given by Zhang Yunxiang et al. (1990, p. 183) results in a volume of 420,000 km^3 of basalt in southwestern China and its fringing areas on the south.

The same authors concluded from their studies of major-element oxides at Emei and Tangfang that the basalts of those two localities "are chemically high-TiO2 transitional quartz tholeiites, although the general lack of orthopyroxene implies a more alkalic classification." Zhang Yunxiang et al. (1990) presented a summary of nearly 400 geochemical analyses of the Emeishan basalt in southwestern China, which appears in terms of weight percent as follows:

SiO_2	47.4	to	48.6
TiO_2	2.32	to	3.92
Al_2O_3	13.4	to	13.5
Fe_2O_3	5.3	to	5.6
FeO	7.7	to	8.4
MnO	0.17	to	0.21
MgO	4.4	to	6.9
CaO	8.2	to	9.1
Na_2O	2.51	to	2.80
K_2O	0.70	to	1.38
P_2O_5	0.22	to	0.32

Low SiO_2, high alkalic content, and high TiO_2, constitute the distinctive features. They conclude, in addition, that most of the Emeishan basalt should be referred to as a weakly alkaline basalt series.

Huang (1932) described the basalt of Emeishan as underlain by Permian limestone with the foraminiferal genus *Neoschwagerina*, and overlain by purplish red Triassic shales. Some 300 km farther south, in the areas of Tangfang and Ertan, the Emeishan basalts continue to lie between Permian limestones with *Neoschwagerina* and Triassic redbeds. Coal measures in these beds contain many *Gigantopteris*, a fernlike genus of plants. Huang concluded that Emeishan basalts may have survived into the Triassic Period in southwestern China, but evidence from many other parts of the world suggests that it is much more likely to be Carboniferous and Permian.

Zhang Yunxiang et al. (1990) do not mention flood basalt in their descriptions, but do call this basalt a typical trap rock. The nature, the geochemistry, and the distribution of Emeishan basalts in greater southwestern China, however, suggest strongly to us that they form a major body of flood basalts.

6.2.5 Wrangellian (Triassic) Flood Basalts of Western and Northwestern North America (Linear)

The "Wrangellia" of Jones et al. (1977), named after the Wrangell Mountains in southeastern Alaska, consists of a major terrane in western and northwestern North America, marked partly by a distinctive Triassic lithostratigraphy, and partly by major effusions of Triassic flood basalts. Wrangellia stretches from southeastern Alaska on the north, to the Washington-Oregon border on the south, a distance of approximately 2,700 km. Compared to its 2,700 km length, the Wrangellian terrane is narrow, even at its wider northern and southern parts. The decided narrowness of the shaded area from the Chichagof Islands to Vancouver Island, however, may be deceptive, because Wrangellian basalts may exist west of the narrow segment under waters of the Pacific Ocean that now lap up against Wrangellia, and wider still if NNW and NNE faults along the present coast have caused major downthrow to the west.

Figure 6.5 shows the position of five thick columns of flood basalt described by Jones et al. (1977) which collectively comprise the Wrangellian flood-basalt province: (1) Wrangel Mountains; (2) Chichagof Island; (3) Queen Charlotte Islands; (4) Vancouver Island; (5) Hells Canyon.

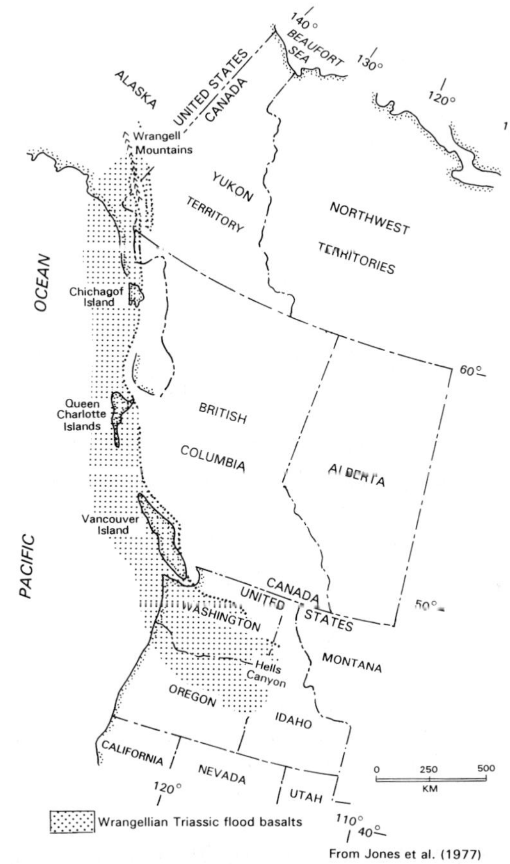

Fig. 6.5. Wrangellian Triassic flood basalts (shaded area) of western and northwestern North America.

Wrangell Mountains, Alaska

Jones et al. (1977) described the flood basalts of this area as "a vast largely subaerial, but locally pillowed basaltic lava field that occurs throughout the Wrangell Mountains and the adjoining eastern Alaska Range and extends eastward into Canada . . ." and that it "consists of intermixed aa and pahoehoe flows between 0.3 and 15 m thick and reaches a cumulate thickness exceeding 3500 m." Olivine-normative tholeiites are present, but the basalt of this area occurs mainly as quartz-normative tholeiites.

Below the flood basalts lie Middle Triassic (Ladinian) strata. Above the column of flood basalts lie Late Triassic (Carnian to Norian) strata. This succession is repeated essentially in each of the four columns that remain to be described. Jones et al. (1977) have stated: "The principal unifying characteristics of Wrangellia are provided by its Triassic strata, which include a thick Middle and Upper Triassic unit mainly composed of tholeiitic basalt . . .."

Chichagof Island and Baranof Island, Alaska
On these two islands within the narrow southern salient of Alaska, low-grade metamorphism has altered the flood basalts into greenstone. A basal fault reduces their thickness.

Queen Charlotte Islands, Canada
Flood basalts having an estimated thickness of 4,000 m crop out widely in the southern part of the Queen Charlotte Islands, on the west coast of Canada. Overlying strata contain the pelecypod *Monotis*, possibly an indicator of Carnian and Norian age (Late Triassic). The age of the top of these flood basalts on the Queen Charlotte Islands is therefore Late Triassic or older. The base of the flood basalts is not exposed.

Vancouver Island, Canada
Muller et al. (1974) noted the presence of effusive basalt throughout Vancouver Island except at its southern tip, adding that "the entire island region, as well as Queen Charlotte Islands consisted of great basaltic lava plateaus . . . the bedded lavas perhaps accumulated as subaerial flood basalts, forming shield volcanoes." Jones et al. (1977) showed a thickness of more than 6,000 m.

Carlisle and Susuki (1974), and Muller et al. (1974), both presented a tripartite division of the flood basalts on Vancouver Island. Their two columns, closely similar, amount to the following: (1) upper basalt flows, pillow lavas and breccias, minor sediments: about 3,000 m; (2) middle pillow breccia, above aquagene tuff: 600 to 1,100 m; and (3) lower pillow lava: about 2,500 m.

The column of flood basalts on Vancouver Island ranges from the late Middle Triassic Ladinian Stage at bottom to the early Late Triassic Carnian Stage at top (Carlisle and Susuki, 1974; Jones et al., 1977). The corresponding span of time is 5 m.y.

Hells Canyon and Wallowa Mountains, Oregon
In Oregon, the rocks of Wrangellian affinities consist of 2,000 to 2,500 m of basaltic flows, breccias, and aquagene tuffs, with occasional sedimentary intercalations. As on Vancouver Island, Daonella is found near the base of the basalt, and intercalating Carnian sediments near its top indicate an age of late Middle Triassic to early Late Triassic.

6.2.6 Kirkpatrick Basalt-Ferrar Dolerite (Jurassic), Antarctica (Linear)

Effusive tholeiitic basalts on this southern continent occur in three areas of outcrop: (1) southern Antarctica; (2) south-central Antarctica; and (3) far northern Antarctica (Fig. 6.6). The name given to the effusives in the two southern areas is the Kirkpatrick Basalt. Elliot (1970, 1972) has described them as "tholeiitic flood basalt" whereas Kyle (1980), and Faure et al. (1982), have used the terms "tholeiitic lava flow" and "lava flow" respectively.

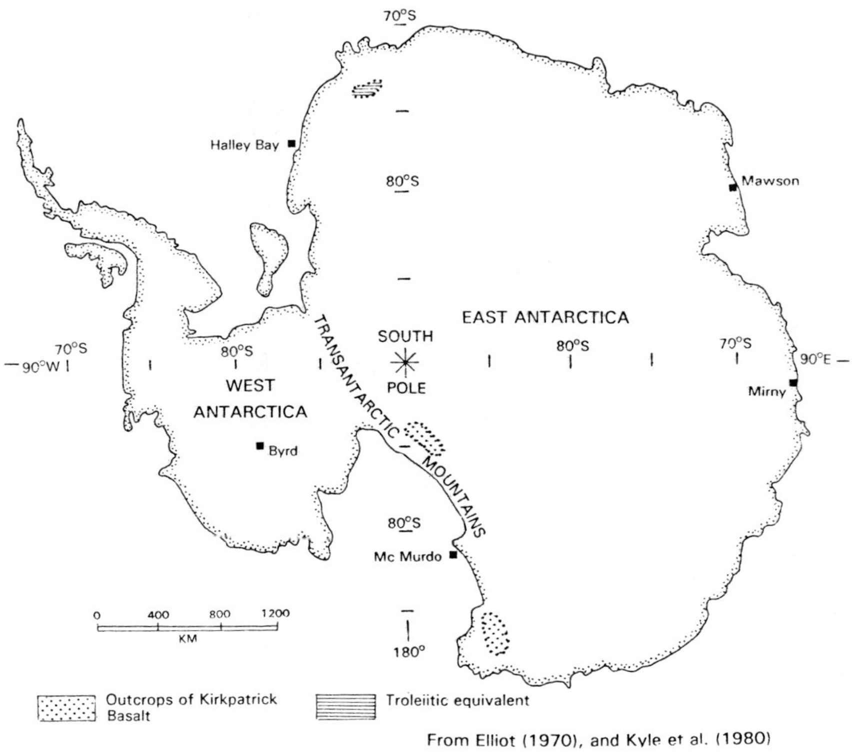

Fig. 6.6. Outcrops of effusive Kirkpatrick Basalt in the Transantarctic Mountains and a presumed tholeiitic equivalent in the north.

The effusives of far northern Antarctica, described by Faure et al. (1979), differ geochemically from those of the Kirkpatrick Basalt. Although lower in SiO_2 and K_2O, and higher in TiO_2, Al_2O_3, MgO, and CaO, they still occupy a position on "A-F-M" diagrams that corresponds to a "tholeiitic basalt."

These tholeiitic basaltic effusives have yielded radiometric ages of 182 to 162 Ma in far northern Antarctica and 179 to 164 Ma in the other two areas (Faure et al., 1979, 1982). These data suggest that the three small shaded areas of Figure 6.6 are remnants of a single province with a north-south length of nearly 4,000 km. With an arbitrary width of 100 km applied to that distance, the total area involved would be nearly 400,000 km^2. This area could be significantly greater if accompanying dolerites are interpreted as feeders to now-eroded flood basalts.

The spread of radiometric ages from 182 to 162 Ma for the Antarctic basalts is equivalent to an interval from early Middle Jurassic to Middle Late Jurassic. The Jurassic effusions on Antarctica take their place in a great "Gondwanan" spread that includes the Cretaceous and Jurassic effusions in southern Africa and Cretaceous effusions in southern Brazil. This great spread, in fact, is one that has given rise to a veritable outpouring of literature on Mesozoic dynamic associations, especially those papers that presage continental drift and the existence of an ancient supercontinent.

6.2.7 Karroo Flood Basalts (Jurassic) of East Southern Africa (Ovate)

Mesozoic basalts in southern Africa outcrop intermittently from the Atlantic Ocean to the Indian Ocean (Fig. 6.7). Along latitude 25° S, they span from ocean to ocean for a distance of 2,000 km.

All basalts on Duncan's (1987) map of southern Africa appear to be flood basalts. The basalts of Namibia are younger and more sparsely exposed than those toward the Indian Ocean. As a result, those exposed closer to the Indian Ocean have received much more attention in the literature. The regions thus involved consist of parts of South Africa, Botswana, Zimbabwe, and Mozambique (Fig. 6.7).

There is consensus that present outcrops are relics of a once continuous cover of basaltic effusions. Flood basalts of the east, in that context, would have covered an area of approximately 1 million km^2. Thicknesses, as in all large areas of basaltic effusion, vary considerably. Thicknesses as great as 9,000 m may occur in the border areas of southern Mozambique, and elsewhere may reach several hundred meters. Assuming an average thickness of 200 m, the total volume of flood basalts in east southern Africa might have amounted originally to 200,000 km^3.

The Karroo flood basalts are essentially tholeiitic. Values of SiO_2 as low as 48 to 49 percent by weight are sufficiently undersaturated to appear abnormal compared to the flood basalts of other areas. An opposite abnormality also occurs, however, with SiO_2 values in some samples rising as high as 53 percent.

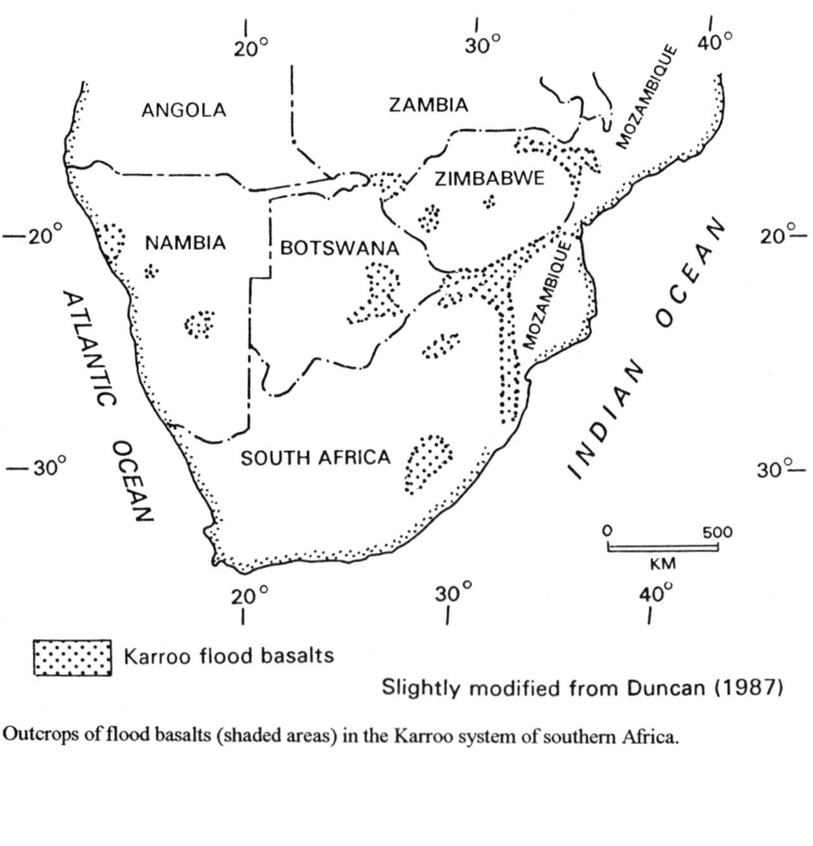

Fig. 6.7. Outcrops of flood basalts (shaded areas) in the Karroo system of southern Africa.

For the Karroo flood basalts on the east side of southern Africa, the table that follows contains our summary of the geochemistry of the major elements provided by Basalt Volcanism Study Project (1981):

SiO_2	48.3	to	52.9
TiO_2	0.93	to	2.76
Al_2O_3	11.3	to	15.8
Fe_2O_3	1.5	to	7.3
FeO	4.5	to	9.9
MnO	0.14	to	0.20
MgO	3.9	to	7.7
CaO	8.8	to	11.2
Na_2O	2.04	to	2.77
K_2O	0.10	to	1.63
P_2O_5	0.08	to	0.55

Radiometric age studies of the Karroo basalts have not always identified the type of basalt analyzed. An age of 180 Ma along the border between Zimbabwe and southern Mozambique (Fitch and Miller, 1981) appears to refer to flood basalts. Ages along the border between South Africa and southern Mozambique are 190 to 180 Ma.

Bristow and Saggerson (1983), in their comprehensive study of Karroo volcanism, postulated the emplacement of a number of carbonatite complexes around 205 Ma in southeast Zimbabwe. Thereafter, the "focus of volcanism" spread eastward and southward to a culmination of "extensive basaltic lavas and more localized rhyolitic flows." Their basaltic lavas are likely to be flood basalts, and the "culmination" shared with localized rhyolites lasted 13 million years, from 190 to 177 Ma! The time interval represented by these two radiometric dates extend from Early Jurassic to Middle Jurassic.

6.2.8 Malaita Island (Cretaceous) (Ovate)

Malaita Island, one of the four northernmost Solomon islands shown on Figure 6.8, lies approximately 600 km SSE of the Ontong Java Plateau. In a monograph on the geology and geophysics of this island, Kroenke (1972) reported basaltic lavas below Late Cretaceous mudstones in outcrop. Kroenke noted that the basaltic lavas show pillow structure, and that they are not rich in sodium. The evidence did not, however, enable Kroenke to distinguish between an early Late Cretaceous and pre-Late Cretaceous effusion of basalts on Malaita Island.

Kroenke's results permit only a suggestion that the basalts on Malaita Island are Cretaceous and that they are flood basalts, although it is conceivable that they are older than Cretaceous (i.e., Jurassic). The Cretaceous thus could have been a period during which the effusion of flood basalts occurred across an unusually wide belt of the area now covered by equatorial waters of the Ontong-Malaita sector. The cross-latitudinal spread of such a belt, taking the distance from the north edge of the shaded area on Figure 6.8 to Malaita Island, is at least 1,400 km.

6.2.9 Ontong Java Plateau (Cretaceous) (Ovate)

The Ontong Java Plateau lies under Pacific oceanic waters to the northeast and east-northeast of New Guinea. The form of the plateau is shown most clearly by the 4,000-meter-below-sea-level (mbsl) bathymetric contour. Where the line of the equator crosses this plateau, the width

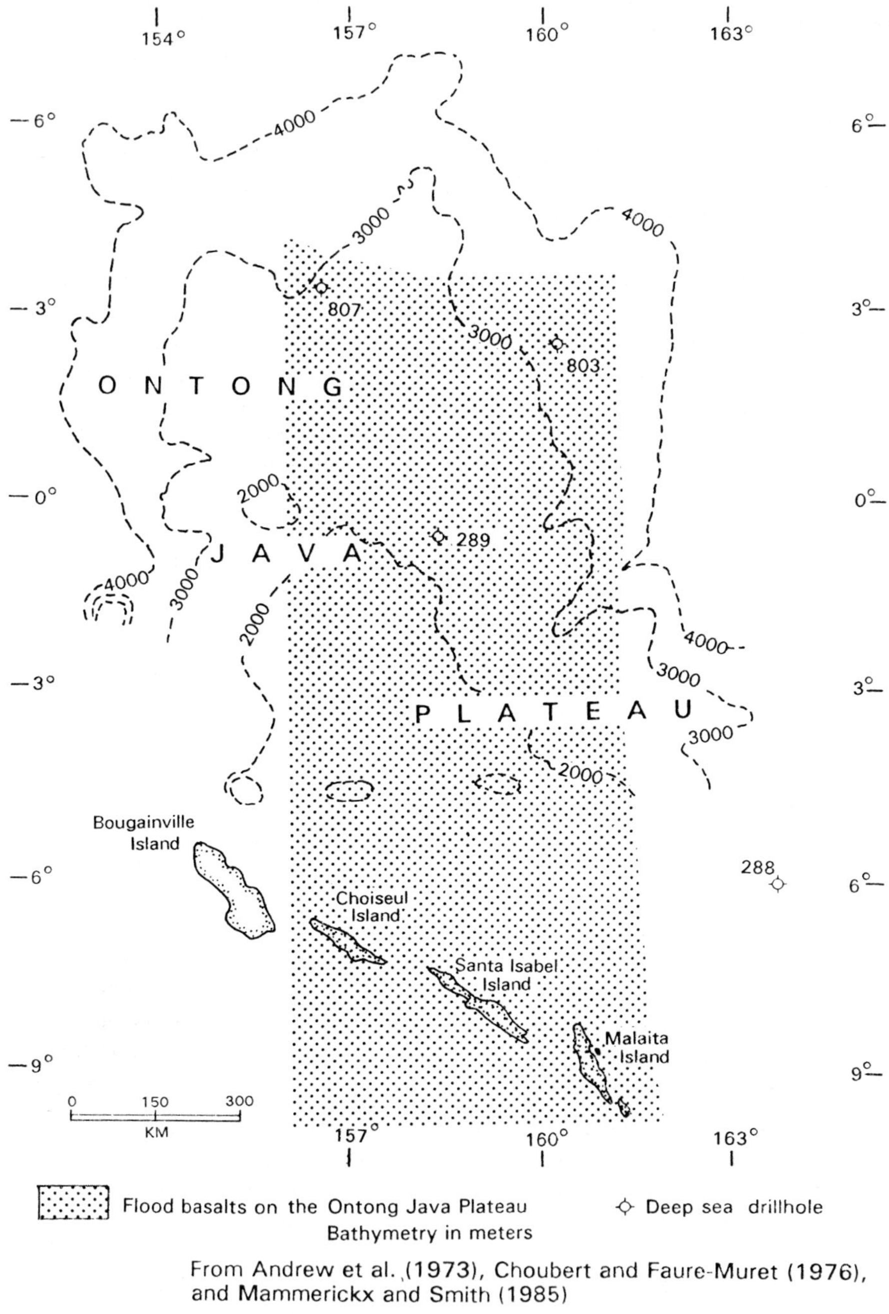

Fig. 6.8. Shaded area shows presumed extent of flood basalts on the Ontong Java Plateau and south.

from 4,000 mbsl on the west to 4,000 mbsl on the east amounts to approximately 900 km. The length at longitude 158° E amounts to approximately 1,100 km (Fig. 6.8). A marked extension to the east at latitude 1° S brings the total area of the Ontong Java Plateau to at least 1,000,000 km^2.

Information concerning flood basalts comes from Kroenke and Berger (1990), who reported on the results of deep-sea drilling. As a result of such drilling, these authors described the Ontong Java Plateau as ". . . the world's largest oceanic flood basalt plateau." This plateau rose for the most part "by deep-water widespread effusion of massive flood basalts during Aptian time" (later Early Cretaceous). At drillholes 803 and 807, the rock of the flood basalts is "predominantly olivine-bearing. . . erupted from submarine vents . . .." At drillhole 289, Andrews et al. (1973) reported extrusive basalt immediately below Aptian limestone. They indicated only that this basalt is fine grained, and that, in their words, it is a "Tholeiitic plateau basaltic lava flow . . .." Their statements, while ambiguous, seem sufficient to justify our assumption that the basalt of drillhole 289 is an Aptian flood basalt.

Drillholes 289, 803, and 807 demarcate a triangle of Aptian flood basalts whose area amounts to approximately 60,000 km^2. How far these Aptian flood basalts spread beyond the triangle must remain for the present only surmise. The bathymetry of the Ontong Java Plateau, however, leads us to outline an expansion of the flood basalts as shown by the shaded area on Figure 6.8. Such an expansion brings the total to approximately 400,000 km^2, and we tentatively classify this province as ovate.

6.2.10 Paraná Flood Basalts (Cretaceous) of Greater Southern Brazil (Linear)

Flood basalts in the Serra Geral Formation of the Paraná basin lie mainly in southern Brazil, with the remainder in fringing areas of Paraguay, Argentina, and Uruguay (Fig. 6.9). We estimate their areal extent on the basis of the map of Amaral et al. (1966), at a little more than 700,000 km^2, which differs considerably from the 1,200,000 km^2 given by Almeida (1986). Almeida, however, was citing Maack (1952), whose estimate has been widely quoted in the literature.

Exploration wells in the Paraná basin have drilled through unusually thick columns of basalt, at least one well showing more than 1,500 m of basalt. Leinz et al. (1966), from their isopachs of the flood basalts, found an average of 650 m for the basin. Given an area of 1,200,000 km^2, the volume of Paraná flood basalts would amount to 780,000 km^3. The volume corresponding to our own area of 700,000 km^2 would amount to 450,000 km^3, still sufficient to equal or exceed the volumes of most major continental flood basalts. At latitudes 22° S and 23° S, Vetter and Fodor (1983) reported that wells drilled offshore show basaltic material as far east as 125 km east of the present shoreline (Figure 6.9). There is, however, no data to show the continuity of flood basalt from southern Brazil to the offshore wells.

Almeida (1986) regarded the basalts of the Paraná magmatic province as predominantly tholeiitic, but offered no chemical analysis as supporting evidence. The following summary of major-element oxides by weight percent for a "tholeiitic" suite come from Bellieni et al. (1988):

SiO_2	48.7	to	50.5
TiO_2	1.37	to	2.25
Al_2O_3	13.3	to	14.7
Fe_2O_3	5.5	to	5.7
FeO	6.8	to	8.6

MnO	0.19	to	0.21
MgO	3.9	to	7.1
CaO	8.7	to	10.8
Na_2O	2.30	to	2.53
K_2O	0.35	to	1.15
P_2O_5	0.16	to	0.33

Peate et al. (1988) examined cuttings of basalt and sediments recovered from four wells located between São Paulo on the east and the border with Paraguay on the west. Drilled thicknesses of Paraná basalts in these four wells ranges from 933 m to 1,723 m. Among other interpretations made in their study, Peate et al. (1988) interpreted 18 sills in the basalt. The presence of these sills reduces the range of drilled thickness of flood basalts in this north-central part of the Paraná basin to 375 m to 887 m. Geochemical analyses of cuttings from the four wells are from depths above the sills. They refer, for that reason, to flood basalts. For the major item of SiO_2, the numbers are 49.5 to 54.6 by weight percent. Comparable numbers for outcrops, as given above, are 48.7 to 50.5. Peate et al. (1988) do not comment on the possibility that 54.6 is a number higher than usual for the definition of tholeiite.

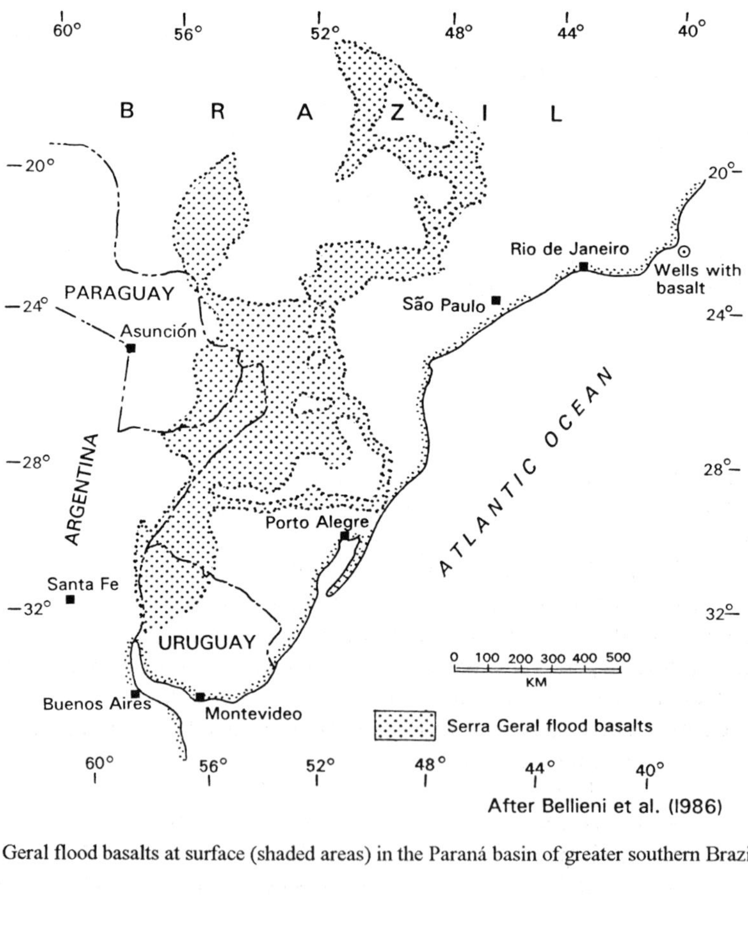

Fig. 6.9. Serra Geral flood basalts at surface (shaded areas) in the Paraná basin of greater southern Brazil.

Amaral et al. (1966) recorded radiometric ages of Paraná flood basalts by K-Ar over a cross-latitudinal distance of approximately 800 km. Their numbers point to the onslaught of volcanism at 150 Ma with later strong peaks clustered around 120 Ma. Peate et al. (1988) considered that K-Ar evidence pointed to a clustering from 135 to 115 Ma "with a preferred age of 130 Ma." The 130 Ma of Peate et al. corresponds to the earlier half of the Early Cretaceous (Valanginian-Hauterivian boundary). The 120 Ma of Amaral et al. corresponds to the beginning of the later half of the Early Cretaceous (Barremian-Aptian boundary).

In the wells drilled between latitudes 22° S and 23° S, in South Atlantic waters 75 km to 125 km east of the present shoreline, basalts from cores yielded ages from 124 to 112 Ma (Vetter and Fodor, 1983). The number 124 Ma corresponds to the earlier half of the Early Cretaceous (Hauterivian-Barremian boundary): that of 112 Ma to the later half of the Early Cretaceous (Aptian-Albian boundary).

The radiometric data of Amaral et al. and Peate et al. suggest that the effusion of flood basalts in the Paraná magmatic province took place in Early Cretaceous time.

6.2.11 Deccan Flood Basalts (Late Cretaceous-Tertiary), Western Peninsular India (Ovate)

Figure 6.10 shows the position and the outline of Deccan flood basalts. Along latitude 22° S, they appear as far as 1,400 km inland from the Indian Ocean. Along longitude 76° E, the flood basalts continue almost uninterrupted for 800 km. The total area covered by these flood basalts, presented in many studies of the literature, amounts to 500,000 km^2. With regard to volume, West (1985) gave a figure of "more than 250,000 cubic kilometers."

Mahoney (1984) pointed out the presence of at least five continental flood-basalt provinces on the "Indian subcontinent." These range in age from the middle Proterozoic to the early Tertiary, with the Deccan occurrence not only the youngest but also by far the most extensive. Petrologically, it is uniform and tholeiitic. From Mahoney's study, a summary of variations in major elements by weight percent follows:

SiO_2	46	to	53
TiO_2	1.0	to	3.5
Al_2O_3	9.5	to	15.8
FeO	10.3	to	15.4
MgO	3.5	to	12.2
CaO	7.0	to	12.5
Na_2O	1.4	to	3.3
K_2O	0.2	to	2.9
P_2O_5	0.1	to	0.6

Enrichment in Fe and Ti is characteristic. Values of FeO greater than 15 weight percent, and of TiO_2 greater than 3 weight percent, have been obtained from a number of localities. Mg-numbers (MgO/FeO) tend to be correspondingly depressed at such localities, also substantially less than in typical midocean ridge basalts (MORB). Values of MgO greater than 8 percent by weight and Mg-numbers greater than 58 percent by weight occur in the lower flows of the coastal region.

The nature and the relationships of rocks in the Deccan have been known since well before the advent of this century. As early as 1867, Blanford (1867a, 1867b) reported with considerable expertise on the area and the age of Deccan basalts, which he called traps, and

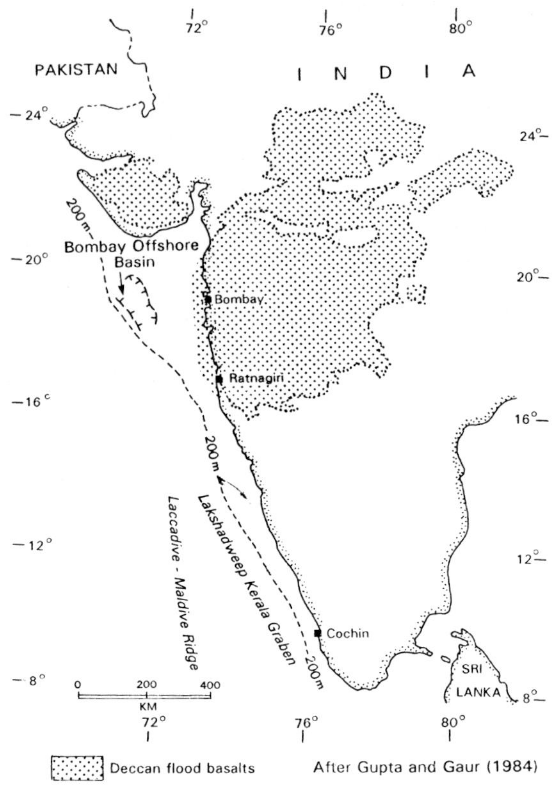

Fig. 6.10. The Deccan flood basalts (shaded areas), western peninsular India.

their intercalated sediments, intertrappean horizons. Since Blanford's time, work on the basalts of Deccan has concentrated largely on the geochemistry of the major elements, although the 1980s witnessed a surge of interest in the rare-earth elements (REE). However, the density of sampling has remained comparatively low. The pace of investigation has been slowed in part by the large areas involved as well as the metasomatism of the basalt itself.

The Deccan Traps, like the Siberian Traps, were emplaced by a shifting eruptive center(s) (Subbarao, 1988; Mitchell and Widdowson, 1991). Radiometric dates obtained from them range from 111 to 4.2 Ma (Agarwal and Rama, 1976; Alexander, 1981; Saxena, 1986), but most of the traps were emplaced in the time span 102-31 Ma (Alexander, 1981).

Some authors point to the possibility that tholeiitic magma poured out over coastal Deccan to form a column presumably 2,000 m thick in as little as 2 million years. Courtillot et al. (1988), for their part, argued for an interval of time as brief as only 1 million years. Either of these two figures supports a thesis of catastrophic volcanism. Duncan and Pyle (1988) visualized an average rate across Deccan of 1 km^3 per year in a regime consisting of periods of more rapid production of lava separated by quieter periods. Cox (1988a, b) favored a similar idea, expressed as periods of "repose," evidenced either by intercalated lateritic horizons or by intercalated more normal sedimentary horizons. However, a large body of radiometric (111-4.2 Ma) and stratigraphic data (lateritic horizons and intercalated sedimentary sections) suggest that the Deccan Traps were emplaced over a long geological time span.

6.2.12 Flood Basalts of the Argentinian Foreland (Late Cretaceous-Tertiary) (Linear)

East of the Chilean-Argentinian Andes, the foreland contains a major terrane of flood basalts, whose occurrences and extent (shaded area) appear in Figure 6.11. The northern end of the terrane lies at about 36° S latitude, and the southern end at about 52° S latitude, for a length of approximately 2,000 km. An average width of 250 km gives the terrane of flood basalts a total area of approximately 500,000 km^2. Volume overall, however, may be comparatively small, because thicknesses of the flood basalts in the Patagonian sector rarely exceed hundreds of meters. A factor that reduces volume further is a perceptible eastward thinning, sufficient to have prevented these flood basalts from spreading as far eastward as the shoreline of the present Atlantic Ocean (Fig. 6.11).

In the Patagonian sector, Butler et al. (1991) described the flood basalts as ". . . a sequence of flat-lying Late Cretaceous through Tertiary basaltic and andesitic rocks exposed east of the Andes in dissected mesetas of Chilean and Argentine Patagonia." An earlier study in the Patagonian sector by Ramos et al. (1982) described Late Cretaceous and Cenozoic magmatism east of the Andes as "a series of alkali plateau basalts." Butler et al. (1991) also characterized

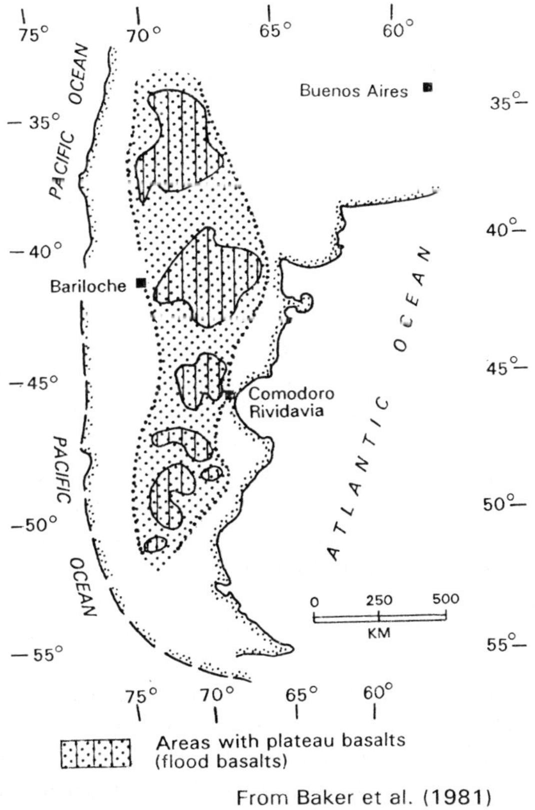

Fig. 6.11. Flood basalts and plateau basalts in Argentina.

Patagonian flood basalts also as alkaline. An average of their figures for SiO_2, however, amounts to 51.5 percent by weight. Their analyses follows:

SiO_2	51.5
TiO_2	1.7
Al_2O_3	16.9
Fe_2O_3	9.8
MnO	0.2
MgO	4.2
CaO	7.5
Na_2O	3.9
K_2O	1.5
P_2O_5	0.9

Ramos et al. (1982), by means of a graph, showed that the flows of basalt in Patagonia, although spasmodic, were repetitive, not only throughout the Late Cretaceous Epoch and the Tertiary but also up to almost historical time.

6.2.13 Brito-Arctic (Tertiary-Quaternary) Basaltic Province (Linear)

From the southernmost point of significant Tertiary basaltic effusion in northeastern Ireland, to a relatively similar effusion at Cap Dyer on Baffin Island far to the north, lie more than 3,000 km of present land and sea. As part of that long span, the subcontinent of Greenland lies athwart the Tertiary basaltic igneous province, with its own total length of 2,500 km (Fig. 6.12), and with a northern edge at more than 80° N latitude.

Rampino and Stothers (1988) reported lower Tertiary flood basalts across northwestern Britain, northern Ireland, and the Faeroe Islands. They also noted "rocks that are possibly equivalent" in East Greenland, in West Greenland, and on Baffin Island. "Together these rocks form the Brito-Arctic Basalt Province. The total area covered is about 100,000 km^2, but it was probably much greater before continental rifting and subsequent erosion."

For Iceland and its marine environs alone, the data given by Einarsson (1960) suggest an original area of lower Tertiary flood basalts not less than 150,000 km^2. If we add to this number the shaded areas of Figure 6.12 for East Greenland and West Greenland, the total rises to about 300,000 km^2. Extrapolation, perhaps beyond permissible limits, would permit an original continuity of lower Tertiary flood basalt from Greenland to Iceland and a total area of about 800,000 km^2. More territory must be added to include the impressive lower Tertiary flood basalts of the Faeroe Islands southeast of Iceland; thus a total of 1,000,000 km^2 is possible.

Rampino and Stothers (1988) analyzed radiometric K-Ar and Ar-Ar results from 135 samples, and from their statistical analysis they estimated that "Brito-Arctic flood basalts began erupting at 62±3 Ma, at about the time of the initial rifting of the Norwegian Sea." This postulated beginning of effusion corresponds to the lower Paleocene Danian Stage, the lowest stage of the Tertiary Period.

The Basalt Volcanism Study Project (1981), citing Evans et al. (1983), has written: "Volcanism began in this province in the Late Cretaceous with eruptions around or just before 65-66 m.y. years ago, and the major events are considered to have terminated between 60 and 56 m.y. ago. Minor activity may have continued until the Eocene . . . ." The total interval of flood-basalt activity, using these data but more recent geological time scale, would be from approximately the Cretaceous-Tertiary boundary to the latest Paleocene/earliest Eocene.

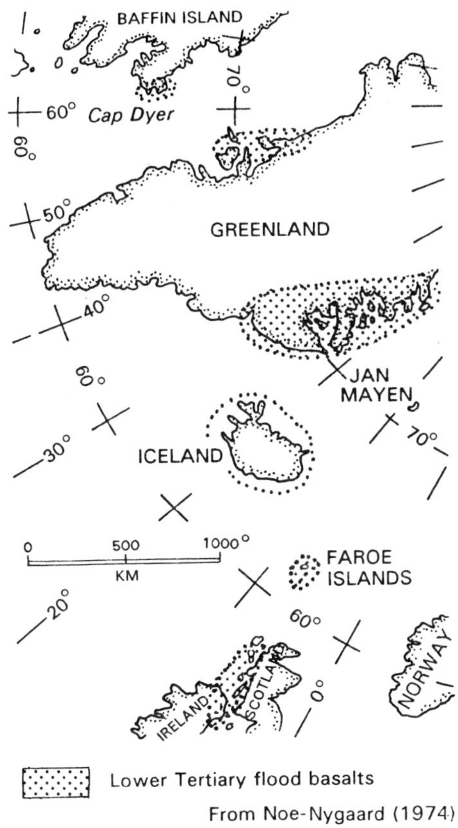

Fig. 6.12. Lower Tertiary flood basalts (shaded areas) in the Brito-Arctic basaltic province.

6.2.14 Syrian-Arabian-Greater Ethiopian Magmatic Province (Tertiary-Quaternary (Ovate)

The shaded portion of Figure 6.13 shows the area over which basaltic effusion took place throughout the Cenozoic Era across most of Ethiopia and across its fringing areas of Yemen, Djibouti, Somalia, Kenya, and Sudan. Distances across this magmatic province are considerable. A nearly north-south line from the Ethiopian-Saudi Arabian border on the Red Sea to meeting points in Ethiopia, Somalia, and Kenya measures approximately 1,400 km. An east-west line along latitude 10° N measures approximately 1,100 km. The result is an area of 1,400,000 km^2 for the greater Ethiopian Cenozoic basaltic province.

Blanford, as long ago as 1869, recognized the Ethiopian highlands as a province of great basaltic effusions, and, even at that time, he compared their "enormous mass and persistent horizontality" to those of the Deccan flood basalts of India. Mohr (1983) observed that "The Ethiopian volcanic province is dominated by a Tertiary basaltic pile, fed from fissures aligned along developing pairs of opposed, nascent continental margins. The Ethiopian and Yemen margins run either side of the Red Sea, and within Ethiopia the plateau margins face each other across southern Afar and the rift valley." The rift valley mentioned by Mohr is the Ethiopian

Rift, which in simplified form appears in Figure 6.13, as a feature trending NNE to Addis Ababa (Berhe et al., 1987).

In contrast to this NNE trend is the NW trend of the submarine rift valley of the present Red Sea. This kind of contrast in trend typifies the swinging track of the whole Afro-Arabian system of rift valleys. These, well documented in the literature, reach nearly 6,000 km from Mozambique on the south to Syria on the north. In Ethiopia, the typical antithetic form of rifting exemplified by the Ethiopian Rift ceases northeast and east of Addis Ababa. The western fault of the rift turns northward, whereas the eastern fault turns eastward. The resulting divergence causes the rift, as such, to lose identity.

Mohr (1983) was concerned with a lack of standards in the nomenclature of Ethiopian basalts, and with a tendency toward a greater use of chemical criteria than of mineralogical criteria. For Ethiopia, the criterion of "transitional basalt" is one "that falls within ± 1 wt % alkalis" from the line that divides alkalic basalt from tholeiite. The following table contains Mohr's numbers for variations in major elements by weight percent in each of three mineralogical categories, with samples mainly from the northern half of the western plateau:

	Andesine tholeiite	Transitional basalt slightly alkalic	Alkalic basalt
SiO_2	50.1 to 50.4	46.9 to 48.7	46.1 to 47.2
TiO_2	2.7 to 3.3	2.9 to 4.8	2.7 to 3.0
Al_2O_3	13.8 to 14.7	10.9 to 15.4	14.8 to 15.1
Fe_2O_3	4.0 to 4.4	n.s. and 6.2	4.9 to 5.9
FeO	7.9 to 9.1	8.5 to 11.9	6.8 to 8.3
MnO	0.19 to 0.24	0.23 and n.s.	0.20 to 0.21
MgO	5.0 to 5.7	6.4 to 7.3	6.5 to 7.0
CaO	9.3 to 10.3	10.6 to 10.9	10.5
Na_2O	3.0 to 3.2	2.5 to 3.0	3.3
K_2O	0.69 to 1.2	0.63 to 1.2	1.1 to 1.3
P_2O_5	0.34 to 0.56	0.50 to 0.59	0.62 to 0.82

Berhe et al. (1987) approached the characterization of Ethiopian basalts from a mixed viewpoint: of numerical status by radiometric age, and of non-numerical status by qualitative description. To the gist of their tabulation of these two factors we add our own factor of chronostratigraphy. The resulting tabulation follows:

NW & Central E	W and SW	Million years	Chronostratigraphy
	alkalic olivine basalts	15 to 7	middleMiocene(Langhian Serravalian boundary) to late Miocene (late Tortonian)
alkalic olivine basalts, high Al_2O_3, low TiO_2		25 to 13	latest Oligocene (latest Chattian) to middle Miocene (middle Serravalian)
	slightly alkalic to tholeiitic	30 to 21	early-late Oligocene boundary (Rupelian-

	basalts		Chattian boundary) to early Miocene (early Burdigalian)
slightly alkalic to tholeiitic		30 to 26	early-late Oligocene boundary (Rupelian-Chattian boundary) to late Oligocene (middle Chattian)
tholeiitic to slightly alkalic "transitional" basalts		34 to 30	early Oligocene (late Rupelian) to early-late Oligocene boundary (Rupelian-Chattian boundary)
	slightly alkalic to "transitional basalts	40 to 30	middle-late Eocene boundary (Bartonian-Priabonian boundary) to early-late Oligocene boundary (Rupelian-Chattian boundary)
	alkalic to slightly alkalic basalts	49 to 45	early middle Eocene (early Lutetian) to late middle Eocene (late Lutetian)
alkalic to slightly alkalic basalts, low Al_2O_3 high TiO_2		54 to 42	late early Eocene (late Ypresian) to late middle Eocene (early Bartonian)

In the same study, Berhe et al. divided Ethiopian Tertiary basalts into flood basalts and "central type volcanics," that is, into tholeiites and alkalic olivine basalts. Flood basalts range in age from 54 Ma to 21 Ma, and "central type volcanics" from 25 to 7 Ma.

Chiesa et al. (1989) studied the Tertiary flood basalts of Yemen, across the present Red Sea from Ethiopia proper (Fig. 6.13). There they distinguished two cycles of activity. The older cycle, represented by a thick plateau, lasted from 30 to 26 Ma, identical in duration to one of the intervals given by Berhe et al. (1987) for central and northwestern Ethiopia. The younger cycle, in the central and southern part of the plateau which is distinctive for its less alkalic basaltic lavas, lasted from 23 to 19 Ma (early Miocene, Aquitanian to early Burdigalian). Effusion of basaltic lava in Yemen continued after this younger cycle, but on a much smaller scale, with concentrations of activity around 10 Ma and 5 Ma, that is, late Miocene (early Tortonian) and early Pliocene (early Zanclean).

From the tabulation of Berhe et al. (1987), and from the description of Chiesa et al. (1989), the duration of Tertiary flood basalts in the greater Ethiopian magmatic province was from 54 to 5 Ma. The corresponding chronostratigraphic interval is from early Eocene (late Ypresian) to early Pliocene (early Zanclean).

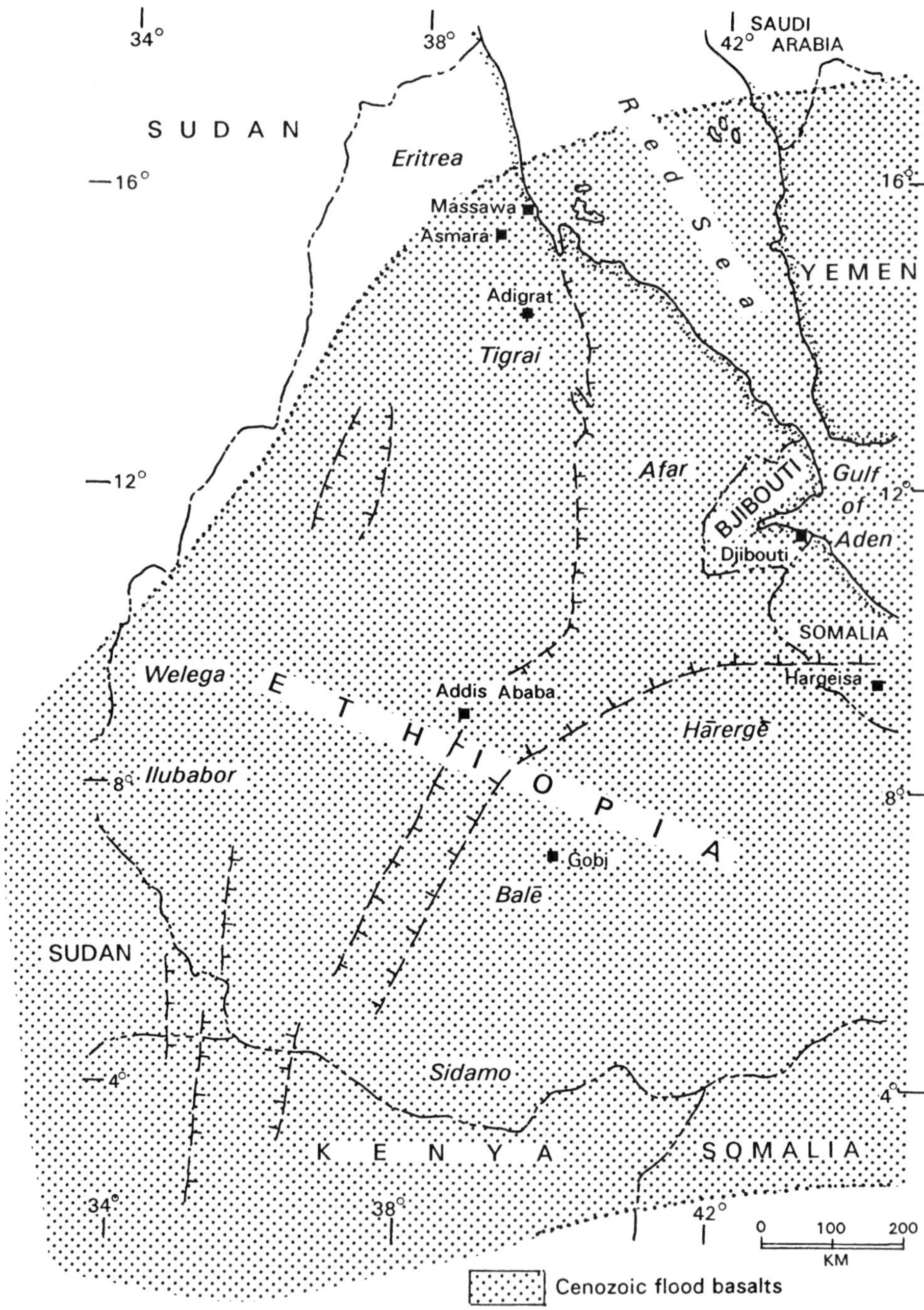

Fig. 6.13. Cenozoic basaltic effusives (shaded areas) in the greater Ethiopian magmatic province.

6.2.15 Tertiary Flood Basalts of Central America (Ovate)

From the border with Mexico southeastward to the northern half of Costa Rica, an important characteristic of Tertiary volcanism in Central America is the dominance of eruption over effusion, and a consequent abundance of ignimbrites in volcanics at outcrop (Fig. 6.14). In Nicaragua, McBirney and Williams (1965) recognized an older and a younger volcanic series. In the older volcanic series, basalts occur together with andesites, dacites, and their pyroclastics, in a group whose radiometric age is at least 15 Ma, chronostratigraphically equivalent to middle Miocene (Langhian-Serravallian). In the younger volcanic series, Weyl (1980) reported basalts together with "andesitic basalts" at "6.65" Ma, that is, late Miocene to latest Miocene (Tortonian-Messinian boundary).

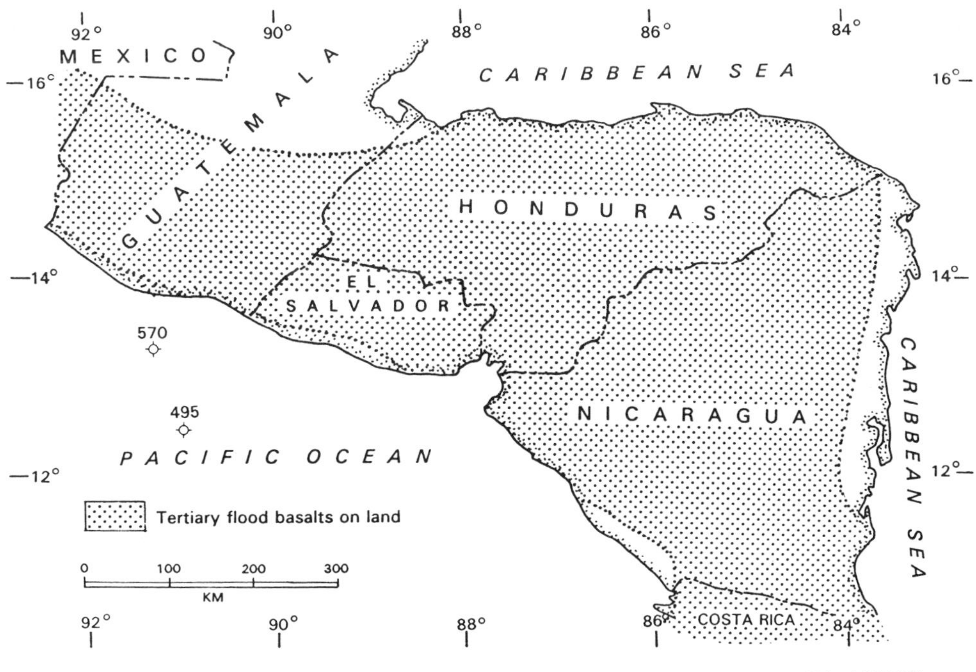

Fig. 6.14. Tertiary flood basalts on land (shaded areas) in Central America.

For Honduras, Weyl (1980) described the volcanics as "chiefly of andesitic and basaltic lavas and their pyroclastics together with a few acidic intercalations." This sequence has yielded a radiometric age of 28 Ma, or late Oligocene (early Chattian). For Guatemala, Weyl (1980) noted that "alternation of mafic and felsic products" corresponds to a similar alternation in Nicaragua. The alternation here includes porphyritic basalt and other basalt, both "calc-alkaline in character." In the Pacific Basin off Guatemala, 150 km SSW of the Guatemalan-El Salvadoran border, basalt occurs at the bottom of Drillhole 495 of the Deep Sea Drilling Program (Aubouin et al., 1982). Most of the 18 m penetrated are described as "phyric" basalt, in which olivine and plagioclase are recognizable as phenocrysts. The cover of this basalt consists of middle Miocene manganiferous chalk and chert.

Drilling to date in the Pacific and Caribbean, whether near to Central America or far distant, points to the potential for considerable volumes of effusive basalt. Post-effusion epeirogenesis may have removed large amounts, but with the whole of the Tertiary epoch in context, the sum of successive flows originally may have been sufficient to cover extremely large stretches of eastern Pacific and western Caribbean waters.

Drillholes 155 and 504 also have yielded evidence of basaltic volcanism farther south. Drillhole 155, located 550 km southeast of the Nicaraguan-Costa Rican border, reached basalt below middle Miocene chalk. Fragments of chalk within the basalt, with no evidence of metasomatism, suggested to van Andel et al. (1973) that the basalt at this location "is a flow rather than an intrusive body." A Miocene flood basalt thus becomes a possibility at Hole 155.

Drillhole 504, 1,100 km south of the Nicaraguan-Costa Rican border, established the presence of a record-breaking column of mafic rock, more than 1,075. 5 m long (Emmermann, 1985). The top 571 m of this column consists of "pillow basalts and brecciated pillow material with some interlayered thin lava flows (1 to 2 m) and occasional intercalations of massive cooling units up to 15 m thick." The underlying 209 m consists of "pillow basalts, massive units, and dikes, the latter becoming progressively more abundant with depth." The bottom 295.5 m consists "solely of dikes and massive cooling units, and which is interpreted as part of a sheeted dike complex." Emmermann characterized both pillow lavas and dikes as "olivine tholeiites," while Anderson et al. (1985) noted that the top of the mafic rock lies below late Miocene sediment. Neither Emmermann nor Anderson et al. suggested a profound disconformity or unconformity between sediment and underlying basalt. The top of the basalt at Drillhole 504 thus could be as young as Miocene.

Additional information on the column of mafic rock in Drillhole 504 is geochemical, and from data in Emmermann (1985) we summarize numbers for major-element oxides by weight percent as follows:

SiO_2	45.7	to	51.1
TiO_2	0.56	to	1.51
Al_2O_3	12.9	to	17.6
Fe_2O_3	1.37	to	4.11
FeO	4.8	to	8.8
MnO	0.15	to	0.39
MgO	7.7	to	9.9
CaO	11.0	to	13.4
Na_2O	1.53	to	2.45
K_2O	0.01	to	0.08
P_2O_5	0.04	to	0.18

In central Guatemala, Meyerhoff (1967) found a 100 km-wide sedimentary section in the north which grades into a volcanic section farther south. On the Caribbean side, Meyerhoff's studies suggest the possible disappearance of the volcanics between latitudes 15° S and 11° S, slightly before reaching the east coast of Nicaragua (Fig. 6.14).

Evidence cited from the various authors mentioned above leads to the conclusion that effusion of basalt across Central American land during 65 million years of Tertiary time covered approximately 300,000 km². This area, compared to 200,000 km² for the classic Tertiary flood basalts of Columbia River in the northwestern United States, indicates that the Tertiary flood basalts of present Central American constitute an effusive volcanic event of considerable significance.

6.2.16 Columbia River Flood Basalts (Tertiary-Quaternary), Northwestern United States (Ovate)

Russell (1893) described the volcanics of the Columbia River region as ". . . the most important geological series in the Northwest, and one which it will be convenient to designate by a specific name. As the region it occupies is drained almost entirely by the Columbia river, we venture to name it the 'Columbia lava'." Almost a hundred years later, current usage includes terms such as "Columbia River volcanism," "Columbia River basalts," "Columbia River Basalt Group," and "Columbia River flood basalt province." Figure 6.15 shows the extent of lavas onshore.

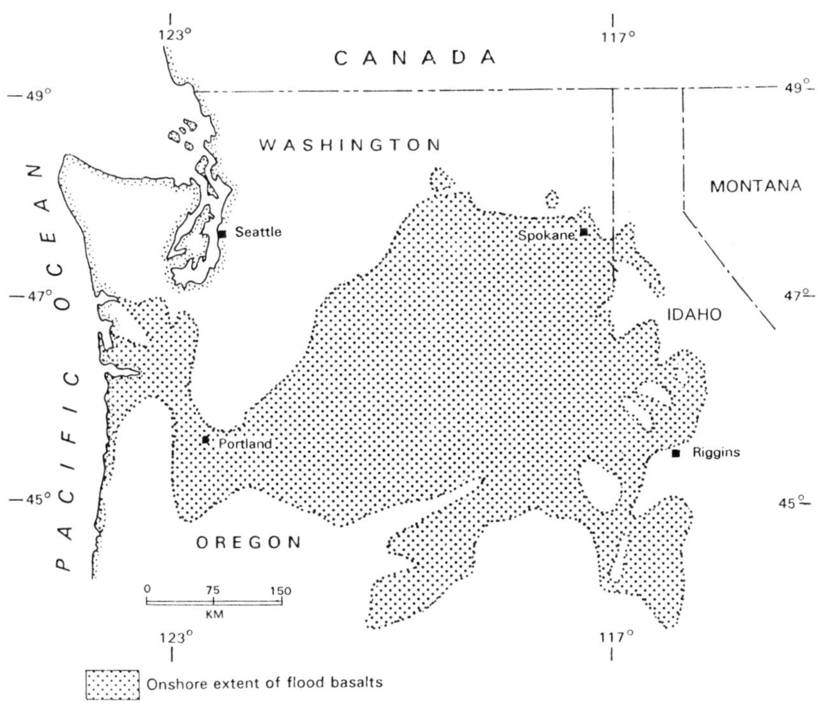

Fig. 6.15. Onshore extent (shaded areas) of flood basalts in the Columbia River region, northwestern United States.

Russell also wrote that the borders of his "Columbia lava" had "never been traced," but he quoted a contemporary estimate of 200,000 mi^2 for its area. Modern estimates vary to some degree. An area of 200,000 km^2, given by Rockwell International (1979), appears also in Hooper (1982). Tolan et al. (1989) calculated a "new" estimate of area from digitized isopach maps "of the 34 units that define the CRBG" (Columbia River Basalt Group). They submitted a figure of 163,721 km^2 for area, "far smaller than previously accepted estimates." The digitization of isopachs, while not necessarily more exact than other methods, is likely to command attention.

The Basalt Volcanism Study Project (1981), in an overview of the Columbia River Basalt Group, described that group as a "tholeiitic flood basalt sequence." Within that sequence, the Grande Ronde Basalt, which fills almost the whole area of outcrop in this igneous province, provides representative geochemical values for tholeiites, which we summarize from Swanson and Wright (1981) in terms of weight percent of major-element oxides, as follows:

SiO_2	52	to	56
TiO_2	1.8	to	2.7
Al_2O_3	13.9	to	14.4
FeO	11.3	to	12.3
MnO		0.2	
MgO	3.4	to	5.2
CaO	6.9	to	9.1
Na_2O	2.8	to	3.4
K_2O	1.0	to	2.0
P_2O_5	0.3	to	1.4

Rockwell International (1979) characterized the Columbia Plateau as "the Earth's youngest recognized flood-basalt province," forming between 16.5 and 6 Ma, two dates taken from Watkins and Baksi (1974) and McKee et al. (1977). Rampino and Stothers (1983) statistically analyzed 116 whole-rock K-Ar ages and 14 Ar-Ar ages. Their analysis resulted in dates of 18 to 16 Ma for the onset of flood-basalt activity in the Columbia River igneous province. The chronostratigraphic equivalent is early Miocene (Burdigalian). The interval of 16.5 and 6 Ma quoted above for the total time of activity covers early Miocene (Burdigalian) to latest Miocene (Messinian).

Watkins and Baksi (1974) concluded that "eruption" (effusion) of the Ronde Grande Basalt was complete by 14 Ma. If their conclusions are correct, the greatest volume of effusions occurred in less than 3 million years after onset, a comparatively rapid rate.

6.2.17 Mid-Ocean Ridge Flood Basalts (Linear)

Basalts below present oceanic waters cover an area that at the very least is more than 20 times greater than those covered by continental flood basalts. The percentage of flood basalts in the ocean consequently is a matter of great geological importance, and although there is a tremendous volume of literature on the geochemistry of midocean ridge basalts (MORB), there are relatively few studies of The extent and general character of oceanic flood basalts. Our purpose herein is only to mention the presence of midocean ridge flood basalts and note that they deserve extensive examination in future studies.

MacDougall (1988) considered differences and similarities in MORB (midocean ridge basalts) and CFB (continental flood basalts), but principally in terms of geochemistry. The differences, for MacDougall, appear on a basis of small amounts of radioactive material. Whether differences of this kind are significant appears to us open to question. On the side of similarities, MacDougall noted that both MORB and CFB are tholeiitic, yet their equality in this broad respect did not lead him to conclude that flood basalts are present in MORB.

Yoder (1988), in a general discussion of basalt as magma, assumed almost didactically that "basaltic floods" are as much an element of oceanic regimes as they are an element of continents. Two of his statements bring out his approach clearly: (1) " . . . the oceanic floods [i.e., the basaltic floods] appear to be . . . effective world-wide along the entire mid-ocean ridge system, currently 60,000 km long." (2) "The so-called flood basalts are not the result of an unusual set of circumstances. They occur both on continents and ocean floors as periodic instabilities in the pervasive plastic layer at the base of the lithosphere plates . . .." Yoder's second statement leaves no doubt that his own term of "basaltic floods" and Tyrrell's 1937 term of "flood basalts" are identical. He leaves no doubt that flood basalts represent an important element in and under the waters of the world's present oceans.

6.3 The Use of Geochemistry in Identifying Flood Basalts

6.3.1 INTRODUCTION

Geochemical/petrochemical studies of igneous rocks for many decades were restricted to (1) studies of the bulk chemistry (major compounds only) of each rock type, and (2) deviations from the "norm" determined for each rock type. High-pressure equipment and high-temperature studies were conducted in the search for the chemical phases and eutectics of rock melts. Such studies were invaluable in determining the origins of various rock types, and led to many classical papers, especially the Yoder and Tilley (1962) and Yoder (1976) treatises on the origin and generation of basalt.

With the advent of plate tectonics, petrochemistry was used increasingly as a supplement to traditional methods of identifying tectonic environments. The assumption was made that each tectonic environment had its own petrochemical "signature." When major-element studies failed to bear out this assumption, however, increasing attention was given to minor (trace) and rare earth elements. Regrettably, nearly all large-scale petrochemical research concentrated on the basalts (e.g., the NASA-sponsored Basaltic Volcanism Study Project published in 1981), and other rock types have failed to receive anything like the attention that the basalts received. As an inevitable consequence, many conclusions were made on the basis of basalt geochemistries alone. Our points are: (1) that a great deal of research---many decades, in fact---will be necessary before sound conclusions regarding the chemical "signature" of tectonic environments will or can be soundly based; and (2) even though the more silicic magma types are in very large part aggregates of crustal compounds and processes, they too have important scientific "messages" to impart. It is too early to reach final conclusions based only on basalt data.

The results of minor and rare-earth element studies, however, have been helpful, for they document in part the history of each sample with the use of spidergrams (Fig. 6.16). They also discriminate easily between midocean-ridge basalts and other basalts, although this already was possible from major element data alone. At first it was believed---and still is by many workers---that spidergrams would make it possible to identify each of the several tectonic environments in which basalts, andesites, and other rocks occur. However, as we document below, the ability of spidergrams to discriminate among most tectonic settings is doubtful without much additional information, partly from isotope data and, in the long run, with the aid of actual field data.

An important step that must be taken now is to standardize the *order* in which the trace and rare-earth elements appear on a spidergram (Fig. 6.16). Second, there is no consistency about which elements are included or excluded (Fig. 6.16), and this problem also must be resolved. Too often elements important to an interpretation are omitted on spidergrams. Finally, there is no consistency about which material is used for "normalizing" element plots. Currently some are chondrite-normalized; some are normalized against an idealized midocean-ridge basalt composition; and many are normalized against the composition of an hypothetical primordial mantle, a practice which, as Thompson et al. (1983) have noted, introduced unnecessary subjectivity into interpretations.

6.3.2 Basalt Magmas

According to Yoder (1976, 1988), the overwhelming choice of parental material from which basaltic liquid is to be derived is garnet peridotite. There are several reasons for this choice: (1) garnet peridotite is found in nodules among rocks that originated at levels deeper than anything known (as determined from geothermometry and geobarometry; Finnerty and Boyd, 1984; Yoder, 1988); (2) it has the appropriate seismic velocities to be a major rock type in the mantle (Anderson, 1987a, b); (3) it is one of the few rock types demonstrated to yield a basaltic liquid in the laboratory; and (4) it yields a residuum with the mineralogy and the relative abundances of other common rock types found among nodules (garnet harzburgite, harzburgite, dunite; Mathias et al., 1970).

Garnet peridotite, according to Yoder (1976, 1988), melts in a eutectic-like fashion. At a depth equivalent to 40 kb (ca. 130 km), the eutectic composition is 3% olivine, 47% clinopyroxene, 47% garnet, and 3% orthopyroxene. No matter what the proportions of these four minerals in garnet peridotite, all eutectic liquids will be of basaltic composition.

The volume of eutectic liquid is fixed primarily by the volume of olivine assumed to be in the parent material. If the average amount for nodules of 60% olivine is used for calculation, as much as 30% basaltic liquid can be derived from the parent material. (If the amount of olivine is smaller, probably larger amounts of basalt can form.) In the 60% case, the major constituents of the liquid will not change until a phase in the eutectic is lost after the 30% of liquid is reached. More advanced melting would yield picritic basalt. Presnall and Hoover (1987), however, studied 1,700 analyses of midocean-ridge basalt glasses, but did not find one picritic basalt. From this fact, one may conclude that melting is terminated by the fractional melting process; that is, the midocean-ridge liquids were extruded before 30% melting was achieved.

It is important to be aware that the concentration of incompatible trace elements* changes greatly in this basaltic liquid, depending on their relative partition coefficients, initial concentrations, and dilution rates. In the midocean-ridge basalts, the volume of incompatible

*"Incompatible elements are those most likely to be transported by melts and other fluids passing through the mantle. They are, therefore, the elements most likely to preserve evidence of mantle enrichment and depletion processes in their relative abundances." (Fitton et al., 1991, p. 13702).

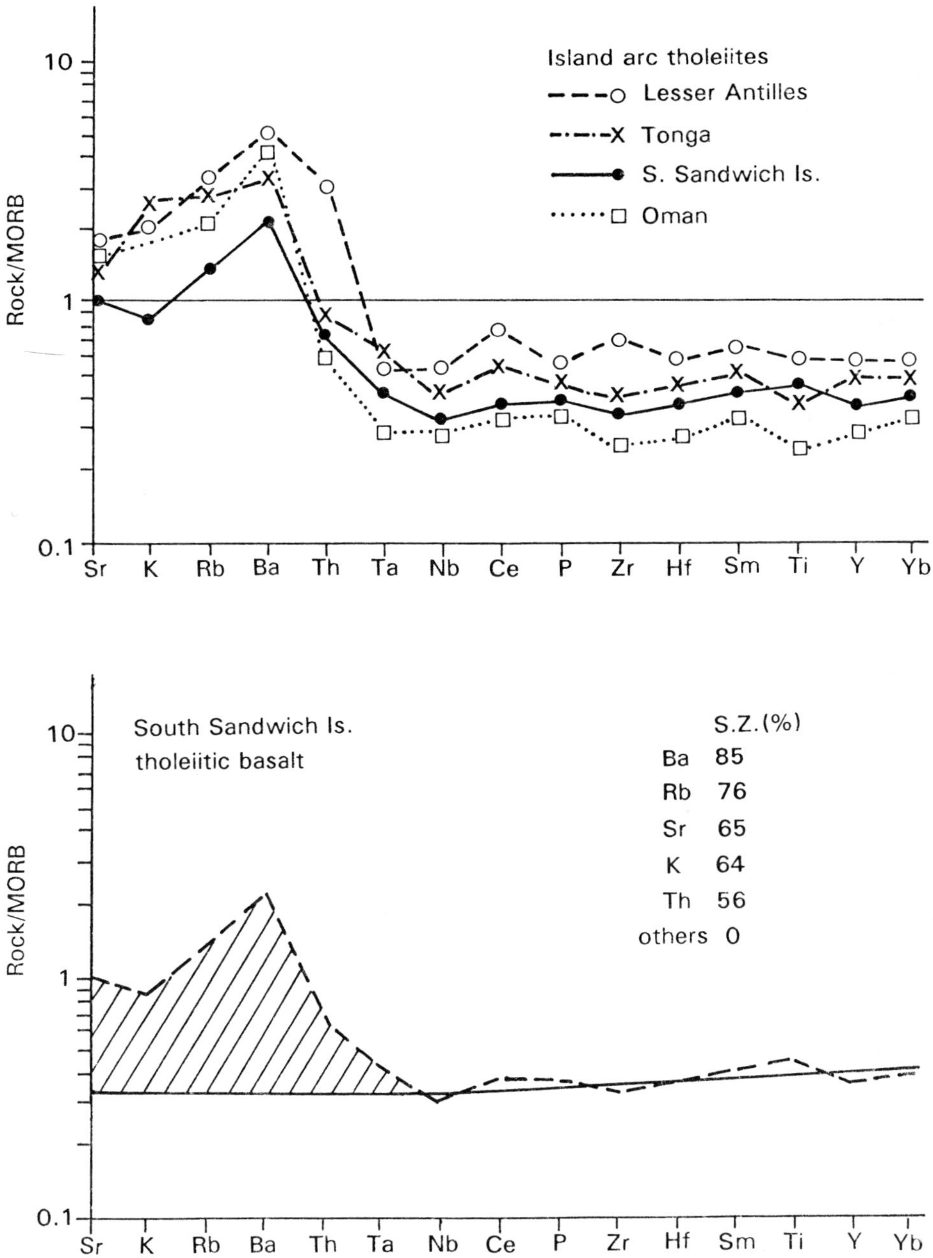

Fig. 6.16. Geochemical patterns of island arc tholeiites (a) and the interpretation of the pattern for the South Sandwich basalt (b). According to Pearce (1993), in (a) the Tonga sample is from Falcon Island and the Lesser Antilles sample from St. Vincent. In (b) s.z. (%) = percentage of the subduction component in the basalt for the elements listed. From Pearce (1983).

minor elements is very small, a fact that suggests that the parental material has already undergone some partial melting and loss of liquid, but still retains parts of all major melt phases (Yoder, 1988).

Several processes involved in the emplacement of magmas in the crust complicate the above picture. The composition of surface samples from rocks that were molten and under high pressure is not necessarily that of the parental liquid at depth. This is true because (1), as the liquid rises, internal reaction relations take place that successively eliminate olivine and orthopyroxene (O'Hara and Yoder, 1967; Kushiro and Yoder, 1974; Yoder, 1988). Hence the composition of the basalt may be altered considerably during its rise from ca. 130 km; (2) of heat loss; (3) the change in pressure further changes the liquid composition; and (4) the rise of the melt produces a change in the stable phases within the liquid.

The reasons for the differences between continental flood basalts and midocean-ridge basalts are related in part to the above factors, but differences in the thickness of the lithosphere clearly must exert an important influence as well (Yoder, 1988). The penetration of an old, thicker, continental massif by basalt melt is clearly more difficult than that of the much thinner oceanic lithosphere, although the rising magma rises in the same way under both lithospheres, following the Peach-Köhler climb force (Newton's Law of Gravity; Weertman and Weertman, 1964; Weertman, 1971; Gretener, 1969; Corry, 1988; Walker, 1989) and stops when the level of neutral buoyancy has been reached (Walker, 1989). The longer---or slower---the rise beneath the continental crust, the greater the fractionation, as reflected in the more iron-rich character of the continental lavas (Wilkinson, 1981; Yoder, 1988). Deep-seated magma segregation beneath the continents provides for more alkalic parental magmas, a greater range of enrichments, and a greater variation that depends on repose time, interactions with the continental crust, and the rates of ascent. The bimodal character of so many continental flood basalts implies the presence for periods of time of mu ltiple magma chambers.

6.3.3 Studies of Minor and Rare Earth Elements

When studies of major elements and compounds revealed difficulties in discerning chemical signatures peculiar to each tectonic environment, research began to focus on studies of minor (trace) elements, rare-earth elements, and chemical isotopes. Although a high degree of success has been claimed for such studies, the facts tell quite a different story. Indeed, it is a poor reflection on the state of current geoscientific research that the eagerness of some researchers to satisfy preconceived hypotheses and models has led some into publishing material that is scientifically sound. Minor (trace) element, rare-earth element and chemical isotopes studies are summarized for the following environments:

Midocean-Ridge Basalts (Ocean-Floor Volcanism)

Midocean ridge basalts are severely depleted in minor and rare-earth abundances, unlike most other basalt types. Moreover, midocean-ridge basalts have isotope abundances that are smaller than in other basalt types, and the isotope ratios in midocean-ridge basalts are distinctive. Thus studies of minor and rare-earth elements confirmed what already was known from studies of major elements and compounds, specifically, that midocean-ridge basalts are easily identified. The participants in the Basalt Volcanism Study Project (1981, p. 132) wrote that "Early studies emphasized the unique and consistent chemical composition of ocean-floor basalts, and in particular their tholeiitic characteristics: constant silica values, low potassium abundances, low

K/Na ratios, low magmaphile content and high K/Rb ratios (e.g., Engel et al., 1965; Kay et al., 1970). More recent studies . . . have shown that chemical and petrological diversity at a single [sampling] site is frequently as great as that previously identified for ocean-floor basalts in general . . .. Nevertheless, using chemical discriminants, [midocean-ridge] basalts are readily distinguishable from basalts erupted in other tectonic regimes. . ., with the possible exception of 'primitive' tholeiitic basalts from island arcs."

Ocean-Island Basalts (Oceanic Intraplate Volcanism)

These are the basalts of oceanic plateaus and linear island and seamount chains. Examples include the Iceland Plateau and the Hawaiian-Emperor Ridge. The participants in the Basalt Volcanism Study Project (1981, p. 161) wrote, "Erupted basalts are dominated by enormous volumes of tholeiites, chemically distinct from mid-ocean ridge tholeiites. . .. These basalts may be oversaturated quartz tholeiites, slightly undersaturated olivine tholeiites, or picrites. In the simplest model, the late stage . . . activity is dominated by extrusion of alkaline series basalts which form a thin cap over the tholeiitic base . . .." The study-project participants subsequently (p. 190) wrote, "Relative to [midocean-ridge basalt] sources, the island sources have higher total iron, TiO_2 and K_2O. Lower Al_2O_3 contents may be an intrinsic property as well; however, garnet retention in the source, which is compatible with the [rare-earth element] data, could also account for this feature.

"Incompatible element abundances are 2-5 times chondrites ([midocean-ridge basalt] sources are 1-2 times chondrites) and Rb/Sr ratios are ca. 0.03 as opposed to 0.01 in [midocean ridge-basalt] sources."

Continental Flood Basalts (Continental Intraplate Volcanism)

Chemically there is no distinction between continental flood basalts and basalts formed in a continental rift (Basalt Volcanism Study Project, 1981). The project participants wrote (p. 59), "It has long been recognized that 'average' or 'typical' continental flood basalts have a distinctive major element composition range relatively high in Si, Fe, Ti, and K. Pearce et al. (1977) found, for example, a fairly clear statistical distinction between continental basalts (including plateau basalts) and others on an MgO-FeO-Al_2O_3 diagram. However, the major basalt plateaus actually show a wide range of compositions, overlapping with most of the other principal basalt types."

Regarding continental rift volcanism, the Basalt Volcanism Study Project (1981, p. 108) scientists wrote, "The volcanic rocks in continental rifts range from extensive flood basalt provinces (Columbia Plateau, Keweenawan of Lake Superior, Deccan Traps, Paraná Basin, Newer Basalts of southeastern Australia) to sporadically distributed small centers of volcanic activity associated with largely amagmatic graben (Rhinegraben)."

Volcanic Arc Basalts ("Subduction" Basalts)

Volcanic arc basalts are erupted in two fundamentally different settings, ensimatic (western Pacific, Caribbean Sea, Scotia Sea) and ensialic (e.g., Andes and other places in the eastern Pacific). They are host to a complete range of eruptive rocks from basalt to rhyolite. The Basalt Volcanism Study Project (1981, p. 193) participants wrote that "Basalts form an important component of 'island arc' volcanic suites. The most important distinction of these from other suites of basaltic rocks lies in their tectonic setting, being associated with the great suite of 'orogenic' lavas erupted at convergent plate boundaries. Two subsets may be distinguished:

'island arc' and 'continental margin' volcanic suites. It is not clear that any fundamental genetic difference exists between these two types. There is overlap in all chemical and mineralogical parameters. These voluminous rocks, which comprise andesites, dacites and rhyolites in addition to basalts [Fig. 6.17], are second in volumetric importance only to Mid-Ocean Ridge Basalts (MORB) in the terrestrial volcanic environment."

Island Arc Basalts--According to the Basalt Volcanism Study Project (1981, p. 210) scientists, "The distinction from other basalts is somewhat . . . difficult to make . . .. . . .the island arc basalt compositions grade into those of marginal basin basalts . . .. The following generalizations may be used to distinguish island arc basalts from [midocean-ridge basalts], with the proviso that any one of these criteria may be mimicked in another environment. In comparison with [midocean-ridge basalts], many arc basalts have higher Al_2O_3, K_2O, Sr and Ba and lower TiO_2, K/Cs and K/Rb ratios. [Rare-earth element] patterns and other trace element abundances show complete overlap and do not serve to discriminate arc basalts from [midocean-ridge basalts] or marginal-basin basalts."

Continental Margin Volcanic Arcs--The Basalt Volcanism Study Project (1981, p. 1000) reported that, ". . . it is clear that these [volcanics] differ fundamentally from island arc volcanics. The proportion of sialic rocks is generally much higher (as in the Andes or Sumatra . . .), and the isotopic signatures of Sr, Pb, Nd (and O and H) are consistently displaced in a direction suggesting significant crustal interaction." In contrast to these continental arcs, the ensimatic arcs of the western Pacific contain much higher percentages of basaltic rocks.

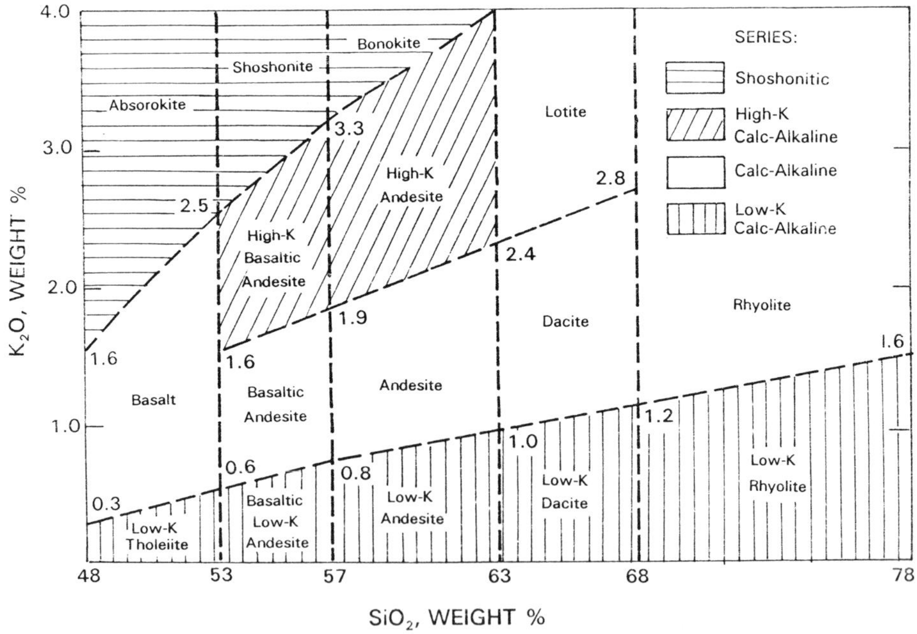

Fig. 6.17. Classification scheme for island arc volcanic rocks. From Basalt Volcanism Study Project (1981, p. 193).

6.4 Geochemical Comparisons among Basalts Erupted in Different Tectonic Settings

6.4.1 MIDOCEAN RIDGE BASALTS

As we have noted, the midocean-ridge basalts are distinguished readily from nearly all other basalt types. It is not possible to discriminate geochemically, however, between the basalts of the oceans and the basalts of the marginal seas. A few rare 'primitive' basalts in the volcanic arcs also may have midocean-ridge basalt geochemical signatures.

6.4.2 OCEAN ISLAND AND CONTINENTAL FLOOD BASALTS

Tholeiitic and alkalic basalts characterize these two closely related settings. A dramatic example, the Cameroon Line, was studied by Norry and Fitton (1983). The Cameroon Line is the only intraplate alkaline volcanic province to cross a continental margin (Fitton, 1987). This line is characterized by a chain of Tertiary-Recent volcanoes that extend from the island of Pagalu in the Atlantic, through Cameroon and northern Nigeria, into the interior of Africa (Fitton, 1987). The geochemistry along the Cameroon line is illustrated on Figure 6.19. Norry and Fitton (1983, p. 14) wrote, "The Cameroon line (Figure 4 [our Fig. 6.19]) is unique among within plate volcanic provinces in that it straddles a continental margin. Half the line lies in the Atlantic ocean . . . . The other half has been erupted on to continental crust . . . . The line has been active volcanically for the last 65 Ma . . .. The volcanoes are largely composed of alkalic basalt, trachyte, phonolite and alkali rhyolite. The only exception is the nephelinite

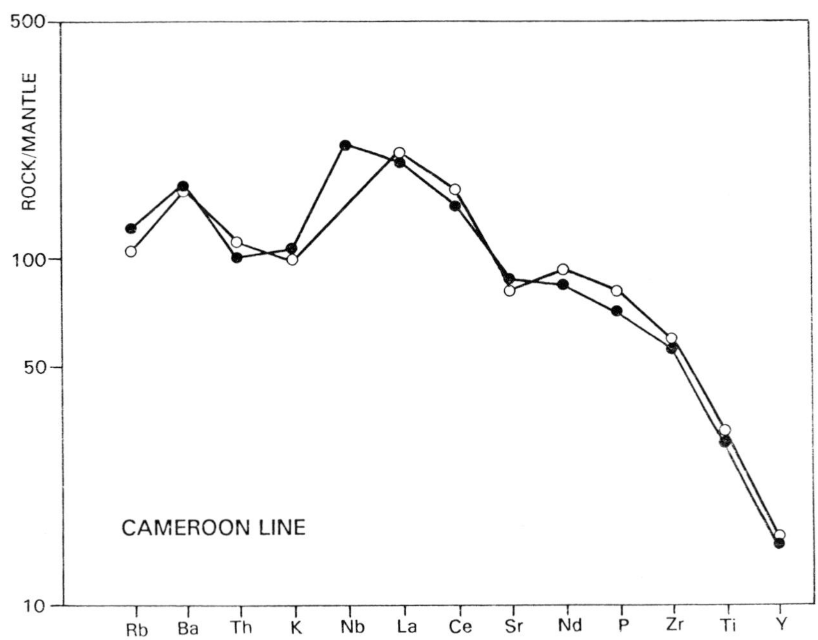

Fig. 6.18. Incompatible element 'mantle' normalized diagram for alkali basalts from the Cameroon line. Solid circle = continental sector; open circle = oceanic sector. From Norry and Fitton (1983).

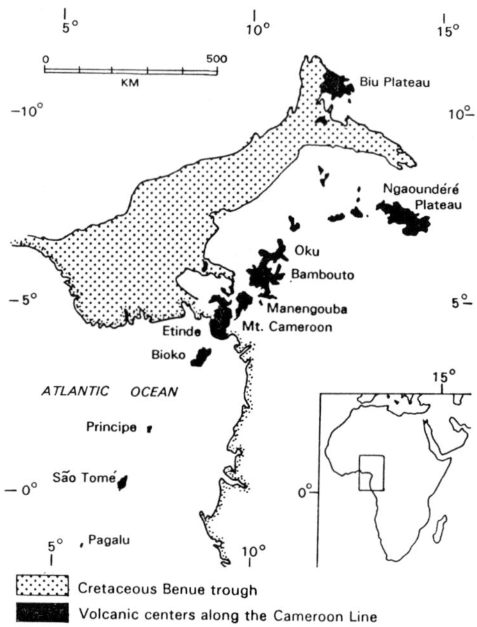

Fig. 6.19. Map of the volcanic centers along the Cameroon line (black). The stippled area is the Cretaceous Benue trough. From Norry and Fitton (1983).

volcano Etinde. This is, therefore, an ideal area in which to compare oceanic and continental within plate volcanism.

"A total of 144 oceanic and 134 continental alkali basalts . . . collected from the main volcanic centers on the line (Figure 4 [our Fig. 6.19]) have been analyzed for a range of incompatible elements using X-ray fluorescence techniques. All the samples were analyzed in Edinburgh against the same batch of standards to ensure internal consistency with the results. The oceanic and continental mean compositions are plotted, normalized to primordial mantle, in Figure 5 [our Fig. 6.18], and it is clear that alkali basalts from the two environments are virtually identical. It follows that the two sets of basalts had a similar source and they evolved through similar partial melting and fractionation processes."

Thompson et al. (1983, p. 172) made several pertinent observations concerning the relations among midocean-ridge, ocean-island, and continental island basalts. They wrote, "Alkalic magmatism is a . . . very persistent feature of continental flood provinces . . .. Several large examples such as Ethiopia . . . and the North Atlantic Tertiary (Brito-Arctic) . . . are substantially alkalic and there is a complete gradation into totally alkalic magmatism in such areas as the East African and Rio Grande Rifts. Even in the largest tholeiitic provinces, such as Karroo and Deccan, there are minor occurrences of alkalic lavas forming integral parts of the magmatism . . .. Within the ocean basins, the alkalic magmas are associated with [ocean-island basalt] tholeiites and <u>not</u> with [midocean-ridge basalts]. The Hawaiian volcanoes and the Iceland-Reykjanes Ridge are obvious examples of this association. Indeed, . . . [ocean-island basalt] magmas may form as much as 25% of the total oceanic crust.

"The spidergrams of representative [ocean-island basalts] and continental alkalic lavas are essentially indistinguishable [Fig. 6.20A] and we agree with Norry and Fitton (1983) that the two groups are elementally and isotopically the same . . .. The spidergrams of [ocean-island basalt] tholeiites are closely similar to those of the [continental flood basalts] [compare Figs. 6.20A and 6.20B]. We therefore propose that [ocean-island basalts], rather than [midocean-ridge basalts], should be compared and contrasted with [continental flood basalts].

"We know of no cases of magmas erupted in a continental setting with the precise elemental and isotopic composition of [midocean-ridge basalts]. Nevertheless, there are several sorts of [continental flood basalts] which closely resemble [midocean-ridge basalts] . . . at first sight. Thompson et al. (1983) then give as an example the Brito-Arctic basalt province of Baffin Island, Svartenhuk (West Greenland), Skye, and elsewhere in that province, noting (p. 173) that "The spidergrams of isotopically-uncontaminated examples of these lavas closely resemble [ocean-island basalt] magmas from the Reykjanes Ridge and similar localities. . .."

6.4.3 Continental Flood and Volcanic-Arc Basalts

We quoted the Basalt Volcanism Study Project, (1981, p. 210) as having stated that volcanic arc basalts are very difficult to tell from the other basalt types. Thompson et al. (1983, p. 173, 175) enlarge upon this topic. They query, "How may continental flood basalts be distinguished from the products of subduction-related magmatism?" Then they continue, as follows: "At first sight this seems to be a ludicrous topic. A volcanologist studying current igneous activity could hardly fail to distinguish a fissure-erupted basalt plateau, in an area of regional crustal extension, from a row of andesite volcanoes above a Benioff zone. Nevertheless, a serious problem arises when dealing with ancient, deeply eroded swarms of sills or dykes. In these cases the original plate-tectonic environment may be far from clear and the surface manifestations of the volcanism long since eroded away.

"Is it, therefore, easy to distinguish chemically between [continental flood basalts] and subduction-related basalts (using samples only from unambiguous tectonic environments)? Far from it, the two groups are *extremely* hard to tell apart. The majority of Mg-rich (> 8% MgO) subduction-related basalts are silica-saturated (quartz-tholeiites, magnesian andesites or boninites), whilst *Mg-rich* [continental flood basalts] are silica-undersaturated . . .. But there is a wide zone of normative overlap between the two groups and their related basaltic andesites have almost identical ranges (Ewart , 1982). Their spidergrams and ranges of isotopic ratios are very similar (e.g., Sun, 1980; Basalt Volcanism Study Product, 1981) and this likeness remains if only the subduction-related basalts are considered."

6.4.4 The Lebombo or East Karroo Continental-Flood Basalts

To get a firm idea of just how *much* variation in trace and rare-earth elements can take place, we cite the example of the southern Lebombo (Karroo) flood basalts. Cox (1983, p. 156-157) wrote, "Finally, it is interesting to note the relationships of Karroo lavas in the Ti-Y-Zr diagram used originally by Pearce and Cann (1973) to discriminate between within plate basalts and other types. As data have accumulated this has become a complex question. . . .it is clear that northern Lebombo [Karroo] rocks plot in the within plate field as expected, but southern Lebombo [Karroo] and Lesotho [Karroo] rocks plot firmly outside it in the fields of calc-alkali basalts, ocean floor basalts and low-K tholeiites. . .. Some south Lebombo rocks (e.g., primitive Rooi Rand dolerites, also a few of the lavas) have trace element patterns very similar to island

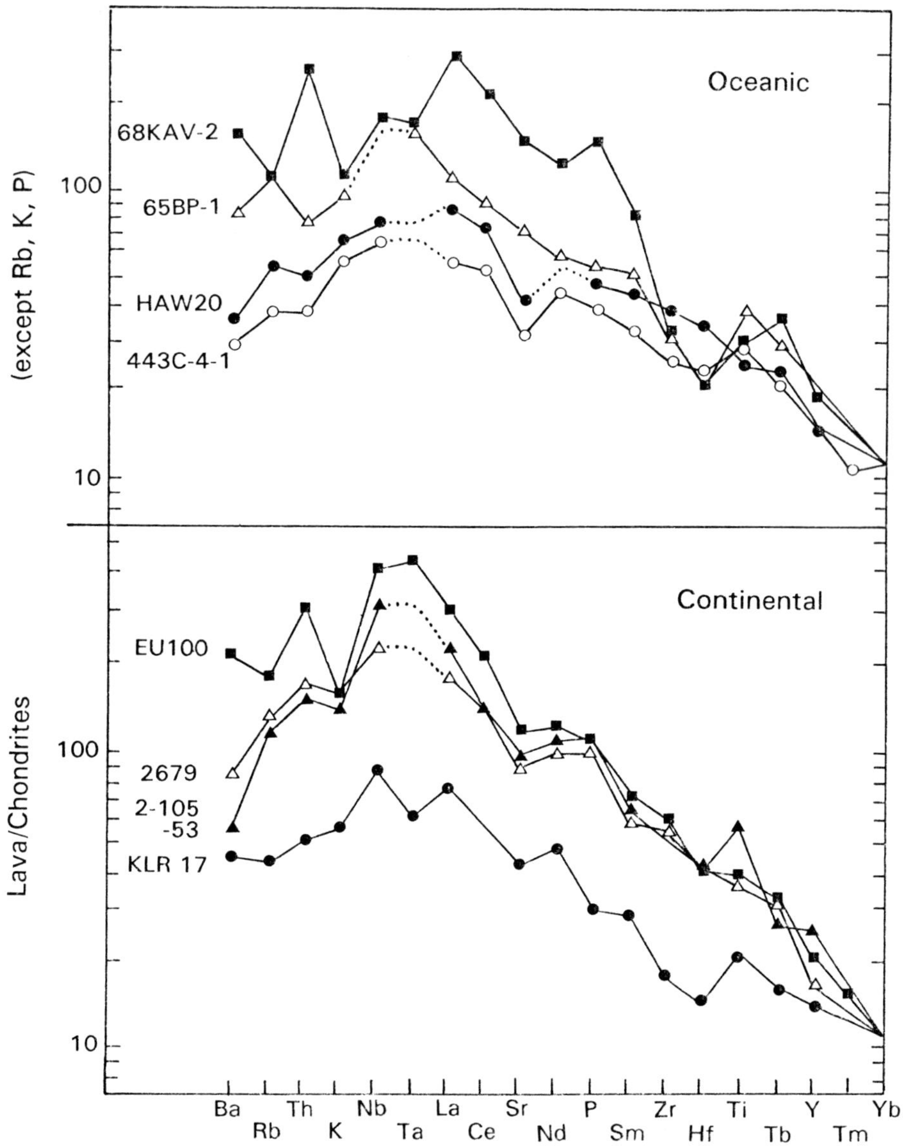

Fig. 6.20. Spidergrams of representative oceanic and continental alkalic and strongly alkalic lavas. From Thompson et al. (1983): "Key to samples: 68KAV-2, nepheline melilitite and 65BP-1, basanite, Oahu, Hawaii (Clague and Frey, 1982); HAW20, alkali olivine basalt, Kohala, Hawaii (Basalt Volcanism Study Project, 1981); 443C-4-1, alkali olivine basalt, Suiko, Emperor seamounts (Bence et al., 1980); EU100, melilitite, Hegau, Germany (unpublished data); 2679, basanite, Mt. Shadwell, Australia (Frey et al., 1978); 2-105-53, basanite, Dry Valley, Antarctica (Kyle, 1980); KLR 17, alkali olivine basalt, Gregory Rift, East Africa (Baker et al., 1977). The irregularities in the spidergram of melilitite 68KAV-2 are discussed by Clague and Frey (1982), who conclude that they were caused by various minor phases remaining stable within the upper-mantle residuum during the genesis of this magma. Note similar irregularities in the Ba-Nb sections of the spidergrams of mililitite EU 100 and the basanites."

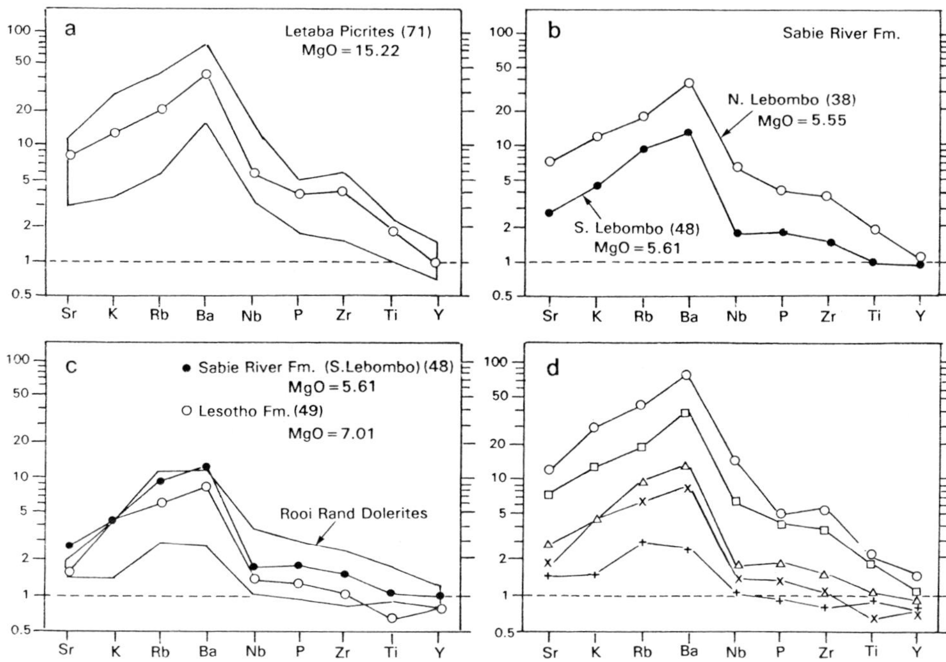

Fig. 6.21. From Cox (1983): "(a) Incompatible element concentrations of picrite basalts from the Letaba Formation in southeast Zimbabwe and the northern part of the Lebombo monocline. All values are normalized against typical MORB of Pearce et al. (1981). Solid line represents average of 71 samples from the northern Lebombo, Nuanetsi and Tuli. Shaded area represents total range shown by the suite. Average MgO is 15.22 weight per cent. (b) Incompatible elements for lavas of the Sabie River Formation. Open circles labelled N. Lebombo represent average of 38 samples (average MgO = 5.55 weight per cent) from Nuanetsi, the northern part of the Lebombo monocline, and from the Sabie River section (north-central Lebombo). Solid circles represent average of 48 samples (average MgO = 5.61 weight per cent) from Zululand, Swaziland, and the Komati River section (south-central Lebombo). (c) Incompatible elements for the Lesotho Formation (open circles). This is the average of 49 samples (average MgO = 7.01 weight per cent). S. Lebombo lavas from (b) are shown for comparison. Shaded area represents field of Rooi Rand dolerites from Cox and Clifford (1982). (d) Summary diagram. Open circles = most enriched rocks from Letaba Formation (upper edge of shaded area in (a)); squares = Sabie River Formation, northern area; triangles = Sabie River Formation, southern area; diagonal crosses = Lesotho Formation; plus signs = most primitive Rooi Rand dolerites (lower edge of shaded area in (c)). "

arc rocks . . . [Fig. 6.21]." Nor is this a minor example that can be dismissed as "atypical" or "unusual" or "unique." (The southern Lebombo province embraces a volume well in excess of 300,000 km^3 of continental flood basalts, flood rhyolites, and ignimbrites [Dingle et al., 1983; Bristow and Armstrong, 1989].) We discuss several examples in the paragraphs that follow.

6.4.5 Peruvian Andes

Figures 6.22, 6.23, and 6.24 show respectively the spidergrams published by Atherton et al. (1985a, 1985b) for the Berriasian (lowermost Cretaceous; Puente Piedra Group), Albian (uppermost Lower Cretaceous; Casma Group), and Tertiary (Calipuy Group). We review briefly their findings.

According to Atherton et al. (1985a, p. 52), "The trace element patterns shown in Figure 6.11 [Fig. 6.22 herein] are revealing. Firstly the rather variable nature of the rocks is clearly shown. Secondly, the distinction between V17 and V18 and the rest is very evident. These two rocks show a clear enrichment relative to [midocean-ridge basalts] from Sr to Yb, at least, strong but decreasing enrichment in Ba, Th, Ta and Nb and an almost linear decrease in enrichment from P to Yb with an average slight Ti enrichment relative to Y and Yb. Cr and Ni are strongly depleted. These characteristics indicate marked alkaline or within-plate components, while the enrichment in low ionic potential elements and marked depletion of Cr and Ni indicate an island-arc tholeiite component, although Y and Yb are relatively undepleted relative to tholeiitic mid-ocean ridge basalt. The patterns are best interpreted as 'anomalous' arc magmas with characteristics of alkaline basalts (Pearce, 1982). De Long et al. (1975) noted that sodic alkaline lavas occur at specific sites in arc systems, e.g., along lateral edges of subduction zones, or where a fracture perpendicular to the trench is being subducted. Similar anomalous magmas have been found close to continents during the initial stages of back arc spreading or back arc rifting . . .." In a subsequent passage, the authors wrote (p. 55), "Cr/Y relations show a clear fractionation type trend to low Cr contents, the most unfractionated rocks lying within the overlap of [midocean-ridge basalt], within-plate basalt and volcanic arc fields." Farther along in their text, the authors state (p. 55), the "The Puente Piedra rocks all lie in the volcanic arc field . . .." which suggest to us some confusion in their interpretation.

Regarding the overlying Albian Casma Group volcanics (Fig. 6.23), Atherton et al. (1985a, p. 55) wrote that "Cr v. Ce/Sr plots of Casma rocks . . . lie in the volcanic arc basalt field and the [midocean ridge basalt] field." Then they wrote (p. 55), "Trace element patterns of the Casma rocks [our Fig. 6.23] show a pattern very similar to [island-arc tholeiites] with low K, high Ba and Ce and P peaks. However, the compatible elements Sc, Cr and Ni are significantly higher than [island-arc tholeiites]. This plus the new M/FM of some of the basalts indicates a non-arc component. In these respects and others . . ., the Casma rocks show [midocean-ridge basalt] type characteristics and are very similar to basalts from the back-arc spreading centre in the East Scotia Sea . . .." Then the authors comment (p. 56), "The variation in chemistry in back arc marginal basins or spreading centers is so great that it is hazardous to designate basalts as marginal basin or back-arc rocks on chemistry alone. Many have similarities with those generated at mid-ocean ridges while others have characteristics more readily associated with island arc and calc-alkaline basalts." These authors cannot discriminate among tectonic environments without information from a source other than the chemistry.

Atherton et al. (1985b), in a separate but parallel study of the overlying Tertiary Calipuy Group (Fig. 6.23), reported that the group consists of basalt (14%), basaltic andesite (31%),

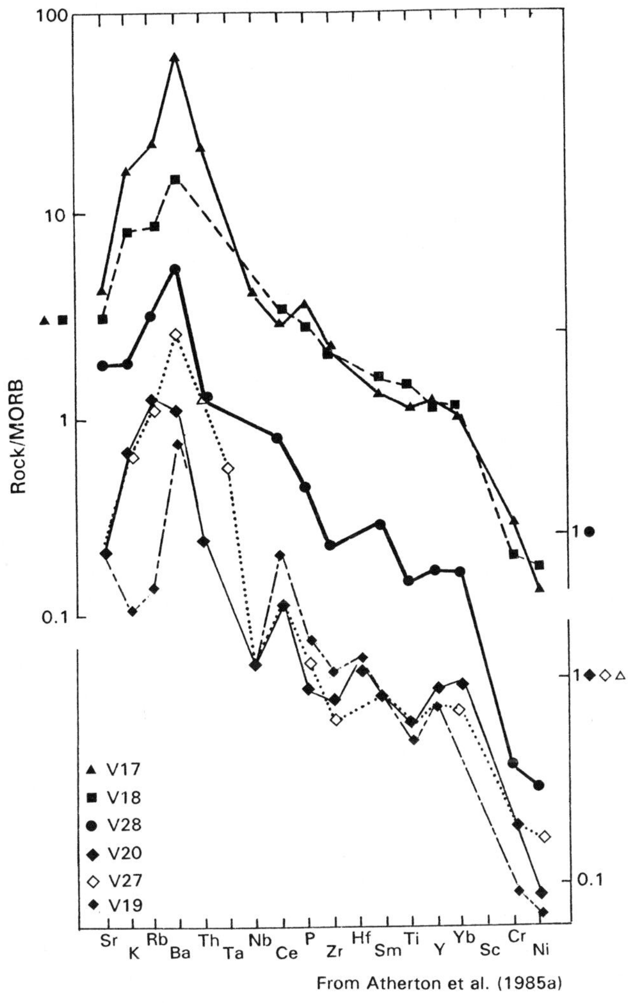

Fig. 6.22. Mid-ocean ridge basalt-normalized trace-element patterns for typical basalts from the (a) Albian Casma Group and (b) Berriasian Puente Piedra Group.

andesite (21%), dacite (22%), and rhyolite (12%). The rocks range from calc-alkaline to tholeiitic. Because of the presence of the siliceous lavas, andesite through rhyolite, the authors had no difficulty in assigning the entire Calipuy Group to a continental volcanic arc setting, e.g. (p. 277), "The multi-element plots [our Fig. 6.23] show a clear volcanic arc pattern . . .." Yet, in detail, it is nearly impossible to separate Figure 6.23 from Figure 6.22. Both show peaks at Rb-Ba-Th, Ce, Sm, and Yb and troughs at Sr, Ta-Nb, Zr-Hf, and Ti. The principal difference between the Puenta Piedra Group (Fig. 6.22) and the Calipuy Group (Fig. 6.24) is that the former consists almost entirely of basalt and basaltic andesite, whereas in the Calipuy Group, the full spectrum of volcanic-arc rocks has been preserved. In our view, it is not possible to assign a tectonic environment on the basis of spidergrams alone.

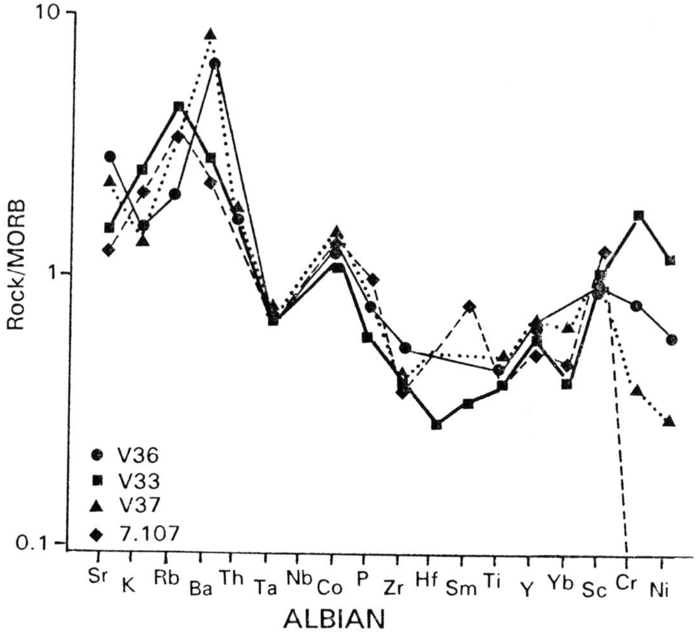

From Atherton et al. (1985a)

Fig. 6.23. Normalized trace-element patterns for the Albian Casma Group.

6.4.6 Pigafetta Basin, Western Pacific Ocean

The Ocean Drilling Program (ODP) Hole 801C recovered the first rocks of Jurassic age encountered in the deep Pacific Basin. The hole, drilled in 5,674 m of water east of the Marianas Arc, penetrated Upper Jurassic strata and reached total depth in basalts that yielded a radiometric date of 158 Ma (Callovian). The basalt is overlain by fossiliferous deep-sea Callovian strata (Floyd et al., 1991).

The basalt section showed unusual chemistry for a site located in 5,674 m of water. The lower basalt was normal tholeiitic midocean-ridge basalt, but the upper flows and sills were ocean-island basalt, an alkali basalt. Floyd et al. (1991) explain the anomalous presence of alkali basalt in this location by sea-floor spreading, i.e., the alkali basalt formed on the midocean ridge and reached its present location by sea-floor spreading.

Figures 6.25 and 6.26 show the chondrite-normalized rare-earth-element plot for the basalts, and the variations in Zr/Y and Zr in the same basalts (Floyd et al., 1991). The rare-earth element plots clearly differentiate between the alkali and tholeiitic basalts. The former are enriched in rare-earth elements; the latter are depleted in rare-earth elements.

We wish to emphasize that the seafloor in the deepest parts of an ocean basin can yield geochemical surprises. Midocean-ridge basalts are not the only chemical variety that is present at such depths. Hence the presence of alkali basalt does not automatically mean an ocean-island environment.

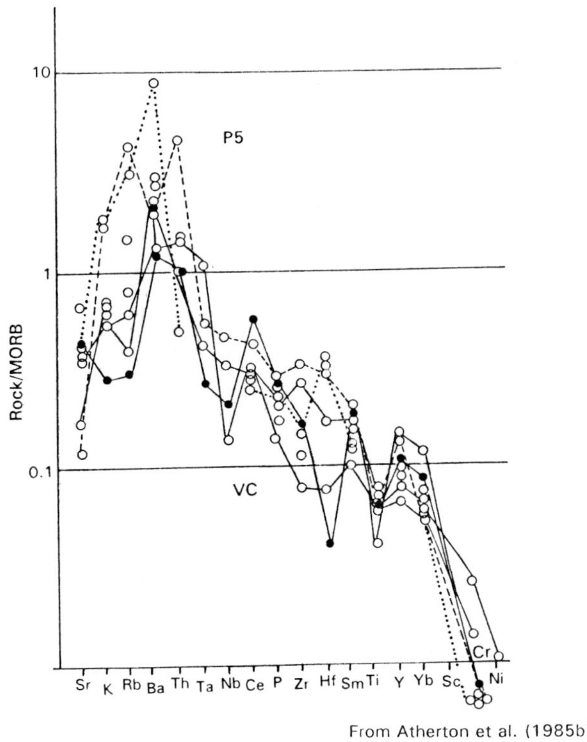

Fig. 6.24. Multi-element plot (rock/MORB) for selected basalts of the Tertiary Calipuy Group: high-K and intermediate-K basalts (P5); low-K basalts (VC).

6.4.7 Conclusion

Our examination of the literature on basalt rocks has led us to conclude that geochemistry is useful in distinguishing between midocean-ridge basalts and other basalts. This is true of bulk geochemistry, major-element geochemistry, and minor (trace) element and rare-earth element geochemistry. Geochemical techniques are not yet capable of discriminating among the various tectonic settings other than that of the midocean ridge. Exceptions to this statement do exist, but only in areas where the investigator has exceptional knowledge of the field relations among the various igneous units that he/she is investigating. Geochemical techniques *are* useful, however, in deciphering the chemical histories of the various igneous units, subject once again to the proviso that field relations among the various units being studied are well understood.

6.5 Duration of Individual Basalt Floods

6.5.1 Introduction

The length of time during which a particular basalt flooding episode lasts differs greatly among the various flood-basalt provinces. Some, such as the Siberian flood-basalt province, have been

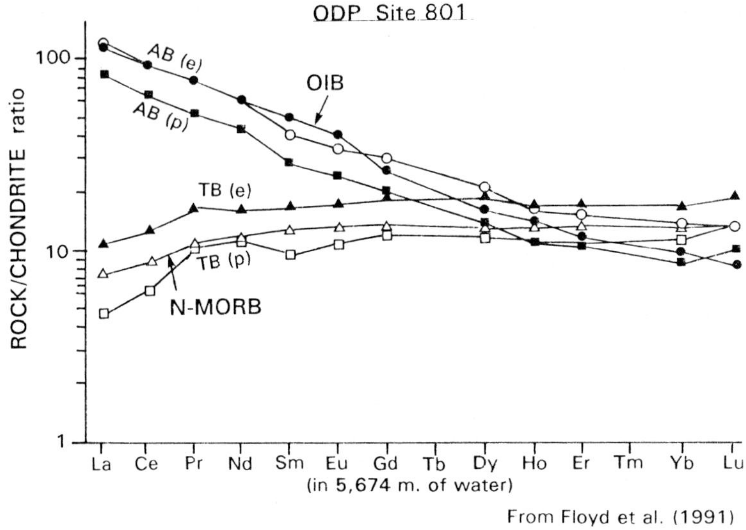

Fig. 6.25. Chondrite-normalized rare-earth element plot for Jurassic basalts from ODP Site 801, Hole 801C.

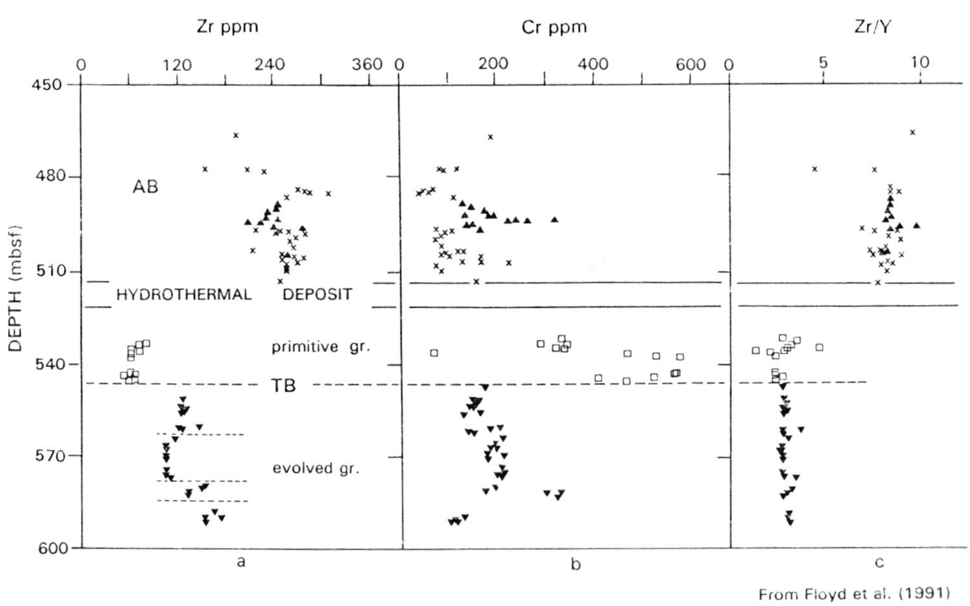

Fig. 6.26. Variations of Zr/Y, Zr and Cr in Jurassic oceanic basalts at ODP Hole 801C with depth.

active more than 200 Ma. Others---the Wrangellian province, for example---probably completed their flood activity in 5 Ma or less. Even in flood-basalt provinces of long duration, the largest volume of basalts may have been extruded in one, or perhaps two or three relatively short bursts. A close relationship seems to exist between times of tectogenesis and times of major basalt flooding.

6.5.2 Flood-Basalt Provinces of Long Duration

Radiometric and/or paleontologic constraints are available for only a few flood-basalt provinces. Therefore, we mention only places where good dating is available. The radiometric data are summarized on Table 6.1.

Siberian Flood-Basalt Province

Kuznetsov and Naumov (1975) summarized the radiometric data available at that time from the Siberian Platform region. Zhuravlev (1986) described the westward subsurface extension of this province beneath the West Siberian Basin. The radiometric dates range from 283 to 70 Ma on the Siberian Platform and from 254 to 175 Ma beneath the West Siberian Basin. The locus of maximum flood-basalt activity shifted from time to time, with an overall shift toward the northeast and southeast (Kuznetsov and Naumov, 1975). Baksi (1989), without stating his grounds, dismissed all radiometric dates in this province as "suspect." He alleged that the petrographic reports accompanying all of the "suspect" dates show that the "material dated was badly altered . . ." (p. 13). Baksi may have been unaware that much of the dated material does come from fairly unaltered surface exposures and diamond cores. More importantly, he also made no comments about the thousands of subsurface occurrences in the West Siberian Basin where unaltered cores have been used widely for dating.

Recently, in an effort to find possible causes of mass faunal and floral extinctions at different times in Phanerozoic history, numerous papers have appeared advocating short-lived massive flood-basalt eruptions in many of the known flood-basalt provinces and linking these massive eruptions to mass extinctions. One of these papers was by Renne and Basu (1991), who claimed that most of the Siberian Traps were erupted in a period not exceeding 2.5 Ma, despite the mass of undoubted evidence to the contrary. The samples that Renne and Basu (1991) dated came from two localities in the northwestern part of the Siberian Platform where most of the traps were emplaced within the span of 40 Ma in a region of rapidly shifting eruptive centers.

Deccan Flood-Basalt Province

The Deccan Traps were extruded mainly during the time interval 111-4.2 Ma (Agarwal and Rama, 1976; Alexander, 1981; Saxena, 1986). Most of the traps were emplaced in the time span 102-31 Ma (Alexander, 1981), that is, over 71 Ma. Here again, as in the Siberian province, repeated attempts are being made to compress the time of flood volcanism into a very brief time span that encompasses the Cretaceous-Tertiary (K-T) boundary, currently placed at 65 Ma (Berggren et al., 1995) and the alleged mass extinctions at the K-T boundary. Hence the usual condensed time spans alleged are in the range 68-61 Ma (Weaver and Tarney, 1983; Duncan and Pyle, 1988; Baksi and Farrar, 1991; Vandamme et al., 1991). These attempts to set the time of the Cretaceous-Tertiary boundary are somewhat disturbing, particularly because it has been shown that nearly 75% of the Cretaceous extinctions took place before the time of the Cretaceous-Tertiary boundary (Jaeger, 1986). Moreover, such attempts altogether dismiss the many reliable dates reported by Agarwal and Rama (1976), Alexander (1981), and more than a half dozen other specialists. The workers who insist on the 68-61 Ma age span invariably dismiss dates that are at odds with their ideas as the results of analytical error, rock alteration, and the like; they do not appear to present solid evidence to support such claims. The Deccan Traps, like the Siberian Traps, were emplaced by a shifting eruptive center (Subbarao, 1988; Mitchell and Widdowson, 1991).

6.5.3 Flood-Basalt Provinces of Short Duration

Columbia River Flood-Basalt Province
The Columbia River Basalt was extruded between 17.5 and 6 Ma (Hooper and Reidel, 1989). Baksi (1989) has presented data which he interprets to mean that most of the basalts were extruded between 17.5 and 15.5 Ma. Most workers do agree (e.g., Carlson and Hart, 1988) that the vast bulk of the basalts had been extruded by 13.5 Ma. Thus the time of maximum basalt extrusion endured nearly 4 Ma.

Wrangellian Flood-Basalt Province
This province commonly is referred to as the "Wrangellia Terrane" (Jones et al., 1977) and recently Panuska (1990) referred to it as a "world class flood basalt province." This province, containing about 1,000,000 km^3 of basalt in a narrow rift valley some 2,700 km long, erupted in an incredibly short period of time between the end of middle Ladinian (latest Middle Triassic) and the beginning of late Carnian (early Late Triassic) times. Total elapsed time probably did not exceed 10 million years and may have been only a small part of this time span (Jones et al., 1977).

6.5.4 Conclusion

We have discussed several flood-basalt provinces which were active during periods that ranged from more than 210 Ma (long duration) to less than 12 Ma (short duration). We have found no evidence to suggest that there are any time controls or any rules of thumb that guide the length of time during which a flood-basalt province may remain active. Nor is there a relationship between type of flood-basalt province and the duration of its extrusion. For example, the Columbia River province is ovate while the Wrangellian province is linear; yet the two endured for approximately the same lengths of time. Reports that the Deccan and Siberian flood-basalt provinces were in fact of very short duration are based on a lack of information. In fact, information adequate to determine the "lifespans" of most flood-basalt provinces, including the Siberian province, is not yet available.

6.6 Flood-Basalt Provinces and Frequency in Geologic Time

As we observed near the beginning of this chapter, the commonly used textbooks of physical geology, structural geology, and geotectonics rarely list more than 10 to 20 flood-basalt provinces. However, the magnificent review of basalts by the participants in the Basalt Volcanism Study Project (1981) mentions or figures not less than 56 flood-basalt provinces and

45 additional provinces of dike swarms which the project participants thought might have fed flood-basalt provinces that have since been removed by erosion.

The participants in the Basalt Volcanism Study Project (1981) concurred on a large number of phenomena that characterize flood-type volcanism. However, they showed considerable confusion, ambivalence, and lack of agreement on which, and what type of, provinces should or should not be described as flood-basalt volcanism. This confusion and ambivalence manifest themselves with respect to the differences between flood-basalt provinces and continental rift-related provinces. Additionally, they used interchangeably the terms "flood basalt," "plateau basalt," "continental rift volcanism," and "hot-spot volcanism."

We summarize here briefly their overall remarks on the ages of flood-basalt activity. They wrote that (1) most flood provinces are less than 200 Ma; (2) no major flood-basalt activity took place in the interval 1,100-200 Ma (yet on pages 718-719, they list eight provinces within this time span, two of which are huge---the Siberian Flood-Basalt Province and the Emeishan Flood-Basalt Province); (3) reasonably well-preserved remnants of flood provinces are known from the time interval 2,150-1,100 Ma; and (4) a few poorly preserved remnants are present in the geological record to 3,760 Ma (Basalt Volcanism Study Project, 1981, p. 717-720). Yet, on page 41, the same authors state that flood-basalt provinces older than 1,200 Ma are unknown.

The participants in the Basalt Volcanism Study Project (1981) have for the first time, to the best of our knowledge, provided solid evidence that flood-basalt volcanism is a phenomenon that has persisted since the beginning of---or since very early in---the Earth's history. However, we have not seen any convincing evidence to support the claim by Rampino and Stothers (1988), and a similar claim by White and McKenzie (1989), that flood-basalt volcanism is periodic, with large outpourings every 32 to 30 Ma. We suspect, but cannot prove, that flood volcanism is triggered by tectogenic (orogenic) pulses that are episodic. In our opinion, the available evidence all but demonstrates an endogenic origin. Which of the various possible endogenic causes is the correct one must await the careful sampling and dating of thousands of more carefully located igneous-rock samples in every major flood-basalt province.

Yoder (1988, p. 140) wrote that "Great basaltic 'floods' have appeared on the continents throughout geologic time (Table 1)," but showed on his Table 1 none older than 1,200 ± 50 Ma. He also (p. 151) made it clear that he regards midocean-ridge and other oceanic basalts as flood basalts, as have a number of earlier workers (e.g., Press and Siever, 1974). We concur absolutely with their interpretation. We also concur with the participants of the Basalt Volcanism Study Project (1981) that evidence of the existence of flood provinces extends back in time to at least 3,760 Ma, and very likely to the Earth's earliest (but nowhere preserved) history. Interestingly, most of what Press and Siever (1974), Yoder (1988), and we concur in what was anticipated by the pioneer work of Engel et al. (1965).

6.7 Non-Basalt Flood Volcanism in Flood-Basalt Provinces

The bimodal nature of many flood-basalt provinces has been known and stressed for many years (Basalt Volcanism Study Project, 1981). Time seems not to be a major factor (the idea being that, the longer an underlying magma chamber is present, the more the magma will interact with the continental crust above it). The most important factor may be the crustal stress state.

Estimates of the volume of non-basaltic rocks in a given flood-basalt province are difficult to find. Accordingly in Table 6.2 we have left many blank spaces and the percentages that we have supplied are poorly documented except in local areas.

The Keweenawan-Midcontinent province contains up to 25% rhyolite and rhyolitic ignimbrite, a figure based on data summarized in the Basaltic Volcanism Study Project (1981, p. 62). This figure applies only to the North Shore Volcanic subprovince; it is much lower in the other subprovinces, but such rocks are present in all of them, probably 5% or more.

The Karroo province extends from Namibia to Mozambique, across much of southern Africa. The Drakenberg Lavas consist mainly of different types of mafic lava, including picritic basalt, in the Karroo basin. However, along the Lebombo monocline paralleling the eastern coast of South Africa, as we pointed out in a preceding section, the volcanics can contain up to 50% rhyolite and rhyolitic ignimbrite. In Namibia on the Atlantic coast, the Entendeka lavas include near the top a number of trachytic lavas and related ignimbrites, but these comprise only a small (amount unknown) percentage of the Entendeka section (Haughton, 1969; Dingle et al., 1983).

The Siberian province seems to be mainly mafic, with limited amounts of andesite, trachyandesite, trachyte, and dacite. Basaltic tuffs abound, in places dominating the section. Sills and dikes abound in certain areas (Zolotukhin and Al'mukhamedov, 1988).

The Paraná (Serra Geral) flood basalt of Argentina, Brazil, Paraguay, and Uruguay has an important silicic mode. As in the Karroo province, the silicic mode is developed best close to the Atlantic continental margin. In the northern third of the province, silicic lavas at the top of the section comprise only 0.3 percent (by volume) of the province; in the central part, silicic lavas comprise 10%; in the southern part, they comprise 35%. The silicic rocks are mainly andesite, dacite, rhyodacite, and rhyolite. For the province as a whole, andesite constitutes 7% of the total volume of lavas and sills, whereas rhyolite and rhyodacite comprise 3% (Piccirillo et al., 1988).

The Ethiopian flood-basalt province also is bimodal. The flood basalts in the southern part of the province are mildly alkalic; their equivalents in the northern part of the plateau, however, are tholeiitic (Mohr and Zanettin, 1988). Flood trachyte, rhyolitic ignimbrite, and alkalic rhyolite are present in the north; trachyte, trachytic ignimbrite, rhyolite, and rhyolitic ignimbrite are present in the south. As we interpret Mohr and Zanettin's (1988) thickness tables, silicic rocks in some places comprise up to 30-35% of the total column.

The Snake River Plain flood province consists approximately 65% of rhyolite and rhyolitic ignimbrite, and 35% of mafic rocks, mainly basalt; the rhyolitic rocks are calc-alkaline, one of the indicators of an island-arc origin. However, the Snake River Plain province is truly bimodal, and probably is more closely related to the Oregon/Modoc Plateau province on the southwest than to the Columbia River Basalt province to which it is linked repeatedly. The Snake River lavas, in a general way, are oldest in the southwest and youngest in the Yellowstone area on the northeast (Bonnichsen, 1989). Of all the provinces we have discussed or listed to this point, the Snake River Plain is the first province in which silicic lavas considerably outnumber mafic lavas---approximately 46,000 km^3 of rhyolite and rhyolitic ignimbrite compared with about 24,000 km^3 of basalt.

We believe that the evidence from these examples demonstrates convincingly that there is a complete gradation from all-basalt and basaltic andesite flood provinces to bimodal provinces containing mainly rhyolite and ignimbrite. Hence, there are basalt floods and rhyolite floods.

In northern Mexico, the Sierra Madre Occidental occupies an area of at least 400,000 km^2

Table 6.2. Estimates of volumes of non-basaltic rocks in selected flood-basalt provinces. Blank spaces are left to show poor documentation of these rocks. See Table 10 for information on paleontological and radiometric ages of each province.

Age	Flood-Basalt Province	Bimodal		Petrograph of Nonbasalt Mode	Source
		Yes	No		
Archean	Ventersdorp Supergroup				
Proterozoic-Early Cambrian	Antrim Plateau				
	Cape Smith				
	Coppermine River				
	Keweenawan (Mid-Continent)	X		Rhyolite, rhyolitic ignimbrite, 5-25%, qtz lahte, trachy-andesite, andesite	Basalt Volcanism Study Program (1981)
	Willorran				
Late Paleozoic-Middle Jurassic	Drakensberg (Karoo)	X		Rhyolite, rhyolitic, ignimbrite, In east, up to 50%	Haughton (1969); Dingle et al. (1983)
	Emeishan				
	Franz Josef				
	High-Middle Atlas				
	Siberian	X		Basaltic andesite, andesite, trachy-andesite, trachyte - <2%; tuff very extensive	Zolotukhin and Al'mukhamedor (1988)
	Venzuela-Colombian Andes				
	Wrangellia				

and contains more than 1,000,000 km^3 of extrusive rocks and their related volcanic necks, dikes, and sills (Fig. 6.27). According to Wark et al. (1990, p. 1555), "The Sierra Madre volcanic field is the largest remnant of a broad and extensive belt of continental-margin, calc-alkalic magmatism that was active along the western edge of North America. Covering much of western Mexico, the field includes the Sierra Madre Occidental . . . proper and parts of the Mexican Basin and Range province farther east . . .. Exposed rocks are dominated by rhyolite ash-flow tuffs that were extruded from large caldera structures mostly between ca. 36 and 27 Ma, across the width of the entire volcanic field . . .. The volumetric predominance of these ash-flow tuffs has led to recognition of the [Sierra Madre Occidental] as the world's largest rhyolite-

Late Jurassic-Cretaceous	Axel Heiberg				
	Entendeks	X			Dingle et al. (1983)
	Labrador Shelf				
	Paraná (Serra Geral)	X			Piccirillo et a.. (1983)
	Rajmahal				
	Svalbard				
Late Cretaceous-Tertiary	Deccan				
	Madagascar				
	Patagonian		X		
Tertiary-Quaternary	Arabian				
	Brito-Arctic				
	Chilcotin				
	Columbia		X		
	Ethiopian	X		Rhyolite, Rhyolitic ignimbrite ,trachyte, trachy-ignimbrite	Mohr and Zannetti (1988)
	Matagalpa				Weyl
	Snake River-Yellowstone	X		Rhyolite, Rhyolitic ignimbrite	
	Yakutat				

dominated volcanic province" (Fig. 6.28).

The base of the section studied by Wark et al. (1990) is ca. 38 Ma (Wark, 1991), and consists of andesite interbedded with rhyolite ash-flow tuffs that are 38-35 Ma. These are overlain by rhyolite, rhyolite ash-flow tuffs, trachyte, and trachyandesite. Andesite and basaltic andesite appear only as a very minor constituent in the section, which is 34-29 Ma. At the top, the rhyolitic rocks intertongue and are capped by basaltic andesite of Cameron et al.'s (1989) "southern cordilleran basaltic andesite" suite, which are 29-24 Ma, and include flows of basaltic andesite, trachybasalt, and basalt. Volcanic activity lasted until 17 Ma, the youngest of the basaltic andesites. Rhyolite ash-flow tuffs are interbedded everywhere with the basaltic andesite, whether in the Sierra Madre Occidental or on the foreland east and north of it (Cameron et al., 1989; Wark et al., 1990; Wark, 1991).

Thus, from 38 Ma until 17 Ma, a truly bimodal column of extrusive rocks accumulated in northern Mexico and adjoining parts of the United States, with rhyolite at one end, basaltic andesite at the other, and very little rock of intermediate compositions. It is this section that many workers refer to as the "orogenic andesite suite." Spidergrams of these rocks appear in Figures 6.29 through 6.35, and should be compared with our other spidergrams (e.g., Figs. 6.17, 6.18, 6.20A, 6.20B). Note that a spidergram of the Grande Ronde volcanic field of the Columbia River flood-basalt province is shown for comparison. A number of workers have called the basaltic andesite of the region an example of flood basalt. Cameron et al. (1989, p. 7822) observed that the mafic rocks of the "southern cordilleran basaltic andesite" suite plot in the field of Hawaiian alkalic basalt, then concluded (p. 7835), "In our view, [southern cordilleran basaltic andesite] represents arc-related magmas erupted in an extensional regime." Yet both the Hawaiian basalts and the Grande Ronde basalts of the Columbia River province are recognized widely as being flood basalts!

We believe that these basalts of the "southern cordilleran basaltic andesite" suite are flood basalts. And if they are flood basalts, then we have demonstrated that the same mechanism that leads to continental and oceanic basalt outpourings also produces the "orogenic andesite suite".

The Okhotsk-Chukotka Volcanic Belt, a linear belt of Cretaceous volcanics, is similar to the Sierra Madre Occidental. It extends 3,000 km from the mouth of Uda Bay (northwestern

Fig. 6.27. From Cameron et al. (1989): "Cenozoic “back arc” basaltic provinces (stippled pattern) of western North America. CP = Columbia Plateau; OP = Oregon Plateau; SCORBA = Southern Cordilleran Basaltic Andesite; and MS= Mojave-Sonora megashear. According to Cameron et al. (1989), from which this figure is taken, SCORBA may be more widespread, extending into southern California, northward into Nevada and Colorado, and farther south into Mexico."

Sea of Okhotsk) to the Bering Sea almost at St. Lawrence Island. It seems to have every type of volcanic from andesite through rhyolite. Basalts are scarce. Soviet geologists either ignore it or say that it is the remnant of a volcanic arc.

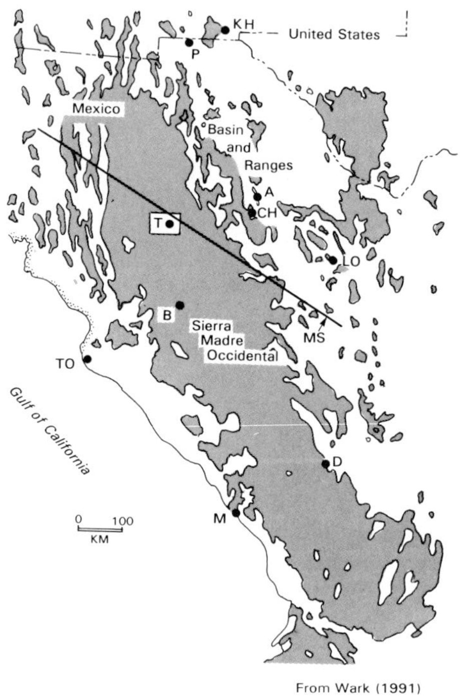

Fig. 6.28. Distribution of Tertiary volcanic rocks in western Mexico (shaded areas). Locality symbols: A=Aldama; B=Batopilas; CH=Chihuahua; D=Durango; LO=La Olivina; T=Tomochic; TO=Topolobampoo; KH=Kilbourne Hole; M=Mazatlan; P=Potrillo; MS=Mojave-Sonora Megashear. See text for discussion.

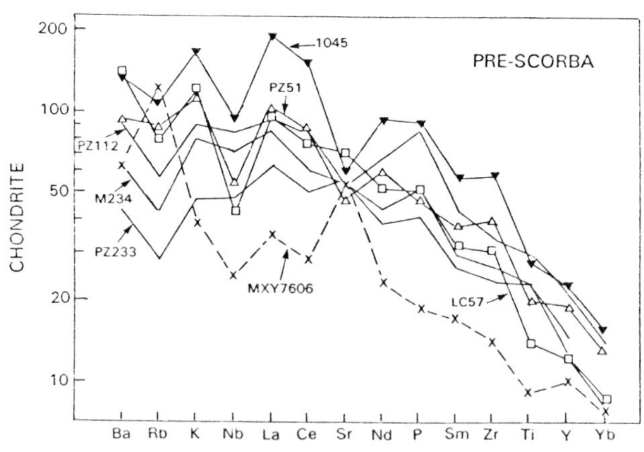

Fig. 6.29. Spidergram of pre-SCORBA (Southern CORdilleran Basaltic Andesite).

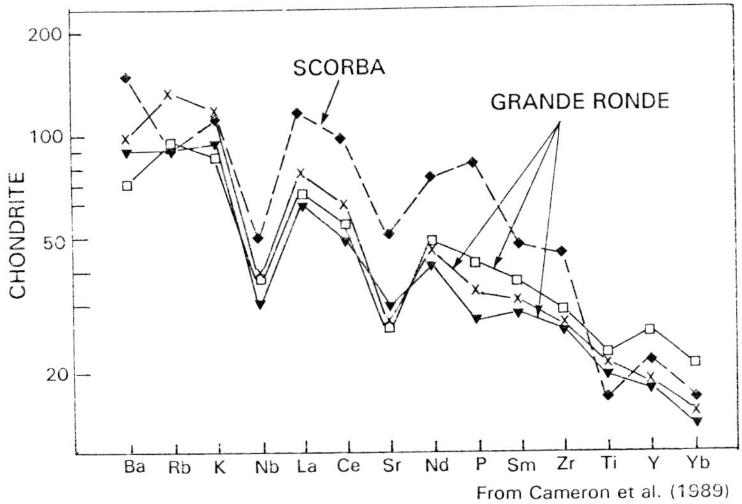

Fig. 6.30. Spidergram of Grande Ronde volcanic field of the Columbia River flood-basalt province. SCORBA shown for comparison.

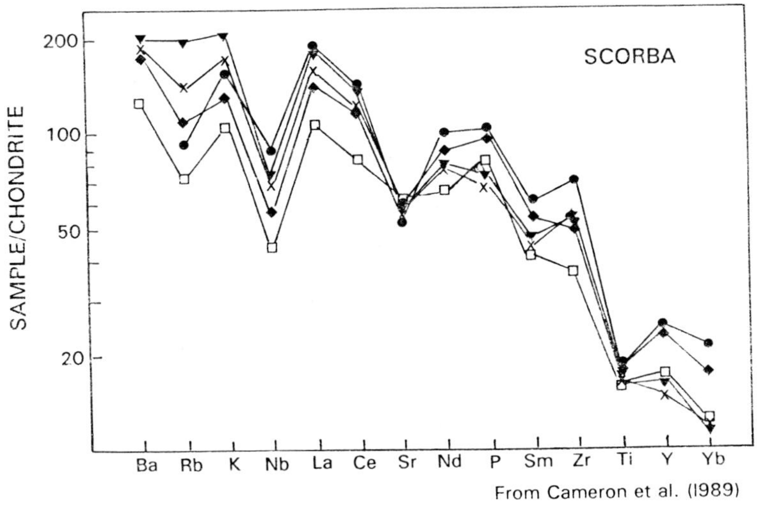

Fig. 6.31. Spidergram of SCORBA.

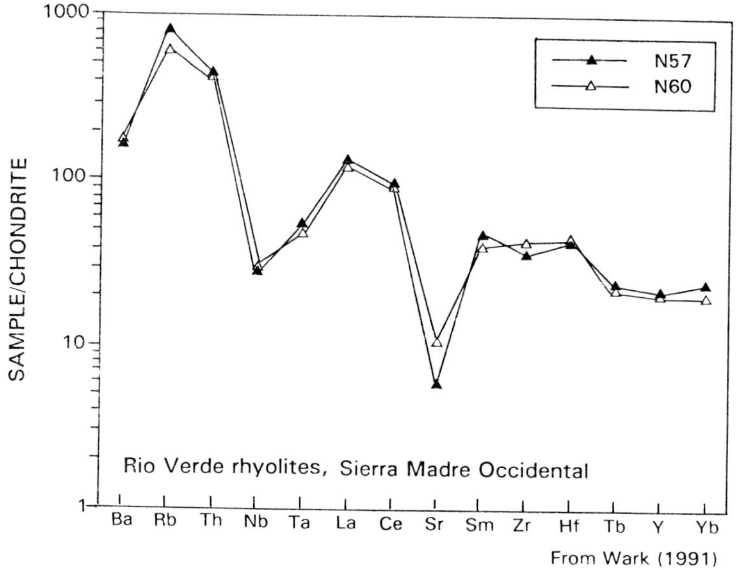

Fig. 6.32. Spidergram of Rio Verde rhyolites, Sierra Madre Occidental.

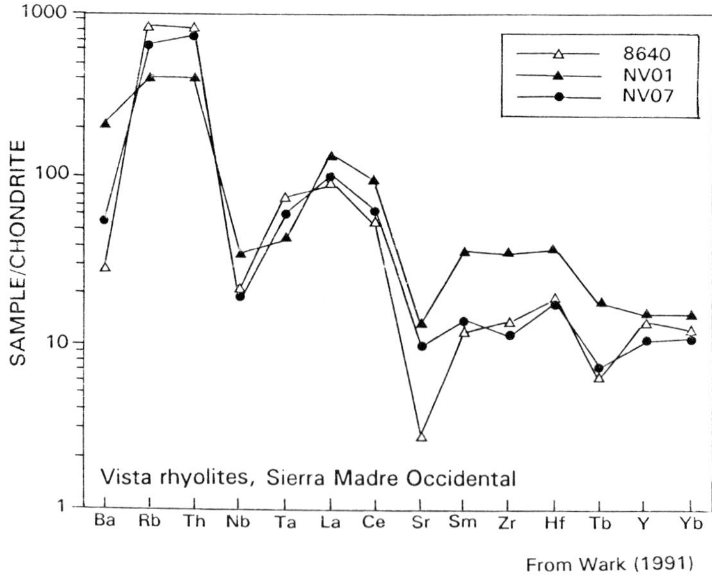

Fig. 6.33. Spidergram of Vista rhyolites, Sierra Madre Occidental.

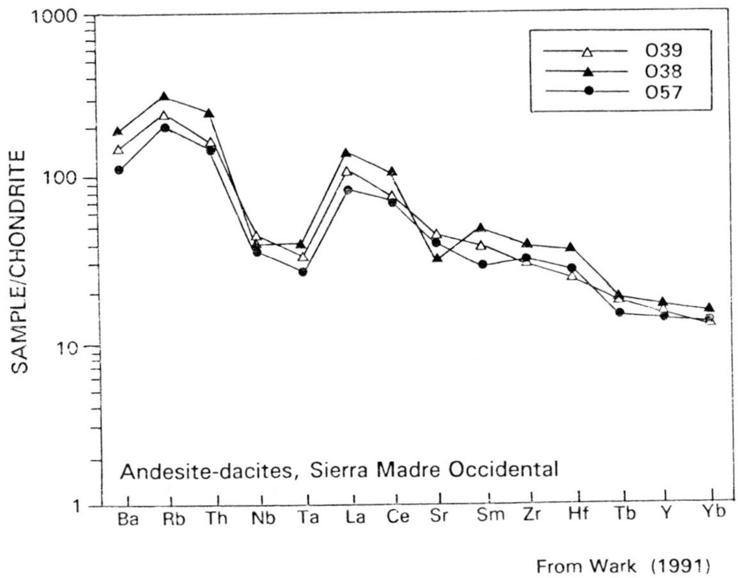

Fig. 6.34. Spidergram of andesites and dacites, Sierra Madre Occidental.

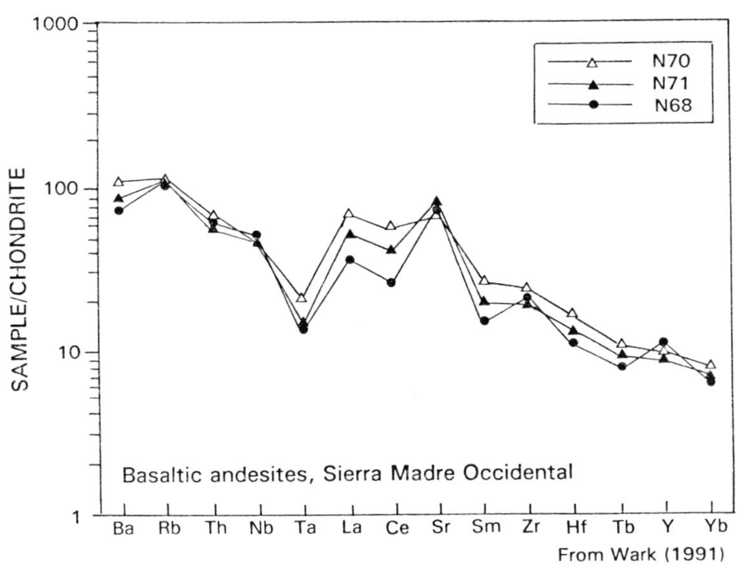

Fig. 6.35. Spidergram of basaltic andesites, Sierra Madre Occidental.

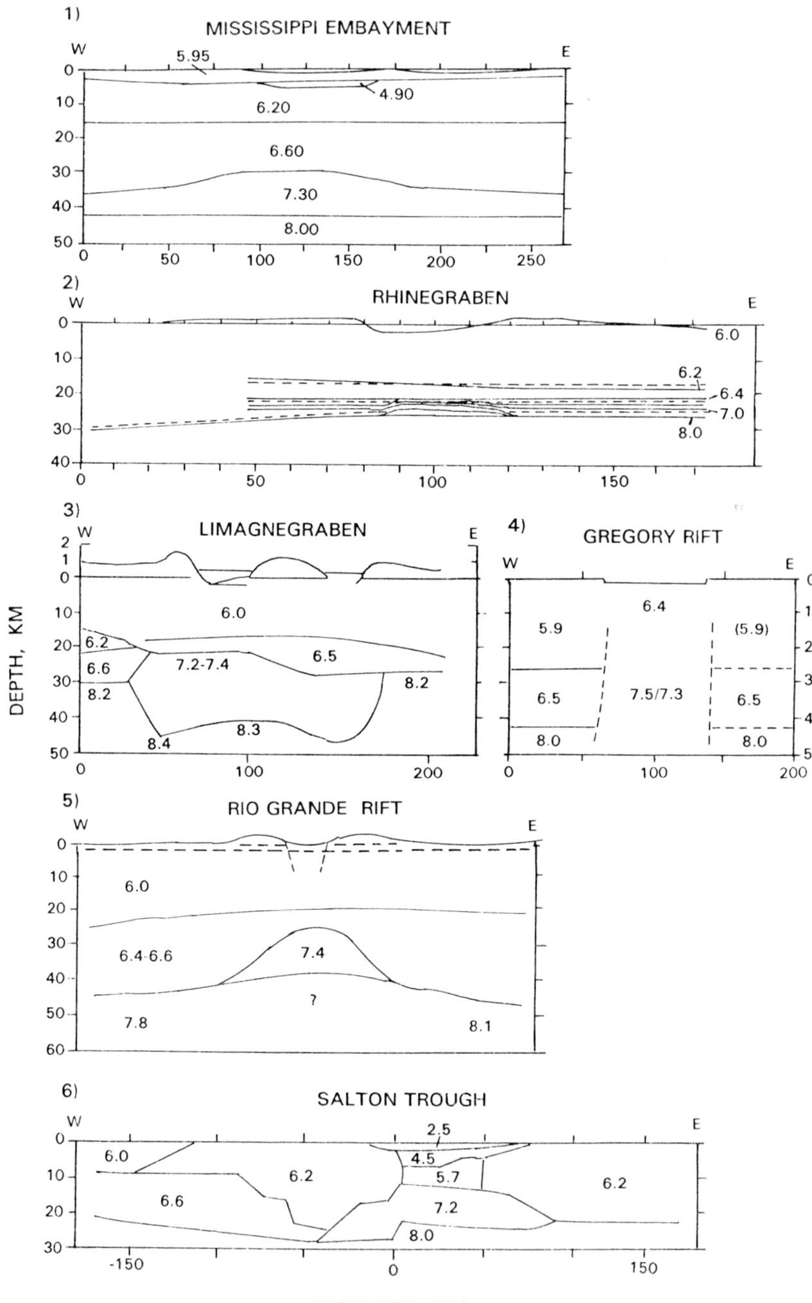

Fig. 6.36. Comparison of crustal velocity structure of the Mississippi Embayment (1) with the Rhine Graben (2), Limagnegraben (3) and Gregory Rift, East Africa, Rio Grande Rift (5), and Salton Trough (6). Note presence of anomalous lower crust with P-wave velocities in the 7.0-7.7 km/s range in all of the rifts.

6.8 Flood Basalts or Magma Floods?

Although we advocate the continued use of the term "flood basalt," it is clear that another term is needed to describe floods of andesite, dacite, and rhyolite. For future studies, we suggest the all-encompassing term *magma floods*. In this way, we can include all of the various lava types, dikes, necks, and sills. It is a term that even embraces situations such as the Ferrar Dolerite of Antarctica and the network of sills and dikes of the Amazon basin.

6.9 Surge-Tectonics Origin of Magma Floods

In the preceding pages we have referred to the presence of several flood-basalt provinces around the world, and have shown that some flood provinces include large volumes of silicic rocks, usually rhyolite and/or dacite. We have also shown by the northern Mexican example that flood basalts can interfinger with the andesite orogenic suite. In addition, we have presented evidence that spidergrams are not more effective at identifying the tectonic setting than bulk chemistry. The available evidence has led us to the conclusion that the same mechanism causes volcanism in the midocean ridges, linear island and seamount chains, oceanic plateaus, island arcs, and continental intereriors. We next attempt an explanation of our conclusion.

Many attempts have been made to explain flood volcanism in the framework of the plate-tectonics hypothesis. The two principal explanations involve (1) hot spots, or mantle plumes and (2) an extraterrestrial cause (e.g., an asteroid impact).

Extraterrestrial causes have been proposed by Alt et al. (1988), who applied this hypothesis to the Columbia River flood-basalt province. A major problem with this concept is that it does not explain linear flood-basalt provinces such as the Keweenawan (Mid-Continent) rift and Wrangellia. Furthermore, Mitchell and Widdowson (1991) pointed out that impact and shock phenomena should be present in the area surrounding the Columbia River province if it resulted from extraterrestrial action, but they are entirely absent.

Mantle diapirism or asthenosphere upwelling constitutes the hot-spot or mantle-plume hypothesis (Morgan, 1971) used widely in tectonic models today. Recent literature on mantle plumes include works by Cox (1988a, b), Rampino and Stothers (1988), Richards et al. (1989), White and McKenzie (1989), Hooper (1990), Duncan and Richards (1991), Larson (1991), and Mitchell and Widdowson (1991). Hot spots are often portrayed as diapiric bodies that are essentially cylindrical, mushrooming plumes. While this might account for isolated volcanoes, it does not account for the massive ovate and linear flood basalt provinces found in many parts of the world..

Mantle upwelling also has been invoked by many writers to explain the presence of long, linear continental rifts (e.g., Mooney et al., 1983), which are, for the most part, similar to one another. The Keweenawan (Mid-Continent) rift is 75 to 150 km wide and at least 2,400 long. The Wrangellian province is 75 to 180 km wide and at least 3,300 km long. The Okhotsk-Chukotka volcanic belt is 80 to 160 km wide and 3,000 km long. If the Sikhote-Alin volcanic belt is a southern extension of the Okhotsk-Chukotka belt, then the total length exceeds 4,500 km. The Rio Grande rift is 40 to 90 km wide and 1,200 km long. The East African rift is of the order of 100-150 km wide and, between Afar and the mouth of the Zambezi River, is 3,500 km long. The Baykal rift and its extensions toward the west and east are 100-160 km wide; the rift is 2,000 km long. The Rhine graben is 20-30 km wide and 280 km long. It is notable that the two short rifts (the Rio Grande and the Rhine) are also the narrowest, less than 100 km wide. In contrast, the Mid-Atlantic Ridge-rift system is about 13,500 km long and 800 to 2,000 km wide.

As we noted in Chapters 3 and 4, Mooney et al. (1983) observed that all active rifts studied by them have an anomalous lower crust with P-wave velocities in the 7.0 to 7.7 km/s range (Fig. 6.36). Olsen (1983) obtained the identical result, as did Krylov et al. (1979) and Sychev (1985) for the Baykal rift. In our review of the literature on rifts, we found no exceptions; *all rifts have a pod of 7.0-7.7 km/s material beneath the rift axis*. Fuchs (1974) believed that this pod of anomalous lower crustal material houses the mechanism that causes rifting. It is interesting to note that all midocean ridges have a pod of 7.0-7.7 km/s as well (Ewing and Ewing, 1959; Talwani et al., 1965). (Furthermore, each island arc and foldbelt also has a pod of 7.0-7.7 km/s material that pinches out from the center of the arc or foldbelt (e.g., Kushiro, 1987, for the Japan arc; Marillier et al., 1989, for the Appalachians.)

Figure 3.6 is a cross section across the Baykal rift, from Krylov et al. (1979) and Sychev (1985). Years of refraction work have shown the Lake Baykal is underlain at about 32 km by a pod that is connected to the deeper asthenosphere. The shallow pod contains a low-velocity zone that presumably is a partial melt. The pod extends the full length of the rift. It is, in short, a channel containing partly molten magma and an excellent example of one of our surge channels. Were it to burst, we believe that it would produce another linear flood-basalt province.

According to our surge tectonic hypothesis, magma in surge channels moves both vertically and horizontally. When two surge channels come in contact, their magmas join together. If they are oriented at an appreciable angle to one another, we believe that the result is a "collision". These "collisions" are responsible for the eruption of round or ovate flood-basalt provinces worldwide.

CHAPTER 7

CONCLUSIONS

We have proposed a new hypothesis of global tectonics in this book, one that is different and will be considered unorthodox by many scientists and non-scientists alike. However, we believe that current tectonic hypotheses cannot adequately explain the increasing volume of data being collected by both old and new technologies. We believe that the hypothesis of surge tectonics does explain these data sets, in a way that is simple and more accurate.

The major points of the surge-tectonics hypothesis can be summarized as follows:

1. All linear to curvilinear mesoscopic and megascopic structures and landforms observed on Earth (and similar features seen on Mars, Venus, and the moons of Jupiter, Saturn and Uranus), and all magmatic phenomena are generated, directly or indirectly, by surge channels. The surge channel is the common denominator of geology, geophysics, and geochemistry.

2. Surge channels formed and continue to form an interconnected worldwide network in the lithosphere. They contain fluid to semifluid magma, or mush, differentiated from the Earth's asthenosphere by the cooling of the Earth. All newly differentiated magma in the asthenosphere must rise into the lithosphere. The newly formed magma has a lower density and therefore, is gravitationally unstable in the asthenosphere. It rises in response to the Peach-Köhler climb force to its level of neutral buoyancy (that is, to form a surge channel).

3. Lateral movements in the Earth's upper layers are a response to the Earth's rotation. Differential lag between the more rigid lithosphere above and the (more) fluid asthenosphere below causes the fluid, or mushy, materials to move relatively eastward.

4. Surge channels are alternately filled and emptied. A complete cycle of filling and emptying is a geotectonic cycle. The geotectonic cycle takes place along this sequence of events:

a. Contraction of the strictosphere is always underway, because the Earth is cooling;

b. The overlying lithosphere, which is already cool, does not contract, but adjusts its basal circumference to the upper surface of the shrinking strictosphere by large-scale thrusting along lithosphere Benioff zones and normal-type faulting along the strictosphere Benioff zones.

c. Thrusting of the lithosphere is not a continuous process, but occurs when the lithosphere's underlying dynamic support fails. When the weight of the lithosphere overcomes combined resistance of the asthenosphere and Benioff zone friction, lithosphere collapse begins in an episodic fashion. Hence, tectogenesis is episodic.

d. During anorogenic intervals between lithosphere collapses, the asthenosphere volume increases slowly as the strictosphere radius decreases and decompression of the asthenosphere begins.

e. Decompression is accompanied by rising temperature, increased magma generation, and lowered viscosity in the asthenosphere, which gradually weakens during the time intervals between collapses.

f. During lithosphere collapse into the asthenosphere, the continentward (hanging wall) sides of the lithosphere Benioff zones override (obduct) the ocean floor. The entire lithosphere buckles, fractures, and founders. Enormous compressive stresses are created in the lithosphere.

g. When the lithosphere collapses into the asthenosphere, the asthenosphere-derived magma in the surge channels begins to surge intensely. Where volume of magma in the channels exceeds volumetric capacity, and when compression in the lithosphere exceeds the strength of the lithosphere that directly overlies the surge channels, the surge-channel roofs rupture along the cracks that comprise the fault-fracture-fissure system generated before the rupture. Rupture is bivergent and forms continental rifts, foldbelts, strike-slip zones, and midocean rifts. We call such bilaterally deformed belts kobergens.

h. Once tectogenesis is completed, another geotectonic cycle or subcycle sets in, commonly within the same belt.

5. Movement in the surge channel during the taphrogenic phase of the geotectonic cycle is parallel with the channel. It is also very slow, not exceeding a few centimeters per year. Flow at the surge-channel walls is laminar as evidenced by the channel-parallel faults, fractures, and fissures observed at the Earth's surface (Stoke's Law). Such flow also produced the more or less regular segmentation observed in tectonic belts.

6. Tectogenesis has many styles. Each reflects the rigidity and thickness of the overlying lithosphere. In ocean basins where the lithosphere is thinnest, massive basalt flooding occurs. At ocean-continent transitions, eugeosynclines with alpinotype tectogenesis form. In continental interiors where the lithosphere is thicker, either germanotype foldbelts or continental rifts are created.

7. During the geotectonic cycle, and within the eugeosynclinal regime, the central core (crest of the surge channel) evolves from a rift basin to a tightly compressed alpinotype foldbelt. Thus a rift basin up to several hundred kilometers wide narrows through time until it is a zone no more than a few kilometers wide that is occupied by a streamline (strike-slip) fault zone (e.g., the San Andreas fault). Then as compression takes over and dominates the full width of the surge-channel crest, the streamline fault zone is distorted until it and the adjacent rocks are severely metamorphosed. If the underlying, and now deformed, surge channel still contains any void space, the overlying rocks may collapse into it, and through this process of *Verschluckung* (engulfment) become a *Verschluckungszone*.

8. The Earth above the strictosphere resembles a giant hydraulic press that behaves according to Pascal's Law. A hydraulic press consists of a containment vessel, fluid in that vessel, and a switch or trigger mechanism. In the case of the Earth, the containment vessel is the interconnected surge-channel system; the fluid is the magma in the channels; and the trigger mechanism is worldwide lithosphere collapse into the asthenosphere when that body becomes too weak to sustain the lithosphere dynamically. Thus tectogenesis may be regarded as surge-channel response to Pascal's Law.

9. Surge channels, active or inactive, underlie nearly every major feature of the Earth's surface, including all rifts, foldbelts, metamorphic belts, and strike-slip zones. These belts are roughly bisymmetrical, have linear surface swaths of faults, fractures, and fissures, and belt-parallel stretching lineations. Aligned plutons, ophiolites, mélange belts, volcanic centers, kimberlite dikes, diatremes, ring structures and mineral belts are characteristic. Zoned metamorphic belts are also characteristic. In some areas, linear river valleys, flood basalts, and/or vortex structures may be present. A lens of 7.8-7.0 km/s material always underlies the belt.

10. Active surge channels are most easily recognized by the presence of high heat flow (Fig. 2.26), microseismicity, lines of thermal springs, small negative Bouguer gravity anomalies, and a 7.8-7.0 km/s lens of material that is transparent in the center or throughout.

11. Inactive surge channels possess a linear positive magnetic anomaly, a linear Bouguer positive gravity anomaly, and a linear, lens-shaped pod of 7.8-7.0 km/s material that is reflective throughout.

12. A surge-tectonics approach to geodynamics provides a new means for determining the origin of the Earth's features and their evolution through time, for analyzing regions prone to earthquakes and volcanism, and for predicting the location and formation of mineral deposits throughout the globe.

APPENDIX

PHYSICAL PRINCIPLES MENTIONED IN THE TEXT

Bernoulli's Theorem

In a stream of liquid, the sum of the elevation head, the pressure head, and the velocity head (1) remains constant along any line of flow provided that no work is done by or upon the liquid in the course of its flow and (2) decreases in proportion to the energy lost in viscous flow.

Hooke's Law

The stress within an elastic solid up to the elastic limit is proportional to the strain responsible for it.

Level of Neutral Buoyancy

In the lithosphere, this is the level at which a rising column of magma no longer can rise because below that level the country rock is denser than the magma, and above that level the country rock is less dense than the magma (see Walker, 1989). This level also has been called the level of gravitational stability.

Navier-Coulomb Maximum Shear-Stress Theory

We do not treat this concept in detail or with mathematics. Those interested in details are referred to Jaeger (1962, p. 75-80). The Navier-Coulomb maximum shear-stress theory states that failure (by shear) of a material occurs when the maximum shear stress equals the shear strength of the material. Compressive shear strength is always greater than tensile shear strength. Because of this, shear caused by compressive stress must be at an angle of less than 45° to the direction of that stress.

In any three-dimensional system, there are three principal stresses. If this three-dimensional system is the Earth's lithosphere, one of the three principal stresses is essentially vertical. Thus three cases----three types of faulting----arise according to whether the vertical stress direction is the greatest, intermediate, or the least (Jaeger, 1962).

Case 1, Thrust faults----As shown on Figure A.1a (from Meyerhoff et al.,1992b) the vertical principal stress is the least in magnitude and the other two principal stresses are compressive. The planes of fracture pass through the direction of the intermediate principal stress and make angles of less than 45° with the direction of the greatest compressive (horizontal) stress. This case is illustrated in nature by Figures 2.36, 3.1 and 3.18, which show the lithosphere Benioff zones dip at an angle less than 45° to the Earth's surface (see Scheidegger and Wilson, 1950).

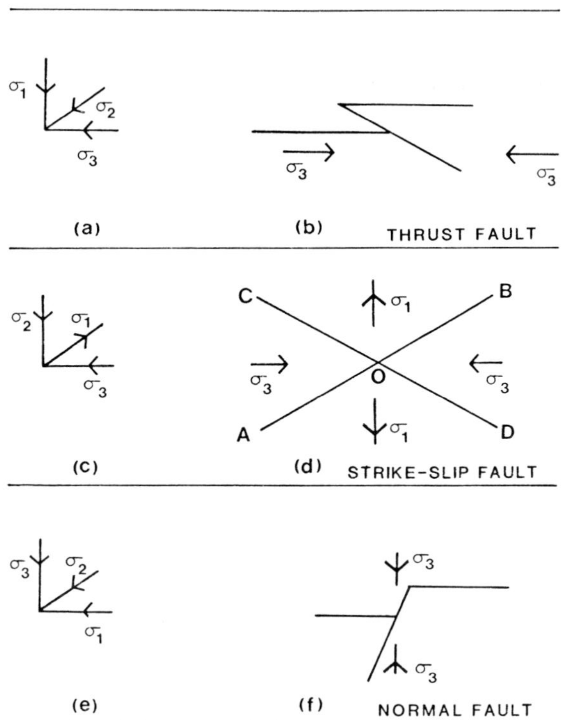

From Jaeger (1962)

Fig. A.1. Three examples of Navier-Coulomb maximum shear-stress theory: in each case, fracture takes place at an angle >45° with the direction of maximum principal stress. s_1 in each diagram is the direction of least principal stress. s_2 is the direction of intermediate principal stress. s_3 is the direction of maximum principal stress. (**a**) and (**b**) describe thrust faulting; (**c**) and (**d**) describe strike-slip faulting; (**e**) and (**f**) describe normal faulting. From Jaeger (1962, p. 75-80).

Case 2, Streamline—strike-slip faults----As shown on Figure A.1c (from Meyerhoff et al.,1992b), the vertical principal stress is the intermediate stress. Of the other two principal stresses, one will be a compression and the other will be small and may even be a tension. In this case, failure can take place on either of two vertical planes (AOB, COD) that are equally inclined at angles less then 45° to it.

Case 3, Normal faults----As shown on Figure A.1e (from Meyerhoff et al., 1992b), the vertical principal stress is the greatest in magnitude. This state can be expected mainly at considerable depths (Jaeger, 1962, p. 80). Failure will take place at an angle of less then 45° to the vertical, that is, at an angle greater than 45° to the Earth's surface. Figures 2.36, 3.1 and 3.18 illustrate natural examples in which the strictosphere Benioff zone dips at an angle greater than 45° to the Earth's surface.

Newton's Laws

All laws, theorems, and principles that involve motion are derived from Newton's three laws of motion and the law of gravity.

First Law of Motion----A body at rest remains at rest and a body in motion remains in uniform motion in a straight line unless acted upon by an external force.

Second Law of Motion----The rate of change of the velocity of a body, that is, its acceleration, is equal to the resultant of all external forces acting on the body divided by the mass of the body, and is in the direction of the straight line in which the force acts (that is, in the same direction as the resultant force).

Third Law of Motion----To every action there is always opposed an equal reaction; that is, the mutual actions of two bodies upon each other are always equal and in the opposite direction; or, for every force, there is an equal and opposite force or reaction.

Law of Gravitation----Every particle of matter in the universe attracts every other particle with a force that is directly proportional to the produce of the masses of the particles and inversely proportional to the square of the distance between them.

Pascal's Law, Theorem, or Principle

Pressure applied to a confined liquid at any point is transmitted undiminished through the fluid in all directions and acts upon every part of the conf ning vessel at right angles to its interior surfaces and equally upon equal areas.

Peach-Köhler Climb Force

Although the following works best in mafic lavas because of their low viscosities, the statements apply to more silicic lavas as well. However, the relatively lower viscosity of mafic lavas makes it easier for them to move through the lithosphere and seek a position of gravitational stability (Walker, 1989).

In the case of a vertical crack filled with magma, the upward force exerted by the crack is a result of the lower density of the magma filling the crack (Corry, 1988), Weertman (1971, p. 1177) explained the process.

"The force of (p - p') g^V [where p is the density of the country rock at a given depth; p' is the magma density; g is the acceleration due to gravity; and V is the magma volume per unit of crack width] experienced by the crack is identical to the Archimedian buoyancy that would act on the crack if it were a solid of density p' embedded in a liquid of density p. There is a fundamental difference, however, between these two forces. For the true Archimedian force the body force p'g^V differs from the net force pg^V exerted on the embedded solid by the hydrostatic pressure. In the case of the liquid-filled crack, the body force p'g^V exerted on the liquid within the crack must be, and is, exactly balanced by hydrostatic forces exerted against the crack walls. "There is no net force exerted on the liquid within the crack." The Peach-Köhler climb force is produced by a gradient in elastic strain energy within the solid (rock body). The origin of the true and the pseudo-Archimedian forces is the same, namely, the gravity field. "The crack will stop propagating, and thus the magma rise will stop, when the magma reaches its level of neutral buoyance, which is the level below which the country rock is denser than the magma, and above which the country rock is less dense than the magma" (Walker, 1989).

Poiseuille's Law

The velocity of flow of a liquid through a capillary tube varies directly as the pressure and the fourth power of the diameter of the tube, and inversely as the length of the tube and the coefficient of viscosity.

Stokes's Law and Linear Geologic Structures

Stokes's Law, one of many expressions of Newton's Second Law of Motion, is little used by geologists, despite its fundamental importance. Because the law usually is thought of as a description of the flow of solids through air and water, it almost never is regarded as describing the interactions between any flowing medium and a solid substance (e.g., the flow of lava through lava tubes). One critic of an earlier version of this manuscript wrote that our use of Stokes's Law ". . . only confuses a discussion of the well-known streamline patterns of viscous flow." This critic apparently did not know that streamline patterns of viscous flow (in glaciers, for example) are commonly cited in physics texts as examples of Stokes's Law (e.g., Blatt, 1983)! Nor did he realize that such patterns of streamline flow are what led in the first place to the 1845 formulation of this law by Sir George G. Stokes (e.g., Sears et al., 1974).

Stokes's Law may be stated most simply as follows: "The force required to move a sphere through a given viscous fluid at a low uniform velocity is directly proportional to the velocity and radius of the sphere" (*Webster's Third New International Dictionary*, 1971, p. 2248). Under the heading of Stokes's Law, Sears et al. (1974, p. 225-226) wrote that "When an ideal fluid of zero viscosity flows past a sphere, or when a sphere moves through a stationary fluid, the streamlines form a perfectly symmetrical pattern around the sphere. . .." If the fluid is viscous, however, there will be a viscous drag on the sphere. Moreover, if the surface of the sphere consists of alternate parallel bands of rough and smooth texture, differential (viscous) drag will take place tween the fluid that flows over the rough-textured surfaces and that which flows over the smooth-textured surfaces. Then the streamlines are accentuated close to the sphere's surface by the difference in fluid velocities close to that surface, and are more appropriately termed sliplines. Flow close to a solid surface is called laminar flow, which has been defined as ". . . streamline flow in a viscous fluid near a solid boundary" (*Webster's Third New International Dictionary*, 1971, p. 1267). The important point is that streamlines, or sliplines are parallel with the direction of the laminar flow.

Many features related closely to laminar flow have been described by petrographers, such as Balk (1937), Knopf and Ingerson (1938), and many others. Many of the structures that we discuss in this paper are analogous to the flow and flow-line structures of Balk (1937). Thus, in geological terms, there are whole families of linear and curvilinear structures, at all scales (e.g., stretching lineations; *schlieren*; strike-slip faults; linear faults, fracture systems of the midocean ridges; several types of linear ridges), which are the products, directly or indirectly, of flow close to and parallel with them. This statement is simply another way of expressing Stokes's Law, or Newton's Second Law of Motion.

For example, the long linear faults, fractures, and fissures of Figure 3.28 that parallel the axes of the Mid-Atlantic Ridge and East Pacific Rise respectively demonstrate that the motion----in any case, the last motion----beneath these ridges was parallel with, not orthogonal to them. Likewise, the jagged tension crack of Figure 3.29 clearly formed as a consequence of motions orthogonal to the overall trend of the crack. The type of fracture that forms in a solid substance is a clear reflection of the direction of the principal stress that deformed it.

Another reviewer of this manuscript wrote, "To me the linear features along a midocean ridge are more plausibly explained by the breaks and cracks of up-and-down shiftings that would result from moving apart the two sides, aided by the forcing apart by magma moving in both directions from a point of injection." This explanation, also favored by Macdonald et al. (1987), requires that a whole series of "points of injection" and "collision zones" between adjacent but opposite magma flows be present along each midocean ridge. Points of injection should be characterized by annular and/or radial structures many kilometers in diameter that alternate along strike with zones of transverse structures, including pressure ridges and like features. We have complete sonograph coverage of several lengthy segments of midocean ridges, including some 600 km of continuous coverage by the U.S. Geological Survey along the Juan de Fuca Ridge. There are no such structures; the only interruptions along strike are the ridge-transverse (transform fault) fault zones (see Figs. 9, 13, 18, 22 of Meyerhoff et al., 1992a). The only other types of interruptions along strike are “overlapping spreading centers”, which appear to be nothing more than giant eddies (see Figs. 16-18 of Meyerhoff et al., 1992a) that also are produced by ridge-parallel flow.

The same critic also wrote that ". . . it seems to me that *en énchelon* bundles of fissures and faults could form just as well by pulling apart as the two sides move orthogonally away from the ridge." Figure 8 in Meyerhoff et al. (1992a) shows such en echelon structures on a small scale; Figure 3.76---an overlapping spreading center---is an en echelon structure on a large scale. As Olson and Pollard (1989) showed, such structures cannot form unless they parallel the direction of flow, as illustrated by Figures 13 and 15 of Meyerhoff et al. (1992a).

The en echelon structures are incipient vortices, in the terminology of fluid dynamics (Sears et al., 1974, p. 226; Tritton, 1988); they are the vortex structures of Meyerhoff et al. (1992a, b) and of this publication. As explained by Sears et al. (p. 226), "When the velocity of a fluid flowing in a tube exceeds a certain critical value (which depends on the properties of the fluid and the diameter of the tube) the nature of the flow becomes extremely complicated. Within a very thin layer adjacent to the tube walls, called the *boundary layer*, the flow is still laminar. The flow velocity in the boundary layer is zero at the tube walls and increases uniformly through the layer . . .. Beyond the boundary layer, the motion is highly irregular. Random circular local currents called *vortices* develop within the fluid, with a large increase in the resistance to flow. Flow of this sort is called *turbulent*." The properties of boundary layers and vortical, eddy, and related structures of incipient and full turbulence are discussed in detail by Tritton (1988). Most of the structural features of the midocean ridges are a consequence of ridge-parallel flow, a conclusion that seems to be well documented, not just because of the linear structures, en echelon phenomena, and related structures we have mentioned, but also because of their near-identity with structures produced both in lava tubes (Yamagishi, 1985) and artificial tubes (Tritton, 1988), in which the flow, by definition, parallels the strike of the tubes. Tritton (1988), in fact, presented at least 39 photographs and line drawings of laboratory experiments in which some of the major structural features of midocean ridges, rift zones, and foldbelts are duplicated (e.g., his figs. 4.4, 17.20, 18.3, and 22.12).

Vortex street

A body towed through a fluid creates a band of oppositely flowing eddies behind it (Tritton, 1988). This band is a vortex street. (In nontechnical language, this band is a wake.) A series of immobile bodies that project into the fluid stream produce the same effect. In the case of the Earth, its rotation theoretically can cause eastward migration of the mushy part of the asthenosphere. Where mushy asthenosphere passes between two cratons rooted to the strictosphere, a vortex street might be created.

BIBLIOGRAPHY

* *Sources used for data compilation of paleotectonic surge channel maps for southeastern Asia.*

Ádám, A., 1983, EM induction in Finland and general crustal physics, *in* Hjelt, S.E., ed., The development of the deep geoelectric model of the Baltic shield, pt. 2. Proceedings of the First Project Symposium, Oulu (Finland), 15-18/11/1983: University of Oulu, Department of Geophysics, Rept. no. 8, p. 1-16.

Ádám, A., 1987, Are there two types of conductive anomaly (CA) caused by fluid in the crust?: Physics of the Earth and Planetary Interiors, v. 45, p. 209-215.

Ádám, A., Duma, G., Gutceursch, R., Verõ, J., And Wallner, A., 1986, Periadriatic lineament in the Alps studied by magnetotellurics: Journal of Geophysics, v. 59, p. 103-111.

Ádám, A., Landy, K., and Nagy, Z., 1989, New evidence for the distribution of the electric conductivity in the earth's crust and upper mantle in the Pannonian basin as a "hotspot": Tectonophysics, v. 164, no. 2-4, p. 361-368.

Adams, J., and Basham, P., 1991, The seismicity and seismotectonics of eastern Canada, *in* Slemmons, D. B., Engdahl, E. R., Zoback, M. D., and Blackwell, D. D., eds., Neotectonics of North America: Boulder, CO, Geological Society of America, Decade of North American Geology, Decade Map Volume, p. 261-276.

Agarwal, J.K., and Rama, S.N.I., 1976, Chronology of Mesozoic volcanics of India: Indian Academy of Sciences, Proceedings, v. 84, p. 157-179.

Agocs, W. B., Meyerhoff, A. A., and Kis, K., 1992, Reykjanes Ridge: quantitative determinations from magnetic anomalies, *in* Chatterjee, S., and Hotton, N. III, eds., New concepts in global tectonics: Lubbock, TX, Texas Tech University Press, p. 221-238.

*Ahmad, T., and Bhat, M. I., 1987, Geochemistry and petrogenesis of the Mandi-Darla volcanics, northwestern Himalayas: Precambrian Research, v. 37, p. 231-256.

Aihara, A., 1989, Paleogeothermal influence on organic metamorphism in the neotectonics of the Japanese Islands: Tectonophysics, v. 159, no. 3/4, p. 291-305.

Aleinikoff, J.N., Dusel-Bacon, C., Foster, H.L., and Nokleberg, W.J., 1987. Lead isotopic finger-printing of tectono-stratigraphic terranes, east-central Alaska: Canadian Journal of Earth Sciences, v. 24, no. 10, p. 2089-2098.

*Alexander, J.B., 1968, The geology and mineral resources of the Bentong area, Pahang District: Geological Survey of West Malaysia, 250 p.

Alexander, P.O., 1981, Age and duration of Deccan volcanism: K-Ar evidence: Geological Society of India Memoir no. 3, p. 244-258.

Allen, C.R., Luo Zhuoli, Qian Hong, Wen Xueze, Zhou Huawei, and Huang Weishi, 1989, Segmentation and recent rupture history of the Xianshuihe fault, southwestern China, *in* Schwartz, D. P., and Sibson, R. H., eds., Fault segmentation and controls of rupture initiation and termination: United States Geological Survey Open-File Report 89-315, p. 10-30.

Allen, J.R.L., Dineley, D.L., and Friend, P.F., 1967, Old Red sandstone basins of North America and northwest Europe, *in* Oswald, D.H., ed., International symposium on the Devonian System, Calgary, 1967, v. 1: Alberta Society of Petroleum Geologists, p. 69-98.

*Almashoor, S.S., and Tjia, H.D., 1987, A prominent fault across the Malaysia-Thai boundary; preliminary report: Geological Society of Malaysia Newsletter, v. 13, no. 2, p. 35-37.

Almeida, F.F.M. de, 1986, Distribução regional e relações tectônicas de magmatismo pós-paleozóico no Brasil: Revista Brasileira de Geociências, v. 16, no. 4, p. 325-349.

Al'mukhamedov, A.I., and Zolotukhin, C.V., 1991, Kaynozoyskiye trappy Dekana, 1: osnovnyye cherty petrologii: Geologiya I Geofizika no. 5, p. 19-27.

Alt, D., Sears, J.W., and Hyndman, D.W., 1988, Terrestrial maria: the origins of large basalt plateaus, hotspot tracks, and spreading ridges: Journal of Geology, v. 96, no. 6, p. 647-662.

*Amantov, V.A., Blagonravov, Yu.A., Borsakovskiy, Yu.A., Durante, M.V., Zonenshayn, L.P., Luvsandanzan, B., Matrosov, P.S., Suyetenko, O.D., Filippova, I.B., and Khasin, R.A., 1970, Osnovnyye cherty stratigrafii Paleozoya Mongol'skoy Narodnoy Respubliki, *in* Luvsandanzan, B., and Marinov, N. A., eds., Stratigrafiya I tektonika Mongol'skoy Narodnoy Respubliki, Sovmestnaya Sovetsko-Mongol'skaya Geologicheskaya Ekspeditsiya, Tyudy vyp. 1: Moscow, Izdatel'stvo "Nauka," p. 8-63.

Amaral, G., Cordani, U.G., Kawashita, K., et al., 1966, Potassium-argon dates of basaltic rocks from southern Brazil: Geochimica et Cosmochimica Acta, v. 30, no. 2, p. 159-189.

*Amiscaray, E., and Tan, M., 1992, Semirara Island Group: a part of the north Palawan block? (Abstract), *in* Second International Symposium on Gondwana Dispersion and Asian Accretion--Geological Evolution of Eastern Tethys, Abstracts: Kyoto, Kochi and Shirokawa (Japan), IGCP Project 321, p. 8.

Ampferer, O., 1906, Über das Bewegungsbild von Faltengebirgen: Jahrbuch der Kaiserlich-Königlichen Geologischen Reichsanstalt, v. 56, no. 3-4, p. 539-622.

Ampferer, O., 1941, Gedanken über das Bewegungsbild des atlantischen Raumes: Sitzungsberichte der Akademie der Wissenschaften in Wien, matematisch-naturwissenschaftliche Klasse, Bd. 150, p. 19-35.

Ampferer, O., and Hammer, W., 1911, Geologischer Querschnitt durch die Östalpen vom Allgäu zum Gardasee: Jahrbuch der Kaiserlich-Königlichen Geologischen Reichsanstalt, v. 61, no. 3-4, p. 531-710.

Anderson, D.L., 1987a, Thermally induced phase changes, lateral heterogeneity of the mantle, continental roots, and deep slab anomalies: Journal of Geophysical Research, v. 91, no. B13, p. 13968-13980.

Anderson, D.L., 1987b, The depths of mantle reservoirs, *in* Mysen, B.O., ed., Magmatic processes: physicochemical principles: University Park (PA), Geochemical Society, special publication no. 1, p. 3-12.

Anderson, D.L., Tanimoto, T., and Zhang Yushen, 1992a, Plate tectonics and hotspots: the third dimension: Science, v. 256, no. 5064, p. 1645-1651.

Anderson, D.L., Zhang Yushen, and Tanimoto, T., 1992b, Plume heads, continental lithosphere, flood basalts and tomography, *in* Storey, B. C., Alabaster, T., and Pankhurst, R. J., eds., Magmatism and the causes of continental break-up: London, Geological Society of London Special Publication no. 68, p. 99-124.

Anderson, R.N., Honnorez, J., Becker, K., and Scientific Party, 1985, Site 504B, *in* Initial Reports, DSDP, 83: Washington, D.C., p. 5-181.

Anderson, T.H., and Schmidt, V.A., 1983, The evolution of Middle America in the Gulf of Mexico-Caribbean Sea region during Mesozoic time: Geological Society of America Bulletin, v. 94, no. 8, p 941-966.

Andrews, J.E., Burns, R.E. Churkin, M., Jr., Davies, T. A., Dumitrica, P., Edwards, A.R., Galehouse, J.S., Kennett, J.P., Packham, G.H. , Van der Lingen, G.J., 1973, Deep Sea Drilling Project: Leg 21; Tasman Sea-Coral Sea (Preliminary Results) in Oceanography of the South Pacific, 1972: Wellington, N. Z. Comm. Unesco, p. 185-199.

Anfiloff, V., 1988, Polycyclic rifting--an interpretation of gravity and magnetics in the North West Shelf, *in* Purcell, P. G., and Purcell, R. E., eds., The North West Shelf, Australia: Perth, Petroleum Exploration Society of Australia, p. 443-455.

Anfiloff, V., Barlow, B.C., Murray, A.S., Denham, D., and Sandford, R., 1976, Compilation and production of the 1976 1:5 000 000 gravity map of Australia: BMR Journal of Australian Geology and Geophysics, v. 1, no. 4, p. 273-276.

Angeheister, G., Bögel, H., Gebrande, H., Schmidt-Thomé, P., and Zeil, W., 1972, Recent investigations of surficial and deeper crustal structures of the eastern and southern Alps: Geologische Rundschau, Bd. 61, Hft. 2, p. 349-395.

Annells, R.N., 1973, Proterozoic flood basalts of eastern Lake Superior: the Keweenawan volcanic rocks of the Mamainse Pointe area, Ontario: Geological Survey of Canada, Paper 72-10, 51 p.

Anonymous, 1988, Precambrian layers drilled in Kansas: EOS, v. 69, no. 31, p. 753.

Antipov, M.P., Zharkov, S.M., Kozhenov, V.Ya., and Pospelov, I.I., 1990, Structure of the Mid-Atlantic Ridge and adjacent parts of the abyssal plain at latitude 13° N: International Geology Review, v. 32, no. 5, p. 468-478.

Arabasz, W.J., Pechmann, J.C., and Brown, E.D., 1992, Observational seismology and the evaluation of earthquake hazards and risk in the Wasatch front area, Utah: Reston, VA, U. S. Geological Survey Professional Paper 1500-D, 36 p.

Archanjo, C.J., Olivier, P., and Bouchez, J.-L., 1992, Plutons granitiques du Seridó (NE du Brésil): écoulement magmatique parallèle à la chaîne révélé par leur anisotropie magnétique: Société Géologique de France Bulletin, v. 163, no. 4, p. 509-520.

Argand, E., 1924, La tectonique de l'Asie: Brussels, 13th International Geological Congress, Liège 1922, Comptes Rendus v. 5, p. 171-372 (Engl. translation by A. V. Carozzi, ed., 1977, Tectonics of Asia: New York, Hafner Press, 218 p.)

Argus, D.F., and Gordon, R.G., 1990, Constraints from VLBI on Pacific-North American motion and deformation (Abstract): EOS, v. 71, no. 28, p. 860.

Arldt, T., 1907, Die Entwicklung der Kontinente und ihrer Liebewelt: Leipzig, Christian Hermann Tauchnitz, 729 p.

Armstrong, R.L., 1978, Cenozoic igneous history of the U.S. Cordillera from lat 42° to 49° N, *in* Smith, R. B., and Eaton, G. P., eds., Cenozoic tectonics and regional geophysics of the Western Cordillera: Geological Society of America Memoir 152, p. 265-282.

Arndt, N.T., Nelson, D.R., Compston, W., Trendall, A.F., and Thorne, A.M., 1991, The age of the Fortesque Group, Hamersley basin, Western Australia, from ion microprobe zircon U-Pb results: Australian Journal of Earth Sciences, v. 38, no. 3, p. 261-281.

Artemjev, M.E., and Artyushkov, E.V., 1971, Structure and isostasy of the Baikal rift and the mechanism of rifting; Journal of Geophysical Research, v. 78, no. 5, p. 1197-1211.

*Artyushkov, E.V., 1987, The forces driving plate motion and compression of the crust in fold belts, *in* Fuchs, K., and Froidevaux, C., eds., Composition, structure and dynamics of the lithosphere-asthenosphere system:

Washington, D.C., American Geophysical Union and Geological Society of America Geodynamics Series, v. 16, p. 175-188.

Artyushkov, E.V., 1992, Role of crustal stretching on subsidence of the continental crust: Tectonophysics, v. 215, no. 1-2, p. 187-207.

*ASCOPE, 1981, Tertiary sedimentary basins of the Gulf of Thailand and South China Sea: stratigraphy, structure and hydrocarbon occurrences: Jakarta, ASEAN Council on Petroleum (ASCOPE), 72 p.

*Asfaw, L.M., Bilham, R., Jackson, M., and Mohr, P., 1992, Recent inactivity in African rift: Nature, v. 357, no. 6387, p. 447.

*Ashley, P.M., and Shaw, S.E., 1993, Apparent ages of hydrothermal alteration zones in the North Arm Volcanics, southeast Queensland: Australian Journal of Earth Sciences, v. 40, no. 4, p. 415-421.

Atherton, M.P., Warden, V., and Sanderson, L.M., 1985a, The Mesozoic marginal basin of central Peru: a geochemical study of within-plate-edge volcanism, *in* Pitcher, W.S., Atherton, M.P., Cobbing, E.J., and Beckinsdale, R.D., eds., Magmatism at a plate edge. The Peruvian Andes: Glasgow, Blackie & Son Limited, p. 47-58.

Atherton, M.P., Sanderson, L.M., Warden, V., and McCourt, W.J., 1985b, The volcanic cover: chemical composition and the origin of the magmas of the Calipuy Group, *in* Pitcher, W.S., Atherton, M.P., Cobbing, E.J., and Beckinsale, R.D., eds., Magmatism at a plate edge. The Peruvian Andes: Glasgow, Blackie & Son Limited, p. 273-284.

Aubouin, J., 1977, Méditerranée occidentale; esquisse d'une comparaison de cadre alpin: Société Géologique de France Bulletin, série 7, v. 19, no. 3, p. 421-435.

Aubouin, J., Von Huene, R., Baltuck, M., Arnott, R., Bourgois, J., Filewicz, M., Helm, R., Kvenvolden, K. A., Lienert, B., McDonald, T., McDougall, K., Ogawa, Y., Taylor, E., and Winsborough, B., 1982, Subduction sans accretion; la marge pacifique du Guatemala; premiers resultats du Leg 84 du Deep Sea Drilling Project (janvier-fevrier 1982): Comptes Rendus des Seances de l'Academie des Sciences, Serie 2, v. 294, no. 13, p. 803-812.

Auden, J.B., 1981, India's former crustal neighbours: Indian National Science Academy Proceedings, part A, v. 47, no. 6, p. 588-630.

*Audley-Charles, M.G., 1968, The geology of Portuguese Timor: London, Geological Society of London Memoir 4, 76 p.

*Audretsch, F.C., d' Kluiving, R.B., and Oudemans, W., 1966, Economic geological investigation of NE Vogelkop (Western New Guinea): Koninklijk Nederlands Geologisch Mijnbouwkundig Genootschap Verhandelingen, Geologische Serie no. 23, 151 p.

Aumento, F., and Loncarevic, B.D., 1969, The Mid-Atlantic Ridge near 45° N., III. Bald Mountain: Canadian Journal of Earth Sciences, v. 6, no. 1, p. 11-23.

Baker, B.H., 1987, Outline of the petrology of the Kenya rift alkaline province, *in* Fitton, J.G., and Upton, B.G.J., eds., Alkaline igneous provinces Geological Society of London Special Publication no. 30, p. 293-311.

Baker, B.H., Goles, G.G., Leeman, W.P., and Lindstrom, M.M., 1977, Geochemistry and petrogenesis of a basalt-benmoreite-trachyte suite from the southern part of the Gregory Rift, Kenya: Contrib. Mineral. Petrol. Beitr. Mineral. Petrol., v. 64, no 3, p 303-332.

Baker, P.E., Rea, W.J., Skarmeta, J., Caminos, C., and Rex, D.C., 1980, Igneous history of the Andean Cordillera and Patagonian Plateau around latitude 46° S: Royal Society of London, Philosophical Transactions, series A, v. 303, no. 1474, p. 105-149.

Baksi, A.,1989, Elucidating the time of initiation and duration of volcanism for various Mesozoic-Tertiary flood basalt provinces (Abstract): New Mexico Bureau of Mines and Mineral Resources Bulletin 131, p. 13.

Baksi, A., 1990, Timing and duration of Mesozoic-Tertiary flood-basalt volcanism: EOS, v. 71, no. 49, p. 1835-1836, 1840.

Baksi, A., and Farrar, E., 1991, $^{40}Ar/^{39}Ar$ dating of the Siberian Traps, USSR: evaluation of the ages of the two major extinction events relative to episodes of flood-basalt volcanism in the USSR and the Deccan Traps, India: Geology, v. 19, no. 5, p. 461-464.

Balk, R., 1937, Structural behavior of igneous rocks: New York, Geological Society of America Memoir 5, 177 p.

Balkwill, H.R., 1978, Evolution of Sverdrup basin, Arctic Canada: American Association of Petroleum Geologists Bulletin, v. 62, no. 6,. p. 1004-1028.

Balkwill, H.R., and Fox, F.G., 1982, Incipient rift zone, western Sverdrup basin, Arctic Canada, *in* Embry, A.F., and Balkwill, H.R., eds., Arctic geology and geophysics: Canadian Society of Petroleum Geologists, Memoir 8, p. 171-187.

Ballard, R.D., and Moore, J.G., 1977. Photographic atlas of the Mid-Atlantic Ridge rift valley: New York, Springer-Verlag, 114 p.

*Bandyopadhyaya, S., Subramanyam, M.R., and Sharma, P.N., 1973, The geology and mineral resources of the Andaman and Nicobar Islands: Geological Survey of India Records, v. 105, no. 2, p. 25-68.

Banks, M.R., 1989, Igneous rocks: introduction, *in* Burrett, C.F., and Martin, E.L., eds., Geology and mineral resources of Tasmania: Geological Society of Australia, Special Publication 15, p. 375-378.

*Bannert, D., and Helmcke, D. 1983, Structural map of Burma, scale: 1:2,000,000, 1 sheet, in Bender, F., 1983, Geology of Burma: Berlin, Gebrüder Borntraeger, 293 p.

Banno, S., and Nakajima, T., 1992, Metamorphic belts of Japanese Islands: Annual Review of Earth and Planetary Sciences, v. 20, p. 159-179.

Baragar, W.R.A., 1972, Coppermine River basalts: geological setting and interpretations *in* Wanless, R.K., and Loveridge, W.D., eds., Rubidium-strontium isochron age studies, Report 1: Geological Survey of Canada Paper 72-73, p. 21-24.

Baragar, W.R.A., 1977, Volcanism of the stable crust, *in* Barager, W.R.A., Coleman, L.C., and Hall, J.M., eds., Volcanic regimes in Canada: Geolo-Association of Canada, Special Paper no. 16, p. 377-405.

Baragar, W.R.A., and Donaldson, J.A., 1973, Coppermine and Dismal Lakes map-areas: Geological Survey of Canada, Paper 71-39, 20 p.

Barrell, J., 1914, The status of hypotheses of polar wanderings; Science, v. 40, n. ser., p. 333-340.

Barrows, A.G., Kahle, J.E., and Beeby, D.J. 1985, Earthquake hazards and tectonic history of the San Andreas fault zone, Los Angeles County, California: California Division of Mines and Geology Open-File Report 85-10, 139 p.

*Barsukov, V.L., Basilevsky, A.T., Burba, G.A., and 27 other authors, 1986, The geology and geomorphology of the Venus surface as revealed by the radar images obtained by Veneras 15 and 16, *in* Proceedings of the Sixteenth Lunar and Planetary Science Conference, Pt. 2: Journal of Geophysical Research, v. 91, no. B-4, p. D378-D398.

Basalt Volcanism Study Project, 1981, Basalt volcanism on the terrestrial planets: New York, Pergamon Press, Inc., 1286 p.

Basu, A.R., Renne, P.R., DasGupta, D.K., Teichmann, F., and Poreda, R.J., 1993, Early and late alkali igneous pulses and a high-^{3}He plume origin for the Deccan flood basalts: Science, v. 261, no. 5123, p. 902-906.

Bayer, R., Carozzo, M.G., Lanza, R., Miletto, M., and Rey, D., 1989, Gravity modelling along the ECORS-CROP vertical seismic reflection profile through the western Alps: Tectonophysics, v. 162, no. 3/4, p. 203-218.

Beall, G.H., Hurley, P.M., Fairbairn, H.W., and Pinson, W.H., Jr., 1963, Comparison of Kar and whole-rock Rb-Sr dating in New Quebec and Labrador: American Journal of Science, v. 261, no. 6, p. 571-580.

Beck, S.L., and Christensen, D.H., 1991, Rupture process of the February 4, 1965, Rat Islands earthquake: Journal of Geophysical Research, v. 96, no. B-2, p. 2205-2221.

*Beckinsale, R.D., Suensilpong, S., Nakapadungrat, S., and Walsh, J.N., 1979, Geochronology and geochemistry of granite magmatism in Thailand in relation to a plate tectonic model: Geological Society of London Journal, v. 136, pt. 5, p. 529-540.

Bellieni, G., Piccirillo, E.M., Comin, C.P., Melfi, A.J., and Da Roit, P., 1988, Mineral chemistry of continental stratoid volcanics and related intrusives from the Paraná Basin (Brazil), *in* Piccirillo, E.M., and Melfi, A.J., eds., The Mesozoic floor volcanism of the Paraná Basin; petrogenetic and geophysical aspects: Univ. Trieste, Ist. Miner. e Petrogr. Trieste, Italy, p. 73-92.

Beloussov, V.V., 1948. Obshchiye zakonomevnosti geotektonickes-kogo protsessa: Akademiya a Nauk SSSR, Izvestya, Seriya Geologiya, no. 5, p. 67-87.

Beloussov, V. V., 1980, Geotectonics: Berlin, Springer-Verlag, 330 p.

Beloussov, V.V., 1981, Continental endogenous regimes: Moscow, Mir Publishers, 295 p.

Beloussov, V.V., 1989, Osnovy geotektoniki: Moscow, Izdatel'stva. Nedra, 382 p.

Beloussov, V.V., 1992. Endogenic regimes and the evolution of the tectonosphere, *in* Chatterjee, S., and Hotton, N., III, eds., New Concepts in Global Tectonics. Lubbock, Texas Tech University Press, p. 411-420.

Belyy, V.F., Gel'men, M.L., and Paraketsov, K.V., 1989, Mesozoic volcanism and structural development in the northeast USSR: International Geology Review, v. 31, no. 5, p. 455-468.

Belyy, V.F., Yefimova, A.F., and Paraketsov, K.V., 1966, Lower Cretaceous of Okhotsk-Chukotka belts: International Geology Review, v. 8, no. 10, p. 1226-1236.

*Bénard, F., Muller, C., Létouzey, J., Rangin, C., and Tahir, S., 1990, Evidence of multiphase deformation in the Rajang-Crocker Range (northern Borneo) from Landsat imagery interpretation: geodynamic implications: Tectonophysics, v. 183, no. 1-4, p. 321-339.

Bence, A.E., Taylor, S.R., and Fisk, M., 1980, Major--and trace-element geochemistry of basalts from Ojin, Nintoku, and Suiko seamounts of the Emperor Seamount chain; DSDP-IPOD Leg 55, *in* Shambach, J., Jackson, E. D., et al., Initial Reports of the Deep Sea Drilling Project, Leg 55: Washington D. C. (Govt. Printing Office), p. 599-605.

*Bender, F., 1983, Geology of Burma: Berlin, Gebrüder Borntraeger, 193 p.

Benioff, H., 1949, Seismic evidence for the fault origin of oceanic deeps: Geological Society of America Bulletin, v. 60, no. 12, p. 1837-1856.

Benioff, H., 1954, Orogenesis and deep crustal structure--additional evidence from seismology: Geological Society of America Bulletin, v. 65, no. 5, p. 385-400.

Benn, K., Sawyer, E.W., and Bouchez, J.-L., 1992, Orogen parallel and transverse shearing in the Opatica belt, Quebec: implications for the structure of the Abitibi subprovince: Canadian Journal of Earth Sciences, v. 29, no. 11, p. 2429-2444.

Bercovici, D,., Schubert, G., Glatzmaier, G. A., 1989, Three-dimensional spherical models of convection in the Earth's mantle: Science, v. 244, no 4907, p. 950-955.

Berendsen, P., Doveton, J. H., Gerhard, L. C., Newell, K. D., Steeples, D. W., Watney W. L., Borcherding, R., 1988, Preliminary geologic report of the Texaco 1 Poersch borehole: Oil and Gas Journal, v. 86, no. 44, p. 48-52, 54.

Bergantino, R. N., 1973-1974, Physiographic diagram of the North Atlantic, Charts NA-6, NA-7, NA-9 and NA-10, U. S. Naval Oceanographic Office Charts, scale 1:10,000,000.

Berggren, W.A., Kent, D.V., Swisher, C.C., III, and Aubry, M.-P., 1995, A revised geochronology and chronostratigraphy, in Geochronology Time Scales and Global Stratigraphic Correlation: SEPM Special Publication No. 54, p. 129-212.

Berhe, S.M., Desta, B., Nicoletti, M., Teferra, M., 1987, Geology, geochronology and geodynamic implications of Cenozoic magmatic province in W. and SE Ethiopia: Journal of the Geological Society, London, v. 144, no. 2, p 213-226

Berrocal, J., Fernandes, C., Bueno, A., Seixas, N., and Bassini, A., 1993, Seismic activity in Monsuaba (State of Rio de Janeiro), Brazil, between 1988 December and 1989 February: Geophysical Journal International, v. 113, no. 1, p. 73-82.

Bertrand, M., 1897, Structure des Alpes françaises et récurrence de certains facies sédimentaires: Sixth International Geological Congress, Lausanne 1894, Comptes Rendus, pt. 3, p. 161-177.

Besairie, H., 1972, Géologie de Madagascar, I. Les terrains sédimentaires: Tananarive, Annales Géologiques de Madagascar, fasc. no. 25, 463 p. + 89 plates.

Bevier, M.L., 1983, Regional stratigraphy and age of Chilcotin Group basalts, south-central British Columbia: Canadian Journal of Earth Sciences, v. 20, no. 4, p. 515-524.

Bevier, M.L., 1989, A lead and strontium study of the Anahim volcanic belt, British Columbia: additional evidence for widespread suboceanic mantle beneath western North America: Geological Society of America Bulletin, v. 101, no. 7, p. 973-981.

Bevins, B.E., Lees, G.J., and Roach, R.A., 1991, Ordovician bimodal volcanism in SW Wales: geochemical evidence for petrogenesis of the silicic rocks: Geological Society of London, Journal, v. 148, p. 4, p. 719-729.

*Bhat, M.I., 1984, Abor volcanics: further evidence for the birth of the Tethys Ocean in the Himalayan segment: Geological Society of London Journal, v. 141, pt. 4, p. 763-775.

*Bhat, M.I., 1987, Spasmodic rift reactivation and its role in the pre-orogenic evolution of the Himalayan region: Tectonophysics, v. 134, no. 1-3, p. 103-127.

*Bhat, M.I., Le Fort, P., and Ahmad, T., 1994, Bafliaz volcanics, NW Himalayas: origin of a bimodal, tholeiite and alkali basalt suite: Chemical Geology, v. 114, no. 3-4, p. 217-234.

Bibee, L.D., Shor, G.G., Jr., and Lu, R.S., 1980, Inter-arc spreading in the Mariana Trough: Marine Geology, v. 35, no. 1-3, p. 183-197.

Billings, M.P., 1972, Structural geology, 3rd ed., Englewood Cliffs (NJ), Prentice-Hall, Inc. 606 p.

Bills, B.G., and Fischer, M.A., 1992, A spatial domain Stokes flow model for the gravity and topography of the middle latitudes of Venus: Journal of Geophysical Research, v. 97, no. E-11, p. 18285-18294.

Biq Ching Chang, Shyu, C.T., Chen, J.C., and Boggs, S., Jr., 1985, Taiwan: geology, geophysics, and marine sediments, *in* Nairn, A.E.M., Stehli, F.G., and Uyeda, S., eds., The ocean basins and margins, v. 7A, The Pacific Ocean: New York, Plenum Press, p. 503-550.

Birch, F., Roy, R.F., and Decker, E.R., 1968, Heat flow and thermal history in New England and New York, *in* Zen, E.A., White, W.S., Hadley, J.B. and Thompson, J.B., Jr., eds., Studies of Appalachian geology, northern and maritime: New Interscience Publishers, p. 437-451.

Birkenmajer, K., 1981, The geology of Svalbard, *in* Nairn, A.E.M., Churkin, M., Jr., and Stehli, F.G., eds., The ocean basins and margins, v. 5. The Arctic Ocean: New York, Plenum Press, p. 265-329.

Blackett, P.M.S., 1956, Lectures on rock magnetism: Jerusalem, The Weizmann Science Press of Israel, 131 p.

*Blackinton, J.G., Hussong, D.M., and Kosalos, J., 1983, First results from a combination side scan sonar and seafloor mapping system (SeaMARC II): Houston, Offshore Technology Conference, OTC 4478, p. 307-311.

Blakely, R.J., and Jachens, R.C., 1990, Volcanism, isostatic residual gravity, and regional tectonic setting of the Cascade volcanic province: Journal of Geophysical Research, v. 95, no. 12, p. 19,439-19,451.

Blanchard, M.-C., Jameison, R.A., and More, E.B., 1984, Late Devonian-Early Carboniferous volcanism in western Cape Breton Island, Nova Scotia: Canadian Journal of Earth Sciences, v. 21, no. 7, p. 162-714.

Blanford, W.T., 1867a, On the geology of a portion of Cutch: Geological Survey of India, Memoir 6, p. 17-38.

Blanford, W.T., 1867b, On the traps and intertrappean beds of western and central India: Geological Survey of India, Memoir 6, p. 137-162.

Blasius, K.R., Cutts, J.A., Guest, J.E., and Masursky, H., 1977, Geology of the Valles Marineris: first analysis of imaging from the Viking 1 orbiter primary mission: Journal of Geophysical Research, v. 82, no. 28, p. 4067-4091.

Blatt, F.J., 1983, Principles of physics: Boston, Allyn and Bacon, Inc., 815 p.

Bloomer, S.H., Stern, R.J., Fisk, E., and Geschwind, C.H., 1989, Shoshonitic volcanism in the northern Mariana arc, 1. Mineralogic and major and trace element characteristics: Journal of Geophysical Research, v. 94, no. B4, p. 4469-4496.

Bluck, B.J., Haughton, P.D.W., House, M.R., Selwood, E.B., and Tunbridge, I.P., 1988, Devonian of England, Wales and Scotland, *in* McMillan, N.J., Embry, A.F., and Glass, D.J., eds., Devonian of the world. Proceedings of the Second International Symposium on the Devonian System, v. I: Regional syntheses: Canadian Society of Petroleum Geologists, Memoir 14, p. 305-324.

Blümling, P., and Prodehl, C., 1983, Crustal structure beneath the eastern part of the Coast Ranges (Diablo Range) of central California from explosion seismic and near earthquake data: Physics of the Earth and Planetary Interiors, v. 31, p. 313-326.

Blümling, P., Mooney, W.D., and Lee, W.H.K., 1985, Crustal structure of the southern Calaveras fault zone, central California, from seismic refraction investigations: Seismological Society of America Bulletin, v. 75, no. 1, p. 193-209.

Boardman, S.J., ed., 1983, Revolution in the earth sciences. Advances in the past half-century: Dubuque (Iowa), Kendall/Hunt Publishing Company, 385 p.

Bodri, B., and Jessop, A.M., 1989, Geothermal model of the continental margins of eastern Canada: Tectonophysics, v. 164, no. 2-4, p. 139-150.

Bois, C., Pinet, B., and Roure, F., 1989, Dating lower crustal features in France and adjacent areas from deep seismic profiles, *in* Mereu, R. F., Mueller, S., and Fountain, D. M., eds., Properties and processes of earth's lower crust: American Geophysical Union Geophysical Monograph no. 51, p. 17-31.

Bondi, H., and Gold, T., 1955: Month. Noth. Roy, Astron. Soc., v. 115, p. 41.

Bonini, W.E., Loomis, T.P., and Robertson, J.D., 1973, Gravity anomalies, ultramafic intrusions, and the tectonics of the region around the Strait of Gibraltar: Journal of Geophysical Research, v. 78, no. 8, p. 1372-1382.

Bonnichsen, B., 1989, The nature of the silicic volcanism in the Snake River Plain, Idaho, U.S.A. (Abstract): New Mexico Bureau of Mines and Mineral Resources Bulletin 131, p. 27.

Bosshard, E., and Macfarlane, D.J., 1970, Crustal structure of the western Canary Islands from seismic refraction and gravity data: Journal of Geophysical Research, v. 75, no. 26, p. 4901-4918.

Bott, M.P.H., 1971, The interior of the earth; New York, St. Martin's Press, 316 p.

Bowin, C., 1976, Caribbean gravity field and plate tectonics: Geological Society of America Special Paper 169, 79 p.

Bowin, C., Purdy, G.M., Johnston, C., Shor, G., Lawver, L., Hartono, H.M.S., and Jezek, P., 1980, Arc-continent collision in Banda Sea region: American Association of Petroleum Geologists Bulletin, v. 64, no. 6, p. 868-915.

Bowin, C., Warsi, W., and Milligan, J., 1982, Free-air gravity anomaly atlas of the world: Geological Society of America Map and Chart Series MC-46, 5 p., 74 sheets, scale, 1:4,000,000.

Boyd, W.W., 1991, Crustal evolution in the Pensacola Mountains: inferences from chemistry and petrology of the igneous rocks and nodule-bearing lamprophyre dykes, *in* Thomson, M.R.A., Crame, J.A., and Thomson, J.W., eds., Geological evolution of Antarctica: Cambridge University Press, p. 557-561.

Boyer, S.E., and Elliott, D., 1982, Thrust systems: American Association of Petroleum Geologists Bulletin, v. 66, no. 9, p. 1196-1230.

Braile, L.W., 1989, Crustal structure of the continental interior, *in* Pakiser, L.C., and Mooney, W.D., eds., Geophysical framework of the continental United States: Geological Society of America Memoir 172, p. 285-315.

Breen, N.A., Silver, E.A., and Roof, S., 1989, The Wetar back arc thrust belt, eastern Indonesia: the effect of accretion against an irregularly shaped arc: Tectonics, v. 8, no. 1, p. 85-98.

Bridgwater, D., Sutton, J., and Watterson, J., 1974, Crustal downfolding associated with igneous activity: Tectonophysics, v. 21, no. 1-2, p. 57-77.

Bristow, J.W., and Armstrong, R.A., 1989, An emplacement and petrogenetic model for the high temperature ash flows of the Jozini Formation, South Africa (Abstract): New Mexico Bureau of Mines and Mineral Resources Bulletin 131, p. 31.

Bristow, J.W., and Saggerson, E. P., 1983, A general account of Karoo vulcanicity in Southern Africa: Geologische Rundschau, v. 72, no. 3, p 1015-1059.

Brodie, K.H., and Rutter, E.H., 1987, Deep crustal extensional faulting in the Ivrea zone of northern Italy: Tectonophysics, v. 140, no. 2-4, 193-212.

Brönnimann, P., and Pardo, G., 1956, Jurassic-Cretaceous stratigraphy of the carbonate rocks of northern Las Villas Province, Cuba, *in* Resúmenes de los trabajos presentados: XX Congreso Geológico Internacional, México 1956, p. 328-329.

Brooks, C., James, D.E., and Hart, S.R., 1976, Ancient lithosphere: its role in young continental volcanism: Science, v. 193, no. 4258, p. 1086-1094.

Brousse, R., and Bellon, H., 1983, Réflexions chronologiques et pétrologiques sur le volcanism associé au développement des rifts de France, *in* Popoff, M., and Tiercelin, J.-J., eds., Rifts et fossés anciens: Pau (France), Centres de la Recherche d'Exploration-Production Bulletin, Elf-Aquitaine, v. 7, no. 1, p. 409-424.

*Brown, D.A., Campbell, K.S.W., and Crook, K.A.W., 1968, The geological evolution of Australia and New Zealand: Oxford, Pergamon Press, 409 p.

Brown, R.L., and Tippett, C.R., 1978, The Selkirk fan structure of the southeastern Canadian Cordillera: Geological Society of America Bulletin, v. 89, no. 4, p. 548-558.

Brown, M., D'-Lemos, R.S., Strachan, R.A., 1992. Transpression and strike-slip faulting; their role in the generation, segregation, ascent and emplacement of felsic magmas, *in* 29th International Geological Congress (Kyoto, Japan, August 24-September 3, 1992), Abstracts, p. 518.

Browne, S.E., and Fairhead, J.D., 1983, Gravity study of the central African rift system: a model of continental disruption, 1. The Ngaoundere and Abu Gabra rifts: Tectonophysics, v. 94, no. 1-4, p. 187-203.

Brune, J.N., 1969, Surface waves and crustal structure, *in* Hart, P.J., ed., The earth's crust and upper mantle: American Geophysical Union Geophysical Monograph 13, p. 230-242.

Brunel, M., 1986, Ductile thrusting in the Himalayas: shear sense criteria and stretching lineations: Tectonics, v. 5, no. 2, p. 247-265.

Bucher, W.H., 1933, The deformation of the earth's crust: Princeton, NJ, Princeton University Press, 518 p.

Bucher, W.H., 1947, Problems of earth deformation illustrated by the Caribbean Sea basin: New York Academy of Sciences Transactions, series II, v. 9, no. 3, p. 98-116.

Bucher, W.H., 1955, Deformation in orogenic belts, *in* Poldervaart, A., ed., Crust of the earth (a symposium): New York, Geological Society of America Special Paper 62, p. 343-368.

Bucher, W.H., 1956, Role of gravity in orogenesis: Geological Society of America Bulletin, v. 67, no. 10, p. 1295-1318.

*Bull, J.M., Martinod, J., and Davy, P., 1992, Buckling of the oceanic lithosphere from geophysical data and experiments: Tectonics, v. 11, no. 3, p. 537-548.

Bultitude, R.J., 1976, Flood basalts of Early Cambrian age in Northern Australia, *in* Johnson, R.W., ed., Volcanism in Australasia: New York, Elsevier Publishing Company, p. 1-20.

*Bunopas, S., 1983, Paleozoic succession in Thailand, *in* Nutalaya, P., ed.-in-chief, Proceedings of the workshop on stratigraphic correlation of Thailand and Malaysia, September 8-10, 1983, v. 1, Technical papers: Bangkok, Geological Society of Thailand and Geological Society of Malaysia, p. 39-76.

*Bunopas, S., and Vella, P., 1978, Late Palaeozoic and Mesozoic structural evolution of northern Thailand: a plate tectonic model, *in* Nutalaya, P., ed., Proceedings of the Third Regional Conference on Geology and Mineral Resources of Southeast Asia, Bangkok, November 14-18, 1978 (GEOSEA III): Bangkok, Asian Institute of Technology, p. 133-140.

Burchfiel, B.C., and Davis, G.A., 1968, Two-sided nature of the Cordilleran orogen and its tectonic implications: 23rd International Geological Congress, Prague 1968, Repts. v. 3, p. 175-194.

Burchfiel, B.C., and Davis, G.S., 1975, Nature and controls of Cordilleran orogenesis, western United States: extensions of an earlier synthesis: American Journal of Science, v. 275-A, p. 363-396.

Bureau of Mineral Resources, 1976, Bouguer gravity map of Australia: Canberra, Bureau of Mineral Resources, 1 sheet, scale, 1:5,000,000.

Burkley, L.A., 1976, Geochronology of the central Venezuelan Andes: unpublished Ph.D. dissertation, Case Western Reserve University, 150 p.

*Burri, P., 1989, Hydrocarbon potential of Tertiary intermontane basins in Thailand, *in* Thanasuthipitak, T., and Ounchanum, P., ed., Proceedings of the International Symposium on Intermontane Basins: Geology and Geophysics: Chiang Mai (Thailand), Chiang Mai University, Faculty of Science, p. 3-12.

Burton, C.K., 1967, Graptolite and tentaculite correlations and palaeogeography of the Silurian and Devonian in the Yunnan-Malaya geosyncline: Palaeontological Society of Japan Transactions and Proceedings, new series, v. 65, p. 27-46.

Butler, R.F., Gehrels, G.E., McClelland, W.C., May, S.R., and Klepacki, D., 1989, Regional tilt rather than large-scale displacement: Geology, v. 17, no. 8, p. 691-694.

Butler, R.F., Herve, F., Munizaga, F., Beck, M.E., Jr., Burmester, R.F., Oviedo, E.S., 1991, Paleomagnetism of the Patagonian plateau basalts, southern Chile and Argentina: Journal of Geophysical Research, ser. B, Solid Earth and Planets, v. 96, no. 4, p. 6023-6034.

Cady, J.W., 1991, Aeromagnetic map of Alaska from latitude 65°-68° N., longitude 141°-162° W.: color-shaded relief: U. S. Geological Survey Geophysical Investigations Map GP-992, scale: 1:500,000, 2 sheets.

Cahen, L., Snelling, N.J., Delhal, J., and Vail, J.R., 1984, The geochronology and evolution of Africa: Oxford, Clarendon Press, 512 p.

Cameron, K.L., Nimz, G.J., Kuentz, D., Niemeyer, S., and Gunn, S., 1989, Southern Cordilleran basaltic andesite suite, southern Chihuahua, Mexico: a link between Tertiary continental arc and flood basalt magmatism in North America: Journal of Geophysical Research, v. 94, no. B-6, p. 7817-7840.

Camp, V.E., and Roobol, M.J., 1989, The Arabian continental alkali basalt province: Part I. Evolution of Harrat Tahat, Kingdom of Saudi Arabia: Geological Society of America Bulletin, v. 101, no. 1, p. 71-95.

Camp, V.E., and Roobol, M.J., 1992, Upwelling asthenosphere beneath western Arabia and its regional implications: Journal of Geophysical Research, v. 97, no. B-11, p. 15255-15271.

Camp, V.E., Roobol, M.J., and Hooper, P.R., 1991, The Arabian continental alkali basalt province: Part II. Evolution of Harrats Khaybar, Ithnayn, and Kura, Kingdom of Saudi Arabia: Geological Society of America Bulletin, v. 103, no. 3, p. 363-391.

Cardwell, R.K., Isacks, B.L., and Karig, D.E., 1980, The spatial distribution of earthquakes, focal mechanism solutions, and subducted lithosphere in the Philippine and northeastern Indonesian islands, *in* Hayes, D.E., ed., The tectonic and geological evolution of southeast Asian seas and islands: American Geophysical Union Geophysical Monograph 23, p. 1-35.

Caress, D.W., Burnett, M.S., and Orcutt, J.A., 1989, Tomographic imaging of the magma chamber at 12°50' N on the East Pacific Rise (Abstract): EOS, v. 70, no. 15, p. 455-456.

Carey, S.W., 1976, The expanding earth: Amsterdam, Elsevier Publishing Company, 488 p.

Carlisle, D.., and Susuki-Takeo, 1974, Emergent basalt and submergent carbonate-clastic sequences including the Upper Triassic Dilleri and Welleri Zones on Vancouver Island: Canadian Journal of Earth Science, v. 11, no. 2, p. 254-279.

Carlson, R.W., and Hart, W.K., 1988, Flood basalt volcanism in the northwestern United States, *in* Macdougall, J.D., ed., Continental flood basalts: Dordrecht, Kluwer Academic Publishers, p. 35-61.

Carter, E.K., and Brooks, J.H., 1960, Precambrian, north-western Queensland, *in* Hill, D., and Denmead, A.K., eds., The geology of Queensland: Melbourne University Press, p. 21-62.

Cathles, L.M., III, 1975, The viscosity of the earth's mantle: Princeton, NJ, Princeton University Press, 386 p.

*Cazenave , A., Valette, J.J., and Boucher, C., 1992, Positioning results with DORIS on SPOT2 after first year of mission: Journal of Geophysical Research, v. 97, no. B-5, p. 7109-7119.

Čermák, V., and Rybach, L., eds., 1979, Terrestrial heat flow in Europe: New York, Springer-Verlag, 328 p.

*Chang [Chengfa], Pan Yusheng, and Sun Yiying, 1989, The tectonic evolution of Qinghai-Tibet Plateau: a review, *in* Şengör, A. M. C., ed., Tectonic evolution of the Tethyan region: Dordrecht, Kluwer Academic Publishers, p. 415-476.

Chao, Y.T., 1929, Geological notes in Szechuan: Geological Society of China, Bulletin, v. 8, no. 2. (Not seen.)

*Charlton, T.R., Kaye, S.J., Samodra, H., and Sardjono, 1991, Geology of the Kai Islands: implications for the evolution of the Aru trough and Weber basin, Banda arc, Indonesia: Marine and Petroleum Geology, v. 8, no. 1, p. 6269.

*Charlton, T.R., Smet, M.E.M. de, Samodra, H., and Kaye, S.J., 1991, The stratigraphic and structural evolution of the Tanimbar Islands, eastern Indonesia: Journal of Southeast Asian Earth Sciences, v. 6, no. 3-4, p. 343-358.

Chatterjee, S., and Hotton, N. III, 1986, The paleoposition of India: Journal of Southeast Asian Earth Sciences, v. 1, no. 3, p. 145-189.

*Chaudhuri, J.N. Bhadra, Helfmeier, H.U., and Galbert, F., 1993, The REE-bearing minerals of intercratonic extensional zone in Chhotonagpur continental plate of eastern Indian Shield: Journal of Southeast Asian Earth Sciences, v. 8, no. 1-4, p. 279-285.

Chen, C., 1980, Non-marine setting of petroleum in the Sungliao basin of northeastern China: Journal of Petroleum Geology, v. 2, no, 3, p. 233-264.

Chen Guoda, 1989, Tectonics of China: Oxford, Pergamon Press, 258 p.

Chen Guoda, 1992, Historistic-causationist geotectonics: Changsha (Hunan), Geotectonica et Metallogenia, v. 16, no. 1-2, p. 1-98.

*Chen Haihong, Sun Shu, Li Jiliang, Wang Qingchen, Peng Haibo, Xu Shutong, and Hsü, K. J., 1993, A preliminary study on tectonics of the Xuefeng Mountains, western Hunnan: Scientia Geologica Sinica, v. 28, no. 3, p. 201-210.

Chen Zongji, 1987, Geodynamics and tectonic evolution of the Panxi rift: Tectonophysics, v . 133, no. 3/4, p. 287-304.

Cheng, Q., Park, K.H., Macdougall, J.D., Zindler, A., Lugmair, G.W., Staudigel, H., Hawkins, J., and Lonsdale, P., 1987, Isotopic evidence for a hotspot origin of the Louisville Seamount chain, *in* Keating, B. H., Fryer, P.,

Batiza, R., and Boehlert, G. W., eds., Seamounts, islands, and atolls: American Geophysical Union Geophysical Monograph 43, p. 283-296.

Cherkasov, G.N., 1979, Sledy posletriasovpgo vulkanizma na Sibirskoy platforme: Novosibirsk, Geologiya i Geofizika, no. 4, p. 154-159.

Cherkis, N.Z., Fleming, H.S., and Brozena, J.M., 1989, Bathymetry of the South Atlantic Ocean - 3° to 40° S, GSA Map and Chart Series, MCH-069, scale: 1:5,737,447 at the equator.

Chi Shunliang, and Luo Mingjin, 1992, Tidal energy flow in solid earth: Acta Seismologica Sinica (English Edition), v. 5, no. 4, p. 877-881.

Chiarenzelli, J.R., and McLelland, J.M., 1991, Age and regional relationships of granitoid rocks of the Adirondack Highlands: Journal of Geology, v. 99, no. 4, p. 571-590.

Chiesa, S., Civetta, L., De Fino, M., La Volpe, L., and Orsi, G., 1989, The Yemen Trap Series; genesis and evolution of a continental flood basalt province: Journal of Volcanology and Geothermal Research, v. 36, no. 4, p. 337-350.

*Chikov, B.M., and Nguyen Chong Iyem, 1992, Deformatsiono-metamorficheskaya skhema tektoniki Severnogo V'yetnama: Novosibirsk, Geologiya i Geofizika, no. 5, p. 44-56.

*Chinbunchorn, N., Pradidtan, S., and Sattayarak, N., 1989, Petroleum potential of Tertiary intermontane basins in Thailand, *in* Thanasuthipitak, T., and Ounchanum, P., eds., Proceedings of the International Symposium on Intermontane Basins: Geology and Resources: Chiang Mai (Thailand), Chiang Mai Univeristy, Faculty of Science, p. 29-42.

*Chinese Academy of Geological Sciences, 1979, Seismo-structural map of China: Beijing, Geological Publishing House, 1 sheet, scale, 1:4,000,000 + explanatory booklet, 11 p.

Choi, D.R., 1993, On the Korea-Kyushu-Palau Ridge system: Hokuriku Geology Institute (Japan), Report no. 3, p. 123-132.

*Choubert, G., and Faure-Muret, A., coordinators, 1976-1985, Geological world atlas: Paris, United Nations Educational, Scientific and Cultural Organization (UNESCO), Commission for the Geological Map of the World, scale, 1:10,000,000, 21 sheets.

Choukroune, P., 1989, The ECORS Pyrenean deep seismic profile reflection data and the overall structure of an orogenic belt: Tectonics, v. 8, no. 1, p. 23-39.

Christensen , U.R., and Yuen, D.A., 1984, The interaction of a subducting lithospheric slab with a chemical or phase boundary: Journal of Geophysical Research, v. 89, no. 6, p. 4389-4402.

Clague, D.A., and Frey, F.A., 1982, Petrology and trace element geochemistry of the Honolulu Volcanics, Oahu; implications for the oceanic mantle below Hawaii: Journal of Petrology, v. 23, no. 3, p. 447-504.

Cloos, E., and Dietz, C., eds., 1954, Conversation with the earth, by Hans Cloos: New York, Alfred A. Knopf, 413 p.

Cloos, H., and Cloos, E., 1927, Das Strömungsbild der Wolkenberg: Zeitschrift für Vulkanologie, v. 11, p. 93-95.

Clowes, R.M., Cook, F.A., Green, A.G., Keen, C.E., Ludden, J.N., Percival, J.A., Quinlan, G.M., and West, G.F., 1992, Lithoprobe: new perspectives on crustal evolution: Canadian Journal of Earth Sciences, v. 29, no. 9, p. 1813-1864.

*Cluzel, D., 1992, Ordovician bimodal magmatism in the Ogcheon belt (South Korea): intracontinental rift-related activity: Journal of Southeast Asian Earth Sciences, v. 7, no. 2-3, p. 195-209.

Cobbing, E.J., Mallick, D.I.J., Pitfield, P.E.J., and Teoh, L.H., 1986, The granites of the Southeast Asian tin belt: Journal of the Geological Society of London, v. 143, part 3, p. 537-550.

Collet, LW., 1927, The structure of the Alps, 2nd ed.: London, E. Arnold, 289 p. (reprinted: 1974, The structure of the Alps, 2nd ed.: Huntington, New York, Robert E. Krieger Publishing Company, 304 p.)

Condie, K.C., 1982, Early and middle Proterozoic supracrustal successions and their tectonic setting: American Journal of Science, v. 282, no. 3, p. 341-357.

Coney, P.J., 1980, Cordilleran metamorphic core complexes: an overview, *in* Crittenden, M.D., Jr., Coney, P.J., and Davis, G.H., eds., Cordilleran metamorphic core complexes: Geological Society of America Memoir 153, p. 7-31.

Cooper, M.A., Collins, D.A., Ford, J., Murphy, F. X., Trayner, P.M., and O'Sullivan, M., 1986, Structural evolution of the Irish Variscides: Geological Society of London Journal, v. 153, pt. 1, p. 53-61.

*Cooper, M.A., Herbert, R., and Hill, G.S., 1989, The structural evolution of Triassic intermontane basins in Thailand: *in* Thanasuthipitak, T., and Ounchanum, P., eds., Proceedings of the International Symposium on Intermontane Basins: Geology and Resources: Chiang Mai (Thailand), Chiang Mai University, Faculty of Science, p. 231-242.

Corry, C.E., 1988, Laccoliths; mechanics of emplacement and growth: Boulder, CO, Geological Society of America Special Publication 220, 110 p.

Costain, J.K., and Bollinger, G.A., 1991, Correlations between streamflow and intraplate seismicity in the central Virginia, U.S.A., seismic zone: evidence for possible climatic controls: Tectonophysics, v. 186, no. 1-2, p. 193-214.

Courtillot, V., Vandamme, D., Besse, J., Jaeger, J. J., 1988, Deccan volcanism at the Cretaceous-Tertiary boundary, *in* Global catastrophes in Earth history; an interdisciplinary conference of impacts, volcanism, and mass mortality: LPI-Contribution, v. 673, p. 31-32.

Courtney, R.C., and White, R.S., 1986, Anomalous heat flow and geoid across the Cape Verde Rise: evidence for dynamic support from a thermal plume in the mantle: Royal Astronomical Society Geophysical Journal, v. 87, p. 815-867.

Coward, M.P., 1982, Surge zones in the Moine thrust zone of NW Scotland: Journal of Structural Geology, v. 4, p. 247-256.

Coward, M.P., and Smallwood, S., 1984, An interpretation of the Variscan tectonics of SW Britain: Geological Society of London Special Publication no. 14, p. 89-102.

Cox, A., and Hart, R.B., 1986, Plate tectonics: how it works: Boston, Blackwell Scientific Publications, Inc., 389 p.

Cox, K.G., 1983, The Karoo province of southern Africa: origin of trace element enrichment patterns, *in* Hawkesworth, C.J., and Norry, M.J., eds., Continental basalts and mantle xenoliths: Nantwich (U.K.), Shive Publishing Limited, p. 139-157.

Cox, K.G., 1988a, The Karoo province, *in* Macdougall, J.D., ed., Continental flood basalts: Dordrecht, Kluwer Academic Publishers, p. 239-271.

Cox, K.G., 1988b, Inaugural address, *in* Subbarao, K.V., ed., Deccan flood basalts: Geological Society of India Memoir 10, p. xv-xxii.

Cox, K.G., 1989, The role of mantle plumes in the development of continental drainage patterns: Nature, v. 342, no. 6252, p. 873-877.

Cox, K. G., and Clifford, P., 1982, Correlation coefficient patterns and their interpretation on three basaltic suites: Contributions to Mineralogy and Petrology, v. 79, no. 3, p. 268-278.

Crane, K., and O'Connell, S., 1983, The distribution and implications of heat flow from the Gregory rift in Kenya: Tectonophysics, v. 94, no. 1-4, p. 253-275.

Crawford, A.J., and Hilyard, D., 1990, Geochemistry of late Proterozoic flood basalts, Adelaide geosyncline, South Australia, *in* Jago, J.B., and Moore, P.S., eds., The evolution of a late Precambrian-early Paleozoic rift complex: the Adelaide geosyncline: Geological Society of Australia, Special Publication no. 16, p. 49-67.

Crone, A.J., and Haller, K.M., 1989, Segmentation of Basin-and-Range normal faults: examples from east-central Idaho and southwestern Montana, *in* Schwartz, D. P., and Sibson, R. H. eds., Fault segmentation and controls of rupture initiation and termination: United States Geological Survey Open-File Report 89-315, p. 110-130.

Crovelli, R.A., 1992, Glossary of terms for chaos, fractals, and dynamics: Reston, VA, U. S. Geological Survey Open-File Report 92-19, 30 p.

*Cull, J.P., 1989, Geothermal models and mantle rheology in Australia: Tectonophysics, v. 164, no. 2-4, p. 107-117.

Curray, J.R., Shor, G.G., Jr., Raitt, R.W., and Henry, M., 1977, Seismic refraction and reflection studies of crustal structure in the eastern Sunda and western Banda arcs: Journal of Geophysical Research, v. 82, no. 17, p. 2479-2489.

Dagis, A.S., and Kazakov, A.M., 1984, Stratigrafiya, litologiya i tsiklichnost' triasovykh otlozheniy Severa Sredney Sibiri: Novosibirsk, Izdatel'stvo Nauka, Sibirskoye Otdeleniye, 177 p.

Dainty, A.M., Keen, C.E., Keen, M.J., and Blanchard, J.E., 1966, Review of geophysical evidence on crust and upper-mantle structure on the eastern seaboard of Canada, *in* Steinhart, J. S., and Smith, T. J., eds., The earth beneath the continents: Washington, D.C., American Geophysical Union Geophysical Monograph 10, p. 349-369.

Daly, R.A., 1925, The geology of Ascension Island: Philadelphia, American Academy of Arts and Sciences Proceedings, v. 60, p. 3-124.

Daly, R.A., 1940, Strength and structure of the earth: New York, Prentice-Hall, Inc., 434 p.

Dana, J.D., 1847, Geological results of the earth's contraction in consequence of cooling: American Journal of Science, 2nd series, v. 3, p. 176-188; v. 4, p. 88-92.

Dana, J.D., 1896, Manual of geology, 4th edition: New York, American Book Company, 1087 p.

Danilov, V.I., and Budnikov, I.V., 1989, Paleobotanicheskoye obosnovaniye korrelyatsii verkhnepaleozoyskikh otlozheniy Vilyuyskoy sineklizy i Zapadnogo Verkhoyan'ya, *in* Dagis, A. S., and Dubatolov, V. N., eds., Verkhniy Paleozoy i Trias Sibiri: Novosibirsk, Akademiya Nauk SSSR, Sibirskoye Otdeleniye, Instituta Geologii i Geofiziki Trudy, v. 732, p. 20-31.

Darwin, G., 1887, Note on Mr. Davison's paper on the straining of the earth's crust in cooling: Royal Society of London Philosophical Transactions, series A, v. 178, p. 242-249.

*Datta, A.K., 1983, Geological evolution and hydrocarbon prospects of Rajasthan basin, *in* Bhandari, L. L., Venkatachala, B. S., Kumar, R., Nanjunda Swamy, S., Garga, P., and Srivastava, D. C., eds., Petroliferous basins of India: Dehra Dun, Petroleum Asia Journal, v. 6, no. 4, p. 93-100.

Davis, A.S., and Plafker, G., 1986, Eocene basalts from the Yakutat terrane: evidence for the origin of an accreting terrane in southern Alaska: Geology, v. 14, no. 11, p. 963-966.

Davis, D.W., Blackburn, C.E., and Krogh, T.E., 1982, Zircon Upb ages from the Wabigoon-Manitou Lakes region, Wabigoon subprovince, northwest Ontario: Canadian Journal of Earth Sciences, v. 19, no. 2, p. 254-266.

Davis, G.A., and Lister, G.S., 1988, Detachment faulting in continental extension; perspectives from the southwestern U.S. Cordillera, *in* Clark, S.P., Jr., Burchfiel, B.C., and Suppe, J., eds., Processes in continental lithospheric deformation: Geological Society of America Special Paper 218, p. 133-159.

Davis, G.H., 1984, Structural geology of rocks and regions: New York, John Wiley & Sons, 492 p.

Davison, C., 1887, On the distribution of strain in the earth's crust resulting from secular cooling; with special reference to the growth of continents and the formation of mountain chains: Royal Society of London Philosophical Transactions, series A, v. 178, p. 231-242.

de Jong, K.A., and Scholten, R., eds., 1973, Gravity and tectonics: New York, John Wiley & Sons, 502 p.

De Long, S.E., Hodges, F.N., and Arculus, R.J., 1975, Ultramafic and mafic inclusions, Kanaga Island, Alaska, and the occurrence of alkaline rocks in island arcs: Journal of Geology, v. 83, no. 6, p. 721-736.

*de Moustier, C., and Kleinrock M. C., 1986, Bathymetric artifacts in Sea Beam data: how to recognize them and what causes them: Journal of Geophyscial Research, v. 91, no. B-3, p. 3407-3424..

dePolo, C.M., Clark, D.G., Slemmons, D.B., and Aymard, W.H., 1989, Historical Basin and Range Province surface faulting and fault segmentaion, *in* Schwartz, D. P., and Sibson, R. H., eds., Fault segmentation and controls of rupture initiation and termination: United States Geological Survey Open-File Report 89-315, p. 131-162.

Decker, R.W., 1989, The Hawaiian-Emperor chain, *in* Bally, A. W., and Palmer, A. R., eds., The geology of North America; an overview: Boulder, CO, Geological Society of America, Decade of North American Geology, v. A, p. 291-295.

Dehler, S.A., and Clowes, R.M., 1988, The Queen Charlotte Islands refraction project; Part I, The Queen Charlotte fault zone: Canadian Journal of Earth Sciences-Journal Canadien des Sciences de la Terre, v. 25, no. 11, p. 1857-1870.

Deichmann, N., and Rybach, L., 1989, Earthquakes and temperatures in the lower crust below the northern Alpine foreland of Switzerland, *in* Mereu, R. F., Mueller, S., and Fountain, D. M., eds., Properties and processes of earth's lower crust; American Geophysical Union Geophysical Monograph no. 51, p. 197-213.

Denham, D., Alexander, L.G., and Worotnicki, G., 1979, Stresses in the Australian crust: evidence from earthquakes and in-situ stress measurements: BMR Journal of Australian Geology and Geophysics, v. 4, no. 4, p. 289-295.

Dennis, J.G., ed., 1982, Orogeny: Benchmark papers in geology, v. 62: Stroudsburg (PA), Hutchinson Ross Publishing Co., 379 p.

*Derrick, G.M., 1979, General geology, *in* BMR earth science atlas: Canberra, Bureau of Mineral Resources, 2 sheets (1 sheet of text and map sheet), scale, 1:10,000,000.

Descartes, R., 1644, Principia philosophiae: Amsterdam, Elsevier Publishing Company, 310 p.

Detrick, R.S., and Watts, A.B., 1979, An analysis of isostasy in the world's oceans, 3. Aseismic ridges: Journal of Geophysical Research, v. 84, no. B-7, p. 3637-3653.

Dewey, J.F., 1988a, Extensional collapse of orogens: Tectonics, v. 7, no. 6, p. 1123-1139.

Dewey, J.F., 1988b, Tectonic evolution of Asia (Abstract): Geological Society of Malaysia Newsletter, v. 14, no. 5, p. 205.

Dewey, J.F., Hempton, M.R., Kidd, W.S.F., Saroglu, F., and Şengör, A.M.C., 1986, Shortening of continental lithosphere: the neotectonics of eastern Anatolia--a young collision zone, *in* Coward, M.P., and Ries, A.C., eds., Collision tectonics: Geological Society of London Special Publication no. 19, p. 3-36.

Dick, T., 1838, Celestial Scenery: Scotland, 184 p.

Dicke, R.H., 1957, Principle of equivalence and the weak interactions: Rev. Mod. Phys., v. 29, no. 3, p. 355-362.

*Dickins, J.M., Shah, S.C., Archbold, N.W., Jin Yugan, Liang Dingyi, and Liu Benpei, 1993, Some climatic and tectonic implications of the Permian marine faunas of peninsular India, Himalayas and Tibet, *in* Findlay, R. H., Unrug, R., Banks, M. R., and Veevers, J. J., eds., Gondwana Eight: assembly, evolution and dispersal: Rotterdam, A. A. Balkema, p. 333-343.

Dietz, R.S., 1961, Continent and ocean basin evolution by spreading of the sea floor: Nature, v. 190, no. 4779, p. 854-857.

Dietz, R., 1962, Ocean-basin evolution by sea-floor spreading, *in* Continental Drift, Chapter10. New York, Academic Press (Internat. Geophysics Ser., v. 3), p. 289-298

Dillon, W.P., and Sougy, J.M.A., 1974, Geology of West Africa and Canary and Cape Verde Islands, *in* Nairn, A.E.M., and Stehli, F.G., eds., The ocean basins and margins, v. 2. The North Atlantic: New York, Plenum Press, p. 315-390.

Dimitrijević, M.D., 1974, The Dinarides: a model based on the new global tectonics, *in* Janković, S., ed., Metallogeny and concepts of the geotectonic development of Yugoslavia: Belgrade, Faculty of Mining and Geology, Belgrade University, Department of Economic Geology, p. 141-176.

Dimroth, E., 1978, Région de la fosse du Labrador (56°30'-57°15'): Ministère Richesses Naturelles du Québec, Rapport de la Géologie RG-193, 396 p.

Dingle, R.V., Siesser, W.G., and Newton, A.R., 1983, Mesozoic and Tertiary geology of southern Africa: Rotterdam, A.A. Balkema, 375 p.

Director-General, 1985, Geological map of peninsular Malaysia, 8th ed.: Kuching, Geological Survey of Malaysia (Malaysia Jabatan Penyiasatan Kajibumi), 2 sheets, scale: 1:500,000.

Divis, A.F., 1980, The petrology and tectonics of recent volcanism in the central Phillippine Islands, *in* Hayes, D.E., ed., The tectonic and geologic evolution of southeast Asian seas and islands: American Geophysical Union Geophysical Monograph 23, p. 127-144.

Dixon, J.E., Fitton, J.E., and Frost, R.T.C., 1981, The tectonic significance of post-Carboniferous igneous activity in the North Sea basin, *in* Illing, L.V. and Hobson, G.D., eds., Petroleum geology of the continental shelf of north-west Europe: London, Heyden & Sons Ltd., p. 121-137.

Dohr, G., 1970, Digital recording of reflections from great depth, *in* Yevropeyskaya Seysmologicheskaya Komissiya, 10th, Trudy, vol. 1, p. 337-351 (include. Russian summary). Akad. Nauk SSSR, Sov. Geofiz. Kom. Moscow, 1970.

Doig, R., 1983, Rb-Sr isotopic study of Archean gneisses north of the Cape Smith fold belt, Ungava, Quebec: Canadian Journal of Earth Sciences, v. 20, no. 5, p. 821k-829.

Doig, R., 1987, Rb-Sr geochronology and metamorphic history of Proterozoic to early Archean rocks north of the Cape Smith fold belt, Quebec: Candian Journal of Earth Sciences, v. 24, no. 4, p. 813-825.

Donnell, L H., 1941, Stress concentrations due to elliptical discontinuities in plates under edge forces, *in* Contributions to applied mechanics and related subjects, Theodore von Kármán Anniversary Volume: Pasadena (CA), California Institute of Technology, p. 293-309.

Donovan, R.N., and Meyerhoff, A.A., 1982, Paleomagnetic evidence for a large (ca. 2,000 km) sinistral offset along the Great Glen fault during Carboniferous time: comment: Geology, v. 10, no. 11, p. 604-605.

Dostal, J., Baragar, W.R.A., and Dupuy, C., 1983, Geochemistry and petrogenesis of basaltic rocks from Coppermine River area, Northwest Territories: Canadian Journal of Earth Sciences, v. 20, no. 5, p. 684-698.

Douglas, R.J.W., ed., 1970, Geology and economic minerals of Canada: Geological Survey of Canada, Economic Geology Report no. 1, 838 p.

*Dow, D.B., 1968, A geological reconnaissance in the Nassau Range, West New Guinea: Geologie en Mijnbouw, v. 47, no. 1, p. 37-46.

*Dow, D.B., Robinson, G.P., and Ratman, N., 1985, New hypothesis for formation of Lengguru foldbelt, Irian Jaya, Indonesia: American Association of Petroleum Geologists Bulletin, v. 69, no. 2, p. 203-214.

Dragašević, T., 1974, Contemporary structure of the earth's crust and upper mantle on the territory of Yugoslavia, *in* Janković, S., ed., Metallogeny and concepts of the geotectonic development of Yugoslavia: Belgrade, Faculty of Mining and Geology, Belgrade University, Department of Economic Geology, p. 73-87.

Drake, C.L., and Nafe, J.E., 1968, The transition from ocean to continent from seismic refraction data, *in* Knopff, L., Drake, C.L., and Hart, P.J., eds., The crust and upper mantle in the Pacific area: American Geophysical Union Geophysical Monograph no. 12, p. 174-186.

Drazin, P.G., and Reid, W.H., 1981, Hydrodynamic stability: Cambridge, Cambridge University Press, 525 p.

Drewry, G.E., Ramsay, A.T.S., and Smith, A.G., 1974, Climatically controlled sediments, the geomagnetic field, and trade wind belts in Phanerozoic time: Journal of Geology, v. 82, no. 5, p. 531-553.

Druecker, M.D., and Gay, S.P., Jr., 1987, Mafic dyke swarms associated with Mesozoic rifting in eastern Paraguay, South America, *in* Halls, H.C., and Fahrig, W.F., eds., Mafic dyke swarms: Geological Association of Canada, Special Paper 34, p. 187-193.

Duncan, A.R., 1987, The Karoo igneous province; a problem area for inferring tectonic setting from basalt geochemistry: Journal of Volcanology and Geothermal Research, v. 32, no. 1-3, p. 13-34.

*Duncan, R.A., and Hargraves, R.B., 1990, $^{40}Ar/^{39}Ar$ geochronology of basement rocks from the Mascarene Plateau, the Chagos Bank, and the Maldives Ridge, *in* Duncan, R. A., Backman, J., et al., Proceedings of the Ocean Drilling Program, v. 115, Scientific results: College Station (TX), Ocean Drilling Program, p. 43-51.

Duncan, R.A., and Pyle, D.G., 1988, Rapid eruption of the Deccan flood basalts, western India, *in* Subbarao, K. V., ed., Deccan flood basalts: Bangalore, Geological Society of India Memoir no. 10, p. 1-9.

Duncan., R.A. and Richards, M.A., 1991, Hot spots, mantle plumes, flood basalts, and true polar wander: Reviews of Geophysics, v. 26, no. 1, p. 31-50.

Dunkelman, T.J., Karson, J.A., and Rosendahl, B.R., 1988, Structural style of the Turkana rift, Kenya: Geology, v. 16, no. 3, p. 258-261.

Dunn, P.R., and Brown, M.C., 1969, North Australian plateau volcanics, *in* Brown, D.A., ed., Proceedings of specialists' meeting held at Canberra, 25-31 May 1968: Geological Society of Australia, Special Publication no. 2, p. 117-122.

Dupré, B., Schiano, P., Polvé, M., and Joron, J.-L., 1994, Variability: a new parameter which emphasizes the limits of extended rare earth diagrams: Société géologique de France Bulletin, v. 165, no. 1, p. 3-13.

du Toit, A. L., 1937, Our Wandering Continents, an Hypothesis of Continental Drifting. Edinburgh, Oliver and Boyd, 366 p.

Dziewonski, A.M., and Anderson, D.L., 1984, Seismic tomography of the earth's interior: American Scientist, v. 72, no. 5, p. 438-494.

Dziewonski, A.M., and Woodhouse, J.H., 1987, Global images of the earth's interior: Science, v. 236, no. 4797, p. 37-48.

Easton, R.M., 1983, Crustal structure of rifted continental margins: geological constraints from the Proterozoic rocks of the Canadian shield: Tectonophysics, v. 94, no. 1-4, p. 371-390.

*Easton, W.H., and Melendres, M.M., Jr., 1963, First Paleozoic fossil from Philippine Archipelago: American Association of Petroleum Geologists Bulletin, v. 47, no. 10, p. 1871-1873.

Eaton, G.P., 1986, A tectonic redefinition of the Southern Rocky Mountains: Tectonophysics, v. 132, no. 1-3, p. 163-193.

Eaton, G.P., 1987, Topography and origin of the Southern Rocky Mountains and Alvarado Ridge, *in* Coward, M.P., Dewey, J.F., and Hancock, P.L., eds., Continental extensional tectonics: Geological Society of London Special Publication no. 28, p. 355-369.

Eaton, G.P., Wahl, R.R., Prostka, H.J., Mabey, D.R., and Kleinkopf, M.D., 1978, Regional gravity and tectonic patterns: their relation to late Cenozoic epeirogeny and lateral spreading in the Western Cordillera, *in* Smith, R.B., and Eaton, G.P., eds., Cenozoic tectonics and regional geophysics of the Western Cordillera: Geological Society of America Memoir 152, p. 51-91.

Ebinger, C.J., 1989, Tectonic development of the western branch of the East African rift system: Geological Society of America Bulletin, v. 101, no. 7, p. 885-903.

*Ebinger, C J., Bechtel, T.D., Forsyth, D.W., and Bowin, C.O., 1989a, Effective elastic plate thickness beneath the East African and Afar Plateaus and dynamic compensation of the uplifts: Journal of Geophysical Research, v. 94, no. B-3, p. 2883-2901.

*Ebinger, C.J., Deino, A.L., Drake, R.E., and Tesha, A.L., 1989b, Chronology of volcanism and rift basin propagation: Rungwe volcanic province, East Africa: Journal of Geophysical Research, v. 94, no. B-11, p. 15785-15803.

*ECAFE-UNESCO, 1971, Geological map of Asia and the Far East, 2nd ed. (revised): Paris, United Nations Economic Commission for Asia and the Far East and United Nations Economic, Scientific and Cultural Organization, 4 sheets, scale, 1:5,000,000.

Edgar, N.T., Ewing, J.I., and Hennion, J., 1971, Seismic refraction and reflection in Caribbean Sea: American Association of Petroleum Geologists Bulletin, v. 55, no. 6, p. 833-870.

Eguchi, T., 1984, Seismotectonics around the Mariana Trough: Tectonophysics, v. 102, no. 1-3, p. 33-52.

Egyed, L., 1956, Determination of changes in the dimensions of the Earth from paleogeographical date: Nature, v. 178, p. 534.

Egyed, L., 1969, The slow expansion hypothesis, *in* Runcorn, S. K., ed., The application of modern physics to the earth and planetary interiors: London and New York, Wiley-Interscience, p. 65-75.

Einarsson, Tr., 1960, The plateau basalt areas in Iceland, *in* Askelsson, Jr., Bödvarsson, G., Einarsson, Tr., Kjartansson, G., and Thorarinsson, S., eds., On the geology and geophysics of Iceland; 21st International Geological Congress, Norden 1960: Guide to Excursion No. H-2, p. 5-20

Élie de Beaumont, J.B.A.L.L., 1831, Researches on some of the revolutions which have taken place on the surface of the globe: Philosophical Magazine, v. 10, p. 241-264. (English translation of the original French article published in 1829.)

Elliot, D.H., 1970, Jurassic basalts of the central Transantactic Mountains, Antarctica, *in* Gilmour, E.H., and Stradling, D., eds., Proceedings of the Second Columbia River Basalt Symposium, Eastern Washington State College, Cheney, Washington, March 1969: Cheney, Washington, Eastern Washington State College Press, p. 301-325.

Elliot, D.H., 1972, Major oxide chemistry of the Kirkpatrick Basalt, central Transantarctic Mountains, *in* Adie, R.J., ed., Antarctic geology and geophysics: Oslo, Universitetsforlaget, p. 413-418.

Ellis, M., and Watkinson, A.J., 1987, Orogen-parallel extension and oblique tectonics: the relation between stretching lineations and relative plate motions: Geology, v. 15, no. 11, p. 1022-1026.

Ellsworth, W.L., and Koyanagi, R.Y., 1977, Three-dimensional crust and mantle structure of Kilauea Volcano, Hawaii: Journal of Geophysical Research, v. 82, no. 33, p. 5379-5394.

Embley, R W., and Wilson, D.S., 1992, Morphology of the Blanco trqansform fault zone-- NE Pacific: implications for its tectonic evolution: Marine Geophysical Researches, v. 14, no. 1, p. 25-45.

Emery, K.O., and Niino, H., 1968, Stratigraphy and petroleum prospects of Korea Strait and the East China Sea: Economic Commission for Asia and the Far East, CCOP Technical Bulletin, v. 1, p. 13-27.

Emmerman, R., 1985, Basement geochemistry, Hole 504B, *in* Anderson, R.N., Honnorez, J., Adamson, A.C., et al., Initial Reports of the Deep Sea Drilling Project, 83: Washington, D.C. (Govt. Printing Office), p. 183-199.

Engel, A.E.J., Engel, C.G., and Havens, R.G., 1965, Chemical characteristics of oceanic basalts and the upper mantle: Geological Society of America Bulletin, v. 76, no. 7, p. 719-734.

Ervin, P., and McGinnis, L.D., 1975, Reelfoot rift: reactivated precursor to the Mississippi embayment: Geological Society of America Bulletin, v. 86, no. 9, p. 1287-1295.

Eva, C., Cattaneo, M., and Merlanti, F., 1988, Seismotectonics of the central segment of the Indonesian arc: Tectonophysics, v. 146, no. 1-4, p. 241-259.

*Evans, C.A., Hawkins, J.W., and Moore, G.F., 1983, Petrology and geochemistry of ophiolitic and associated volcanic rocks on the Talaud Islands, Molucca Sea collision zone, northeast Indonesia, *in* Hilde, T.W.C., and Uyeda, S., eds., Geodynamics of the western Pacific-Indonesian region: Washington, D.C., American Geophysical Union-Geological Society of America Geodynamics Series, v. 11, p. 159-172.

Ewart, A., 1982, The mineralogy and petrology of Tertiary-recent orogenic volcanic rocks: with special reference to the andesitic-basaltic compositional range, *in* Thorpe, R.S., ed., Andesites. Orogenic andesites and related rocks: New York, John Wiley & Sons, p. 25-95.

Ewing, J., and Ewing, M., 1959, Seismic-refraction measurements in the Atlantic Ocean basins, in the Mediterranean Sea, on the Mid-Atlantic Ridge, and in the Norwegian Sea: Geological Society of America Bulletin, v. 70. no. 3, p. 291-317.

*Fallon, F.W., and Dillinger, W.H., 1992, Crustal velocities from geodetic very long baseline interferometry: Journal of Geophysical Research, v. 97, no. B-5, p. 7129-7136.

Faure, G., Bowman, J.R., and Elliot,D.H., 1979, The initial $^{87}Sr/^{86}Sr$ ratios of the Kirwan volcanics of Dronning Maud Land: comparison with the Kirkpatric Basalt, Transantarctic Mountains: Chemical Geology, v. 26, no. 1/2, p. 79-90.

Faure, G., Pace, K.K., and Elliot, D.H., 1982, Systematic variations of $^{87}Sr/^{86}Sr$ ratios and major element concentrations in the Kirkpatrick Basalt of Mount Falla, Queen Alexandra Range, Transantarctic Mountains, *in* Craddock, C., ed., Antarctic geoscience: Madison, Wisconsin, The University of Wisconsin Press, p. 715-713.

Fedotov, S.A., 1973, Deep structure under the volcanic belt of Kamchatka, *in* Coleman, P.J., ed., The western Pacific: island arcs, marginal seas, geochemistry: University of Western Australia Press, p. 247-254.

Feng, R., and McEvilly, T.V., 1983, Interpretation of seismic reflection profiling data for the structure of the San Andreas fault zone: Seismological Society of America Bulletin, v. 73, no. 6, p. 1701-1720.

Filloux, J.H., 1982, Magnetotelluric experiment over the ROSE area: Journal of Geophysical Research, v. 87, no. B10, p. 8364-8378.

Finlayson, D.M., Leven, J.H., and Wake-Dyster, K.D., 1989, Large-scale lenticles in the lower crust under an intra-continental basin in eastern Australia, *in* Mereu, R. F., Mueller, S., and Fountain, D. M., eds., Properties and processes of earth's lower crust: Washington, D.C., American Geophysical Union Geophysical Monograph 51, p. 3-16.

Finnerty, A.A., and Boyd, F.R., 1984, Evaluation of thermobarometers for garnet peridotites: Geochimica et Cosmochimica Acta, v. 48, p. 15-27.

Fitch, F.J., and Miller, J.A., 1981, Dating Karoo igneous rocks by the conventional K-Ar and (40) Ar/(39) Ar age spectrum methods: Geocongress '81; South African Geodynamics Project, v. 1, p. 78-79.

Fitton, J.G., 1987, The Cameroon Line, West Africa: a comparison between oceanic and continental alkaline volcanism, *in* Fitton, J.G., and Upton, E.G.J., eds., Alkaline igneous rocks: Geological Society Special Publication 30, p. 273-291.

Fitton, J.G., James, D., and Leeman, W.P., 1991, Basic magmatism associated with late Cenozoic extension in the western United States: compositional variations in time and space: Journal of Geophysical Research, v. 96, no. B-8, p. 13693-13711.

Flack, C.A., and Warner, M.R., 1990, Three-dimensional mapping of seismic reflections from the crust and upper mantle, northwest of Scotland: Tectonophysics, v. 173, no. 1-4, p. 469-481.

Fletcher, J.B., Sbar, M.L., and Sykes, L.R., 1978, Seismic trends and travel-time residuals in eastern North America and their tectonic implications: Geological Society of America Bulletin, v. 89, no. 11, p. 1656-1676.

Flint, R.F., and Skinner, B.J., 1977, Physical geology, 2nd ed., New York, John Wiley & Sons, Inc., 594 p. + 5 appendices, glossary, and index.

Floyd, P.A., Castillo, P.R., and Pringle, M., 1991, Tholeiitic and alkalic basalts of the oldest Pacific Ocean crust: Terra Nova, v. 3, no. 3, p. 257-264.

*Foose, R.M., 1977, Structural lineaments and tectonics of the Mediterranean basin, *in* Biju-Duval, B., and Montadert, L., eds., Structural history of the Mediterranean basins: Paris, Éditions Technip, p. 221-232.

*Foose, R.M. 1985, Geological information from satellite surveys of the Mediterranean region, *in* Stanley, D. J., and Wezel, F.-C., eds., Geological evolution of the Mediterranean basin: Berlin, Springer-Verlag, p. 33-53.

Ford, A.B., and Kistler, R.W., 1981, K-Ar age, composition, and origin of Mesozoic mafic rocks related to the Ferrar Group, Pensacola Mountains, Antarctica: New Zealand Journal of Geology and Geophysics, v. 23, no. 3, p. 371-390.

Ford, R.J., 1989, Cretaceous alkaline rocks, *in* Burrett, C.F., and Martin, E.L., eds., Geology and mineral resources of Tasmania: Geological Society of Australia, Special Publication no. 15, p. 381-383.

Francis, D., Ludden, J., and Hynes, A., 1983, Magma evolution in a Proterozoic rifting environment: Journal of Petrology, v. 24, pt. 4, p. 556-582.

Francis, E.H., 1988, Mid-Devonian to Early Permian volcanism: Old World, *in* Harris, A. L., and Fettes, D. J., eds., The Caledonian-Appalachian orogen: London, Geological Society of London Special Publication no. 38, p. 573-584.

Frei, W., Heitzmann, P., Lehner, P. and Valasek, P., 1989, Die drei Alpentraversen von NFP 20: Swiss Association of Petroleum Geologists and Engineers Bulletin, v. 55, no. 128, p. 13-43.

Frey, F.A., Green, D. H., and Roy, S. D., 1978, Intergrated models of basalt petrogenesis; a study of quartz tholeiites to olivine melilitites from Southeastern Australia utilizing geochemical and experimental petrological data: Journal of Petrology, v. 19, no. 3, p. 463-513.

Froitzheim, N., Stets, J., and Wurster, P., 1988, Aspects of western High Atlas tectonics, *in* Jacobshagen, V.H., ed., The Atlas system of Morocco. Studies on its geodynamic evolution: New York, Springer-Verlag, p. 219-244.

Fryer, B.J., 1972, Age determinations in the circum-Ungava geosyncline and evolution of Precambrian banded iron formations: Canadian Journal of Earth Sciences, v. 9, no. 4, p. 652-663.

*Fuchs, G., 1975, Contributions to the geology of the north-western Himalayas: Wien (Vienna), Abhandlungen der Geologischen Bundesanstalt, Bd. 32, 59 p.

*Fuchs, G., and Gupta, V. J., 1979, The significance of the Tanol Formation of the western Himalaya, *in* Gupta, V. J., ed., Upper Palaeozoics of the Himalayas: Delhi, Hindustan Publishing Corporation (I), p. 154-161.

Fuchs, K., 1974, Geophysical contributions to taphrogenesis, *in* Illies, H., and Fuchs, K., eds., Approaches to taphrogenesis: Stuttgart, Schwitzerbart, p. 420-432.

Fuchs, K., and Landisman, M., 1966, Detailed crustal investigation along a north-south section through the central part of western Germany, *in* Steinhart, J. S., and Smith, T. J., eds., The earth beneath the continents: Washington, D.C., American Geophysical Union Geophysical Monograph 10, p. 433-452.

Fuis, G.S., and Kohler, W.M., 1984, Crustal structure and tectonics of the Imperial Valley region, California, *in* Rigsby, C.A., ed., The Imperial Basin--tectonics, sedimentation and thermal aspects: Society of Economic Geologists and Paleontologists, Pacific Section Book 40, p. 1-13.

Fuis, G.S., Zucca, J.J., Mooney, W.D., and Milkereit, B., 1987, A geologic interpretation of seismic-refraction results in northeastern California: Geological Society of America Bulletin, v. 98, no. 1, p. 53-65

Fukao, Y., 1992, Seismic tomogram of the earth's mantle: geodynamic implications: Science, v. 258, no. 5082, p. 625-630.

Fukao, Y., Obayashi, M., Inoue, H., and Nenbai, M., 1992, Subducting slabs stagnant in the mantle transition zone: Journal of Geophysical Research, v. 97, no. B-4, p. 4809-4822.

Fullagar, P.D., and Butler, J.R., 1979, 325 to 265 m.y.-old granitic plutons in the Piedmont of the southeastern Appalachians: American Journal of Science, v. 279, no. 2, p. 161-185.

Furon, R., 1949, Sur les trilobites draguées à 4225 m de profondeur par le Talisman (1883): Paris, Académie des Sciences, Compte Rendu, v. 258, no. 19, p. 1509-1510.

Furumoto, A.S., Webb, J.P., Odegard, M.E., and Hussong, D.M., 1976, Seismic studies on the Ontong Java Plateau, 1970: Tectonophysics, v. 34, no. 1/2, p. 71-90.

Fúster, J.M., Fernandez-Santin, S., and Sagredo, J., 1968a, Geología y vulcanología de las Islas Canarias. Lanzarote: Madrid, Consejo Superior de Investigaciones Científicas, Instituto "Lucas Mallada," 177 p.

Fúster, J.M., Cendrero, A., Gastesi, P., Ibarrola, E., and López-Ruiz, J., 1968b, Geología y vulcanología de las Islas Canarias. Fuerteventura: Madrid, Consejo Superior de Investigaciones Científicas, Instituto "Lucas Mallada," 239 p.

*Gansser, A., 1964, Geology of the Himalayas: London, Interscience Publishers, 289 p.

Gansser, A., 1968, The Insubric Line, a major geotectonic line: Schweizerische Mineralogische und Petrographische Mitteilungen, Bd. 48, p. 123-143.

Gansser, A., 1977, The great suture zone between Himalaya and Tibet: a preliminary account, *in* Écologie et géologie de l'Himalaya, Colloques Internationaux no. 268, 7-10 Déc. 1976: Paris, Centre National de la Recherche Scientifique, p. 181-191.

Gansser, A., 1981, The geodynamic history of the Himalaya, *in* Gupta, H. K., and Delany, F. M., eds., Zagros-Hindu Kush-Himalaya: geodynamic evolution: Washington, D.C., American Geophysical Union and Geological Society of America Geodynamics Series, v. 3, p. 111-121.

*Gansser, A., 1983a, Geology of the Bhutan Himalaya: Basel, Birkhauser Verlag, Schweizerischen Naturforschenden Gesellschaft Denkschriften, Bd. 96, 181 p.

Gansser, A., 1983b, The morphogenic phase of mountain building, *in* Hsü, K. J., ed., Mountain building processes: London, Academic Press, p. 221-228.

*Gansser, A., 1991, Facts and theories on the Himalayas: Eclogae Geologicae Helvetiae, v. 84, no. 1, p. 33-59.

Garfunkel, Z., 1966, Problems of wrench faults: Tectonophysics, v. 3, no. 5, p. 457-473.

Gass, I.G., Smith, P.J., and Wilson, C.L., eds., 1972, Understanding the earth: Cambridge, The M.I.T. Press, 383 p.

*Gaulier, J. M., Le Pichon, X., Lyberis, N., Avedik, F., Geli, L., Moretti, I., Deschamps, A., and Hafez, S., 1988, Seismic study of the crust of the northern Red Sea and Gulf of Suez: Tectonophysics, v. 153, no. 1-4, p. 55-88.

*Ge, G. G., Zagruzina, I. A., Zyong, Nguyen Van, and Tkhi, Fan Chyong, 1979, Novyye dannyye o posdnepaleozoyskom magmatizme V'yetnama: Novosibirsk, Geologiya i Geofizika, no. 10, p. 69-74.

*Geary, E. E., Harrison, T. M., and Heizler, M., 1988, Diverse ages and origins of basement complexes, Luzon, Philippines: Geology, v. 16, no. 4, p. 341-344.

*GEBCO, 1975-1982, General bathymetric chart of the oceans: Ottawa, The Canadian Hydrographic Service, Charts 5-1 through 5-16, scale, 1:10,000,000; Charts 5-17 and 5-18, scale, 1:6,000,000.

Geikie, Sir A., 1902, Text-book of geology, part two: New York, American Dome Library Company, p. 503-976; part three: New York, American Dome Library Company, p. 977-1417.

Geller, C.A., Weissel, J.K., and Anderson, R.N., 1983, Heat transfer and intraplate in the central Indian Ocean: Journal of Geophysical Research, v. 88, no. B-2, p. 1018-1032.

*Genshaft, Yu.S., and Saltykovskiy, A.Ya., 1989, Kontinental'nyy vulkanism, ksenolity i tektonika litosfernykh plit, *in* Beloussov, V. V., ed., Resultaty issledovaniya po mezhdunarodnym geofizicheskim proyektam: Moscow, Akademii Nauk SSSR, Mezhduvedomstvennyy Geofizicheskiy Komitet, Geodinamicheskiye Issledovaniya no. 13, Tektonosfera: yeye stroyeniye i razvitiye, p. 94-105.

Geological Society of America, 1984, Decade of North American geology: geologic time scale: Geological Society of America, Map and Chart Series MC-50, 1 chart.

*Geological Survey Division, 1982, Geological map of Thailand, 1982 edition: Bangkok, Geological Survey Division, Department of Mineral Resources, 2 sheets, scale, 1:1,000,000.

*Geological Survey Division, 1987, Geological map of Thailand: Bangkok, Geological Survey Division, Department of Mineral Resources, 1 sheet, scale, 1:2,500,000.

*George, M.T., Harris, N.B.W., and Butler, R.W.H., 1993, The tectonic implications of contrasting granite magmatism between the Kohistan island arc and the Nanga Parbat-Haramosh massif, Pakistan Himalaya, *in* Treloar, P. J., and Searle, M. P., eds., Himalayan tectonics: London, Geological Society of London Special Publication no. 74, p. 173-191.

*Ghosh, D., Das, L.K., Ghatak, T.K., Saha, D.K., and Bose, R.N., 1993, Morphotectonic configuration of cratonic Gondwana depocentres of eastern India: Tectonophysics, v. 223, no. 3-4, p. 423-438.

Gilbert, G.K., 1877, Geology of the Henry Mountains, Utah: United States Geographical and Geological Survey of the Rocky Mountain Region, 170 p.

Gill, A.E., 1982, Atmosphere-ocean dynamics, International Geophysics Series, v. 30: New York, Academic Press, 662 p.

Girdler, R.W., 1965, The formation of new oceanic crust, in a symposium on continental drift: Royal Society (London) Philosophical Transactions, series A, v. 258, no. 1088, p. 123-136.

*Glaessner, M.F., Llewellyn, K.M., and Stanley, G.A.V., 1950, Fossiliferous rocks of Permian age from the Territory of New Guinea: Australian Journal of Science, v. 13, no. 1, p. 24-25.

Glangeaud, L., 1957, Essai de classification géodynamique des chaînes et des phénomènes orogèniques: Révue de Géographie Physique et de Géologie Dynamique, séries 2, v. 1, fasc. 4, p. 200-220.

Glangeaud, L., 1959, Classification géodynamique des chaînes de montagnes. 9.---Chaînes intracratoniques: 1. Structure et embryologie des cratons (croute sialique): Révue de Géographie Physique et de Géologie Dynamique, séries 2, v. 1, fasc. 4, p. 197-204.

Glenn, M.F., 1970, Introducing an operational multi-beam array sonar: International Hydrographic Review, v. XLVII, no. 1, p. 35-39.

Glennie, K.W., 1986, Development of N. W. Europe's southern Permian gas basin, *in* Brooks, J., Goff, J. C., and Hoorn, B. van, eds., Habitat of Palaeozoic gas in N. W. Europe: London, Geological Society of London Special Publication no. 23, p. 3-22.

Gold, T., 1955, Instability of the earth's axis of rotation. Nature, v. 175, no. 4456, p. 526-529.

Goldfinger, C., Kulm, L. D., Yeats, R. S., Appelgate, B., Mackay, M. E., and Cochrane, G. R., 1991, Active strike-slip faulting and folding of the Cascadia plate boundary and forearc in central and northern Oregon: U. S. Geological Survey Open-File Report 91-441-S,. 42 p.

Goldreich, P., and Toomre, A., 1969, Some remarks on polar wandering: Journal of Geophysical Research, v. 74, no. 10, p. 2555-2567.

Golonetskiy, S.I., and Misharina, L.A., 1978, Seismicity and earthquake focal mechanisms in the Baikal rift zone: Tectonophysics, v. 45, no. 1, p. 71-85.

González de Juana, C., Iturralde de Arozena, J.M., and Cadillat, X.P., 1980, Geología de Venezuela y de sus cuencas petrolíferas: Caracas, Ediciones Foninves, 407 p.

Goodacre, A., 1991, Continental Drift: Nature, v. 351, no. 6351, p. 261.

Goodwin, A.M., 1991, Precambrian geology. The dynamic evolution of the continental crust: London, Academic Press, 666 p.

Goodwin, E.B., and Thompson, G.A., 1988, The seismically reflective crust beneath highly extended terranes: evidence for its origin and extension: Geological Society of America Bulletin, v. 100, no. 10, p. 1616-1626.

*Goosens, P.J., 1978, The metallogenic provinces of Burma: theory definitions, geologic relationships and extension into China, India and Thailand, *in* Nutalaya, P., ed., Proceedings of the Third Regional Conference on Geology and Mineral Resources of Southeast Asia, Bangkok, November 14-18, 1978 (GEOSEA III): Bangkok, Asian Institute of Technology, p. 431-492.

*Gordin, V.M., 1989, Magnitnoye pole okeanov i gipoteza vayna-met'usa, *in* Beloussov, V. V., ed., Tektonosfera yeye stroyeniye i razvitiye: Moscow, Akademii Nauk SSSR, Mezhduvedomstvenniy Geofizicheskiy Komitet, Geodinamicheskiye Issledovaniya no. 13, p. 129-158.

Gordon, R.G., and Stein, S., 1992, Global tectonics and space geodesy: Science, v. 256, no. 5055, p. 333-342.

Gorshkov, A.G., and Lukashevich, I.P., 1989, Computing of magma chamber temperatures in rift zones of the world ocean: Tectonophysics, v. 159, no. 3-4, p. 337-346.

Gorshkov, G.S., 1973, Petrochemistry of volcanic rocks in the Kuril Island arc with some generalizations on volcanism, *in* Coleman, P.J., ed., The western Pacific: island arcs, marginal seas, geochemistry: University of Western Australia Press, p. 459-467.

Goslin, J., and Sibuet, J.C., 1975, Geophysical study of the easternmost Walvis Ridge, south Atlantic: Geological Society of America Bulletin, v. 86, no. 12, p. 1713-1724.

Gossler, J., and Kind, R., 1996, Seismic evidence for very deep roots of continents: Earth and Planetary Science Letters, v. 138, p. 1-13.

Gough, D.I., 1984, Mantle upflow under North America and plate dynamics: Nature, v. 311,no. 5985, p. 428-433.

Gough, D.I., 1989, Magnetometer array studies, earth structure, and tectonic processes: Review of Geophysics, v. 17, no. 1, p. 141-157.

*Govindan, A., 1984, Stratigraphy and sedimentation of East Godavari sub-basin, Andhra Pradesh, *in* Biswas, S. K., convener, Petroliferous basins of India--II: Dehra Dun, Petroleum Asia Journal, v. 7, no. 1, p. 132-146.

Grabau, A. W., 1931, The Permian of Mongolia--A report on the Permian fauna of the Kisu Honguer Limestone of Mongolia and its relations to the Permian of other parts of the world, *in* Natural history of Central Asia, v. IV: New York, American Museum of Natural History, 665 p. + p. i-xlii.

Graham, J.W., 1949, The stability and significance of magnetism in sedimentary rocks: Journal of Geophysical Research, v. 54, no. 2, p. 131-167.

Gramberg, I.S., and Smyslov, A.A., 1986, The map of heat flow and hydrothermal mineralization in the world ocean: Moscow, Ministry of Geology of the USSR, 6 sheets, scale, 1:20,000,000.

Grand, S.P., 1987, Tomographic inversion for shear velocity beneath the North American plate: Journal of Geophysical Research, v. 92, no. B-13, p. 14065-14090.

Gravity Anomaly Map Committee, 1987, Gravity anomaly map of North America: Boulder (CO), Geological Society of America, Continent-Scale Map-002, 5 sheets, scale, 1:5,000,000.

Greeley, R., 1982, The Snake River plain, Idaho: representative of a new category of volcanism: Journal of Geophysical Research, v. 87, no. B-4, p. 2705-2712.

Green, D.H., Hibberson, W.O., and Jaques, A.L., 1979, Petrogenesis of mid-ocean ridge basalts, *in* McElhinny, M. W., ed., The earth: its origin, structure and evolution: London, Academic Press, p. 265-299.

Green, J.C., 1982, Geology of Keweenawan extrusive rocks, *in* Wold, R.J., and Hinze, W.J., eds., Geology and tectonics of the Lake Superior basin: Geological Society of America Memoir 156, p. 47-55.

Green, W.L., 1875, Vestiges of the molten globe as exhibited in the figure of the earth's volcanic action and physiography, part I: London, 59 p.

Green, W.L., 1887, Vestiges of the molten globe as exhibited in the figure of the earth's volcanic action and physiography, part II. The earth's surface features and volcanic phenomena: Honolulu, 337 p.

Gregory, J.W., 1899, The plan of the earth and its causes: Geographical Journal, v. 13, p. 225-251.

Gregory, J.W., 1901, The plan of the earth and its causes: American Geologist, v. 27, p. 100-119, 134-147.

Gregory, J.W., ed., 1929, The structure of Asia: London, Methuen, 227 p.

Gretener, P.E., 1969, On the mechanics of the intrusion of sills: Canadian Journal of Earth Sciences, v. 6, no. 6, p. 1415-1419.

Gries, R., 1983, North-south compression of Rocky Mountain foreland structures, *in* Lowell, J.D., and Gries, R., eds., Rocky Mountain foreland basins and uplifts: Rocky Mountain Association of Geologists, p. 9-32.
Griffith, A.A., 1920, Philosophical Transactions of the Royal Society of London, A 221, p. 163.
Griggs, D., 1939, A theory of mountain building: American Journal of Science, v. 237, no. 9, p. 611-650.
Grow, J.A., 1973, Crustal and upper mantle structure of the central Aleutian arc: Geological Society of America Bulletin, v. 84, no. 7, p. 2169-2191.
Gunn, B.M., 1962, Differentiation in Ferrar dolerites, Antarctica: New Zealand Journal of Geology and Geophysics, v. 5, p. 820-863.
Gupta, H.K., 1993, The deadly Latur earthquake: Science, v. 262, no. 5140, p. 1666-1667.
Gupta, M.L., and Gaur, V. K., 1984, Surface heat flow and probable evolution of Deccan volcanism: Tectonophysics, v. 105, no. 1-4, p. 309-318.
Gutenberg, B., and Richter, C.F., 1949, Seismicity of the earth and associated phenomena: Princeton, Princeton University Press, 273 p.
Gutenberg, B., and Richter, C.F., 1954, Seismicity of the earth and associated phenomena: 2nd ed.: Princeton, Princeton University Press, 310 p.
Haarmann, E., 1930. A theory of mountain building: American Journal of Science, v. 237, no. 9, p. 611-650.
*Hahn, L., 1984, The Indosinian orogeny in Thailand and adjacent areas, *in* Buffetaut, E., Jaeger, J.-J., and Rage, J.-C., organizers, Paléogéographie de l'Inde, du Tibet et de sud-est Asiatique: confrontation des données paléontologiques avec les modèles géodynamics: Société Géologique de France Mémoire no. 147 (nouvelle série), p. 71-82.
*Haile, N.S., 1969, Geosynclinal theory and the organizational pattern of the Northwest Borneo geosyncline: Geological Society of London Quarterly Journal, v. 124, pt. 2, p. 171-194.
*Haile, N.S., 1970, Notes on the geology of the Tambelan, Annambas [sic] and Bunguran (Natuna) Islands, Sunda shelf, Indonesia, including radiometric age determinations: Economic Commission for Asia and the Far East, CCOP Technical Bulletin, v. 3, p. 55-89.
*Haile, N.S., 1974, Borneo, *in* Spencer, A. M., ed., Mesozoic-Cenozoic orogenic belts: London, Geological Society of London Special Publication no. 4, p. 333-347.
*Haile, N.S., Barber, A. J., and Carter, D. J., 1979, Mesozoic cherts on crystalline schists in Sulawesi and Timor: Geological Society of London Journal, v. 136, pt. 1, p. 65-70.
Haile, N.S., Stauffer, P. H., Krishnan, D., Lim, T. P., and Ong, G. B., 1977, Palaeozoic redbeds and radiolarian chert: reinterpretation of their relationships in the Bentong and Raub areas, West Pahang, peninsular Malaysia: Geological Society of Malaysia Bulletin, no. 8, p. 45-59.
Hales, A.L., 1936, Convection currents in the mantle: Royal Astronomical Society Monthly Notices, Geophysical Supplement, v. 3, p. 372-379.
Hales, A.L., 1969, Gravitational sliding and continental drift: Earth and Planetary Science Letters, v. 6, p. 31-34.
*Hall, J.M., and Ryall, P. J. C., 1977, Paleomagnetism of basement rocks, Leg 37, *in* Aumento, F., Melson, W. G., et al., Initial reports of the Deep Sea Drilling Project, v. 37: Washington, D. C., U. S. Government Printing Office, p. 425-448.
*Hall, R., and Nichols, G.J., 1990, Terrane amalgamation in the Philippine Sea margin: Tectonophysics, v. 181, no. 1-4, p. 207-222.
*Hall, R., Audley-Charles, M.G., Banner, F.T., Hidayat, S., and Tobing, S.L., 1988a, Basement rocks of the Halmahera region, eastern Indonesia: a Late Cretaceous-early Tertiary arc and fore-arc: Geological Society of London Journal, v. 145, pt. 1, p. 65-84.
*Hall, R., Audley-Charles, M.G., Banner, F.T., Hidayat, S., and Tobing, S.L., 1988b, Late Palaeogene-Quaternary geology of Halmahera, eastern Indonesia: initiation of a volcanic arc: Geological Society of London Journal, v. 145, pt. 4, p. 577-590.
Haller, J., 1979, Himalayan orogenesis in perspective, *in* Verma, P. K., ed., Metamorphic rock sequences of the eastern Himalaya: Calcutta, K. P. Bagchi & Co., p. I-xxxiii.
Halls, H.D., 1978, The late Precambrian central North American rift system--A survey of recent geological and geophysical investigrations, *in* Neumann, E.R., and Rambert, I.B., eds., Tectonics and geophysics of continental rifts, v. 2: Dordrecht, Kluwer Academic Publishers, p. 111-123.
Ham, W.E., Denison, R.E., and Merritt, C.A., 1964, Basement rocks and structural evolution of southern Oklahoma: Oklahoma Geological Survey Bulletin 95, 302 p.
Hamilton, D.A., and Mereu, R.F., 1993, 2-D tomographic imaging across the North American Mid-Continent Rift system: Geophysical Journal International, v. 112, no. 3, p. 344-358.
Hamilton, R.M., and Johnston, A.C., 1990, Tecumseh's prophecy; preparing for the next New Madrid earthquake; a plan for an intensified study of the New Madrid seismic zone: Reston, VA, U. S. Geological Survey, Geological Survey Circular, 30 p.
Hamilton, W.B., 1963, Tectonics of Antarctica, *in* Childs, O. E., and Beebe, B. W., eds., Backbone of the Americas--

tectonic history from pole to pole: Tulsa, OK, American Association of Petroleum Geologists Memoir 2, p. 4-15.

*Hamilton, W. B., 1979, Tectonics of the Indonesian region: Reston, VA, U. S. Geological Survey Professional Paper 1078, 345 p.

Hamilton, W. B., 1988, Laramide crustal shortening, *in* Schmidt, C. J., and Perry, W. J., Jr., eds., Interaction of the Rocky Mountain foreland and the Cordilleran thrust belt: Boulder, CO, Geological Society of America Memoir 171, p. 27-39.

Hansen, V.L., 1988, A model for terrane accretion: Yukon-Tanana and Slide Mountain terranes, northwest North America: Tectonics, v. 7, no. 6, p. 1167-1177.

*Harbury, N.A., and Kallagher, H. J., 1991, The Sunda outer-arc ridge, north Sumatra, Indonesia: Journal of Southeast Asian Earth Sciences, v. 6, no. 3-4, p. 463-476.

Harding, A.J., Orcutt, J.A., Kappus, M.E., Vera, E.E., Mutter, J.C., Buhl, P., Detrick, R.S., and Brocher, T.M. 1989, Structure of young oceanic crust at 13° N on the East Pacific Rise from expanding spread profiles: Journal of Geophysical Research, v. 94, no. B9, p. 12163-12196.

*Harding, R.R., 1969, Catalogue of age determinations of Australian rocks, 1962-1965: Canberra, Bureau of Mineral Resources, Geology and Geophysics Report 117, 117 p.

Harland, W.B., 1973, Mesozoic geology of Svalbard, *in* Pitcher, M.G., ed., Arctic geology: American Association of Petroleum Geologists, Memoir 19, p. 135-148.

Harland, W.B., Armstrong, R.L., Cox, A.V., Craig, L.E., Smith, A.G., and Smith, D.G., 1990, A geologic time scale 1989: Cambridge, Cambridge University Press, 262 p.

Harland, W. B., Scott, R. B., Auckland, K. A., and Snape, I., 1992, The Ny Friesland orogen, Spitsbergen: Geological Magazine, v. 129, no. 6, p. 679-708.

*Harris, L.B., and Beeson, J., 1993, Gondwanaland significance of lower Palaeozoic deformation in central India and SW Western Australia: Geological Society of London Journal, v. 150, pt. 5, p. 811-814.

Harrison, C.G.A., 1968, Antipodal locations of continents and ocean basins: Science, v. 153, no. 3741, p. 1246-1248.

Harrison, C.G.A., and Douglas, N.B., 1990, Satellite laser ranging and geological constraints on plate motion. Tectonics, v. 9, no. 5, p. 935-952.

*Hartono, H.M.S., and Tjokrosapoetro, S., 1986, Geological evolution of the Indonesian Archipelago: Geological Society of Malaysia Bulletin no. 20, p. 97-136.

Hasegawa, A., Zhao, D., Hori, S., Yamamoto, A., and Horiuchi, S., 1991, Deep structure of the northeastern Japan arc and its relationship to seismic and volcanic activity: Nature, v. 352, no. 6337, p. 683-689.

Hasegawa, H. S., 1991, Four seismogenic environments in eastern Canada: Tectonophysics, v. 186, no. 1-2, p. 3-17.

*Hashimoto, W., and Sato, T., 1973, Geological structure of north Palawan, and its bearing on the geological history of the Philippines, *in* Kobayashi, T., and Toriyama, R., eds., Geology and palaeontology of Southeast Asia, v. 13: Tokyo, University of Tokyo Press, p. 145-161.

Haughton, S.H., 1969, Geological history of southern Africa: Cape Town, Geological Society of South Africa, 535 p.

Haxby, W.F., 1987, Gravity field of the world's oceans: National Geophysical Data Center, National Oceanic and Atmospheric Administration, scale, 1:40,000,000.

Heaman, L.M., LeCheminant, A.N., and Rainbird, R.H., 1990, A Upb baddeleyite study of Franklin igneous events, Canada: Geological Association of Canada, Abstracts with Programs, v. 15, p. A55.

Heezen, B.C., 1960, The rift in the ocean floor: Scientific American, v. 203. no. 4, p. 98-110.

Heezen, B.C., and Tharp, M., 1977, World ocean floor: United States Navy, Office of Naval Research, 1 sheet, scale, 1:46,460,000.

Heezen, B.C., Tharp, M., and Ewing, M., 1959, The floors of the oceans, I: the North Atlantic: Boulder, CO, Geological Society of America Special Paper 65, 122 p.

Hegarty, K.A., and Weissel, J.K., 1988, Complexities in the development of the Caroline plate region, western equatorial Pacific, *in* Nairn, A.E.M., Stehli, F.G., and Uyeda, S., eds., The ocean basins and margins, v. 7B. The Pacific Ocean: New York, Plenum Press, p. 277-301.

Heim, A., 1878, Untersuchungen uber den Mechanismus der Gebirgsbildung, Part 2, Basel.

Heim, A., 1922, Geologie der Schweiz, Bd. II, Die schweizer Alpen, zweite Hälfte: Leipzig, Christian Hermann Tauchnitz, p. 477-1018.

*Hermes, J.J., 1968, The Papuan geosyncline and the concept of geosynclines: Geologie en Mijnbouw, v. 47, no. 2, p. 81-97.

Hershey, O.H., 1903, Structure of the southern portion of the Klamath Mountains, Cal.: American Geologist, v. 31, p. 231-145.

Hershey, O.H., 1906, Some western Klamath stratigraphy: American Journal of Science, 4th series, v. 21, no. 1, p. 58-66.

Hess, H.H., 1962, History of ocean basins, *in* Engel, A.E.J., James, H.L., and Leonard, B.F., eds., Petrologic studies: a volume in honor of H. F. Buddington: Boulder, CO, Geological Society of America, p. 599-620.

Hey, R., and Vogt, P., 1977, Spreading center jumps and sub-axial asthenosphere flow near the Galapagos hotspot: Tectonophysics, v. 37, no. 1-3, p. 41-52.

Hildebrand, R.S., and Baragar, W.R.A., 1991, On folds and thrusts affecting the Coppermine River Group, northwestern Canadian shield: Canadian Journal of Earth Sciences, v. 28, no. 4, p. 523-531.

Hildenbrand, T.G., 1982, Model of the southeastern margin of the Mississippi Valley graben near Memphis, Tennessee, from interpretation of truck-magnetometer data: Geology, v. 10, no. 9, p. 476-480.

Hildenbrand, T.G., Simpson, R.W., Godson, R.H., and Kane, M.F., 1982, Digital colored residual and regional Bouguer gravity maps of the conterminous United States with cut-off wavelengths of 250 km and 1000 km: Reston, VA, U. S. Geological Survey Geophysical Investigations Map GP-953-A, scale, 1:7,500,000, 2 sheets.

Hilgenberg, D.C., 1933, Von wachsenden Erdball: Berlin, Charlotenburg, Giessmann und Bartsch, 55 p.

Hill, D.P., 1969, Crustal structure of the island of Hawaii from seismic-refraction measurements: Seismological Society of America Bulletin, v. 59, no. 1, p. 101-130.

Hill, D.P., Reasenberg, P. A., Michael, A., Arabaz, W. J., Beroza, G., Brumbaugh, D., Brune, J. N., Castro, R., Davis, S., dePolo, D., Ellsworth, W. L., Gomberg, J., Harmsen, S., House, L., Jackson, S. M., Johnston, M. J. S., Jones, L., Keller, R., Malone, S., Munguia, L., Nava, S., Pechmann, J. C., Sanford, A., Simpson, R. W., Smith, R. B., Stark, M., Stickner, M., Vidal, A., Walter, S., Wong, V., and Zollweg, J., 1993, Seismicity remotely triggered by the magnitude 7.3 Landers, California, earthquake: Science, v. 260, no. 5114, p. 1617-1623.

Hills, E.S., 1972, Elements of structural geology, 2nd ed.: New York, John Wiley & Sons, Inc., 502 p.

Hilyard, D., 1990, Willouran basic province: stratigraphy of late Proterozoic flood basalts, Adelaide geosyncline, South Australia, *in* Jago, J.B., and Moore, P.S., eds., The evolution of a late Precambrian-early Paleozoic rift complex: the Adelaide geosyncline: Geological Society of Australia, Special Publication no. 16, p. 35-48.

Hinze, W.J., and Braile, L.W., 1988, Geophysical aspects of the craton: U.S., *in* Sloss, L.L., ed., Sedimentary cover--North American craton: U.S., The geology of North America, v. D-2: Geological Society of America, Decade of North American Geology, p. 5-24.

Hirahara, K., Ikami, A., Ishida, M., and Mikumo, T., 1989, Three-dimensional P-wave velocity structure beneath central Japan: low-velocity bodies in the wedge portion of the upper mantle above high-velocity subducting plates: Tectonophysics, v. 163, no. 1-2, p. 63-73.

Ho, C.S., 1982, Tectonic evolution of Taiwan, Explanatory text of the tectonic map of Taiwan: Taipei, Ministry of Economic Affairs, 126 p.

Hobbs, B.E., Means, W.D., and Williams, P.F., 1976, An outline of structural geology: New York, John Wiley & Sons, 571 p.

Hoffman, P.F., 1989, Precambrian geology and tectonic history of North America, *in* Bally, A.W., and Palmer, A.R., eds., The geology of North America; an overview: Geological Society of America, Decade of North American Geology, v. A, p. 447-512.

Holbrook, W.S., and Kelemen, P. B. 1993, Large igneous province on the US Atlantic margin and implications for magmatism during continental breakup: Nature, v. 364, no. 6436, p. 433-436.

Hollister, L.S., and Crawford, M.L., 1986, Melt-enhanced deformation: a major tectonic process: Geology, v. 14, no. 7, p. 562-566.

*Holloway, N.H., 1982, North Palawan block, Philippines--its relation to Asian mainland and role in evolution of South China Sea: American Association of Petroleum Geologists Bulletin, v. 66, no. 9, p. 1355-1383.

Holmes, A., 1931, Radioactivity and earth movements: Geological Society of Glasgow Transactions, v. 18, pt. 3, p. 559-606.

Honza, E., and Tamaki, K., 1985, The Bonin arc, *in* Nairn, A.E.M., Stehli, F.G., and Uyeda, S., eds., The ocean basins and margins, v. 7A. The Pacific Ocean: New York, Plenum Press, p. 459-502.

Hooper, P.R., 1982, The Columbia River basalts: Science, v. 215, no. 4539, p. 1463-1468.

Hooper, P.R., 1988, The Columbia River basalt, *in* Macdougall, J.D., ed., Continental flood basalts: Dordrecht, Kluwer Academic Publishers, p. 1-33.

Hooper, P.R., 1990, The timing of crustal extension and the eruption of continental flood basalts: Nature, v. 345, no. (17 May), p. 246-249.

Hooper, P.R., and Reidel, S.P., 1989, Dikes and vents feeding the Columbia River basalts, *in* Joseph, N.L., editor-in-chief, Geologic guidebook for Washington and adjacent areas: Washington Division of Geology and Earth Resources Information Circular 86, p. 255-273.

Hopkins, W., 1839, Researches in physical geology, part 1: Royal Society of London, Philisophical Transactions (for 1839), p. 381-432..

Houghton, R.L., Thomas, J.E., Jr., Diecchio, R.J., and Tagliacozzo, A., 1979, Radiometric ages of basalts from DSDP Leg 43, Sites 382 and 385 (New England Seamounts), 384 (J-anomaly), 386 and 387 (central and western Bermuda Rise), *in* Tucholke, B. E., Vogt, P. R., et al., Initial Reports of the Deep Sea Drilling Project, v. 43: Washington, D. C., U. S. Government Printing Office, p. 739-753.

House, M.R., Richardson, J.B., Chaloner, W.G., Allen, J.R.L., Holland, C.H., and Westoll, T.S., 1977, A correlation of the Devonian rocks in the British Isles: Geological Society of London, Special Report no. 8, 110 p.

Howard, K.A., 1991, Intrusion of horizontal dikes: tectonic significance of middle Proterozoic diabas sheets widespread in the upper crust of the southwestern United States: Journal of Geophysical Research, v. 96, no. B7, p. 12461-12478.

Huang Jiqing (Huang, T. K.), 1945, On major tectonic forms of China: Geological Survey of China Memoir, series A, v. 20, 165 p.

Huang Jiqing (Huang Chi-ching; T. K. Huang), 1978, An outline of the tectonic characteristics of China: Eclogae Geologicae Helvetiae, v. 71, no. 3, p. 611-635.

Huang Jiqing, 1984, New researches on the tectonic characteristics of China, *in* Reports of the 27th International Geological Congress, Moscow 1984: Moscow, Vneshtorgizdat, Colloquium 05, Tectonics of Asia, v. 5, p. 13-28.

*Huang Jiqing, and Chen Bingwei, 1987, The evolution of the Tethys in China and adjacent regions: Beijing, Geological Publishing House, 109 p.

Huang Qinghua, 1980, A mechanical model of the whirl structure, *in* Scientific Papers on Geology [for] International Exchange; prepared for the 26th International Geological Congress, Paris, France, v. 1: Beijing, Geological Publishing House, p. 223-232.

Huang, T.K., 1932, The Permian formations of southern China: Geological Survey of China, Geological Memoirs, Series A, no. 10, 140 p.

Hubbert, M.K., 1951, Mechanical basis for certain familiar geologic structures: Geological Society of America Bulletin, v. 62, no. 4, p. 355-372.

Humphreys, E., Clayton, R.W., and Hager, B.H., 1984, A tomographic image of mantle structure beneath southern California: Geophysical Research Letters, v. 11, no. 7, p. 625-627.

Hunziker, J.C., and Zingg, A., 1980, Lower Palaeozoic amphibolite to granulite facies metamorphism in the Ivrea zone (southern Alps, northern Italy): Schweizerische Mineralogische und Petrographische Mitteilungen, Bd. 60, p. 181-213.

Huppert, H.E., Sparks, R.S. J., Whitehead, J.A., and Hallworth, M.A., 1986, Replenishment of magma chambers by light inputs: Journal of Geophysical Research, v. 91, no. B-6, p. 6113-6122.

Hurtig, E., Čermák, V., Haenel, R., and Zui, V., eds., 1981, Geothermal map of Europe: Gotha (East Germany), Verlag Hermann Haack, Geographisch-Kartographische Anstalt, 34 sheets (20 at scale: 1: 2,500,000; 14 at scale, 5,000,000).

Hussong, D.M., and Sinton, J.B., 1983, Seismicity associated with back arc crustal spreading in the central Mariana Trough, *in* Hayes, D.E., ed., The tectonic and geologic evolution of southeast Asian seas and islands; part 2: American Geophysical Union Geophysical Monograph 27, p. 217-235.

*Hutchinson, C.S., 1973, Synthesis of crustal evolution in the Malay Peninsula, *in* Gobbett, D. J., and Hutchinson, C. S., eds., Geology of the Malay Peninsula: New York, Wiley-Interscience, p. 330-334.

*Hutchinson, C.S., 1977, Granite emplacement and tectonic subdivision of peninsular Malaysia: Geological Society of Malaysia Bulletin, v. 9, p. 187-207.

*Hutchinson, C.S., 1983, Multiple Mesozoic Sn-W-Sb granitoids of southeast Asia, *in* Roddick, J. A., ed., Circum-Pacific plutonic terranes: Boulder, CO, Geological Society of America Memoir 159, p. 35-60.

*Hutchinson, C.S., 1988, Stratigraphic-tectonic model for eastern Borneo: Geological Society of Malaysia Bulletin no. 22, p. 135-151.

*Hutchinson, C.S., 1989a, Geological evolution of South-East Asia: Oxford, Clarendon Press, 368 p.

Hutchinson, C.S., 1989b, The Palaeo-Tethyan realm and Indosinian orogenic system of Southeast Asia, *in* Şengör, A. M. C., ed., Tectonic evolution of the Tethyan region: Dordrecht, Kluwer Academic Publishers, p. 585-643.

Hutchinson, D.R., and White, R.S., 1990, Rifts and hot spots--a model for the Proterozoic Midcontinent rift system of North America: U.S. Geological Survey Circular 1060, p. 38-39.

Hyndman, R.D., and Klemperer, S. L., 1989, Lower-crustal porosity from electrical measurements and inferences about composition from seismic velocities: Geophysical Research Letters, v. 16, no. 3, p. 255-258.

Hynes, A., and Francis, D.M., 1982, A transect of the early Proterozoic Cape Smith foldbelt, New Quebec: Tectonophysics, v. 88, no. 1/2, p. 23-59.

Ichikawa, K., Bunopas, S., Khoo, T. T., and Hada, S., 1990, Report of the Fourth International Symposium on pre-Jurassic evolution of eastern Asia, *in* Ichikawa, K., Bunopas, S., Khoo, T. T., and Hada, S., convenors, Pre-Jurassic geologic evolution of eastern continental margin of Asia: Osaka (Japan), IGCP Project 224, Pre-Jurassic evolution of eastern Asia, report no. 5, p. 1-3.

Iliffe, J.E., Lerche, I., and deBuyl, M., 1991, Basin analysis and hydrocarbon generation of the south Mozambique graben using extensional models of heat flow: Marine and Petroleum Geology, v. 8, no. 2, p. 152-162.

Illies, J.H., and Greiner, G., 1979, Holocene movements and state of stress in the Rhinegraben rift system: Tectonophysics, v. 54, no. 1-4, p. 349-359.

Inglis, C.E., 1913, Stresses in a plate due to the presence of cracks and sharp corners: London, Institute of Naval Architecture Transactions, v. 55, p. 219-241.

Irving, E., 1964, Paleomagnetism and its application to geological and geophysical problems: New York, John Wiley & Sons, 399 p.

Irving, E.M., 1975, Structural evolution of the northernmost Andes, Colombia: U.S. Geological Survey Professional Paper 846, 47 p.

Irving, E., and Archibald, D. A., 1990, Bathozonal tilt corrections to paleomagnetic data from mid-Cretaceous plutonic rocks: examples from the Omineca belt, British Columbia: Journal of Geophysical Research, v. 95, no. B-4, p. 4579-4585.

Isacks, B.L., and Barazangi, M., 1977, Geometry of Benioff zones: lateral segmentation and downwards bending of the subducted lithosphere, *in* Talwani, M., and Pitman, W. C. III, eds., Island arcs, deep sea trenches and back-arc basins: Washington, D.C., American Geophysical Union, Maurice Ewing Series 1, p. 99-144.

Iturralde-Vinent, M.A., 1988, Naturaleza geológica de Cuba, 2nd ed.: La Habana, Editorial Científico-Técnica, 248 p.

Iyer, H., and Hitchcock, T., 1989, Upper-mantle velocity structure in the continental U.S. and Canada, *in* Pakiser, L.C., and Mooney, W.D., eds., Geophysical framework of the continental United States: Geological Society of America Memoir 172, p. 681-710.

Jackson, E.D., 1976, Linear volcanic chains on the Pacific plate, *in* Sutton, G.H., Manghnani, M.H., Moberly, R., and McAfee, E.U., eds., The geophysics of the Pacific Ocean basin and its margin: American Geophysical Union Geophysical Monograph 19, p. 319-335.

Jackson, G.D., and Iannelli, T.R., 1981, Rift-related cyclic sedimentation in the Neohelikian Borden basin, northern Baffin Island, *in* Campbell, F.H.A., ed., Proterozoic basins of Canada: Geological Survey of Canada Paper 81-10, p. 269-302.

Jacobs, J.A., 1961, Some aspects of the thermal history of the earth, *in* The earth today: A collection of papers dedicated to Sir Harold Jeffreys: Royal Astonomical Society Geophysical Journal, v. 4, p. 267-275.

Jacobshagen, V., Brede, R., Hauptmann, M., Heinitz, W., and Zylka, R., 1988, Structure and post-Paleozoic evolution of the central High Atlas, *in* Jacobshagen, V. H., ed., The Atlas system of Morocco. Studies on its geodynamic evolution: New York, Springer-Verlag, p. 245-271.

*Jacoby, W.R., 1984, Geophysical aspects of Rhenish Massif uplift and Rhine Graben sibsidence, *in* Proceedings of the 27th International Geological Congress, Moscow 1984, Report v. 8: Amsterdam, VNU Press, p. 73-104.

Jacoby, W.R., and Anderson, A.J., 1984, Mantle convection; gravity, plate motion, convection: Terra Cognita, v. 4, no. 2, p. 151-155.

Jaeger, H., 1986, Die Faunenwende Mesozoikum/Känozoikum--nüchtern betrachtet: Zeitschrift für Geoloische Wissenschaftern, v. 15, p. 629-656.

Jaeger, J.C., 1962, Elasticity, fracture and flow with engineering and geological applications, 2nd ed.: London, Methuen and Company Ltd., 208 p.

Jaeger, J.C., and Cook, N.G.W., 1979, Fundamentals of rock mechanics, 3rd ed.: London, Chapman and Hall, 593 p.

*Jagannathan, C.R., Ratnam, C., Baishya, N.C., and Das Gupta, U., 1983, Geology of the offshore Mahanadi basin, *in* Bhandari, L. L., Venkatachala, B. S., Kumar, R. Nanjunda Swamy, S., Garga, P., and Srivastava, D. C., eds., Petroliferous basins of India: Dehra Dun, Petroleum Asia Journal, v. 6, no. 4, p. 101-104.

Jaillard, E., Soler, P., Carlier, G., and Mourier, T., 1990, Geodynamic evolution of the northern and central Andes during early to middle Mesozoic times: a Tethyan model: Geological Society of London Journal, v. 147, pt. 6, p. 1009-1022.

*Jaiprakash, B.C., Singh, J., and Raju, D.S.N., 1993, Foraminiferal events across K/T boundary and age of Deccan volcanism in Pallakollu area, Krishna-Godavari basin, India: Geological Society of India Journal, v. 41, no. 2, p. 105-117.

Jardetzky, W.S., 1935, Récherches mathematiques sur l'évolution de la terre: Académie Royale Serbe, t. 107, Section Scientifique, mathematique et Naturelle, t. 29, 203 p.

Jardetzky, W.S., 1954, The principal characteristics of the formation of the earth's crust: Science, v. 119, no. 3090, p. 361-365.

Jaupart, C., Mann, J.R., and Simmons, G., 1982, A detailed study of the distribution of heat flow and radioactivity in New Hampshire: Earth and Planetary Science Letters, v. 59, p. 267-287.

Jefferson, C.W., and Parrish, R.R., 1989, Late Proterozoic stratigraphy, Upb zircon ages, and rift tectonics, Mackenzie mountains, northwestern Canada: Canadian Journal of Earth Sciences, v. 26, no. 9, p. 1784-1801.

Jeffreys, Sir H., 1959, The earth: 4th ed.: Cambridge, Cambridge University Press, 420 p.

Jeffreys, Sir H., 1970, The earth, 5th ed.: Cambridge, Cambridge University Press, 525 p.
Jeffreys, Sir H., 1976, The earth, 6th ed.: Cambridge, Cambridge University Press, 574 p.
Jeffreys, Sir H., and Crampin, S., 1970, On the modified Lomnitz Law of damping: Royal Astronomical Society Monthly Notices, v. 147, p. 295-301.
*Jin Yugan, 1985, Permian Brachiopoda and paleogeography of the Qinghai-Xizang (Tibet) Plateau: Palaeontologia Cathayana no. 2, p. 19-71.
Johnson, A.M., and Page, B.M., 1976, A theory of concentric, kink and sinusoidal folding and of monoclinal flexuring of compressible, elastic multilayers, VII. Development of folds within Huasna syncline, San Luis Obispo County, California: Tectonophysics, v. 33, no. 1-2, p. 97-143.
*JOIDES (Joint Oceanographic Institutions' Deep Earth Sampling Program), 1965, Ocean drilling on the continental margin: Science, v. 150, no. 3697, p. 709-716.
Joly, L., 1925, The surface history of the earth: Oxford, The Clarendon Press, 192 p.
Jones, A.G., 1992, Electrical conductivity of the continental lower crust, *in* Fountain, D. M., Arculus, R., and Kay, R. W., eds., Continental lower crust: Amesterdam, Elsevier Science Publishers, p. 81-143.
Jones, D. L., Silberling, N.J., and Hillhouse, J., 1977, Wrangellia--a displaced terrane in northwestern North America: Canadian Journal of Earth Sciences, v. 14, no. 11, p. 2565-2577.
Jordan, T.H., 1975, The continental tectosphere: Reviews of Geophysics and Space Physics, v. 13, no. 1, p. 1-12.
Jordan, T.H., 1978, Composition and development of the continental tectosphere: Nature, v. 274, no. 5671, p. 544-548.
*Juang, W.S., and Chen, C.H., 1992, Geochronology and geochemistry of Penghu basalts, Taiwan Strait and their tectonic significance: Journal of Southeast Asian Earth Sciences, v. 7, no. 2-3, p. 185-193.
Karig, D.E., 1970, Ridges and basins of the Tonga-Kermadec island arc system: Journal of Geophysical Research, v. 75, no. 2, p. 239-254.
Karig, E.E., Anderson, R.N., and Bibee, L.D., 1978, Characteristics of back arc spreading in the Mariana Trough: Journal of Geophysical Research, v. 83, no. B3, p. 1213-1226.
*Karuinakaran, C., Pawde, M. B., Raina, V. K., Ray, K. K., and Saha, S. S., 1964, Geology of South Andaman Island, India, *in* Kohli, G., Krishnaswami, V. S., and Valdiya, K. S., eds., Himalayan and Alpine orogeny: New Delhi, 22nd International Geological Congress, India 1964, Part XI, Section 11, Proceedings (publ. by Temple Press, Calcutta), p. 79-100.
Kashfi, M.S., 1976, Plate tectonics and structural evolution of the Zagros geosyncline, southwestern Iran: Geological Society of America Bulletin, v. 87, no. 10, p. 1486-1490.
Kashfi, M.S., 1988, Evidence for non-collision geology in the Middle East: Journal of Petroleum Geology, v. 11, no. 4, p. 443-460.
*Katili, J.A., 1991, Tectonic evolution of eastern Indonesia and its bearing on the occurrence of hydrocarbons: Marine and Petroleum Geology, v. 8, no. 1, p. 70-83.
*Kavalieris, I., Leeuwen, T.M. van, and Wilson, M., 1992, Geological setting and styles of mineralization, north arm of Sulawesi, Indonesia: Journal of Southeast Asian Earth Sciences, v. 7, no. 2-3, p. 113-129.
Kay, R., Hubbard, N., and Gast, P.W., 1970, Chemical charactersitics and origin of ocean ridge volcanic rocks: Journal of Geophysical Research, v. 75, no. 8, p. 1585-1613.
*Kazakov, I.K., 1986, Dokembriyskiye infrastrukturnyye kompleksy Paleozoid Mongolii: Leningrad, Izdatel'stvo Nauka, Leningradskoye Otdeleniye, 144 p.
Keith, M.L., 1993, Geodynamics and mantle flow: an alternative earth model: Earth-Science Reviews, v. 33, p. 153-337.
Kelly, K.A., and Gille, S.T., 1990, Gulf Stream surface transport at 69° W from the Geosat altimeter: Journal of Geophysical Research, v. 95, no. C-3, p. 3149-3161.
*Kenyon, C.S., and Beddoes, L.R., Jr., eds., 1977, Geothermal gradient map of Southeast Asia: Southeast Asia Petroleum Exploration Society and Indonesian Petroleum Association, 50 p. + 1 map sheet, scale: 1:6,000,000.
Kerr, R.A., 1993, Landers quake's long reach is shaking up seismologists: Science, v. 256, no. 5091, p. 29.
*Ketner, K.B., Kastowo, Subroto Modjo, Naeser, C.W., Obradovich, J.D., Robinson, K., Tatan Suptander, and Wikarno, 1976, Pre-Eocene rocks of Java, Indonesia: Journal of Research of the U. S. Geological Survey, v. 4, no. 5, p. 605-614.
*Khoo, T.T., 1984, The terrane of the Patani Metamorphics: Geological Society of Malaysia Bulletin no. 17, p. 79-95.
*Khoo, T.T., and Hada, S., 1986, Unconformities in peninsular Malaya and their tectonic significance, *in* Ichikawa, K., Bunopas, S., Khoo, T. T., and Hada, S., convenors, Pre-Jurassic geologic evolution of eastern continental margin of Asia: Osaka (Japan), IGCP Project 224, Pre-Jurassic evolution of eastern Asia, Reports and abstracts, p. 88-91.
*Khoo, T.T., and Tan, B.K., 1983, Geological evolution of peninsular Malaysia, *in* Nutalaya, P., ed.-in-chief, Proceedings of the workshop on stratigraphic correlation of Thailand and Malaysia, September 8-10, 1983,

v. 1, Technical papers: Bangkok, Geological Society of Thailand and Geological Society of Malaysia, p. 253-289.

Khudoley, K.M., 1967, Principal features of Cuban geology: American Association of Petroleum Geologists Bulletin, v. 51, no. 5, p. 668-677.

Khudoley, K.M., and Meyerhoff, A.A., 1971, Paleogeography and geological history of Greater Antilles: Boulder, CO, Geological Society of America Memoir 129, 199 p.

Kim, O.J., and Lee, D.S., 1983, Summary of igneous activity in South Korea, *in* Roddick, J. A., ed., Circum-Pacific plutonic terranes: Geological Society of America Memoir 159, p. 87-103.

Kimura, T., 1973, The old 'inner' arc and its deformation in Japan, *in* Coleman, P. J., ed., The western Pacific: island arcs, marginal seas, geochemistry: Nedlands, University of Western Australia Press, p. 255-273.

*Kimura, T., Hayami, I., and Yoshida, S., 1991, Geology of Japan: Tokyo, University of Tokyo Press, 287 p.

Kingma, J.T., 1974, The geological structure of New Zealand: New York, Wiley-Interscience, John Wiley and Sons, 407 p.

Kirikov, V.P., 1988, Devonian Period in evolution of the Russian platform, *in* McMillan, N.J., Embry, A.F., and Glass, D.J., eds., Devonian of the world. Proceedings of the Second International Symposium on the Devonian System, v. I: Regional syntheses: Canadian Society of Petroleum Geologists, Memoir 14, p. 513-525.

*Kiselev, A.I., and Popov, A.M., 1992, Asthenosphere diapir beneath the Baikal rift: petrological constraints: Tectonophysics, v. 208, no. 1-3, p. 287-295.

Kizaki, K., 1986, Geology and tectonics of the Ryukyu Islands: Tectonophysics, v. 125, no. 1-3, p. 193-207.

Klemperer, S. L., 1987, A relation between continental heat flow and the seismic reflectivity of the lower crust: Journal of Geophysics, v. 61, no. 1, p. 1-11.

Klemperer, S. L., 1988, Crustal thinning and nature of extension in the northern North Sea from deep seismic reflection profiling: Tectonics, v. 7, no. 4, p. 803-821.

Klemperer, S.L., Hauge, T.A., Hauser, E.C., Oliver, J.E., and Potter, C.J., 1986, The Moho in the northern Basin and Range province, Nevada, along the COCORP 40° N seismic-reflection transect: Geological Society of America Bulletin, v. 97, no. 5, p. 603-618.

Klemperer, S., and Hobbs, R., 1991, The BIRPS Atlas; deep seismic reflection profiles around the British Isles; Cambridge, Cambridge University Press, 124 p.

Klewin, K.W., and Berg, J.H., 1991, Petroleogy of the Keweenawan Mamainse Point lavas, Ontario; petrogenesis and continental rift evolution: Journal of Geophysical Research, ser. B., Solid Earth and Planets, v. 96, no. 2, p. 457-474.

*Kneupfer, P.L.K., 1989, Implications of the characteristics of end-points of historical surface fault ruptures for the nature of fault segmentation, *in* Schwartz, D. P., and Sibson, R. H., eds., Fault segmentation and controls of rupture initiation and termination: United States Feological Survey Open-File Report 89-315, p. 193-228.

*Knittel, U., and Daniels, U., 1987, Sr-isotopic composition of marbles from the Puerto Galera area (Mindoro, Philippines): additional evidence for a Paleozoic age of a metamorphic complex in the Philippine island arc: Geology, v. 15, no. 2, p. 136-138.

Knopf, E.B., and Ingerson, E., 1938, Structural petrology: Geological Society of America Memoir 6, 270 p.

Kobayashi, K., 1985, Sea of Japan and Ryukyu Trench-back-arc system, *in* Nairn, A.E.M., Stehli, F.G., and Uyeda, S., eds., The ocean basins and margins, v. 7A. The Pacific Ocean: New York, Plenum Press, p. 419-458.

Kober, L., 1921, Der Bau de Erde: Berlin, Gebrüder Borntraeger, 324 p.

Kober, L., 1925, Die Gestaltungsgeschichte der Erde: Berlin, Gebrüder Borntraeger, 200 p.

Kober, L., 1928, Der Bau der Erde, 2nd ed.: Berlin, Gebrüder Borntraeger, 499 p.

Kontorovich, A.E., Nesterov, I.I., Salmanov, F.K., Surkov, V.K., Trofimuk, A.A., and Ervye, Yu. G., 1975, Geologiya nefti i gaza Zapadnoy Sibiri: Moscow, Izdatel'stvo Nedra, 679 p.

*Kravchenko, K.N., 1979, Tectonic evolution of the Tien Shan, Pamir and Karakorum, *in* Farah, A., and De Jong, K. A., eds., Geodynamics of Pakistan: Quetta, Geological Survey of Pakistan, p. 25-48.

Krebs, W., 1975, Formation of southwest Pacific island arc-trench and mountain systems: plate or global-vertical tectonics?: American Association of Petroleum Geologists Bulletin, v. 59, no. 9, p. 1639-1666.

*Krishman, M. S., 1960, Geology of India and Burma, 4th ed.: Madras, Higginbothams (Private) Ltd., 604 p.

Kroenke, L. W., 1972, A possible occurrence of extensive flood basalts within the ocean basins, *in* Cordilleran Section, 68th Annual Meeting: Geological Society of America Abstracts, v. 4, no. 3, p. 185-186.

Kroenke, L. and Berger, W., Co-Chief Scientists ODP Leg 130 Scientific Party, 1990, ODP returns to Ontong Java Plateau: Geotimes, v. 35, no. 9, p. 15-17.

Krummenacher, D., and Noetzlin, J., 1966, Ages isotopiques K/A de roches prelevées dans les possessions françaises de pacifique: Sociéte Geologique de France Bulletin, serie 7, v. 8, p. 173-175.

Krummenacher, D., Dowd, D.H., Duda, V.F. , Cunningham, W.B., Kingery, F.L., and Speidel, W.F., 1972, Potassium-argon ages from zenoliths (sic) and differentiates in coarse-grained rocks from the center of the island of

Tahiti (French Polynesia) (Abstract): Geological Society of America, Abstracts with Programs, v. 4, no. 3, p. 186.

Krylov, S.V., Mishen'kin, B.P., Petrik, G.V., and Seleznev, V.S., 1979, O seismicheskoy modeli verkhov mantii v Baykal'skoy riftovoy zone: Novosibirsk, Geologiya i Geofizika, no. 5, p. 117-129.

Kuchay, V. K., and Yeryemin, G.G., 1990, Astenokanaly perekhodnoy zony ot Pamira k Tyan'-shanyu: Novosibirsk, Geologiya i Geofizika, no. 2, p. 37-46.

*Kumar, S. P., 1983, Geology and hydrocarbon prospects of Krishna Godavari and Cauvery basins, *in* Bhandari, L. L., Venkatachala, B. S., Kumar, R., Nanjunda Swamy, S., Garga, P., and Srivastava, D. C., eds., Petroliferous basins of India: Dehra Dun, Petroleum Asia Journal, v. 6, no. 4, p. 56-65.

*Kushiro, I., 1987, A petrological model of the mantle wedge and lower crust in the Japanese Islands, *in* Mysen, B. O., ed., Magmatic processes: physico-chemical principles. A volume in honor of Hatten s. Yoder, Jr.: The Geochemical Society Special Publication no. 1, p. 165-181.

Kushiro, I., and Yoder, H.S., Jr., 1974, Formation of eclogite from garnet lherzolite: liquidus relations in a portion of the system $MgSiO_3$-$CaSiO_3$-Al_2O_3 at high pressures: Carnegie Institute of Washington Year Book, v. 73, p. 266-269.

Kuznetsov, M. F., and Naumov, V. A., 1975, Trap magmatism and some problems of tectonics in the southeast of the Tunguska syneclise: International Geology Review, v. 17, no. 7, p. 781-785.

Kyle, P. R., 1980, Development of heterogeneities in the subcontinental mantle; evidence from the Ferrar Group, Antarctica: Contributions to Mineralogy and Petrology, v. 73, no. 1, p. 89-104.

Kyle, P.R., Elliot, D.H., and Sutter, J.F., 1981, Jurassic Ferrar Supergroup tholeiites from the Transantarctic Mountains, Antarctica, and their relationship to the initial fragmentation of Gondwana, *in* Cresswell, M.M., and Vella, P., eds., Gondwana Five: Rotterdam, A.A. Balkema, p. 283-387.

Lachenbruch, A.H., and Sass, J.H., 1977, Heat flow in the United States and the thermal regime of the crust, *in* Heacock, J.G., ed., The Earth's crust. Its nature and physical properties: American Geophysical Union Geophysical Monograph 20, p. 626-675.

Lachenbruch, A.H., and Sass, J.H., 1980. Heat flow and energetics of the San Andreas fault zone: Journal of Geophysical Research, v. 85, no. B11, p. 6185-6222.

*Lan, C.-Y., Lee, T., and Lee, C.W., 1990, The Rb-Sr isotopic record in Taiwan gneisses and its tectonic implication: Tectonophysics, v. 183, no. 1-4, p. 129-143.

Lancelot, Y.P., Larson, R.L., et al., 1990, Initial Reports of the Ocean Drilling Program, 129: College Station, TX (Ocean Drilling Program), 488 p.

Landisman, M., Mueller, S., and Mitchell, R.J., 1971, Review of evidence for velocity inversions in continental crust, *in* Heacock, J.G., ed., The structure and physical properties of the earth's crust: American Geophysical Union Geophysical Monograph 14, p. 11-34.

Lanphere, M.A., 1983, $^{40}Ar/^{39}Ar$ ages of basalt from Clubhouse Crossroads test hole #2, near Charleston, South Carolina, *in* Gohn, S.H., ed., Studies related to the Charleston, South Carolina, earthquake of 1886--tectonics and seismicity: U.S. Geological Survey Professional Paper 1313, p. B1-B8.

Larson, R.L., 1991, Latest pulse of earth: evidence for a mid-Cretaceous superplume: Geology, v. 19, no. 6, p. 547-550.

Larson, R.L.,and Chase, C.G., 1972, Late Mesozoic evolution of the western Pacific Ocean: Geological Society of America Bulletin, v. 83, no. 12, p. 3627-3643.

Latin, D.M., Dixon, J.E., Fitton, J.G., and White, N., 1990,Mesozoic magmatic activity in the North Sea basin: implications for stretching history, *in* Hardman, R.F.P., and Brooks, J., eds., Tectonics events responsible for Britain's oil and gas reserves: Geological Society of London, Special Publication no. 55, p. 207-277.

LaTraille, S.L., and Hussong, D.M., 1980, Crustal structure across the Mariana island arc, *in* Hayes, D.E., ed., The tectonic and geologic evolution of southeast Asian seas and islands: American Geophysical Union Geophysical Monograph 23, p. 209-221.

Laubscher, H.P., 1975, Plate boundaries and microplates in Alpine history: American Journal of Science, v. 275, no. 8, p. 865-876.

Laubscher, H.P., 1978, Foreland folding: Tectonophysics, v. 47, no. 3/4, p. 325-337.

Laubscher, H., 1988, Material balance in Alpine orogeny: Geological Society of America Bulletin, v. 100, no. 9, p. 1313-1328.

Laubscher, H., and Bernoulli, D., 1983, History and deformation of the Alps, *in* Hsü, K.J., ed., Mountain building processes: London, Academic Press, p. 169-180.

Laughton, A.S., and Searle, R.C., 1979, Tectonic processes on slow spreading ridges, *in* Talwani, M., Harrison, C. G., and Hayes, D. E., eds., Deep drilling results in the Atlantic Ocean: ocean crust: American Geophysical Union Maurice Ewing Series 2, p. 15-32.

Lawver, L.A., and Williams, D.L., 1979, Heat flow in the central Gulf of California: Journal of Geophysical Research, v. 84, no. B7, p. 3465-3478.

Lawver, L.A., Dalziel, I. W.D., and Sandwell, D.T., 1993, Antarctic plate: tectonics from a gravity anomaly and infrared satellite image: GSA Today, v. 3, no. 5, p. 117-119, 122.

Le Fort, P., 1975, Himalayas: the collided range. Present knowledge of the continental arc: American Journal of Science, v. 275-A, p. 1-44.

*Le Fort, P., 1986, Metamorphism and magmatism during the Himalayan collision, in Coward, M. P., and Ries, A. C., eds., Collision tectonics: London, Geological Society of London Special Publication no. 19, p. 159-172.

*Le Fort, P., 1988, Granites in the tectonic evolution of the Himalaya, Karakoram and southern Tibet, *in* Shackleton, R. M., Dewey, R. F., and Windley, B. F., eds., Tectonic evolution of the Himalayas and Tibet: The Royal Society of London Philosophical Transactions, series A, v. 326, no. 1589, p. 281-298.

*Le Fort, P., Cuney, M., Deniel, G., France-Lanord, C., Sheppard, C. M. F., Upreti, B. N., and Vidal, P., 1987, Crustal generation of the Himalayan leucogranites: Tectonophysics, v. 134, no. 1-3, p. 39-57.

Le Pichon, X., Houtz, R.E., Drake, C.L., and Nafe, J.E., 1965, Crustal structure of the mid-ocean ridges, I. Seismic refraction measurements: Journal of Geophysical Research, v. 70, no. 2, p. 319-339.

Le Pichon, X., Francheteau, J., and Bonnin, J., 1973, Plate tectonics: Amsterdam Elsevier Scientific Publishing Company, 300 p..

Leat, P.T., Jackson, S.E., Thorpe, R.S., and Stillman, C.J., 1986, Geochemistry of bimodal basalt-subalkaline/peralkaline rhyolite provinces within the southern British Caledonides: Geological Society of London, Journal, v. 143, no. 2, p. 259-273.

Lee, M. S., 1981, Geology and metallic mineralization associated with Mesozoic granitic magmatism in South Korea: Mining Geology, v. 31, p. 235-244.

Leeman, W.P., and Fitton, J.G., 1989, Magmatism associated with extension: introductions: Journal of Geophysical Research, v. 94, no. B6, p. 7682-7684.

Legg, M.R., Luyendyk, B.P., Mammerickx, J., Moustier, C. de, and Tyce, R.C., 1989, Sea Beam survey of an active strike-slip fault: the San Clemente fault in the California continental borderland: Journal of Geophysical Research, v. 94, no. B-2, p. 1727-1744.

Lehner, P., and Ruiter, P.A.C. de, 1977, Structural history of Atlantic margin of Africa: American Association of Petroleum Geologists Bulletin, v. 61, no. 7, p. 961-981.

Leinz, V., Bartorelli, A., Sadowski, G.R., et al., 1966, Sôbre o comportamento espacial do trapp basáltico da bacia do Paraná: Sociedade Brasileiro de Geologia, Boletim, v. 15, no. 4, p. 79-91.

*Leo, G.W.,1991, Oliverian domes, related plutonic rocks, and mantling Ammonoosuc Volcanics of the Bronson Hill anticlinorium, New England Appalachians: United States Geological Survey Professional Paper 1516, 92 p.

Lerner-Lam, A.L., and Jordan, T.H., 1987, How thick are the continents?: Journal of Geophysical Research, v. 92, no. B-13, p. 14007-14026.

Létouzey, J., and Kimura, M., 1985, Okinawa Trough genesis: structure and evolution of a backarc basin developed in a continent: Marine and Petroleum Geology, v. 2, no. 2, p. 111-130.

Létouzey, J., Biju-Duval, B., Dorkel, A., Gonnard, R., Kristchev, K., Montadert, L., and Sungurlu, O., 1977, The Black-Sea: a marginal basin. Geophysical and geological data, *in* Biju-Duval, B., and Montadert, L., eds., Structural history of the Mediterranean basins: Paris, Éditions Technip, p. 363-375.

Létouzey, J., Muller, C., and Sage, L., 1988, Structure of sedimentary basins in eastern Asia: Southeast Asia Petroleum Exploration Society Proceedings, v. 8, p. 63-68.

Létouzey, J., Werner, P., and Marty, A., 1990, Fault reactivation and structural inversion. Backarc and intraplate compressive deformations. Example of the eastern Sunda shelf (Indonesia): Tectonophysics, v. 183, no. 1-4, p. 341-362.

Levchencko, O.V., 1989, Tectonic aspects of intraplate seismicity in the northeastern Indian Ocean: Tectonophysics, v. 170, no. 1-2, p. 125-139.

Levin, H.L., 1978, The earth through time: Philadelphia, W.B. Saunders Company, 530 p. + 2 appendices, glossary, and index.

Leybourne, B.D., Villalobos, E.L., Lowrie, A., and Cotten, J.E., Jr., 1995, Geomorphology and surface acoustice characteristics of the South Adriatic basin: evidence for rotation and subsidence: Third Thematic Conference on Remote Sensing for Marine and Coastal Environments, v. 2, p. II-588-II-598.

Li Desheng, and Luo Ming, 1988, The hydrocarbon accumulations of minor Meso-Cenozoic remnant petroliferous basins in south-east China: Acta Petrolei Sinica, v. 9, no. 4, p. 1-6.

Li Siguang (J. S. Lee), 1929, Some characteristic structural types in eastern Asia and their bearing upon the problem of continental movements: Geological Magazine, v. 66, pt. 4, p. 358-473; pt. 5, p. 501-522.

Li Siguang (J. S. Lee), 1939, The geology of China: London, Thomas Murby & Co., 528 p.

Li Siguang (J. S. Lee), 1973, Crustal structure and crustal movement: Scientia Sinica, v. 16, no. 4, p. 519-559.

Lienert, B.R., Whitcomb, J.H., and Phillips, R.J., 1979, Magnetotelluric measurements close to the San Andreas and Garlock faults (Abstract): EOS, v. 60, no. 18, p. 243.

*Liew, T. C., and Page, R. W., 1985, U-Pb zircon dating of granitoid plutons from the west coast province of peninsular Malaysia: Geological Society of London Journal, v. 142, pt. 3, p. 515-526.

Lightfoot, P.C., Naldrett, A.J., Gorbachev, N.S., Doherty, W., and Fedorenko, V.A., 1990, Geochemistry of the Siberian Trap of the Noril'sk area, USSR, with implications for the relative contributions of crust and mantle to flood basalt magmatism: Contributions to Mineralogy and Petrology, v. 104, no. 6, p. 631-644.

*Lim, P.S., 1985, Geological map of Sabah, 3rd ed.: Kuching, Geological Survey of Malaysia (Malaysia Jabatan Penyiasatan Kajibumi), 1 sheet, scale, 1:500,000.

Lin, J., Purdy, G.M., Schouten, H., Sempéré, J.C., and Zervas, C., 1990, Evidence from gravity data for focused magmatic accretion along the Mid-Atlantic Ridge: Nature, v. 344, 12 April, p. 627-632.

*Litvinovskiy, B. A., Artyushkov, Ye. V., and Zanvilevich, A. N., 1989, O prirode magmatizma Mongolo-Zabaykalskogo polyuza: Novosibirsk, Geologiya i Geofizika, no. 2, p. 32-40.

Liu, C.C., and Yu, S.B., 1989, Fast uplifting along the plate boundary in Taiwan (Abstract): EOS, v. 70, no. 15, p. 403.

*Liu Futian, Qu Kexin, Wu Hua, Li Qiang, Liu Jianhua, and Hu Ge, 1989, Seismic tomography of the Chinese continent and adjacent regions: Acta Geophysica Sinica, v. 32, no. 3, p. 281-291.

Liu Guodong, 1987a, MTS studies on the upper mantle conductivity in China: Pure and Applied Geophysics, v. 125, no. 2/3, p. 465-482.

Liu Guodong, 1987b, The Cenozoic rift system of the North China Plain and the deep internal process: Tectonophysics, v. 133, no. 3/4, p. 277-285.

*Liu Jianhua, Liu Futian, Wu Hua, Li Qiang, and Hu Ge, 1989, Three dimensional velocity images of the crust and upper mantle beneath a north-south zone in China: Acta Geophysica Sinica, v. 32, no. 2, p. 142-152.

*Liu Xiangpei, Chen Yinfo, and Wu Yanchang, 1988, On metallogenic conditions and regularities of the Lower and Middle reaches of the Changjiang River: Acta Geological Sinica (English Edition), v. 1, no. 4, p. 435-449.

Lliboutry, L., 1969, Sea-floor spreading, continental drift and lithosphere sinking with an asthenosphere at melting point: Journal of Geophysical Research, v. 74, no. 27, p. 6525-6540.

Lliboutry, L.A., 1971, Rheological properties of the asthenosphere from Fennoscandian data: Journal of Geophysical Research, v. 76, no. 5, p. 1433-1446.

Locardi, E., 1988, The origin of the Apenninic arcs: Tectonophysics, v. 146, no. 1-4, p. 105-123.

Logatchev, N. A., and Florensov, N. A., 1978, The Baikal system of rift valleys: Tectonophysics, v. 45, no. 1, p. 1-13.

Lomnitz, C., 1957, Linear dissipation in solids: Journal of Applied Physics, v. 28, p. 201-205.

Lomnitz, C., 1962, On Andean structure: Journal of Geophysical Research, v. 67, no. 1, p. 351- 363.

Lomnitz, C., 1974, Global tectonics and earthquake risk: Amsterdam, Elsevier Scientific Publishing Company, 320 p.

Lonsdale, P., 1982, Small offsets of the Pacific-Nazca and Pacific Cocos spreading axes (Abstract): EOS, v. 63, no. 45, p. 1108.

Lonsdale, P., 1988, Structural pattern of the Galapagos microplate and evolution of the Galapagos triple junctions: Journal of Geophysical Research, v. 93, no. B-11, p. 13,551-13,574.

Lorenz, J.C., 1988, Synthesis of late Paleozoic and Triassic redbed sedimentation in Morocco, *in* Jacobshagen, V.H., ed., The Atlas system of Morocco: Berlin, Springer-Verlag, p. 139-168.

Lory, Ch. de, 1860, Géologie du Dauphine: Paris (not seen; mentioned by Aubouin [1965] as emended by Curt Teichert [written commun., 1988]).

Lowman, P.D., Jr., 1985, Plate tectonics with fixed continents: a testable hypothesis--I: Journal of Petroleum Geology, v. 8, no. 4, p. 373-388.

Lowman, P.D., Jr., 1986, Plate tectonics with fixed continents: a testable hypothesis--II: Journal of Petroleum Geology, v. 9, no. 1, p. 71-87.

*Lowman, P.D., Jr., 1991, Eastern North America as a convergent plate boundary: a suggested factor in eastern earthquakes (Abstract): Geological Society of America, Abstracts with Programs, v. 23, no. 1, p. 60.

Lowman, P.D., Jr., 1994, Exploring Space-Exploring Earth: The Impact of Space Flight on Geophysics and Geology: First Draft, Goddard Space Flight Center, 185 p.

Lugéon, M., 1902, Les grandes nappes de recouvrement des Alpes du Chablais et de la Suisse: Société Géologique de France Bulletin, sér. 4, t. 1 (1901), p. 723-823.

Luosto, U., Flueh, E.R., Lund, C.E., and Working Group, 1989, The crustal structure along the POLAR profile from seismic refraction investigations: Tectonophysics, v. 162, no. 1/2, p. 51-85.

*Luyendyk, B.P., 1977, Deep sea drilling on the Ninetyeast Ridge: synthesis and a tectonic model, *in* Heirtzler, J. R., Bolli, H. M., Davies, T. A., Saunders, J. B., and Sclater, J. G., eds., Indian Ocean geology and biostratigraphy: Washington, D.C., American Geophysical Union, p. 165-187.

Luyendyk, B.P., and Melson, W. G., 1967, Magnetic properties and petrology of rocks near the crest of the Mid-Atlantic Ridge: Nature, v. 215, no. 5097, p. 147-149.

Lyttleton, R.A., 1982, The earth and its mountains: Chichester (UK), John Wiley & Sons, 206 p.

Ma, C., Ryan, J.W., and Caprette, D., 1989, Crustal dynamics project data analysis--1988: Greenbelt (MD), National Aeronautics and Space Administration, NASA Technical Memorandum 100723, 10 p. + 9 appendices.

*Ma Xingyuan, chief compiler, 1986a, Lithospheric dynamics map of China and adjacent seas: Beijing, Geological Publishing House, scale: 1:4,000,000, 2 sheets.

Ma Xingyuan, 1986b, Explanatory notes for the lithosphere dynamics map of China and adjacent areas: Beijing, Geological Publishing House, 53 p.

Ma Xingyuan, ed.-in-chief, 1989, Lithosphere dynamics atlas of China: Beijing, China Cartographic Publishing House, 68 charts (+ figs. and texts for each).

*Ma Xingyuan, and Wu Daning, 1987, Cenozoic extensional tectonics in China: Tectonophysics, v. 133, no. 3-4, 243-255.

Maack, R., 1952, Die Entwicklung der Gondwana-Schichton sudbrasiliens und ihre Beziehungen zur Karroo-formation südafrikas, *in* Algeria, Algiers, International Geological Congress, 19th, Symposium sur les séries de Gondwana: IUGS, p. 339-372.

*MacDonald, A.S., and Barr, S.M., 1984, The Nan River mafic-ultramafic belt, northern Thailand: geochemistry and tectonic significance: Geological Society of Malaysia Bulletin no. 17, p. 209-224.

MacDonald, G.J.F., 1959, Calculations on the thermal history of the earth: Journal of Geophysical Research, v. 64, no. 11, p. 1967-2000.

MacDonald, G.J.F ., 1963, The deep structure of continents: Reviews of Geophysics, v. 1, no. 4, p. 587-665.

MacDonald, G.J.F., 1964, The deep structure of continents: Science, v. 143, no. 3609, p. 921-929.

MacDonald, G.J.F., 1965a, Geophysical deductions from observations of heat flow, *in* Lee, W. H. K., ed., Terrestrial heat flow: Washington, D.C., American Geophysical Union Geophysical Monograph no. 8, p. 191-210.

MacDonald, G.J.F., 1965b, Continental structure and drift, *in* Blackett, P. M. S., Bullard, Sir E., and Runcorn, S. K., organizers, A symposium on continental drift: Royal Society of London Philosophical Transactions, series A, v. 258, no. 1088, p. 215-227.

Macdonald, K.C., and Fox, P.J., 1983, Overlapping spreading centers: a new kind of accretionary geometry on the East Pacific Rise: Nature, v. 302, no. 5903, p. 55-58.

Macdonald, K.C., Fox, P. J., Perram, L.J., Eisen, M.F., Haymon, R.M., Miller, S.P., Carbotte, S.M., Cormier, M.-H., and Shor, A.N., 1988, A new view of the mid-ocean ridge from the behaviour of ridge-axis discontinuities: Nature, v. 335, no. 6187, p. 217-225.

Macdonald, K.C., Sempère, J.-C., and Fox, P.J., 1984, East Pacific Rise from Siquieros to Orozco fracture zones: along-strike continuity of axial neovolcanic zone and structure and evolution of overlapping spreading centers: Journal of Geophysical Research, v. 89, no. B-7, p. 6049-6069.

Macdonald, K.C., Sempère, J.C., Fox, P.J., and Tyce, R., 1987, Tectonic evolution of ridge-axis discontinuities by the meeting, linking, or self-decapitation of neighboring ridge segments: Geology, v. 15, no. 11, p. 993-997.

MacDonald, W.D., and Opdyke, N.D., 1984, Preliminary paleomagnetic results from Jurassic rocks of the Santa Marta massif, *in* Bonini, W.E., Hargraves, R.G., and Shagam, R., eds., The Caribbean-South American plate boundary and regional tectonics: Geological Society of America, Memoir 162, p. 295-298.

Macdougall, J.D., 1988, Continental flood basalts and MORB; a brief discussion of similarities and differences in their petrogenesis, *in* Continental flood basalts, J.D. Macdouglas, ed.: Boston, Kluwer Academic Publishers, p. 331-341.

Macgregor, A.M., 1951, Some milestones in the Precambrian of Southern Rhodesia: Geological Society of South Africa Proceedings, v. 54, p. xxvii-lxxi.

MacLeod, C.J., and Rothery, D.A., 1992, Ridge axial segmentation in the Oman Ophiolite: evidence from along-strike variations in the sheeted dyke complex, *in* Parson, L. M., Murton, B. J., and Browning, P., eds., Ophiolites and their modern oceanic analogues: Geological Society of London Special Publication no. 60, p. 39-63.

Magnetic Anomaly Map Committee, 1987, Magnetic anomaly map of North America: Boulder (CO), Geological Society of America, Continent-Scale Map-003, 4 sheets, scale, 1:5,000,000.

Mahoney, J.J., 1984, Isotopic and chemical studies of the Deccan and Rajmahal traps, India; mantle sources and petrogenesis: University of California, San Diego, Ph.D. Dissertation, 205 p.

Mahoney, J.J., Storey, M., Duncan, R.A., Spencer, K.J., and Pringle, M.S., 1993, Geochemistry and age of the Ontong Java Plateau, *in* Pringle, M. S., Sager, W. W., Sliter, W. V., and Stein, S., eds., The Mesozoic Pacific: geology, tectonics, and volcanism: Washington, D.C., American Geophysical Union Geophysical Monograph 77, p. 233-261.

Makarenko, G.F., 1983, Trappy c strukture materikov: Moscow, Izdatel'stvo Nauka, 208 p.

Makeyeva, L.I., Vinnik, L.P., and Roecker, S.W., 1992, Shear-wave splitting and small-scale convection in the continental upper mantle: Nature, v. 358, no. 6382, p. 144-147.

Mammerickx, J. and Smith, S.M., 1985, Bathymetry of the northcentral Pacific: Geological Society of America, Map and Chart Series, 52, scale: 1:6,442,194.

Mandelbrot, B.B., 1983, The fractal geometry of nature: New York, W. H. Freeman and Company, 468 p.

Manspeizer, W., DeBoer, J., Costain, J.K., Froelich, A.J., Coruh, C., Olsen, P.E., McHone, G.J., Puffer, J.H., and Prowell, D.C., 1989, Post-Paleozoic activity, *in* Hatcher, R.D., Jr., Thomas, W.A., and Viele, G.W., eds., The geology of North America, v. F-2. The Appalachian-Ouachita orogen in the United States: Geological Society of America, Decade of North American Geology, p. 319-374.

Manspeizer, W., Puffer, J.H., and Cousminer, H.L., 1978, Separation of Morocco and eastern North America: a Triassic-Liassic stratigraphic record: Geological Society of America Bulletin, v. 89, no. 6, p. 901-920.

Marillier, F., Keen, C.E., and Stockmal, G.S., 1989, Laterally persistent seismic characteristics of the lower crust: examples from the northern Appalachians, *in* Mereu, R.F., Mueller, S., and Fountain, D.M., eds., Properties and processes of earth's lower crust: American Geophysical Union Geophysical Monograph no. 51, p. 45-52.

*Markovskiy, A.P., editor-in-chief, 1975, Geologicheskaya karta Evrazii: Moscow, Ministerstvo Geologii SSSR, Vsesoyuznyy Ordena Lenina Nauchno-Issledovatelskiy Geologicheskiy Institut (VSEGEI), 8 sheets, scale, 1:5,000,000.

Marsh, B.D., 1979, Island arc development: some observations, experiments, and speculations: Journal of Geology, v. 87, no. 6, p. 687-713.

Marsh, B.D., and Carmichael, I.S.E., 1974, Benioff zone magmatism: Journal of Geophysical Research, v. 79, no. 8, p. 1196-1206.

Martin, B.D., 1992, Constraints to major right-lateral movements, San Andreas fault system, central and northern California, *in* Chatterjee, S., and Hotton, N., III, eds., New Concepts in Global Tectonics: Lubbock, Texas Tech University Press, p. 131-148.

Martin, G., 1982, Geologie des Küstengebietes von Nordwest-Afrika südlich der Sahara. Neue Erkentnisse aus der Erdölexploration: Giessen, Lenz-Verlag, Giessener Geologische Schriften no. 30, 150 p.

Martin, H., 1975, Structural and palaeogeographical evidence for an upper Palaeozoic sea between southern Africa and South America, *in* Campbell, K. S. W., ed., Gondwana geology. Papers presented at the Third Gondwana Symposium, Canberra, Australia, 1973: Canberra, Australian National University Press, p. 37-51.

Massey, N.W.D., 1982, Mishewawa Lake area, District of Algoma, in Wood, J., White, O. L., Barlow, R. B., Colvine, A. C. Summary of Field Work, 1982: Ontario Geological Survey Misc. Paper, 106, p. 61-64.

Mathias, M., Siebert, J.C., and Rickwood, P.C., 1970, Some aspects of the mineralogy and petrology of ultramafic xenoliths in kimberlite: Contributions to Mineralogy Petrology, v. 26, no. 2, p. 75-123.

*Matsubayashi, O., and Uyeda, S., 1979, Estimation of heat flow in certain exploration wells in offshore areas of Malaysia: Japan, Earthquake Research Institute Bulletin, v. 54, no. 1, p. 31-44.

*Matsubayashi, O., Okubo, Y., Yamazaki, T., Joshima, M., and Miyazaki, T., 1992, Heat flow and Curie point depth map of Japan and adjoining areas, *in* Geological atlas of Japan, 2nd ed.: Kawasaki, Geological Survey of Japan, 1 sheet, scale, 1:5,000,000.

*Mattauer, M., and Collot, B., 1986, Continental subduction, thrusting and strike-slip faulting in the Canadian Cordillera: Societe Geologique de France Bulletin, ser. 8, v. 2, no. 6, p. 899-909.

Maxwell, A.E., Von Herzen, R.P., Andrews, J.A., Boyce, R.E., Milow, E.D., Hsu, K.J., Percival, S.F., and Saito, T., 1970, Summary and conclusions, *in* Initial reports of the Deep Sea Drilling Project, v. 3: Washington, D. C., U. S. Government Printing Office, p. 441-471.

Maxwell, J.C., 1968, The Mediterranean, ophiolites, and continental drift, *in* Johnson, H., and Smith, B.L., eds., The megatectonics of continents and ocean basins: Rutgers University Press, p. 167-193.

Maze, W.B., 1984, Jurassic La Quinta Formation in the Sierra de Perija, northwestern Venezuela: geology and tectonic environment of red beds and volcanic rocks, *in* Bonini, W.E., Hargraves, R.B., and Shagam, R., eds., The Caribbean-South American plate boundary and regional tectonics: Geological Society of America, Memoir 162, p. 263-282.

McBirney, A.R., and Wiliams, H., 1965, Volcanic history of Nicaragua: California Univ. Pubs. Geol. Sci. v. 55, 73 p.

McBride, J.H., Nelson, K.D., and Brown, L.D., 1989, Evidence and implications of an extensive early Mesozoic rift basin and basalt/diabase sequence beneath the southeast coastal plain: Geological Society of America Bulletin, v. 101, no. 4, p. 512-520.

McCaffrey, R., Silver, E.A., and Raitt, R.W., 1980, Crustal structure of the Molucca Sea collision zone, Indonesia, *in* Hayes, D.E., ed., The tectonic and geological evolution of southeast Asian seas and islands: American Geophysical Union Geophysical Monograph 23, p. 161-177.

McCartan, L., and Architzel, R.J., 1988, Heat flow map of the eastern United States: Reston, VA, U. S. Geological Survey Miscellaneous Field Studies Map MF-2057, scale, 1:2,500,000, 2 sheets.

McCartan, L., and Gettings, M.E., 1991, Possible relationship between seismicity and warm intrusive bodies in the Charleston, South Carolina, and New Madrid, Missouri, areas: Reston, VA, U. S. Geological Survey Bulletin 1953, 18 p.

McDougall, I., 1974, Potassium-argon ages on basaltic rocks recovered from DSDP, Leg 22, Indian Ocean, *in* Borch, C. C. von der, Sclater, J. G., et al., Initial reports of the Deep Sea Drilling Project, v. 22: Washington, D. C., U. S. Government Printing Office, p. 377-379.

McDowell, F.W., Lehman, D.H., Gucwa, P.R., Fritz, D., and Maxwell, J.C., 1984, Glaucophane schists and ophiolites of the northern California Coast Ranges and tectonic implications: Geological Society of America Bulletin, v. 95, no. 11, p. 1373-1382.

McElhinny, M.W., ed., 1979, The earth: its origin, structure and evolution: London, Academic Press, 597 p.

McHone, J.G., 1984, Mesozoic igneous rocks of northern New England and adjacent Quebec: summary, description of map, and bibliography of data sources: Geological Society of America, Maps and Chart Series MC-49, map + 5 p.

McHone, J.G., Ross, M.E., and Greenough, J.D., 1987, Mesozoic dyke swarms of eastern North America, *in* Halls, H.C., and Fahrig, W.F., eds., Mafic dyke swarms: Geological Association of Canada, Special Paper 34, p. 279-288.

McKee, E.H., Swanson, D.A., and Write, T.L., 1977, Duration and volume of Columbia River basalt volcanism, Wahington, Oregon, and Idaho: Geological Society of America, Abstracts with Programs, v. 9, p. 463-464.

Megard, F., 1987, Structure and evolution of the Peruvian Andes, *in* Schaer, J.P., and Rodgers, J., eds., The anatomy of mountain ranges: Princeton University Press, p. 179-210.

Mehnert, K.R., 1969, Petrology of the Precambrian basement complex, *in* Hart, P.J., ed., The earth's crust and upper mantle: American Geophysical Union Geophysical Monograph 13, p. 513-518.

Melson, W.G., Hart, S.R., and Thompson, G., 1972, St. Paul's Rocks, equatorial Atlantic: petrogenesis, radiometric ages, and implications on sea-floor spreading, *in* Shagam, R., Hargraves, R. B., Morgan, W. J., Van Houten, F. B., Burk, C. A., Holland, H. D., and Hollister, L. C., eds., Studies in earth and space sciences: Boulder, CO, Geological Society of America Memoir 132, p. 241-272.

Menard, H.W., 1960, The East Pacific Rise: Science, v. 132, no. 3441, p. 1737-1746.

Menard, H.W., 1964, Marine geology of the Pacific: New York, McGraw-Hill Book Company, 271 p.

Menard, H.W., 1986, The ocean of truth. A personal history of global tectonics: Princeton University Press, 252 p.

*Metcalfe, I., 1986, Conodont biostratigraphic studies in Sumatra: preliminary results: Geological Society of Malaysia Bulletin no. 20, p. 243-247.

*Metcalfe, I., 1989, Triassic sedimentation in the central basin of peninsular Malaysia, *in* Thanasuthipitak, T., and Ounchanum, P., eds., Proceedings of the International Symposium on Intermontane Basins: Geology and Resources: Chiang Mai (Thailand), Chiang Mai University, Faculty of Science, p. 173-186.

Meyerhoff, A.A., 1967, Future hydrocarbon provinces of Gulf of Mexico-Carribbean region: Gulf Coast Association Geological Societies Trans., v. 17, p. 217-260.

Meyerhoff, A.A., 1970, Continental drift: implications of paleomagnetic studies, meteorology, physical oceanography, and climatology: Journal of Geology, v. 78, no. 1, p. 1-51.

Meyerhoff, A.A., 1981a, Ordovician (Caradocian) trilobites and graptolites near the Mid-Atlantic Ridge: International Stop Continental Drift Society Newsletter, v. 3, no. 3, p. 1.

Meyerhoff, A.A., 1981b, The oil and gas potential of the Soviet Far East: Beaconsfield (UK), Scientific Press Ltd., 175 p.

Meyerhoff, A.A., 1982, Hydrocarbon resources in Arctic and subarctic region, *in* Embry, A. F., and Balkwill, H. R., eds., Arctic geology and geophysics: Calgary (Alberta), Canadian Society of Petroleum Geologists Memoir 8, p. 451-552.

Meyerhoff, A.A., in press, Surge tectonic evolution of Southeastern Asia: a geohydrodynamics approach: London, Elsevier, 212 p. 76 figs., 1 table.

Meyerhoff, A.A., and Meyerhoff, H.A., 1972, "The new global tectonics": major inconsistencies: American Association of Petroleum Geologists Bulletin, v. 56, no. 2, p. 269-336.

Meyerhoff, A.A., and Meyerhoff, H.A., 1974, Tests of plate tectonics, *in* Kahle, C. F., ed., Plate tectonics--assessments and reassessments: Tulsa, OK, American Association of Petroleum Geologists Memoir 23, p. 43-145.

Meyerhoff, A.A., Agocs, W.B., Taner, I., Morris, A. E. L., and Martin, B. D., 1992a, Origin of midocean ridges, *in* Chatterjee, S., and Hotton, N. III, eds., New concepts in global tectonics: Lubbock, TX, Texas Tech University Press, p. 151-178.

*Meyerhoff, A.A., Kamen-Kaye, M., Chen, C., and Taner, I., 1991, China--stratigraphy, paleogeography and tectonics: Dordrecht, Kluwer Academic Publishers, 188 p.

Meyerhoff, A.A., Meyerhoff, H.A., and Briggs, R.S., Jr., 1972, Continental drift, V: proposed hypothesis of earth tectonics: Journal of Geology, v. 80, no. 6, p. 662-692.

Meyerhoff, A.A., Taner, I., Morris, A.E.L., and Martin, B.D., 1989, Surge tectonics, *in* Chatterjee, S., and Hotton, N. III, eds., New concepts in global tectonics: a discussion meeting sponsored by the Smithsonian Institution and Texas Tech University, 20-21 July, 1989, National Museum of Natural History, Smithsonian Institution, Washington, D.C., Abstracts volume: Lubbock, TX, Texas Tech University Press, p. 25-26.

Meyerhoff, A.A., Taner, I., Morris, A.E. L., Martin, B.D., Agocs, W.B., and Meyerhoff, H.A., 1992b, Surge tectonics: a new hypothesis of earth dynamics, *in* Chatterjee, S., and Hotton, N. III, eds., New concepts in global tectonics: Lubbock, TX, Texas Tech University Press, p. 309-409.

Meyerhoff, H.A., 1954, Antillean tectonics: New York Academy of Sciences Transactions, series II, v. 16, no. 3, p. 149-155.

Meyerhoff, H.A., and Meyerhoff, A.A., 1977, Genesis of island arcs, *in* Geodynamics in south-west Pacific, Symposium International, Noumea, Nouvelle-Calédonie, 27 août-2 septembre 1976: Paris, Éditions Technip, p. 357-370.

Miller, H., Müller, S., and Perrier, G., 1982, Structure and dynamics of the Alps: a geophysical inventory, *in* Berckhemer, H., and Hsü, K., eds., Alpine-Mediterranean geodynamics: American Geophysical Union and Geological Society of America Geodynamics Series, v. 7, p. 175-203.

Milnes, A.G., 1974, Structure of the Pennine zone (central Alps): a new working hypothesis: Geological Society of America Bulletin, v. 85, no. 11, p. 1727-1732.

Milsom, J., 1977, Preliminary gravity map of Seram, eastern Indonesia: Geology, v. 5, no. 10, p. 641-643.

*Milsom, J., 1981, Neogene thrust emplacement from a frontal arc in New Guinea, *in* McClay, K.R., and Price, N.J., eds., Thrust and nappe tectonics: London, Geological Society of London Special Publication no. 9, p. 417-424.

*Minster, J.B., and Jordan, T.H., 1978, Present-day plate motions: Journal of Geophysical Research, v. 83, no. B-11, p. 5331-5354.

*Mitchell, A.H.G., 1986, Mesozoic and Cenozoic regional tectonics and metallogenesis in mainland SE Asia: Geological Society of Malaysia Bulletin no. 20, p. 221-239.

*Mitchell, A.H.G., 1992, Late Permian-Mesozoic events and the Mergui Group nappe in Myanmar and Thailand: Journal of Southeast Asian Earth Sciences, v. 7, no. 2-3, p. 165-178.

*Mitchell, A.H.G., and Leach, T. M., 1991, Epithermal gold in the Philippines: island arc metallogenesis, geothermal systems and geology: London, Harcourt Brace Jovanovich, Publishers (Academic Press), 457 p.

Mitchell, B.J., and Landisman, M., 1971, Geophysical measurements in the southern Great Plains, *in* Heacock, J.G., ed., The structure and physical properties of the earth's crust: American Geophysical Union Geophysical Monograph 14, p. 77-93.

Mitchell, C., and Widdowson, M., 1991, A geological map of the southern Deccan Traps, India and its structural implications: Journal of the Geological Society (London), v. 148, pt. 3, p. 495-505.

Miyashiro, A., 1973, Metamorphism and metamorphic belts: New York, John Wiley and Sons (Halsted Press), 492 p.

Miyashiro, A., Aki, K., and Şengör, A.M.C., 1982, Orogeny: New York, John Wiley & Sons, 242 p.

Miyata, T., 1990, Slump strain indicative of paleoslope in Cretaceous Izumi sedimentary basin along Median Tectonic Line, southwest Japan: Geology, v. 18, no. 5, p. 392-394.

Mohr, P.A., 1976, ENE-trending lineaments of the African rift system, p. 327-326, *in* R.A. Hodgson, S.P. Gray, Jr., and Y. Benjamins, eds., Proceedings of the First International Conference on the new basement tectonics: Utah Geological Assoc. Publication no. 5, 636 p.

Mohr, P.A., 1983, Ethiopian flood basalt province: Nature, v. 303, no. 5918, p. 577-584.

Mohr, P.A., 1987, Patterns of faulting in the Ethiopian rift valley: Tectonophysics, v. 143, nos. 1-3, p. 169-179.

*Mohr, P., and Zanettin, B., 1988, The Ethopian flood basalt province, *in* Macdougall, J.D., ed., Continental flood basalts: Dordrecht, Kluwer Academic Publishers, p. 63-110.

Monger, J. W. H., 1984, Evidence for large-scale horizontal displacements within the North American Cordillera, 27th International Geological Congress, Special session of the International "Lithosphere" Programme, Moscow, v. 27 (IX, Part 1), p. 44.

Moon, F.C., 1987, Chaotic vibrations. An introduction for applied scientists and engineers: New York, John Wiley & Sons, 309 p.

Mooney, W.D., and Braile, L.W., 1989, The seismic structure of the continental crust and upper mantle of North America, *in* Bally, A. W., and Palmer, A. R., eds., The geology of North America; an overview: Boulder, CO, Geological Society of America, Decade of North American Geology, v. A, p. 39-52.

*Mooney, W.D., and Brocher, T.M. 1987, Coincident seismic reflection/refraction studies of the continental lithosphere: a global review: Geophysical Journal of the Royal Astronomical Society, v. 89, no. 1, p. 1-6.

Mooney, W.D., and Meissner, R., 1992, Multi-genetic origin of crustal reflectivity: a review of seismic-reflection profiling of the continental crust and Moho, *in* Fountain, D. M., Arculus, R., and Kay, R. W., eds., Continental lower crust: Amsterdam, Elsevier Science Publishers, p. 45-79.

Mooney, W.D., and Weaver, C.S., 1989, Regional crustal structure and tectonics of the Pacific coastal states: California, Oregon, and Washington, *in* Pakiser, L.C., and Mooney, W.D., eds., Geophysical framework of the continental United States: Geological Society of America Memoir 172, p. 129-161.

Mooney, W.D., Andrews, M.C., Ginzburg, A., Peters, D.A., and Hamilton, R.M., 1983, Crustal structure of the northern Mississippi embayment and a comparison with other continental rift zones: Tectonophysics, v. 94, no. 1-4, p. 327-348.

Moore, A.M.G., Willcox, J.B., Exon, N.F., and O'Brien, G.W., 1992, Continental shelf basins on the west Tasmania margin: Australian Petroleum Exploration Association (APEA) Journal, v. 32, pt. 1, p. 231-250.

Moore, T.C., Jr., Rabinowitz, P.D., Borella, P.E., and Boersma, A., 1984, History of the Walvis Ridge, *in* Moore, T.C., Jr., Rabinowitz, P.D., and others, Initial reports of the Deep Sea Drilling Project, v. 74: Washington (DC), United States Government Printing Office, p. 873-894.

Morgan, W.J., 1968, Rises, trenches, great faults, and crustal blocks: Journal of Geophysical Research, v. 73, no. 6, p. 1959-1982.

Morgan, W.J., 1971, Island chains, aseismic ridges, and plate motions. American Association of Petroleum Geologists Bulletin, v. 55, no. 2, p. 354-355

Morley, L.W., and Larochelle, A., 1964, Palaeomagnetism as a means of dating geological events, *in* Osborne, F. F., ed., Geochronology in Canada: Toronto, Royal Society of Canada Special Publication no. 8, p. 39-51.

Morris, A.E.L., Taner, I., Meyerhoff, H.A., and Meyerhoff, A.A., 1990, Tectonic evolution of the Caribbean region, *in* Dengo, G., and Case, J.E., eds., The geology of North America, v. H. The Caribbean region: Geological Society of America, Decade of North American Geology, p. 423-457.

Morris, W.A., and Park, J.K., 1981, Correlation of upper Proterozoic strata in the Cordillera: paleomagnetism of the Tsezotene sills and the Little Dal lavas, *in* Campbell, F.H.A., ed., Proterozoic basins of Canada: Geological Survey of Canada Paper 81-10, p. 73-78.

Mosmann, R., Falkenhein, F.U.H., Gongalves, A., and Neopmuceno, F. (Filho), 1986, Oil and gas potential of the Amazon Paleozoic basins, *in* Halbouty, M.T., ed., Future petroleum provinces of the world: American Association of Petroleum Geologists Memoir 40, p. 207-241.

Mount, V.S., 1988, State of stress near major strike-slip faults (Abstract): Geological Society of America Abstracts with Programs, v. 20, no. 7, p. A320.

Mount, V.S., and Suppe, J., 1987, State of stress near the San Andreas fault: implications for wrench tectonics: Geology, v. 15, no. 12, p. 1147-1150.

Mueller, A.G., 1991, The Savage Lode magnesian skarn in the Marvel Loch gold-silver mine, Southern Cross greenstone belt, Western Australia. Part 1: Structural setting, petrography, and geochemistry: Canadian Journal of Earth Sciences, v. 28, no. 5, p. 659-685.

Mueller, S., 1983, Deep structure and recent dynamics in the Alps, *in* Hsü, K.J., ed., Mountain building processes: New York, Academic Press, p. 181-199.

Mueller, S., and Ansorge, J., 1989, The crustal structure of western Europe: Annual Review of Earth and Planetary Sciences, v. 17, p. 335-360.

Muir-Wood, R., 1989, Extraordinary deglaciation reverses faulting in northern Fennoscandia, *in* Gregersen, S., and Basham, P.W., eds., Earthquakes at North-Atlantic passive margins: neotectonics and postglacial rebound: Boston, Kluwer Academic Publishers, p. 141-173.

Muller, J.E., Northcote, K. E., and Carlisle, D., 1974, Geology and mineral deposits of Alert Bay-Cape Scott map-area, Vancouver Island, British Columbia: Canadian Geological Survey Paper No. 74-8, 77 p.

Munk, W.H., and MacDonald, G.J.F., 1960, The rotation of the earth. A geophysical discussion: Cambridge, Cambridge University Press, 323 p.

Munk, W.H., and MacDonald, G.J.F., 1975, The rotation of the earth. A geophysical discussion (with corrections): Cambridge, Cambridge University Press, 323 p.

Munson, B.R., Young, D.F., and Okiishi, T.H., 1990, Fundamentals of fluid mechanics: New York, John Wiley & Sons, 843 p.

Murauchi, S., Den, N., Asano, S., Hotta, H., Yoshii, T., Asanuma, T., Hagiwara, K., Ichikawa, K., Sato, T., Ludwig, W.J., Ewing, J.I., Edgar, N.T., and Houtz, R.E., 1968, Crustal structure of the Philippine Sea: Journal of Geophysical Research, v. 73, no. 10, p. 3143-3171.

Murphy, F.X., 1985, The lithostratigraphy and structural geology of the Dungarvan syncline and adjacent areas, Co. Waterford, southern Ireland: unpubl. Ph. D. disser., National University of Ireland, 200 p.

Murphy, F.X., 1990, The Irish Variscides: a fold belt developed within a major surge zone: Geological Society of London Journal, v. 147, pt. 3, p. 451-460.

Murrell, S.A.F., 1964a, The theory of the propagation of elliptical Griffith cracks under various conditions of plane strain or plane stress, part 1: British Journal of Applied Physics, v. 15, p. 1195-1210.

Murrell, S.A.F., 1964b, The theory of the propagation of elliptical Griffith cracks under various conditions of plane strain or plane stress, parts 2 and 3: British Journal of Applied Physics, v. 15, p. 1211-1223.

Murthy, N.G.K., 1987, Mafic dyke swarms of the Indian shield, *in* Halls, H. C., and Fahrig, W. F., eds., Mafic dyke swarms: St. John's, Newfoundland, Geological Association of Canada Special Paper 34, p. 393-400.

Mykura, W., 1960, The lower Old Red sandstone igenous rocks of the Pentland Hills: Geological Survey of Great Britain, Bulletin, v. 16, p. 131-155.

Nagata, T., 1953, Rock-magnetism: Tokyo, Maruzen Co., 232 p.

Nagata, T., 1961, Rock magnetism, revised edition: Tokyo, Maruzen Company Ltd., 350 p.

*Nagibina, N.S., Gatinsky, Yu. G., Grinberg, G.A., Gusey, G.S., and Kovalenko, V.I., 1987, Mesozoic tectonic movements and magmatism in eastern Asia (correlation problem of tectonics and magmatism), *in* Leonov,

Yu. G., and Khain, V. E., eds., Global correlation of tectonic movements: New York, John Wiley & Sons, p. 119-139.

Nakanishi, M., Tamaki, K., and Kobayashi, K., 1992, A new Mesozoic isochron chart of the northwestern Pacific Ocean: paleonmagnetic and tectonic implications: Geophysical Research Letters, v. 19, no. 7, p. 693-696.

Nalivkin, D.V., 1973, Geology of the U.S.S.R.: Edinburgh, Oliver & Boyd, 855 p.

Namson, J.S., and Davis, T.L., 1988, Structural transect of the western Transverse Ranges, California: implications for lithospheric kinematics and seismic risk evaluation: Geology, v. 16, no. 8, p. 675-679.

*Naqvi, S.M., and Rogers, J.J.W., 1987, Precambrian geology of India: New York, Oxford University Press, 223 p.

National Geographic Society, 1992, Atlantic Ocean floor, *in* National Geographic Atlas of the World, revised sixth edition: National Geographic Society, Chart 104, scale: 1:31,860,000 (at the equator).

NAUTILAU Group, 1990, Hydrothermal activity in the Lau basin: EOS, v. 71, no. 18, p. 678-679.

Neev, D., and Hall, J.K., 1984, Mantle-produced counterclockwise vortices along the northern Mediterranean belt (a genetic hypothesis for the Alpine system): Universität Hamburg, Geologische-Paläontologische Institut, Mitteilunger, Festband Georg Knetsch, v. 56, p. 111-127.

Neev, D., Greenfield, L., and Hall, J. K., 1985, Slice tectonics in the eastern Mediterranean basin, *in* Stanley, D. J., and Wezel, F.-C., eds., Geological evolution of the Mediterranean basin: New York, Springer-Verlag, p. 249-269.

Nelson, K.D., 1988, The COCORP atlas, v. 1: Ithaca (NY), Cornell University, Institute for the Study of the Continents, 24 p. + ca. 65 plates.

*Neprochnov, Yu. P., Levchenko, O.V., Merklin, L.R., and Sedov, V.V., 1988, The structure and tectonics of the intraplate deformation area in the Indian Ocean: Tectonophysics, v. 156, no. 1-2, p. 89-106.

Nevin, C.M., 1949, Principles of structural geology, 4th ed.: New York, John Wiley & Sons, Inc., 410 p.

Newhall, C.G., and Dzurisin, D., 1988, Historical unrest at large calderas of the world, v. 2: United States Geological Survey Bulletin 1855, p. 599-1108.

Newton, C. R., 1988, Significance of "Tethyan" fossils in the American Cordillera: Science, v. 242, no. 4877, p. 385-391.

*Nguyen Kim Lap, 1991, The seismotectonics of the Indochina Peninsula: Journal of Southeast Asian Earth Sciences, v. 6, no. 2, p. 55-61.

Nicolas, A., 1987, Asthenosphere structure and anisotropy beneath rifts: Reston, VA, U. S. Geological Survey Circular 956, p. 53-54.

Nicolas, A., 1989, Structures of ophiolites and dynamics of oceanic lithosphere: Dordrecht, Kluwer Academic Publishers, 367 p.

Nishimura, K., 1984, A schematic model of development of active continental margins as inferred from particular features of global-scale geoid undulations: Kyoto University, Disaster Prevention Research Institute Bulletin, v. 34, pt. 4, no. 309, p. 187-210.

Nishimura, K., 1985, On the character of the Meso-Cainozoic tectono-magmatic activity in East Asia: Kyoto University, Disaster Prevention Research Institute Bulletin, v. 35, pt. 3, no. 313, p. 41-54.

Nishimura, K., 1986, A numerical study of mantle tectonic flow as relevant to the Cenozoic structural development of the East Asiatic transition zone: Kyoto University, Disaster Prevention Research Institute Bulletin, v. 36, pts. 3-4, no. 320, p. 113-135.

Noe-Nygaard, A., 1974, Cenozoic to Recent volcanism in and around the North Atlantic Basin, *in* The ocean basins and margins, v. 2, the North Atlantic, A.E.M. Nairn and P.G. Smith, eds: New York City, Plenum Press, p. 391-443.

Norman, T.N., 1984, The role of the Ankara Melange in the development of Anatolia (Turkey), *in* Dixon, J.E., and Robertson, A.H.F., eds., The geological evolution of the eastern Mediterranean: Geological Society of London Special Publication no. 17, p. 441-447.

Norry, M.J., and Fitton, J.G., 1983, Compositional differences between oceanic and continental basic lavas and their significance, *in* Hawkesworth, C.J., and Norry, M.J., eds., Continental basalts and mantle xenoliths: Nantwich (U.K.), Shiva Publishing Limited, p. 5-19.

North, F.K., 1965, The curvature of the Antilles: Geologie en Mijnbouw, v. 44, no. 3, p. 73-86.

Northrop, J. W., III, and Meyerhoff, A. A., 1963, Validity of polar and continental movement hypotheses based on paleomagnetic studies: American Association of Petroleum Geologists Bulletin, v. 47, no. 4, p. 575-585.

Nunn, J.A., and Aires, J.R., 1988, Gravity anomalies and flexure of the lithosphere at the Middle Amazon basin, Brazil: Journal of Geophysical Research, v. 93, no. B1, p. 415-428.

*Nutalaya, P., Campbell, K.V., Macdonald, A.S., Aranyakanon, P., and Suthakorn, P., 1979, Review of the geology of the Thai tin fields: Geological Society of Malaysia Bulletin no. 11, p. 137-159.

*O'Brien, C.A.E., Gray, K.W., and Gillespie, I., 1961, The geological results of exploration for petroleum in western Papua: Geological Society of Australia Journal, v. 8, no. 1, p. 1-133.

Officer, C.B., Ewing, J.I., Hennion, J.F., Harkrider, D.G., and Miller, D.E., 1959, Geophysical investigations in the eastern Caribbean: summary of 1955 and 1956 cruises, *in* Ahrens, L.H., Press, F., Rankama, K., and Runcorn, S.K., eds., Physics and chemistry of the earth, v. 3: New York, Pergamon Press, p. 17-109.

O'Hara, M.J., and Yoder, H.S., Jr., 1967, Formation and fractionation of basic magmas at high pressures: Scottish Journal of Geology, v. 3, p. 67-117.

Oike, K., and Huzita, K., 1988, Relation between characteristics of seismic activity and neotectonics in Honshu, Japan: Tectonophysics, v. 148, no. 1/2, p. 115-130.

Okal, E.A., and Batiza, R., 1987, Hotspots: the first 25 years, in Keating, B. H., Fryer, P., Batiza, R., and Boehlert, G. W. eds., Seamounts, islands, and atolls: American Geophysical Union Geophysical Monograph no. 19, p. 1-11.

O'Keefe, J.A., 1965, Discussion, *in* Blackett, P. M. S., Bullard, E., and Runcorn, S. K., conveners, A symposium on continental drift: Royal Society of London Philosophical Transactions, series A, v. 258, no. 1088, p. 272-275.

*Okubo, Y., 1993, Temperature gradient map of the Japanese Islands: Journal of the Geothermal Research Society of Japan, v. 15, no. 1, p. 1-21.

Olafsson, I., Sundvor, E., Eldholm, O., and Grue, K., 1992, Mre margin: crustal structure from analysis of expanded spread profiles: Marine Geophysical Researches, v. 14, no. 2, p. 137-162.

Oliver, J.E., Dobrin, M., Kaufman, S., Meyer, R., and Phinney, R., 1976, Continuous seismic reflection profiling of the deep basement, Hardeman County, Texas: Geological Society of America Bulletin, v. 87, no. 11, p. 1537-1546.

Olsen, K.H., 1983, The role of seismic refraction data for studies of the origin and evolution of continental rifts: Tectonophysics, v. 94, no. 1-4, p. 349-370.

Olson, J., and Pollard, D.D., 1989, Inferring paleostresses from natural fracture patterns: a new method: Geology, v. 17, no. 4, p. 345-348.

Olson, P., Silver, P.G., and Carlson, R.W., 1990, The large-scale structure of convection in the earth's mantle: Nature, v. 344, no. 6271, p. 109-215.

Opdyke, N.D., and Hekinian, R., 1967, Magnetic properties of some igneous rocks from the Mid-Atlantic Ridge: Journal of Geophysical Research, v. 72, no. 8, p. 2257-2260.

Orowan, E., 1965, Convection in a non-newtonian mantle, continental drift, and mountain building, in a symposium on continental drift: Royal Society (London) Philosophical Transactions, ser. a, v. 258, no. 1088, p. 284-313.

Orowan, E., 1967, Seismic damping and creep in the mantle: Royal Astronomical Society, Geophysical Journal, v. 14, nos. 1-4, p. 191-218.

Ortlieb, L., Ruegg, J.C., Angelier, J., Colletta, B., Kasser, M., and Lesage, P., 1989, Geodetic and tectonic analyses along an active plate boundary: the central Gulf of California: Tectonics, v. 8, no. 3, p. 429-441.

Ouchi, T., Kawakami, H., Nagumo, S., Kasahara, J., and Koresawa, S., 1989, Microseismicity in the middle Okinawa Trough: Journal of Geophysical Research, v. 94, no. B8, p. 10601-10608.

Owen, E.W., 1975, Trek of the oil finders; a history of exploration for petroleum: American Association of Petroleum Geologists Memoir 6, 1647 pp.

Owen, H.G., 1976, Continental displacement and expansion of the earth during the Mesozoic and Cenozoic: Royal Society of London, Philosophical Transactions, series A, v. 281, no. 1303, p. 223-291.

Oxburgh, E.R., and Turcotte, D.L., 1968, Problem of high heat flow and volcanism associated with zones of descending mantle convective flow: Nature, v. 218, no. 5146, p. 1041-1043.

Ozima, M., Takigami, Y., and Kaneoka, I., 1980, $^{40}Ar/^{39}Ar$ geochronological studies on rocks of Deep Sea Drilling Project Sites 443, 445, and 446, *in* Klein, G. D., Kobayashi, K., et al., Initial reports of the Deep Sea Drilling Project, v. 58: Washington, D. C., U. S. Government Printing Office, p. 917-920.

*Page, B.G.N., Bennett, J.D., Cameron, N.R., Bridge, D.M., Jeffery, D.H., Keats, W., and Thaib, J., 1979, A review of the main structural and magmatic features of northern Sumatra: Geological Society of London Journal, v. 136, pt. 5, p. 569-579.

*Page, R.W., 1976, Geochronology of igneous and metamorphic rocks in the New Guinea Highlands: Canberra, Bureau of Mineral Resources, Geology and Geophysics Bull. 162, 117 p.

*Pal, D., 1980, Lithostratigraphy and structure of a part of the Krol belt in Naini Tal region of the Kumaun Himalaya, *in* Saklani, P. S., ed., Structural geology of the Himalaya: New Delhi, Today and Tomorrow's Printers and Publishers, p. 133-143.

*Pandey, O.P. , 1991, Terrestrial heat flow and lithospheric geothermal structure in New Zealand, *in* Čermák, V., and Rybach, L., eds., Terrestrial heat flow and the lithosphere structure: Berlin, Springer-Verlag, p. 338-380.

Panuska, B.C., 1990, An overlooked, world class Triassic flood basalt event: Geological Society of America, Abstracts with Programs, v. 22, no. 7, p. A168.

Papazachos, B.C., and Comninahis, P.E., 1977, Modes of lithospheric interaction in the Aegean area, *in* Biju-Duval,

B., and Montadert, L., eds., Structural history of the Mediterranean basins: Paris, Editions Technip, p. 319-331.

Parkinson, W.D., 1983, Introduction to geomagnetism: Edinburgh, Scottish Academic Press, 433 p.

Parnell, J., 1982, Paleomagnetic evidence for a large (ca. 2,000) sinistral offset along the Great Glen fault during Carboniferous time: comment: Geology, v. 10, no. 11, p. 605.

Pasquarè, G., Poli, S., Vezzoli, L., and Zanchi, A., 1988, Continental arc volcanism and tectonic setting in central Anatolia, Turkey: Tectonophysics, v. 146, no. 1-4, p. 217-230.

Paterson, S.R., Tobisch, O.T., and Morand, V.J., 1989, Eastward thrusting associated with the Wyangala batholith: *surge tectonics* in the Lachlan fold belt?, *in* Australasian tectonics, structural geology and tectonics conference, Kangaroo Island, February 1989: Geological Society of Australia Abstracts, v. 24, p. 116-117.

Pavlenkova, N.I., 1989, Struktura zemnoy i verkhney mantii i tektonika plit, *in* Beloussov, V.V., ed., Tektonosfera: yeye stroyeniye i razvitiye: Moscow, Akademiya Nauk SSSR, Mezhduvedomstvennyy Geofizicheskiy Komitet, Geodinamicheskiye Issledovaniya no. 13, p. 36-45.

Peano, G., 1895, Sopra lo spostamento del polo sulla terra: Accademia Scientifica Torrino, Atti della Classe di Scienze fisiche, matematiche e naturale, t. 30, p. 2-11.

Pearce, J.A., 1982, Trace element characteristics of lavas from destructive plate boundaries, *in* Thorpe, R.S., ed., Andesites. Orogenic andesites and related rocks: New York, John Wiley & Sons, p. 525-548.

Pearce, J.A., 1983, Role of the sub-continental lithosphere in magma genesis at active continental margins, *in* Hawkesworth, C. J., and Norry, M. J., eds., Continental basalts and mantle xenoliths: Nantwich (UK), Shive Publishing Ltd., p. 230-249.

Pearce, J.A., and Cann, J.R., 1973, Tectonic setting of basic volcanic rocks determined using trace element analyses: Earth and Planetary Science Letters, v. 19, p. 290-300.

Pearce, J.R., Morter, W.J., Uytterhoeven, J.B., and Lunsford, J.H., 1981, Crystallographic study of the distribution of cations in Y-type zeolites containing Fe(II) and FE(III): C. S. Faraday Trans. 1, v. 77, no. 4, p. 937-946.

Pearce, T.H., Gorman, B.E., and Birkett, T.C., 1977, The relationship between major element chemistry and tectonic environment of basic and intermediate volcanic rocks: Earth and Planetary Science Letters, v. 36, p. 121-132.

Peate, D.N., Mantovani, M.S. M., Hawkesworth, C.J., 1988, Geochemical stratigraphy of the Parana continental flood basalts: borehole evidence, *in* II Symposium about the evolution of the South Atlantic (IGCP Project No. 242), Part 1: Revista Brasileira de Geociencias, v. 18, no. 2, p. 212-221.

Pêcher, A., 1991, The contact between the Higher Himalaya crystallines and the Tibetan sedimentary series: Miocene large-scale dextral shearing: Tectonics, v. 10, no. 3, p. 587-598.

Pedlovsky, J., 1979, Geophysical fluid dynamics: New York, Springer-Verlag, 624 p.

Pekeris, C.L., 1936, Thermal convection in the interior of the earth: Royal Astronomical Society Monthly Notices, Geophysical Supplement, v. 3, p. 343-368.

Pelletier, B.,and Louat, R., 1989, Seismotectonics and present-day relative plate motions in the Tonga-Lau and Kermadec-Havre region: Tectonophysics, v. 165, no. 1-4, p. 237-250.

Pe-Piper, G., and Jansa, L. F., 1987, Geochemistry of late Middle Jurassic-Early Cretaceous igneous rocks on the eastern North American margin: Geological Society of America Bulletin, v. 99, no. 6, p. 803-813.

Petrushevsky, B.A., 1971, On the problem of the horizontal heterogeneity of the earth's crust and uppermost mantle in southern Eurasia: Tectonophysics, v. 11, no. 1, p. 29-60.

*Pettengill, G.H. , 1977., Orbiter radar mapper experiment, *in* Colin, L., and Hunten, D. M., eds., Pioneer Venus experiment descriptions: Space Science Reviews, v. 20, no. 4, p. 512-515.

*Pettengill, G.H., Ford, P.G., Johnson, W.T.K., Raney, R.K., and Soderblom, L.A., 1991, Magellan: radar performance and data products: Science, v. 252, no. 5003, p. 260-265.

*Pettengill, G.H., Horwood, G.F., and Keller, C.H., 1979, Pioneer-Venus orbiter radar mapper; design and operation: Institute of Electrical and Electronics Engineers, Transactions on Geoscience and Remote Sensing, GE 18, p. 2-32.

Petterson, M.G., Beddoe-Stephens, B., Millward, D., and Johnson, E. W., 1992, A pre-caldera plateau-andesite field in the Borrowdale Volcanic Group of the English Lake District: Geological Society of London Journal, v. 149, pt. 6, p. 889-906.

Phillips, R.P., 1964, Seismic refraction studies in Gulf of California, *in* van Andel, T.H., and Shor, G.G., Jr., eds., Marine geology of the Gulf of California: American Association of Petroleum Geologists Memoir 3, p. 90-121.

Phillips, W.J., and Kuckes, A.F., 1983, Electrical conductivity structure of the San Andreas fault in central California: Journal of Geophysical Research, v. 88, no. B9, p. 7467-7474.

Phinney, R. A., 1986, A seismic cross section of the New England Appalachians: the orogen exposed, *in* Barazangi, M., and Brown, L., eds., Reflection seismology: the continental crust: American Geophysical Union Geodynamics Series, v. 14, p. 157-172.

Phipps, S.P., 1988, Deep rifts as sources for intraplate magmatism in eastern North America: Nature, v. 334, no. 6177, p. 27-31.
Piccirillo, E.M., Melfi, A.J., Comin-Charamonti, P., Bellieni, G., Ernesto, M., Marques, L.S., Nardy, A.J.R., Pacca, I.G., Roisenberg, A., and Stolfa, D., 1988, Continental flood volcanism from the Paraná basin (Brazil), *in* Macdougall, J.D., ed., Continental flood basalts: Dordrecht, Kluwer Academic Publishers, p. 195-238.
Pickering, W.A., 1907, The place of origin of the moon--the volcanic problem: Journal Geology, v. 15, no. 1, p. 23-38.
*Pieters, P. E., Pigram, C. J., Trail, D. S., Dow, D. B., Ratman, N., and Sukamto, R., 1983, The stratigraphy of western Irian Jaya: Bandung, Geological Research and Development Centre Bulletin, no. 8, p. 14-48.
Pilger, R.H., Jr., 1982, The origin of hotspot traces: evidence from eastern Australia: Journal of Geophysical Research, v. 87, no. B3, p. 1825-1834.
Piqué, A., and Michard, A., 1989, Moroccan Hercynides: a synopsis. The Paleozoic sedimentary and tectonic evolution at the northern margin of West Africa: American Journal of Science, v. 289, no. 3, p. 286-330.
*Pitfield, P. E. J., Cobbing, E. J., Mallick, D. I. J., Clarke, M. C. G., Teoh, L. H., 1987, Granite provinces in the southeast Asian tin belt, *in* Horn, M. K., ed., Transactions of the Fourth Circum-Pacific Energy and Mineral Resources Conference, August 17-22, 1986, Singapore: Houston, Circum-Pacific Council for Energy and Mineral Resources, p. 575-589.
Platt, J.P., and Vissers, R.L.M., 1989, Extensional collapse of thickened continental lithosphere: a working hypothesis for the Alboran Sea and Gibraltar arc: Geology, v. 17, no. 6, p. 540-543.
Poliakov, A., and Podladchikov, Y., 1992, Diapirism and topography: Geophysical Journal International, v. 109, no. 3, p. 553-564.
Pol'kin, Ya.I., 1973, Tungusskaya sineklize, *in* Kiparisova, L.D., Radchenko, L.P., and Gorskiy, V.P., eds., Triasovaya sistema: Moscow, Izdatel'stvo Nedra, p. 240-254.
Popov, A.M., 1987, O prichinakh povysheniya elektroprobodnosti v zemnoy kore: Novosibirsk, Geologiya i Geofizika, no. 12, p. 56-64.
Pöschl, T., 1921, Über eine partibulane Lesung des biharmonischen Probleme für den Aussenraum de Ellipse: Zeitschrift für Mathematik, Bol. II, p. 89.
Posgay, K. Hegedus, E., Timar, Z., 1990, The identification of mantle reflections below Hungary from deep seismic profiling. Tectonophysics, v. 173, no. 1-4, p. 379-385.
*Pramumijoyo, S., and Sebrier, M., 1991, Neogene and Quaternary fault kinematics around the Sunda Strait area, Indonesia: Journal of Southeast Asian Earth Sciences, v. 6, no. 2, p. 137-145.
Pratsch, J.-C., 1982, Wedge tectonics along continental margins, *in* Watkins, J. L., and Drake, C. L., eds., Studies in continental margin geology: Tulsa, OK, American Association of Petroleum Geologists Memoir 34, p. 211-220.
Presnall, D.C., and Hoover, J.D., 1987, High pressure phase equilibrium constraints on the origin of mid-ocean ridge basalts, *in* Mysen, B.O., ed., Magmatic processes: physicochemical principles: University Park (PA), Geochemical Society, special publication no. 1, p. 75-89.
Press, F., 1966, Seismic velocities, *in* Clark, S.P., Jr., ed., Handbook of physical constants: Geological Society of America Memoir 97, p. 195-218.
Press, F., and Siever, R., 1974, Planet Earth: San Francisco, W.H. Freeman and Company, 945 p.
Prodehl, C., 1970, Seismic refraction study of crustal structure in the western United States: Geological Society of America Bulletin, v. 81, no. 9, p. 2629-2645.
Prodehl, C., 1979, Crustal structure of the western United States: U.S. Geological Survey Professional Paper 1034, 74 p.
Prodehl, C., and Pakiser, L.C., 1980, Crustal structure of the Southern Rocky Mountains from seismic measurements: Geological Society of America Bulletin, v. 91, no. 3, p. 147-155.
Pszczólkowski, A., 1987, Paleogeography and paleotectonic evolution of Cuba and adjoining areas during the Jurassic-Early Cretaceous: Annales Societatis Geologorum Poloniae, v. 57, p. 127-142.
*Pubellier, M., Quebral, R., Rangin, C., Deffontaines, B., Muller, C., Butterlin, J., and Manzano, J., 1991, The Mindanao collision zone: a soft collision event within a continuous Neogene strike-slip setting: Journal of Southeast Asian Earth Sciences, v. 6, no. 3-4, p. 239-248.
*Pudsey, C.J., Schroeder, R., and Skelton, P.W., 1986, Cretaceous (Aptian/Albian) age for island-arc volcanics, Kohistan, N. Pakistan, *in* Gupta, V. J., ed., Geology of western Himalayas: Delhi, Hindustan Publishing Corporation (I), p. 150-168.
Pushcharovskiy, Yu.M., 1972, Vvedeniye v tektoniku Tikhookeanskogo segmenta zemli: Adademiya Nauk SSSR, Geologicheskiy Institut Trudy, vyp. 234, 222 p.
*Radhakrishna, T., 1987, Collision tectonics in the Himalaya as evidenced by the Indus and Shyok rock assemblages: Tectonophysics, v. 134, no. 1-3, p. 1-16.
Ragland, P.C., and Rogers, J.J.W., eds., 1984, Basalts: a Hutchinson Ross Benchmark book: New York, Van Nostrand Reinhold Company, 429 p.

*Raina, B.N., 1978, A review of the stratigraphy and structure of Lesser Himalaya of Uttar Pradesh and Himachal Pradesh, *in* Saklani, P. S., ed., Tectonic geology of the Himalaya: New Delhi, Today & Tomorrow's Printers and Publishers, p. 79-112.

Raitt, R.W., 1956, Seismic-refraction studies of the Pacific Ocean basin, part I: crustal thickness of the central equatorial Pacific: Geological Society of America Bulletin, v. 67, no. 12, pt. 1, p. 1623-1639.

Ramakrishnan, M., and Anantha Iyer, G.V., 1987, Geochemistry of the Archaean Javanahalli amphibolites and evolution of Chitradurga supracrustal belt by continental rifting, *in* Saha, A.K., ed., Geological evolution of peninsular India. Petrological and structural aspects: Delhi (India), Recent Researches in Geology, v. 13, Hindustan Publishing Co., p. 61-76.

*Ramalingeswara Rao, B., 1992, Seismicity and geodynamics of the low- to high-grade transition zone of peninsular India: Tectonophysics, v. 201, no. 1-2, p. 175-185.

*Ramberg, H., 1952, The origin of metamorphic and metasomatic rocks: Chicago, The University of Chicago Press, 317 p.

*Ramberg, H., 1955, Natural and experimental boudinage and pinch-and-swell: Journal of Geology, v. 63, no. 6, p. 512-526.

*Ramberg, H., 1960, Relationships between length of arc and thickness of ptygmatically filled veins: American Journal of Science, v. 258, no. 1, p. 36-46.

*Ramberg, H., 1962, Contact strain and folding instability of a multilayered body under compression: Geologische Rundschau, v. 51, no. 2, p. 405-439.

*Ramberg, H., 1963a, Experimental study of gravity tectonics by means of centrifuged models: University of Uppsala, Institute of Geology Bulletin, v. 42, p. 1-97.

*Ramberg, H. 1963b, Fluid dynamics of viscous buckling applicable to folding of layered rocks: American Association of Petroleum Geologists Bulletin, v. 47, no. 3, p. 484-505.

*Ramberg, H., 1967, Gravity, deformation and the earth's crust: London, Academic Press, 214 p.

*Ramberg, H., 1968a, Instability of layered systems in the field of gravity, I: Physics of the Earth and Planetary Interiors, v. 1, p. 427-447.

*Ramberg, H., 1968b, Instability of layered systems in the field of gravity, II: Physics of the Earth and Planetary Interiors, v. 1, p. 448-474.

*Ramberg, H., 1972a, Theoretical models of density stratification and diapirism in the earth: Journal of Geophysical Research, v. 77, no. 5, p. 877-889.

*Ramberg, H., 1972b, Mantle diapirism and its tectonic and magmagenetic consequences: Physics of the Earth and Planetary Interiors, v. 5, p. 45-60.

*Ramberg, H., 1973, Model studies of gravity-controlled tectonics by the centrifuge technique, *in* de Jong, K. A., and Scholten, R., eds., Gravity and tectonics: New York, Wiley & Sons, Inc., p. 49-66.

*Ramberg, H., 1981, Gravity, deformation and the earth's crust, 2nd ed.: New York, Academic Press, 452 p.

*Ramberg, H., 1985, The velocity of nappes movement: a correction: Tectonophysics, v. 111, no. 1-2, p. 137-138.

*Ramberg, H., 1986, The stream function and Gauss' principle of least constraint: two useful concepts for structural geology: Tectonophysics, v. 131, no. 3-4, p. 205-246.

*Ramberg, H., 1989, A new numerical simulation method applied to spreading nappes: Tectonophysics, v. 162, no. 3-4, p. 173-192.

*Ramberg, H., 1991, Numerical simulation of spreading nappes, sliding against basal friction: Tectonophysics, v. 188, no. 1-2, p. 159-186.

Ramos, V.A., Niemeyer, H., Skarmeta, J., and Munoz, J., 1982, Magmatic evolution of the Austral Patagonian Andes: Earth-Science Reviews, v. 18, p. 411-443.

Rampino, M.R., and Stothers, R.B., 1988, Flood basalt volcanism during the past 250 million years: Science, v. 241, no. 4866, p. 663-668.

*Rangarao, A., 1983, Geology and hydrocarbon potential of a part of Assam-Arakan basin, *in* Bhandari, L. L., Venkatachala, B. S., Kumar, R., Nanjunda Swamy, S., Garga, P., and Srivastava, D. C., eds., Petroliferous basins of India: Dehra Dun, Petroleum Asia Journal, v. 6, no. 4, p. 127-158.

Rangin, C., 1984, Aspectos geodinámicos de la región noroccidental de México: Universidad Nacional Autónoma de México, Instituto de Geología, Revista, v. 5, no. 2, p. 186-194.

Rangin, C., 1986, Contribution à l'étude géologique du système cordillérain mésozoïque du nord-ouest due Mexique: une coupe de la Basse Californie centrale à la Sierra Madre Occidentale en Sonora: Société Géologique de France, Mémoire no. 148, nouvelle série---1985, 136 p.

*Rangin, C., Bellon, H., Bénard, F., Létouzey, J., Muller, C., and Sanudin, T., 1990, Neogene arc-continent collision in Sabah, northern Borneo (Malaysia): Tectonophysics, v. 183, no. 1-4, p. 305-319.

Ranneft, T. S. M., 1979, Segmentation of island arcs and application to petroleum geology: Journal of Petroleum Geology, v. 1, no. 3, p. 35-53.

Ratschbacher, L., Frisch, W., Neubauer, F., Schmid, S.M., and Neugebauer, J., 1989, Extension in compressional orogenic belts: the eastern Alps: Geology, v. 17, no. 5, p. 404-407.

*Ray, D. K., gen. convenor, 1982, Tectonic map of south and east Asia, 1st ed.: Commission for the Geological Map of the World, Subcommission for South and East Asia, 4 sheets and legend, scale, 1:5,000,000; commentary, 26 p.

Rayleigh, Lord (Strutt, J. W.), 1883, Investigation of the character of the equilibrium of an incompressible heavy fluid of variable density: London Mathematical Society Proceedings, v. 14, p. 170-177.

*Reed, T. B., IV, and Hussong, D., 1989, Digital image processing techniques for enhancement and classification of SeaMARC II side scan sonar imagery: Journal of Geophysical Research, v. 94, no. B6, p. 7469-7490.

Reid, I., Orcutt, J.A., and Prothero, W.A., 1977, Seismic evidence for a narrow zone of partial melting underlying the East Pacific Rise at 21° N: Geological Society of America Bulletin, v. 88, no. 5, p. 678-682.

Reiter, M., and Jessop, A.M., 1985, Estimates of terrestrial heat flow in offshore eastern Canada: Canadian Journal of Earth Sciences, v. 22, no. 10, p. 1503-1517.

*Remy, J.M., 1978, New results on lithostratigraphy and structure of west Nepal Himalaya, *in* Saklani, P. S., ed., Tectonic geology of the Himalaya: New Delhi, Today & Tomorrow's Printers and Publishers, p. 129-149.

Ren Jishun (Jen Chi-shun), Jiang Chunfa, Zhang Zhengkun, and Qin Deyu, 1987, Geotectonic evolution of China: Berlin, Springer-Verlag, 203 p.

Ren Jishun, and Chen Tingyu, 1989, Tectonic evolution of the continental lithosphere in eastern China and adjacent areas: Journal of Southeast Asian Earth Sciences, v. 3, no. 1-4, p. 17-27.

Rénard, V., and Allenou, J.-P., 1979, Sea Beam, multibeam echo-sounding in Jean Charcot: description, evaluation and first results: Monaco, International Hydrographical Review, v. 56, no. 1, p. 35-67.

Renne, P. R., and Basu, A. R., 1991, Rapid eruption of the Siberian traps flood basalts at the Permo-Triassic boundary: Science, v. 253, no. 5016, p. 176-179.

Reston, T.J., 1988, Evidence for shear zones in the lower crust offshore Britain: Tectonics, v. 7, no. 5, p. 929-945.

Revelle, R., 1958, The Downwind Expedition to the southeast Pacific (Abstract): American Geophysical Union Transactions, v. 39, no. 3, p. 528-529.

Revenaugh, J., Jordan, T. H., 1989, A study of mantle layering beneath the western Pacific: Journal of Geophysical Research., ser. B, Solid Earth and Planets, v. 94, no. 5, p. 5787-5813.

Reynolds, P.H., and Clay, W., 1977, Leg 37 basalts and gabbro: K-Ar and $^{40}Ar/^{39}Ar$ dating, *in* Aumento, F., Melson, W. G., et al., Initial reports of the Deep Sea Drilling Project, v. 37: Washington, D. C., U. S. Government Printing Office, p. 629-630.

Reynolds, R.T., Fricker, P.E., and Summers, A.L., 1966, Effects of melting upon thermal models of the earth: Journal of Geophysical Research, v. 71, no. 2, p. 573-582.

Rezanov, I. A., 1962, On continental drift (from paleomagnetic data): International Geology Review, v. 4, no. 10, p. 1118-1134.

Rezanov, I. A., 1968, Paleomagnetism and continental drift: International Geology Review, v. 10, no. 7, p. 765-776.

Rich, J. L., 1951, Origin of compressional mountains and associated phenomena: Geological Society of America Bulletin, v. 62, no. 10, p. 1179-1222.

Richard, M., Bellon, H., Maury, R.C., Barrier, E., and Juang, W.S., 1986, Miocene to recent calc-alkalic volcanism in eastern Taiwan: Tectonophysics, v. 125, no. 1-3, p. 87-124.

*Richards, J. P., and Kerrich, R., 1993, The Porgera gold mine, Papua New Guinea: magmatic hydrothermal to epithermal evolution of an alkalic-type precious metal deposit: Economic Geology, v. 88, no. 5, p. 1017-1052.

*Richards, J. R., and Wilmott, W. F., 1970, K-Ar ages of biotites from Torres Strait: Australian Journal of Science, v. 32, p. 369-370.

Richards, M.A., Duncan, R.A., and Courtillot, V.E., 1989, Flood basalts and hot-spot tracks: plume heads and tails: Science, v. 246, no. 4926, p. 103-107.

Ries, A.C., and Shackleton, R.M., 1990, Structures in the Huqf-Haushi uplift, east central Oman, *in* Robertson, A.H.F., Searle, M.P., and Ries, A.C., eds., The geology and tectonics of the Oman region: Geological Society of London Special Publication 49, p. 653-663.

Ringwood, A. E., 1979, Composition and origin of the earth, *in* McElhinny, M. W., ed., The earth; its origin, structure and evolution: London, Academic Press, p. 1-58.

Ritsema, A.R., 1957, On the focal mechanism of southeast Asian earthquakes, *in* Hodgson, J.H., ed., The mechanics of faulting, with special reference to the fault-plane work: Dominion Observatory (Ottawa) Publications, v. 20, no. 2, p. 341-368.

Ritsema, A.R., 1960, Further focal mechanism studies at De Bilt, *in* Hodgson, J.H., ed., A symposium on earthquake mechanism: Dominion Observatory (Ottawa) Publications, v. 24, no. 10, p. 355-358.

Robbins, S.L., 1968, Gravity and magnetic data over the Calaveras, Hayward, and Silver Creek faults near San Jose,

California, *in* Dickinson, W.R., and Grantz, A., eds., Proceedings of conference on geologic problems of San Andreas fault system: Stanford University Publications, Geological Sciences, v. 11, p. 216-217.

Roberts, A.M., Price, J.D., and Olson, T.S., 1990, Late Jurassic half-graben control on the siting and structure of hydrocarbon accumulations: UK/Norwegian Central graben, *in* Hardman, R.F.P., and Brooks, J., eds., Tectonic events responsible for Britain's oil and gas reserves: Geological Society of London, Special Publication no. 55, p. 229-257.

Roberts, R.J., Crittenden, M.D., Jr., Tooker, E.W., Morris, H.T., Hose, R.K., and Cheney, T.M., 1965, Pennsylvanian and Permian basins in northwestern Utah, northeastern Nevada, and southcentral Idaho: American Association of Petroleum Geologists Bulletin, v. 49, no. 11, p. 1926-1956.

*Robertson, A.H.F., and Degnan, P.J., 1993, Sedimentology and tectonic implications of the Lamayuru complex: deep-water facies of the Indian passive margin, Indus suture zone, Ladakh Himalaya, *in* Treloar, P. J., and Searle, M. P., eds., Himalayan tectonics: London, Geological Society of London Special Publication no. 74, p. 299-321.

Rocci, G., Bronner, G., and Deschamps, M., 1991, Crystalline basement of the West African craton, *in* Dallmeyer, R.D., and Lecorche, J.P., eds., The West African orogens and circum-Atlantic correlatives: Berlin, Springer-Verlag, p. 31-61.

Rockwell International, 1979, Geologic studies of the Columbia Plateau; a status report, October, 1979: Report RHO-BWT-ST-4, 541 p.

Rockwell, T., 1989, Behaviour of individual fault segments along the Elsinore-Laguna Salada fault zone, southern California and northern Baja California: implications for the characteristic earthquake model, *in* Schwqartz, D. P., and Sibson, R. H., eds., Fault segmentation and controls of rupture initiation and termination: United States Geological Survey Open-File Report 89-315, p. 288-308.

Rodgers, J., 1987, The Appalachian geosyncline, *in* Schaer, J.P., and Rodgers, J., eds., The anatomy of mountain ranges: Princeton University Press, p. 241-258.

Rodnikov, A.G., 1988, Correlation between the asthenosphere and the structure of the earth's crust in active margins of the Pacific Ocean: Tectonophysics, v. 146, no. 1-4, p. 279-289.

*Rogerson, R., and Williamson, A., 1986, Age, petrology and mineralisation associated with two Neogene intrusive types in the Eastern Highlands of Papua New Guinea: Geological Society of Malaysia Bulletin no. 20, p. 487-502.

Rosendahl, B.R., Groschel-Becker, H., Meyers, J., and Kaczmarick, K., 1991, Deep seismic reflection study of a passive margin, southeastern Gulf of Guinea: Geology, v. 19, no. 4, p. 291-295.

Rothpletz, A., 1903, Über die Möglichkeit den Genesatz zwischen Contractions - und Expansions - Theorie aufzuheben: Munich, Königliche Bayerische Akademie der Wissenschaften, Matematisch - naturwissenschäftliche Klasse, Sitzungsberichte, Bd. 32, p. 311-325.

Roure, F., and Sosson, M., 1986, Late Jurassic collision between a composite exotic block and the North American continent: a model for the Cordillera building: Société Géologique de France Bulletin, 8e série, v. 2, no. 6, p. 945-959.

Rousseau, A., 1987, A conjectural explanation of some seismic observations in the Mediterranean region. A new hypothesis and its quantitative analysis after comparing the seismicity, the geoid and the earth's surface present motions of the Mediterranean region. A possible key for seismic prediction in this region: Bordeaux, Université de Bordeaux I, Institut Universitaire de Technologie "A," Laboratoire de Génie Civil, Rept., 49 p. + 30 figs.

Rousseau, A., 1992, A new geodynamic model for the seismicity and in situ stresses of the Mediterranean region: Tectonophysics, v. 202, no. 2-4, p. 183-201.

Roussel, J., and Lesquer, A., 1991, Geophysics and the crustal structure of West Africa, *in* Dallmeyer, R. D., and Lécorché, J. P., eds., The West African orogens and circum-Atlantic correlatives: Berlin, Springer-Verlag, p. 9-28.

*Roybarman, A., 1983, Geology and hydrocarbon prospects of West Bengal, *in* Bhandari, L. L., Venkatachala, B. S., Kumar, R., Nanjunda Swamy, S., Garga, P., and Srivastava, D. C., eds., Petroliferous basins of India: Dehra Dun, Petroleum Asia Journal, v. 6, no. 4, p. 51-56.

Royden, L., Horvath, F., and Rumpler, J., 1983, Evolution of the Pannonian basin system, I. Tectonics: Tectonics, v. 2, no. 1, p. 63-90.

Runcorn, S. K., 1965, Changes in the convestion pattern in the earth's mantle and continental drift-evidence for a cold origin of the earth, in a symposium on continetal drift: Royal Society (London) Philosophical Transactions, ser. a, v. 258, no. 1088, p. 228-251.

Rundle, C.C., Brook, M., Snelling, N.J., Reynolds, P.H., and Barr, S.M., 1974, Radiometric age determinations, *in* Davies, T. A., Luyendyk, B. P., et al., Initial reports of the Deep Sea Drilling Project, v. 26: Washington, D. C., U.S. Government Printing Office, p. 513-516.

Rusby, J.S.M., 1970, A long range side-scan sonar for use in the deep sea: Monaco, International Hydrographical Review, v. 47, no. 1, p. 25-34.

Rusby, J.S.M., and Somers, M.L., 1977, The development of the "GLORIA" sonar system from 1970-75, *in* Angel, M. V., ed., Voyage of discovery: Deep-Sea Research, v. 24 (Supplement), p. 611-625.

Rusby, R.I., 1992, GLORIA and other geophysical studies of the tectonic pattern and history of the Easter microplate, southeast Pacific, *in* Parson, L.M., Murton, B.J., and Browning, P., eds., Ophiolites and their modern oceanic analogues: London, Geological Society of London Special Publication no. 60, p. 81-106.

Russell, I.G., 1893, Geologic reconnaissance in central Washington: U.S. Geological Survey, Bulletin 108, 108 p.

Russell, M.J., and Smythe, D.K., 1983, Origin of the Oslo Graben in relation to the Hercynian-Alleghenian orogeny and lithosphere rifting in the North Atlantic: Tectonophysics, v. 94, no. 1-4, p. 457-472.

Russo, R.M. and Silver, P.G., 1995, The Andes deep origins: Natural History, v. 104, no. 2, p. 52-59.

*Rutherford, K. J., and Qureshi, M. K., 1981, Geothermal gradient map of Southeast Asia, 2nd ed.: Singapore, Southeast Asia Petroleum Exploration Society and Indonesian Petroleum Association, scale: 1:10,000,000, 1 sheet.

Rutten, M.G., 1940, On Devonian limestones with Clathrodictyon cf. spatiosum and Heliolites porosus from eastern Borneo: Koninklijke Nederlandsche Akademie van Wetenschappen Proceedings, v. 43, no. 8, p. 1061-1064.

Rutten, M.G., 1969, The geology of western Europe: Amsterdam, Elsevier Publishing Company, 520 p.

Ryan, M.P., 1987, Neutral buoyancy and the mechanical evolution of magmatic systems, *in* Myson, B.O., ed., Magmatic processes: physico-chemical principles: University Park, PA, Geochemical Society Special Publication no. 1, p. 259-287.

Ryan, M.P., 1993, Neutral buoyancy and the structure of mid-ocean ridge magma reservoirs: Journal of Geophysical Research, v. 98, no. B-12, p. 22321-22338.

Rzhonsnitskaya, M.A., 1967, Devonian of the USSR, *in* Oswald, D.H., ed., International symposium on the Devonian System, Calgary, 1967, v. 1: Alberta Society of Petroleum Geologists, p. 331-348.

*Sanderson, G.A., 1966, Presence of Carboniferous in west Sarawak: American Association of Petroleum Geologists Bulletin, v. 50, no. 3, p. 578-580.

Sandwell, D.T., and MacKenzie, K.R., 1989, Geoid height versus topography for oceanic plateaus and swells: Journal of Geophysical Research, v. 94, no. B6, p. 7403-7418.

Sandwell, D.T., and Renkin, M.L., 1988, Compensation of swells and plateaus in the North Pacific: no direct evidence for mantle convection: Journal of Geophysical Research, v. 93, no. B-4, p. 2775-2783.

*Sanford, A.R., and Einarsson, P., 1982, Magma chambers in rifts, *in* Palmasson, G., ed., Continental and oceanic rifts: American Geophysical Union and Geological Society of America Geodynamics Series no. 8, p. 147-168.

Santos, J.O.S., and Loguercio, S.O.C., 1984, A parte meridional do craton Amazônico (escudo Brasil central) e Parecis-Alto Xingu, in Schobbenhaus, C., coordinator, Geologia do Brasil: Brasilia, Departamento Nacional da Produção Mineral, p. 93-127.

Sarewitz, D.R., and Lewis, S.D., 1988, Intra-arc basins in the central Philippines (II): regional extension, incipient sea-floor spreading, and implications for the creation of ophiolites (Abstract): Geological Society of America Abstracts with Programs, v. 20, no. 7, p. A127.

Sattayarak, N., 1983, Review of the continental Mesozoic stratigraphy of Thailand, *in* Nutalaya, P., ed.-in-chief, Proceedings of the workshop on stratigraphic correlation of Thailand and Malaysia, September 8-10, 1983, v. 1, Technical papers: Bangkok, Geological Society of Thailand and Geological Society of Malaysia, p. 127-148.

*Sattayarak, N., Srilulwong, S., and Pum-im, S., 1989, Petroleum potential of the Triassic pre-Khorat intermontane basin in northeastern Thailand, *in* Thanasuthipitak, T., and Ounchanum, P., eds., Proceedings of the International Symposium on Intermontane Basins: Geology and Resources: Chiang Mai (Thailand), Chiang Mai University, Faculty of Science, p. 43-58.

Saul, L.R., 1986, Pacific west coast Cretaceous molluscan faunas: time and aspect of changes, *in* Abbott, P. L., ed., Cretaceous stratigraphy, western North America: Los Angeles, CA, Society of Economic Paleontologists and Mineralogists, Pacific Section, Book 46, p. 131-135.

Savostin, L., Zonenshain, L., and Baranov, B., 1983, Geology and plate tectonics of the Sea of Okhotsk, *in* Hilde, T. W. C., and Uyeda, S., eds., Geodynamics of the western Pacific-Indonesian region: Washington, D.C., American Geophysical Union-Geological Society of America Geodynamics Series, v. 11, p. 189-221.

Sawlan, M.G., 1991, Magmatic evolution of the Gulf of California rift, *in* Dauphin, J.P., and Simoneit, B.R.T., eds., The Gulf and Peninsular province of the Californias: American Association of Petroleum Geologists Memoir 47, p. 301-369.

Saxena, M.N., 1971a, The crystalline axis of the Himalaya: the Indian shield and continental drift: Tectonophysics, v. 12, no. 6, p. 433-447.

Saxena, M.N., 1971b, Geological classification and the tectonic history of the Himalaya: Indian National ScienceAcademy Proceedings, part A, v. 37, no. 1, p. 28-54.

Saxena, M.N., 1978, The Himalaya--a modified arc system and the gravity tectonics, *in* Saklani, P. S., ed., Tectonic geology of the Himalaya: New Delhi, Today and Tomorrow's Printers and Publishers, p. 313-340.

*Saxena, M.N., 1986, Geodynamic synopsis of the Deccan Traps in relation to epochs of volcanic activity of the Indian shield, drift of the subcontinent, and the tectonic development of southern and southeastern Asia: Journal of Southeast Asian Earth Sciences, v. 1, no. 4, p. 205-213.

Saxena, M.N., Gupta, V.J., Meyerhoff, A.A., and Archbold, N.W., 1986, Tectonic and spatial relations between India and Asia since Proterozoic time, *in* Gupta, V. J., ed., Geology of western Himalayas: Delhi, Hindustan Publishing Corporation (I), p. 187-207.

Scheidegger, A.E., 1963, Principles of geodynamics, 2nd ed.: New York, Academic Press Inc., 362 p.

Scheidegger, A.E., 1982, Principles of Geodynamics: Berlin, Springer-Verlag, 3rd ed., 411 p.

Scheidegger, A.E., and Wilson, J.T., 1950, An investigation into possible methods of failure of the earth: Geological Association of Canada Proceedings, v. 3, p. 167-190.

Schwarz, G., and Wigger, P.J., 1988, Geophysical studies of the earth's crust and upper mantle in the Atlas system of Morocco, *in* Jacopshagen, V.H., ed., The Atlas system of Morocco. Studies on its geodynamic evolution: New York, Springer-Verlag, p. 339-357.

Sclater, J.G., Hawkings, J.W., Mammerickx, Jr., and Chase, C.G., 1972, Crustal extension between the Tonga and Lau Ridges: petrologic and geophysical evidence: Geological Society of America Bulletin, v. 83, no. 2, p. 505-517.

Searle, D.L., and Ba Than Haq, 1964, The Mogok Belt of Burma and its relationship to the Himalayan orogeny, *in* Kohli, G., Krishnaswamy, V. S., and Valdiya, K. S., eds., Himalayan and Alpine orogeny: 22nd International Geological Congress, India 1964, Proc. section 11, p. 132-160.

Searle, R.C., Rusby, R.I., Engeln J., Hey, R.N., Zukin, J., Hunter, P.M., LeBas, T.P., Hoffman, H.J., and Livermore, R., 1989, Comprehensive sonar imaging of the Easter Island microplate: Nature, v. 341, 26 October, p. 701-705.

Sears, F.W., Zemansky, M.W., and Young, H.D., 1974, College physics, 4th ed.: Reading (MA), Addison-Wesley Publishing Company, 751 p. + appendices.

Sears, J.W., Hyndman, D.W., and Alt, D., 1990, The Snake River plain, a volcanic hotspot track: Geological Society of America, Abstracts with Programs, v. 22, no. 3, p. 82.

Seeber, L., Armbruster, J.G., and Quittmeyer, R.C., 1981, Seismicity and continental subduction in the Himalayan arc, *in* Gupta, H.K., and Delany, F.M., eds., Zagros-Hindu Kush-Himalaya: geodynamic evolution: Washington, D.C., American Geophysical Union and Geological Society of America Geodynamics Series, v. 3, p. 215-242.

Segawa, J., and Tomoda, Y., 1976, Gravity measurements near Japan and study of the upper mantle beneath the oceanic trench-marginal sea transition zones, *in* Sutton, G.H., Manghnani, M.H., Moberly, R., and McAfee, E.U., eds., The geophysics of the Pacific Ocean basin and its margins: American Geophysical Union Geophysical Monograph 19, p. 35-52.

Seiders, V.M., 1988, Origin of conglomerate stratigraphy in the Franciscan assemblage and Great Valley sequence, northern California: Geology, v. 25, no. 9, p. 783-787.

Selverstone, J., 1988, Evidence for east-west crustal extension in the eastern Alps: implications for the unroofing history of the Tauern window: Tectonics, v. 7, no. 1, p. 87-105.

Şengör, A.M.C., 1979, The North Anatolian transform fault: its age, offset and tectonic significance: Geological Society of London Journal, v. 136, pt. 3, p. 269-282.

Şengör, A.M.C., 1984, The Cimmeride orogenic system and the tectonics of Eurasia: Boulder, CO, Geological Society of America Special Paper 195, 82 p.

Şengör, A.M.C., ed., 1989, Tectonic evolution of the Tethyan region: Dordrecht, Kluwer Academic Publishers, 698 p.

Şengör, A.M.C., 1990, Lithotectonic terranes and the plate tectonic theory of orogeny: a critique of the principles of terrane analysis: *in* Wiley, T.J., Howell, D.G., and Wong, F.L., eds., Terrane analysis of China and the Pacific rim: Houston (TX), Circum-Pacific Council for Energy and Mineral Resources Earth Science Series, v. 13, p. 9-44.

Şengör, A.M.C., and Hsü, K. ., 1984, The Cimmerides of eastern Asia: history of the eastern end of Palaeo-Tethys, *in* Buffetaut, E., Jaeger, J.-J., and Rage, J.-C., organizers, Paléogéographie de l' Inde, du Tibet et du sud-est Asiatique: confrontation des données paléontologiques avec modèles géodynamiques: Société Géologique de France Mémoire 147 (nouvelle série), p. 139-167.

Sevigny, J.H., Cook, F.A., and Clark, E.A., 1991, Geochemical signature and seismic stratigraphic setting of Coppermine basalts drilled beneath the Anderson Plains in northwest Canada: Canadian Journal of Earth Sciences, v. 28, no. 2, p. 184-194.

*Shanker, R., ed.-in-chief, 1991, Geothermal atlas of India: Calcutta, Geological Survey of India, 144 p.

*Sharma, K.K., 1989, Late Mesozoic-Cenozoic magmatism and crustal growth in the Himalaya, *in* Saklani, P. S., ed., Himalayan mountain building: New Delhi, Today & Tomorrow's Printers and Publishers, p. 107-125.

Shaw, H.R., Jackson, E.D., and Bargar, K.E., 1980, Volcanic periodicity along the Hawaiian-Emperor chain: American Journal of Science, v. 280-A, pt. 2, p. 667-708.

Shih, J., and Molnar, P., 1975, Analysis and implications of the sequence of ridge jumps that eliminated the Surveyor transform fault: Journal of Geophysical Research, v. 80, no. 35, p. 4815-4824.

Shimazu, M.S., 1988, Kaynozoyskiy vulkanizm i tektonicheskoye razvitiye Yaponskogo mora i yego obramleniya, *in* Natal'in, B.A., Tuyezov, I.K., and Uyeda, S., eds., Tektonika Vostochno-Aziaskikh okrainykh morey: Moscow, Adademiya Nauk SSSR, Mezhduvedomstvennyy Geofizicheskiy Komitet, Geodinamicheskiye Issledovaniya no. 11, p. 81-88.

Shipboard Scientific Party, 1972, Site 80, *in* Hays, J. D., Cook, H. E., III, et al., Initial reports of the Deep Sea Drilling Project, v. 9: Washington, D. C., U. S. Government Printing Office, p. 401-447.

*Shipboard Scientific Party, 1974, Site 238, *in* Fisher, R L., Bunce, E.T., et al., Initial reports of the Deep Sea Drilling Project, v. 24: Washington, D. C., U. S. Government Printing Office, p. 469-587.

Shurbet, D.H., and Cebull, S.E., 1971, Crustal low-velocity layer and regional extenion in Basin and Range province: Geological Society of America Bulletin, v. 82, no. 11, p. 3241-3243.

*Shuvalov, V.F., 1982, Paleogeografiya i istoriya razvitiya ozernykh sistem Mongolii v yurskoye i melovoye vremya, *in* Martinson, G. G., ed., Mezozoyskiye ozernyye basseyny Mongolii: Leningrad, Izdatel'stvo "Nauka," p. 5-80.

Sial, A.N., 1974, Petrology and tectonic significance of the post-Paleozoic basaltic rocks of northeast Brazil: unpubl. Ph.D. dissertation, University of California at Davis, 405 p.

Silberling, N.J., Jones, D.L., Blake, M.C., Jr., and Howell, D.G., 1987, Lithotectonic terrane map of the western conterminous United States: United States Geological Survey, Miscellaneous Field Studies Map, MF-1874-C, 1 sheet, scale, 1:2,500,000 + 20 p.

Silberling, N.J., Jones, D.L., Monger, J.W.H., and Coney, P.J., 1992, Lithotectonic terrane map of the North American Cordillera: Reston, VA, U.S. Geological Survey Miscellaneous Investigations Series Map I-2176, scale, 1:5,000,000, 2 sheets.

Silver, E.A., and Moore, J.C., 1978, The Molucca Sea collision zone, Indonesia: Journal of Geophysical Research, v. 83, no. B4, p. 1681-1691.

Silver, E.A., Breen, N.A., Prasetyo, H., and Hussong, D.M., 1986, Multibeam study of the Flores backarc thrust belt, Indonesia: Journal of Geophysical Research, v. 91, no. B3, p. 3491-3500.

Simonen, A., and Mikkola, A., 1980, Finland, *in* Geology of the European countries. Denmark, Finland, Iceland, Norway, Sweden: London, Graham and Trotman Ltd., p. 51-126.

*Simpson, G. G., 1943, Mammals and the nature of continents: American Journal of Science, v. 241, no. 1, p. 1-31.

Sinton, J.B., and Hussong, D.M., 1983, Crustal structure of a short length transform fault in the central Mariana Trough, *in* Hayes, D.E., ed., The tectonic and geologic evolution of southeast Asian seas and islands, part 2: American Geophysical Union Geophysical Monograph 27, p. 236-254.

Smith, R.B., 1978, Seismicity, crustal structure, and intraplate tectonics of the interior of the Western Cordillera, *in* Smith, R.B., and Easton, G.P., eds., Cenozoic tectonics and regional geophysics of the Western Cordillera: Geological Society of America Memoir 152, p. 111-114.

Smoot, N.C., 1989, North Atlantic fracture-zone distribution and patterns shown by multibeam sonar: Geology, v. 17, no. 12, p. 1119-1122.

Smoot, N.C., 1990a, Mariana Trough morphology by multi-beam sonar: Geo-Marine Letters, v. 10, p. 137-144.

Smoot, N.C., 1990b, North Atlantic fracture-zone distribution and patterns shown by multibeam sonar: reply: Geology, v. 18, no. 9, p. 912-914.

Smoot, N.C., 1991, The Mariana Trench convergent margin at the Magellan Seamounts: geomorphology: Marine Technology Society '91 Proceedings, v.1, p. 85-91,New Orleans, 11 November.

Smoot, N.C., 1993, Geomorphic effects of seamounts in NW Pacific subduction zones: Geological Society of America Abstracts with Programs, v. 25, A-379, Boston, 28 October.

Smoot, N.C., 1994a, The relation of seamounts to interplate deformation in the North Pacific: EOS, Transactions, American Geophysical Union, v. 75, no. 44, p. 582, San Francisco, 5 December.

Smoot, N.C., 1994b, Plate-wide Pacific trends-orthogonal fracture intersections: EOS, Transactions, American Geophysical Union, v. 75, no. 25, p. 69, Hong Kong, 19 July.

Smoot, N.C., 1995a, The Chinook Trough: a trans-Pacific fracture zone, *in* Proceedings of the Third Thematic Conference on Remote Sensing for Marine and Coastal Environments, v. II, p. 539-550.

Smoot, N.C., 1995b, Mass wasting and subaerial weathering in guyot formation: the Hawaiian and Canary Ridges as examples: Geomorphology, v. 14, p. 29-41.

Smoot, N.C., 1996 (in press), Aligned aseismic buoyant highs, across-trench deformation, clustered volcanoes, and deep earthquakes are not aligned with the current plate-tectonic theory: Geomorphology.

Smooth, N.C., in review, Convergent Margins: NAVOCEANO Technical Report 315, 234 p.

Smoot, N.C. and King, R.E., 1993, The Darwin-Rise demise: Geological Society of America Abstracts with Programs, v. 25, A-375.

Smoot, N.C. and Meyerhoff, A.A., 1995, Tectonic fabric of the North Atlantic Ocean floor: speculation vs. reality: Journal of Petroleum Geology, v. 18, no. 2, p. 207-222.

Smoot, N.C. and Leybourne, B.A., in review, The South Adriatic basin: a vortex structure on the world-encircling vortex street: MTS Journal.

Snider-Pellegrini, A., 1858, La Création et ses Mystères dévoilés. Paris: Franck et Dentu.

Sokolov, V.N., Krasil'shchikov, A.A., Livshitz, Yu. Ya., and Semevsky, D.V., 1973, Structural history of Spitsbergen and adjoining shelves, *in* Pitcher, M.G., ed., Arctic Geology: American Association of Petroleum Geologists, Memoir 19, p. 269-274.

Solomon, S.C., Head, J.W., Kaula, W.M., McKenzie, D., Parsons, B., Phillips, R.J., Schubert, G., and Talwani, M., 1991, Venus tectonics: initial analysis from Magellan: Science, v. 252, no. 5003, p. 297-312.

Solov'yev, V. O., 1966, Permian and Late Cretaceous-early Paleocene volcanism of the south Maritime region: Academy of Sciences of the USSR Doklady (American Geological Institute, English translation, Washington, D.C.), v. 170, p. 63-65.

Somers, M.L., Carson, R.M., Revie, J.A., Edge, R.H., Barrow, B.J., and Andrews, A.G., 1978. GLORIA II -- an improved long range side-scan sonar, *in* Oceanology International 1978, IEEE/IERE sub-conference on offshore instrumentation and communications, Technical Session J: London, BPS Publications Ltd., p. 16-24.

Speed, R., Elison, M.W., and Heck, F.R., 1988, Phanerozoic tectonic evolution of the Great Basin, *in* Ernst, W.G., ed., Metamorphism and crustal evolution of the western United States. Rubey volume 7: Englewood Cliffs (NJ), Prentice Hall, p. 573-605.

*Spizharskiy, T.N., ed., 1966, Tektonicheskaya karta SSSR: Moscow, Ministerstvo Geologii SSSR (VSEGEI), scale, 1:7,500,000, 2 sheets.

*Spizharskiy, T.N., ed., 1977, Paleotektonicheskiye karty SSSR, tom I. Pozdniy proterozoy, kembriy: Moscow, Ministerstvo Geologii SSSR (VSEGEI), 20 sheets, scale, 1:5,000,000 + legend.

*Spizharskiy, T.N., ed., 1979, Paleotektonicheskiye karty SSSR, tom II. Ordovik, silur, devon: Moscow, Ministerstvo Geologii SSSR (VSEGEI), 20 sheets, scale, 1:5,000,000 + legend.

*Spring, L., Bussy, F., Vannay, J.-C., Huon, S., and Cosca, M.A., 1993, Early Permian granitic dykes of alkaline affinity in the Indian High Himalaya of upper Lahul and SE Zanskar: geochemical characterization and geotectonic implications, *in* Treloar, P., J., and Searle, M. P., eds., Himalayan tectonics: London, Geological Society of London Special Publication no. 74, p. 251-264.

*Srinivasan, M.S., and Azmi, R.J., 1976, New developments in the late Cenozoic lithostratigraphy of Andaman-Nicobar Islands, Bay of Bengal, *in* Srinivasan, M. S., ed., VI Indian Colloquium on Micropaleontology and Stratigraphy Proceedings: Varanasi (India), Banaros Hindu University, Department of Geology, p. 302-327.

Stacey, F.D., 1981, Cooling of the earth--a constraint on paleotectonic hypotheses, *in* O'Connell, R. J. and Fyfe, R.S., eds., Evolution of the earth: Washington, D.C., American Geophysical Union and Geological Society of America Geodynamics Series, v. 5, p. 272-276.

Stanley, S.M., 1986, Earth and life through time: New York, W.H. Freeman and Company, 690 p.

Stanton, R.L., 1978, Mineralization in island arcs, with particular reference to the south-west Pacific region. Introductory lecture, William Smith Meeting, Geological Society of London, February 1977: Proceedings of the Australasian Institute of Mining and Metallurgy, v. 268, p. 9-19.

Stanton, R.L., 1994, Ore Elements in Arc Lavas: Oxford, Clarendon Press, 391 p.

St. Onge, M.R., Lucas, S.B., Scott, D.J., and Bégin, N.J., 1989, Evidence for the development of oceanic crust and for continental rifting in the tectonostratigraphy of the early Proterozoic Cape Smith Belt: Geoscience Canada, v. 16, p. 119-122.

Starosel'tsev, V.S., 1980, Tektonicheskoye polozheniye basal'tovykh plato: Geologiya i Geofizika, no. 8, p. 3-10.

*Stauffer, P.H., and Snelling, N., 1977, A Precambrian trondhjemite boulder in Palaeozoic mudstone of N. W. Malaya: Geological Magazine, v. 114, pt. 6, p. 479-482.

Stegena, L., 1990, Correlation between seismicity and horizontal variation of heat flow density in central Europe: Tectonophysics, v. 179, no. 1-2, p. 55-61.

Steinitz, G., and Maze, W.B., 1984, Kar ages on hornblende-andesites from the Sierra de Perija, western Venezuela, *in* Bonini, W.E., Hargraves, R.B., and Shagam, R., eds., The Caribbean-South American plate boundary and regional tectonics: Geological Society of America, Memoir 162, p. 283-286.

Stéphan, J.F., Blanchet, R., Rangin, C., Pelletier, B. Létouzey, J., and Muller, C., 1986, Geodynamic evolution of the Taiwan-Luzon-Mindoro belt since the late Eocene: Tectonophysics, v. 125, no. 1-3, p. 245-268.

*Stephenson, D., and Marshall, T.R., 1984, The petrology and mineralogy of Mt. Popa volcano and the nature of the late-Cenozoic Burma volcanic arc: Geological Society of London Journal, v. 141, pt. 4, p. 747-762.

Stern, T.A., and Davey, F.J., 1989, Crustal structure and origin of basins formed behind the Hikurangi subduction zone, New Zealand, *in* Price, R.A., ed., Origin and evolution of sedimentary basins and their energy and mineral resources: American Geophysical Union Geophysical Monograph 48, p. 73-85.

Stevenson, A.C., 1945, Complex potentials in two-dimensional elasticity: Royal Society of London Proceedings, series A, v. 181, p. 129-179.

Stevenson, D.J., and Turner, J.S., 1979, Fluid models of mantle convection, *in* McElhinny, M. W., ed., The earth: its origin, structure and evolution: London, Academic Press, p. 227-263.

Stewart, J.H., 1978, Basin-range structure in western North America: a review, *in* Smith, R.B., and Eaton, G.P., eds., Cenozoic tectonics and regional geophysics of the Western Cordillera: Geological Society of America Memoir 152, p. 1-31.

Stöcklin, J., 1984, Orogeny and Tethys evolution in the Middle East: an appraisal of current concepts, *in* International Geological Congress Reports, Moscow 1984, Colloquium 05, Tectonics of Asia, v. 5: Moscow, Izdatel'stvo Nauka, p. 65-84.

Stöcklin, J., 1989, Tethys evolution in the Afghanistan-Pamir-Pakistan region, *in* Şengör, A. M. C., ed., Tectonic evolution of the Tethyan region: Dordrecht, Kluwer Academic Publishers, p. 241-264.

Storetvedt, K.M., 1990, The Tethys sea and the Alpine-Himalayan orogenic belt: mega-elements in a new global tectonic system: Physics of the Earth and Planetary Interiors, v. 62, p. 141-184.

Storetvedt, K.M., 1992, Rotating plate: new concept of global tectonics, *in* Chatterjee, S., and Hotton, N., III, eds., New concepts in global tectonics: Lubbock, TX, Texas Tech University Press, p. 203-220.

Strick, E., 1984. Implications of Jeffreys-Lomnitz transient creep: Journal of Geophysical Research, v. 89, no. B1, p. 437-451.

Strong, D.F., 1979, Proterozoic tectonics of northwestern Gondwanaland: new evidence from eastern Newfoundland: Tectonophysics, v. 54, no. 1/2, p. 81-101.

Su Yangzhen, 1988, Devonian paleogeography of northeastern China, *in* McMillan, N.J., Embry, A.F., and Glass, D.J., eds., Devonian of the world. Proceedings of the Second International Symposium on the Devonian System, v. I: regional syntheses: Canadian Society of Petroleum Geologists, Memoir 14, p. 607-618.

Subbarao, K.V., ed., 1988, Deccan flood basalts: Geological Society of India Memoir 10, 393 p.

Suess, E., 1885, Das Antlitz der Erde, Band 1: Wien (Vienna), F. Tempsky, 779 p.

Suess, E., 1901, Das Antlitz der Erde, dritter Band, erste Hälfte: Wien (Vienna), F. Tempsky, 508 p.

Suess, E., 1909, Das Antlitz der Erde: Prague, F. Tempsky.

*Sukhanov, A.L., Pronin, A.A., Burba, G.A., Nikishin, A.M., Kryuchkov, V.P., Basilevsky, A. T., Markov, M.S., Kuzmin, R.O., Bobina, N.N., Shashkina, V. P., Slyuta, E.N., and Chernaya, I.M., 1989, Geomorphic/geologic map of the northern hemisphere of Venus: United States Geological Survey Miscellaneous Investigations Series. Map I-2059, scale, 1:15,000,000, 1 sheet.

*Sulidi-Kondrat'yev, Ye. D., Ivanov, G.M., and Bush, V.A., 1989, Main features of the structure and geodynamics of Laos: Geotectonics, v. 23, p. 543-549.

Sun, S.S., 1980, Lead isotopic study of young volcanic rocks from mid-ocean ridges, ocean islands and island arcs: Royal Society of London Philosophical Transactions, ser. A, v. 297, p. 409-445.

Sun, S.S., 1984, Geochemical characteristics of Archaean ultramafic and mafic volcanic rocks: implications for mantle composition and evolution, *in* Kröner, A., Hanson, G.N., and Goodwin, A.M., eds., Archaean geochemistry. The origin and evolution of the Archaean continental crust: Berlin, Springer-Verlag, p. 25-46.

Sun Ruomei, Liu Futian, and Liu Jianhua, 1991, Seismic tomography in Sichuan: Acta Geophysica Sinica, v. 34, no. 6, p. 708-716.

Sun Ruomei, Zhao Yanlai, and Mei Shirong, 1993, Seismic tomographic image in Bohai Sea and its adjacent area: Acta Geophysica Sinica, v. 36, no. 1, p. 44-54.

Sundvoll, B., Neumann, E.R., Larsen, B., and Tuen, E., Age relations among Oslo Rift magmatic rocks; implications for tectonic and magmatic modelling: Tectonophysics, v. 178, no. 1, p. 67.

Suo Shutian, and Zhou Hanwen, 1992, Transpressive deformation across Tongbo-Dabie orogenic belt: China University of Geosciences Journal, v. 3, no. 1, p. 1-8.

*Suppe, J., 1980, A retrodeformable cross section of northern Taiwan: Geological Society of China Proceedings no. 23, p. 46-55.

*Suppe, J., 1981, Mechanics of mountain building and metamorphism in Taiwan: Geological Society of China Memoir 4, p. 67-89.

Sutter, J.F., and Smith, T.E., 1979, $^{40}Ar/^{39}Ar$ ages of diabase intrusions from Newark trend basins in Connecticut and Maryland: initiation of central Atlantic rifting: American Journal of Science, v. 279, no. 7, p. 808-831.

Sutton, G.H., Maynard, G.L., and Hussong, D.M., 1971, Widespread occurrence of a high-velocity basal layer in the Pacific crust found with repetitive sources and sonobuoys, *in* Heacock, J.G., ed., The structure and physical properties of the earth's crust: American Geophysical Union Geophysical Monograph 14, p. 193-209.

Swanson, D.A., and Wright, T.L., 1981, The regional approach to studying the Columbia River basalt group, *in*

Subbarao, K. V., Sukheswala, R. N., eds., Deccan volcanism and related basalt provinces in other parts of the world: Memoir Geological Society of India, v. 3, p. 58-80.

Swanson, D.A., Wright, T.L., Hooper, P.R., and Bentley, R.D., 1979, Revisions in stratigraphic nomenclature of the Columbia River Basalt Group: U.S. Geological Survey Bulletin 1457-G, 59 p.

Sychev, P. M., 1985, Anomal'nyye zony v. verkhney mantii, mekhanizm obrazovaniya i rol' v razvitii struktur zemnoy kory: akademiya Nauk SSSR, Tikhookeanskaya Geologiya, no. 6, p. 25-35.

Sychev, P.M., and Sharaskin, A.Y., 1984, Heat flow and magmatism in the NW Pacific back-arc basins, *in* Kokelaar, B.P., and Howells, M.F., eds., Marginal basin geology: Geological Society of London Special Publication 16, p. 173-181.

Sykes, LR., 1967, Mechanism of earthquakes and nature of faulting on the mid-ocean ridge: Journal of Geophysical Research, v. 72, no. 8, p. 2131-2153.

Sykes, L.R., 1978, Intraplate seismicity, reactivation of preexisting zones of weakness, alkaline magmatism, and other tectonism postdating continental fragmentation: Review of Geophysics and Space Physics, v. 16, no. 4, p. 621-688.

Talandier, J., 1989, Submarine volcanic activity. Detection, monitoring, and interpretation: EOS, v. 70, no. 18, p. 561, 568-569.

Talandier, J., and Kuster, G.T., 1976, Seismicity and submarine volcanic activity in French Polynesia: Journal of Geophysical Research, v. 81, no. 5, p. 936-948.

Talbot, C.J., and Slunga, R., 1989, Patterns of active shear in Fennoscandia, *in* Gregersen, S., and Basham, P.W., eds., Earthquakes at North-Atlantic passive margins: neotectonics and postglacial rebound: Dordrecht, Kluwer Academic Publishers, p. 441-466.

Talwani, M., Le Pichon, X., and Ewing, M., 1965, Crustal structure of the mid-ocean ridges, 2. Computed model from gravity and seismic refraction data: Journal of Geophysical Research, v. 70, no. 2, p. 341-352.

Tammann, G., 1903, Ueber Aenderungen des Aggregatzustandes bei der Abkühlung eines Weltkörpers: St. Petersburg, Königliche Akademie der Wissenschaften, Permanenton Seimischen Kommission, Verhandlungen Bd. I, Lieferung 2, p. 321-328.

*Tan, B.K., 1976, Tectonic development of peninsular Malaysia, *in* Discussion meeting: geology of the South China Sea area including its continental rim, Abstracts of papers: Ipoh, Geological Society of Malaysia, p. 2.

*Tan, B.K., 1981, The margins of the Central Belt, Peninsular Malaysia (Abstract), *in* Regional geology seminar '81--Abstracts of papers: Geological Society of Malaysia Newsletter, v. 7, no. 2, p. 46-47.

*Tan, B.K., 1984, The tectonic framework and evolution of the Central Belt and its margins, Peninsular Malaysia: Geological Society of Malaysia Bulletin no. 17, p. 307-322.

*Tan, B.K., and Khoo, T.T., 1983, Ultramafic rocks in peninsular Malaysia and their tectonic implications, *in* Fourth Regional Conference on Geology, Mineral and Energy Resources, Southeast Asia, Manila 1981, Proceedings: Manila, Geological Society of the Philippines, p. 259-264.

*Tan, B.K., and Khoo, T.T., 1993, Clinopyroxene composition and tectonic setting of the Bentong-Raub belt, Peninsular Malaysia: Journal of Southeast Asian Earth Sciences, v. 8, no. 1-4, p. 539-545.

*Tan, D.N.K., 1982, Geological map of Sarawak, 1st ed.: Kuching, Geological Survey of Malaysia (Malaysia Jabatan Penyiasatan Kajibumi), 2 sheets, scale, 1:500,000.

Tanaka, K.L., Shoemaker, E.M., Ulrich, G.E., and Wolfe, E.W., 1986, Migration of volcanism in the San Francisco volcanic field, Arizona: Geological Society of America Bulletin, v. 97, no. 2, p. 129-141.

*Taner, I., and Meyerhoff, A.A., 1990, Petroleum at the roof of the world: the geological evolution of the Tibet (Qinghai-Xizang) Plateau, part I: Journal of Petroleum Geology, v. 13, no. 2, p. 157-178; part II: v. 13, no. 3, p. 289-314.

*Tang Hanzhang, ed.-in-chief, 1986, Metamorphic map of China: Beijing, Geological Publishing House, scale, 1:4,000,000, 2 sheets.

*Tang Yaoqing, ed.-in-chief, 1988, Geological map of Qinghai-Xizang (Tibet) Plateau and adjacent areas: Chengdu, Chengdu Institute of Geology and Mineral Resources and Chinese Academy of Geological Sciences, scale, 1:500,000, 6 sheets + explanatory note, 91 p.

Tarakhovskiy, A.N., Fishman, M.V., Shkola, I.V., and Andreichev, V.D., 1983, Vozrast trappov zemli Frantsa-Iosifa, *in* Kavardin, G.I., ed., Prognozirovaniye i otsenka nikelnosnosti novykh rudhnykh rayonov na SevereSibirskoy platformy: Leningrad, Ministerstvo Geologii SSSR, PGO Severmorgeologiya, p. 100-107.

*Tate, R.B., 1991, Cross-border correlation of geological formations in Sarawak and Kalimantan: Geological Society of Malaysia Bulletin no. 28, p. 63-95.

Tate, R.B., and Hon, V., 1991, The oldest rocks in Borneo: a note on the Tuang Formation, west Sarawak and its importance in relation to the presence of a "basement" in west Borneo: Geological Society of Malaysia Newsletter, v. 17, no. 5, p. 221-224.

*Taylor, B., Klaus, A., Brown, G.R., Moore, G.F., Okamura, Y., and Murakami, F., 1991, Structural development of Sumisu rift, Izu-Bonin arc: Journal of Geophysical Research, v. 96, no. B-10, p. 16113-16129.

Taylor, E.M., 1990, Volcanic history and tectonic development of the central High Cascade Range, Oregon (Abstract): EOS, v. 71, no. 9, p. 312.

Taylor, F.B., 1910, Bearing of the Tertiary mountain belt on the origin of the earth's plan: Geological Society of America Bulletin, v. 21, no. 2, p. 179-226.

Taylor, G.I., 1950, The instability of liquid surfaces when accelerated in a direction perpendicular to their planes, I: Royal Society of London Proceedings, ser. A, v. 201, p. 192-196.

Taylor, S.R., 1979, Chemical composition and evolution of the continental crust: the rare earth element evidence, *in* McElhinny, M.W., ed., The earth: its origin, structure and evolution: London, Academic Press, p. 353-376.

Taylor, S.R., 1989a, Geophysical framework of the Appalachians and the adjacent Grenville province, *in* Pakiser, L.C., and Mooney, W.D., eds., Geophysical framework of the continental United States: Boulder, CO, Geological Society of America Memoir 172, p. 317-348.

Taylor, S.R., 1989b, Chemical composition and evolution of the continental crust: the rare earth element evidence, *in* McElhinny, M.W., ed., the earth: its origin, structure and evolution: London Academic Press, p. 353-376.

Taylor, S.R., Töksoz, M.N., and Chaplin, M.P., 1980, Crustal structure of the northeastern United States: contrasts between Grenville and Appalachian provinces: Science, v. 208, no. 4444, p. 595-597.

Teng, L.S., 1992, Geotectonic evolution of Tertiary continental margin basins of Taiwan: Petroleum Geology of Taiwan, no. 27, p. 1-19.

Tesakov, Yu.I., ed., 1990, Stratigrafiya paleozoyskikh otlozheniy yugo-vostoka Zapadno-Sibirskoy plity: Novosibirsk, Akademiya Nauk SSSR, Sibirskoye Otdeleniye, Instituta Geologii i Geofiziki Trudy, v. 766, 216 p.

Tetyayev,M.M., 1934, Osnovy geotektoniki: Leningrad, GONTI, 288 p.

Tezcan, A.K., 1979, Geothermal studies, their present status and contribution to heat flow contouring in Turkey, *in* Čermák, V., and Rybach, L., eds., Terrestrial heat flow in Europe: New York, Springer-Verlag, p. 283-292.

*Thienprasert, A., and Raksaskulwong, M., 1984, Heat flow in northern Thailand: Tectonophysics, v. 103, no. 1-4, p. 217-233.

Thirlwall, M.F., 1981, Implications for Caledonian plate tectonic models of chemical data from volcanic rocks of the British Old Red Sandstone: Geological Society of London, Journal, v. 138, pt. 2, p. 123-128.

Thompson, G.A., and McCarthy, J., 1990, A gravity constraint on the origin of highly extended terranes: Tectonophysics, v. 174, no. 1-2, p. 197-206.

Thompson, G., Bryan, W.B., and Humphris, S.E., 1989, Axial volcanism on the East Pacific Rise, *in* Saunders, A.D., and Norry, M.J., eds., Magmatism in the ocean basins: Geological Society of London Special Publication no. 42, p. 181-200.

Thompson, J.B., Jr., Robinson, P., Clifford, T.N., and Trask, N.J., Jr., 1968, Nappes and gneiss domes in west-central New England, *in* Zen, E. A., White, W. S., Hadley, J. B., and Thompson, J. B., Jr., eds., Studies of Appalachian geology, northern and maritime: New York, Interscience Publishers, p. 203-218.

Thompson, R.N., Morrison, M.A., Dickin, A.P., and Hendry, G.L., 1983, Continental flood basalts...arachnids rule OK?, *in* Hawkesworth, C.J., and Norry, M.J., eds., Continental basalts and mantle xenoliths: Nantwich (UK), Shiva Publishing Ltd., p. 158-185.

*Thuc, Dao Dinh, 1980, Mezozoyskiye vulkanicheskiye kompleksy Severnogo V'yetnama: Novosibirsk, Geologiya i Geofizika, no. 7, p. 62-66.

*Thuc, Pham Van, and Lap, Nguyen Kim, 1981, Seismicity of South-east Asia: Acta Geophysica Polonica, v. 29, no. 1, p. 63-73.

*Tien, Phan Cu, 1991, Stratigraphic correlation of Permian and Triassic in Vietnam, *in* Kotaka, T., Dickins, J. M., McKenzie, K. G., Mori, K., Ogasawara, K., and Stanley, G. D., Jr., eds., Shallow Tethys 3: Sendai (Japan), Saito Ho-on Kai Special Publication no. 3, p. 359-370.

*Tien, Phan Cu, 1993, Upper Carboniferous-Permian volcano-sedimentary formations in Vietnam and adjacent territories, *in* Findlay, R.H., Unrug, U., Banks, M.R., and Veevers, J.J., eds., Gondwana Eight: assembly, evolution and dispersal: Rotterdam, A. A. Balkema, p. 299-306.

*Tikoff, T., and Teyssier, C., 1992, Crustal-scale, en echelon "P-shear" tensional bridges: a possible solution to the batholithic room problem: Geology, v. 20, no. 10, p. 927-930.

Timofeyev, P.P., Udintsev, G.B., Agapova, G.V., Antipov, M.P., Boyko, N.I., Yeremeyev, V. V., Yefimov, V.N., Kurentsova, N.A., and Lyubimov, V.V., 1992, Equatorial segment of the Mid-Atlantic Ridge as a possible structural barrier between the North and South Atlantic: USSR Academy of Sciences Transactions (Doklady), Earth Science Sections, v. 312, no. 3, p. 133-135.

Timoshenko, S. and Goodier, J.N., 1951, Theory of Elasticity: New York, McGraw-Hill, 2nd ed.

Tobin, D.G., Ward, P.L., and Drake, C.L., 1969, Microearthquakes in the rift valley of Kenya: Geological Society of America Bulletin, v. 80, no. 10, p. 2043-2046.

Tobisch, O.T., and Paterson, S.R., 1990, The Yarra granite: an intradeformational pluton associated with ductile thrusting, Lachlan fold belt, southeastern Australia: Geological Society of America Bulletin. v. 102, no. 6, p. 693-703.

Tolan, T.L., Reidel, S.P., Hooper, P.R., Beeson, M.H., Anderson, J.L., Fecht, K.R., and Swanson, D.A., 1989, Revisions to the estimates of the areal extent and volume of the Columbia River Basalt Group, *in* Reidel, S.P., and Hooper, P.R., eds., Volcanism and tectonism in the Columbia River flood-basalt province: Geological Society of America Special Paper 239, p. 1-20.

*Tongkul, F., 1991, Tectonic evolution of Sabah, Malaysia: Journal of Southeast Asian Earth Sciences, v. 6, no. 3-4, p. 395-405.

*Tongue, J.A., 1992, Tomographic study of local earthquake data from the Lake Bogoria region of the Kenya Rift valley: Geophysical Journal International, v. 109, no. 2, p. 249-258.

Tongue, J.A., Maguire, P.K.H., and Young, P.A., V., 1992, Seismicity distribution from temporary earthquake recording networks in Kenya: Tectonophysics, v. 204, no. 1-2, p. 71-79.

Trehu, A.M., Klitgord, K.D., Sawyer, D.S., and Buffler, R.T., 1989, Atlantic and Gulf of Mexico continental margins, *in* Pakiser, L.C., and Mooney, W.D., eds., Geophysical framework of the continental United States: Boulder, CO, Geological Society of America Memoir 172, p. 349-382.

*Trinh Xuan Ben, Nguyen Van Hoai, and Visnepskaya, Yu. E., 1993, The behaviour of rare metals in Piaoac granite massif, Vietnam: Journal of Southeast Asian Earth Sciences, v. 8, no. 1-4, p. 383-386.

Tritton, D.J., 1988, Physical fluid dynamics, 2nd ed.: Oxford, Clarendon Press, 519 p.

Tritton, D.J., and Davies, P.H., 1985, Instabilities in geophysical fluid dynamics, *in* Swinney, H.L., and Gollub, J.P., eds., Hydrodynamic instabilities and the transition to turbulence, 2nd ed.: Berlin, Springer-Verlag, Topics in Applied Physics, v. 45, p. 228-270.

Trümpy, R., 1960, Paleotectonic evolution of the central and western Alps: Geological Society of America Bulletin, v. 71, no. 6, p. 843-907.

Trurnit, P., 1988, Growth and breakup of supercontinents and evolution of oceans and continental margins during the global tectonic megacycles: GeoJournal, v. 17, no. 1, p. 37-73.

Trurnit P., 1991, Space-time analysis of the structural pattern of continents: GeoJournal, v. 25, no. 4, p. 305-358.

Trusheim, F., 1960, Mechanism of salt migration in northern Germany: American Association of Petroleum Geologists Bulletin, v. 44, no. 9, p. 1519-1540.

Tsai, Y.B., 1986, Seismotectonics of Taiwan: Tectonophysics, v. 125, no. 1-3, p. 17-37.

*Tuezov, I.K., 1988, Heat flow map of the Pacific Ocean and the adjacent continents: Khabarovsk, USSR Academy of Sciences, Far East Branch, Institute of Tectonics and Geophysics, 33 p. + 12 sheets, scale, 1:10,000,000.

Turcotte, D.L., and Oxburgh, E.R., 1969, Convection in a mantle with variable physical properties: Journal of Geophysical Research, v. 74, no. 6, p. 1458-1474.

Turcotte, D.L., and Oxburgh, E.R., 1976, Stress accumulation in the lithosphere: Tectonophysics, v. 35, no. 1-3, p. 183-199.

Turner, D.L., and Jarrard, R.D., 1982, K/Ar dating of the Cook-Austral Island chain: a test of the hotspot hypothesis: Journal of Volcanology and Geothermal Research, v. 12, p. 187-220.

Twato, O., 1975, Laramide (Late Cretaceous-early Tertiary) orogeny in the Southern Rocky Mountains, *in* Curtis, B.F., ed., Cenozoic history of the Southern Rocky Mountains: Geological Society of America Memoir 144.

*Tyler, G.L., Ford, P.G., Campbell, D.B., Elachi, C., Pettengill, G.H., and Simpson, R.A., 1991, Magellan: electrical and physical properties of Venus' surface: Science, v. 252, no. 5003, p. 265-270.

Tyrrell, G.W., 1930, The principles of petrology, 2nd ed.: New York, E.P. Dutton and Company Inc. Publishers, 349 p.

Tyrrell, G.W., 1937, Flood basalts and fissure eruption: Bulletin Volcanologique, ser. 2, no. 1, p. 89-111.

Udintsev, G.B., 1972, Tikhiy okean: geomorfologiya i tektonika dna Tikhogo okeana: Moscow, Nauka, 394 p.

Udintsev, G.B., and Kurentsova, N.A., 1993, Igneous activity in the equatorial segment of the Mid-Atlantic Ridge: USSR Academy of Sciences Transactions (Doklady), Earth Science Sections, v. 315, no. 6, p. 165-168.

Udintsev, G.B., Kurentsova, N.A., Pronina, N.V., Smirnova, S.B., and Ushakova, M.G., 1992, Finds of continental rocks and sediments of anomalous age in the equatorial segment of the Mid-Atlantic Ridge: USSR Academy of Sciences Transactions (Doklady), Earth Science Sections, v. 312, no. 3, p. 111-114.

Uliana, M.A., Biddle, K.T., and Cerdan, J., 1989, Mesozoic extension and the formation of Argentine sedimentary basins, *in* Tankard, A.J., and Balkwill, H.R., eds., Extensional tectonics and stratigraphy of the North Atlantic margins: American Association of Petroleum Geologists Memoir 46, p. 599-614.

Umpleby, D.C., 1979, Geology of the Labrador shelf: Geological Survey of Canada Paper 79-13, 34 p.

Urey, H.C., 1952, The planets: their origin and development: New Haven, Conn., Yale University Press, 245 p.

*U.S. Geological Survey, 1981, Altimetric and shaded relief map of Venus: United States Geological Survey Miscellaneous Investigations Servies, Map I-1324, scale, 1:50,000,000, 1 sheet.

*U.S. Geological Survey, 1984, Topographic and shaded relief maps of Venus: United States Geological Survey Miscellaneous Investigations Series, Map I-1562, scale, 1:50,000,000, 2 sheets.

*US/USSR Joint Working Group 1989, Maps of part of the northern hemisphere of Venus. A joint U.S./U.S.S.R.

mapping project: United States Geological Survey Miscellaneous Investigations Series, Map I-2041, scale, 1:15,000,000, 3 sheets.

Utada, M., 1980, Hydrothermal alterations related to igneous activity in Cretaceous and Neogene formations in Japan, *in* Ishihara, S., and Takenouchi, S., eds., Grantic magmas and related mineralization: Tokyo, Society of Mining Geologists of Japan, Mining Geology Special Issue no. 8, p. 67-83.

Utsu, T., 1971, Seismological evidence for anomalous structure of island arcs with special reference to the Japanese region: Reviews of Geophysics and Space Physics, v. 9, no. 4, p. 839-890.

Vallier, T.L., 1974, Volcanogenic sediments and their relation to landmass volcanism and sea floor-continent movements, western Indian Ocean, Leg 25, Deep Sea Drilling Project, *in* Simpson, E.S.W., Schlich, R., et al., Initial reports of the Deep Sea Drilling Project, v. 25: Washington, D.C., U.S. Government Printing Office, p. 515-542.

van Andel, T.H., Heath, G.R., Bennett, R.H., Bukry, J.D., Charleston, S., Cronan, D.S., Dinkelman, M.G., Kaneps, A.G., Rodolfo, K.S., Yeats, R.S., 1973, San Cristobal, Panama to Honolulu, Hawaii; February-March, 1971, *in* Initial Reports of the Deep Sea Drilling Project, Leg 16: Washington, D.C. (Govt. Printing Office), 949 p.

van Bemmelen, R.W., 1931, Magma- und Krusten-Undation: VI Nederland-Indisch Natuurwetenschappen Congress, Bandung, Proceedings, p. 645-653.

van Bemmelen, R.W., 1933, The undation theory of the development of the earth's crust: 16th International Geological Congress, Washington, D.C., Report 2, p. 965-982.

*van Bemmelen, R. W., 1949, The geology of Indonesia, v. IA. General geology of Indonesia and adjacent archipelagoes; v. IB, Portfolio volume: The Hague, Government Printing Office/Martinus Nijhoff, 732 p. + 60-p. portfolio.

van Bemmelen, R.W., 1976, Plate tectonics and the undation model: a comparison: Tectonophysics, v. 32, no. 3/4, p. 145-182.

van der Linden, W.J.M., 1980, Walvis Ridge: a piece of Africa?: Geology, v. 8, no. 9, p. 417-421.

Van der Voo, R., and Scotese, C., 1981, Paleomagnetic evidence for a large (ca. 2,000 km) sinistral offset along the Great Glen fault during Carboniferous time: Geology, v. 9, no. 12, p. 583-589.

Van der Westhuizen, W.A., de Bruiyn, H., and Meintjes, P.G., 1991, the Ventersdorp Supergroup: an overview: Journal of African Earth Sciences, v. 13, no. 1, p. 83-105.

Van Houten, F.B., 1976, Late Variscan nonmarine deposits, northwestern Africa: implications for pre-drift North Atlantic reconstructions: American Journal of Science, v. 276, no. 6, p. 671-693.

Van Houten, F.B., 1977, Triassic-Liassic deposits of Morocco and eastern North America: comparison: American Association of Petroleum Geologists Bulletin, v. 61, no. 1, p. 76-99.

Van Schmus, W.R., and Bowring, S.A., 1980, Chronology of igneous events in the Wopmay orogen, Northwest Territories, Canada: Geological Society of America, Abstracts with Programs, v. 12, no. 7, p. 540.

Van Schmus, W.R., and Hinze, W.J., 1985, The Midcontinent rift system: Annual Review of Earth and Planetary Sciences, v. 13, no. 345-383.

Van Schmus, W.R., Green, J.C., and Halls, H.C., 1982, Geochronology of Keweenawan rocks of the Lake Superior region: a summary, *in* Wold, R.J., and Hinze, W.J., eds., Geology and tectonics of the Lake Superior basin: Geological Society of America, Memoir 156, p. 165-171.

Vandamme, D., Courtillot, V., and Besse, J., 1991, Paleomagnetism and age determinations of the Deccan Traps (India): results of a Nagpur-Bombay traverse and review of earlier work: Reviews of Geophysics, v. 29, no. 2, p. 159-190.

*Vannay, J.-C., and Spring, L., 1993, Geochemistry of the continental basalts within the Tethyan Himalaya of Lahul-Spiti and SE Zanskar, northwest India, *in* Treloar, P.J., and Searle, M.P., eds., Himalayan tectonics: London, Geological Society of London Special Publication no. 74, p. 237-249.

Veevers, J.J., ed., 1984, Phanerozoic history of Australia: Oxford Geological Sciences Series, v. 2, 418 p.

Velasque, P.C., Ducasse, L., Muller, J., and Scholten, R., 1989, The influence of inherited extensional structure on the tectonic evolution of an intracratonic chain: the example of the western Pyrenees: Tectonophysics, v. 162, no. 3-4, p. 243-264.

Vening Meinesz, F.A., 1933, The mechanism of mountain formation on geosynclinal belts: Koninklijke Akademie van Wetenschapperde Amsterdam, Proceedings of the Section of Sciences, v. 36, no. 4, p. 372-377.

Vening Meinesz, F., 1934, Gravity expeditions at sea, 1923-1932, v. 2: Delft, Netherlands Geodetic Commission, 208 p.

Vening Meinesz, F.A., 1954, Indonesian archipelago: a geophysical study: Geological Society of America Bulletin, v. 65, no. 2, p. 143-164.

Verhoogen, J., Turner, F.J., Wahrhaftig, C., and Fyfe, W.S., 1970, The earth. An introduction to physical geology: New York, Holt, Rinehart and Winston, Inc., 748 p.

*Verma, R.K., 1991, Geodynamics of the Indian Peninsula and the Indian plate margin: Rotterdam, A. A. Balkema, 357 p.

Verrall, P., 1989, Speculations on the Mesozoic-Cenozoic tectonic history of the United States, *in* Tankard, A.J., and Balkwill, H.R., eds., Extensional tectonics and stratigraphy of the North Atlantic margins: American Association of Petroleum Geologists Memoir 46, p. 615-631.

Vertlib, M.B., 1978, K opredeleniyu glubin ochagov zemletryaseniy v Pribaykal'ye: Geologiya i Geofizika, no. 9, p. 141-146.

Vetter, S.K., and Fodor, R.V., 1983, Petrology of Early Cretaceous South Atlantic basaltic crust offshore southeastern Brazil: Abstracts, Geological Society of America, v. 15, no. 2, p. 63.

Vine, F.J., and Matthews, D.H., 1963, Magnetic anomalies over oceanic ridges: Nature, v. 199, no. 4897, p. 947-949.

*Visser, W.A., 1968, A geological reconnaissance in the Nassau Range, West New Guinea: discussion: Geologie en Mijnbouw, v. 47, no. 1, p. 47-48.

*Visser, W.A., and Hermes, J.J., 1962, Geological results of the exploration for oil in Netherlands New Guinea: Koninklijk Nederlands Geologisch Mijnbouwkundig Genootschap Verhandlingen, Geologische Serie no. 20, 265 p.

Viswanathan, S., and Chandrasekharam, D., 1981, Geochemical comparison of the Siberian and Deccan traps, *in* Subbarao, K.V., Sukheswala, R.N., eds., Deccan volcanism and related basalt provinces in other parts of the world: Memoir Geological Society of India, v. 3, p. 461-471.

Vlasov, G.M., and Belova, M.B., eds., 1964, Geologiya SSSR, t. 31. Kamchatka, Kuril'skiye i Komandorskiye ostrova, chast' I. Geologicheskoye Opisaniye: Moscow, Nedra, 728 p.

Vogt, P.R., 1974, The Iceland phenomenon: imprints of a hotspot on the ocean crust, and implications for flow below the plates, *in* Kristjansson, L., ed., Geodynamics of Iceland and the North Atlantic area: Dordrecht, D. Reidel Publishing Company, p. 105-126.

Vogt, P.R., Schneider, E.D., and Johnson, G.L., 1969, The crust and upper mantle beneath the sea, *in* Hart, P.J., ed., The earth's crust and upper mantle: American Geophysical Union Geophysical Monograph 13, p. 556-617.

Volterra, V., 1895, Sulla teoria dei moti del polo terrestre: Accademia delle Scienze di Torino, Classe di Scienze Fisiche, Matematiche e Naturali, Atti, v. 30, p. 301-306.

Von Damm, K.L., 1990, Seafloor hydrothermal activity: black smoker chemistry and chimneys: Annual Review of Earth and Planetary Sciences, v. 18, p. 173-204.

Von Herzen, R.P., Cordery, M.J., Detrick, R.S., and Fang, C., 1989, Heat flow and the thermal origin of hot spot swells: the Hawaiian swell revisited: Journal of Geophysical Research, v. 94, no. B10, p. 13783-13799.

von Humboldt, A., 1801, Esquisse d'un tableau géologique de l'Amérique meridionale: Paris, Journal de Physique, Chimie, Histoire Naturelle et Arts, t. 53, p. 30-60.

*Voskresenskiy, I.A., Kravchenko, K.N., Movshovich, E.V., and Sokolov, B.A., 1971, Ocherk geologii Pakistana: Moscow, Izdatel'stvo Nedra, 168 p.

*Wajzer, M.R., Barber, A.J., Hidayat, S., and Suharsono, 1991, Accretion, collision and strike-slip faulting: the Woyla Group as a key to the tectonic evolution of north Sumatra: Journal of Southeast Asian Earth Sciences, v. 6, no. 3-4, p. 447-461.

*Wakhaloo, S.N., 1979, The Panjal volcanics of Kashmir, *in* Gupta, V. J., ed., Upper Palaeozoics of the Himalayas: Delhi, Hindustan Publishing Corporation (I), p. 189-194.

Walker, G.P.L., 1989, Gravitational (density) controls on volcanism, magma chambers and intrusions: Australian Journal of Earth Sciences, v. 36, no. 2, p. 149-165.

Wallace, T.C., 1992, The 1992 Landers earthquake sequence: Arizona Geology, v. 22, no. 3, p. 1-4.

Wallbrecher, E., 1988, A ductile shear zone in the Panafrican basement on the northwestern margin of the West African craton (Sirwa dome, central Anti-Atlas), *in* Jacobshage, V.H., ed., The Atlas system of Morocco. Studies on its geodynamic evolution: New York, Springer-Verlag, p. 19-42.

Walter, A., and Sharpless, S., 1987, Crustal velocity structure of the Sur-Obispo (Franciscan) terrane between San Simeon and Santa Maria, California (Abstract): EOS, v. 68, no. 44, p. 1366.

*Wan Tianfeng, and Zhu Hong, 1991, Tectonic events of late Proterozoic-Triassic in South China: Journal of Southeast Asian Earth Sciences, v. 6, no. 2, p. 147-157.

Wang Chengshan, Zhao Zhichao, Cheng Xueru, and Zhao Wanli, 1993, Volcanic rocks of marginal block type and surge tectonics of continental margin: Chengdu College of Geology Journal, v. 20, no. 4, p. 55-66.

*Wang Erchie, and Chu, J.J., 1988, Collision tectonics in the Cenozoic orogenic zone bordering China, India and Burma: Tectonophysics, v. 147, no. 1-2, p. 71-84.

*Wang Hongzhen, chief comp., 1985, Atlas of the palaeogeography of China; Institute of Geology, Chinese Academy of Geological Sciences, Wuhan College of Geology: Beijing, Cartographic Publishing House, xiii p., 143 p. (charts), 85 p. (Chinese text), 28 p. (English text) + 25 p. (appendix).

*Wang Hongzhen, and Liu Benpei, 1980, Historical geology: Beijing, Science Press, 352 p.

*Wang Jingming, 1987, The Fenwei rift and its recent periodic activity: Tectonophysics, v. 133, no. 3-4, p. 257-275.

Wang Jun, Huang Shangyao, Huang Geshan, and Wang Jiyang, 1986, Basic characteristics of the earth's temperature distribution in southern China: Acta Geological Sinica (Trial English Edition), v. 60, no. 3, p. 91-106.

Wang Liankiu, Zhai Bin, Zhu Weifang, Cai Yuangji, and Li Tongjin, 1980, Characteristics and melting experiments of granites in southern China, *in* Ishihara, S., and Takenouchi, S., eds., Granitic magmas and related mineralization: Tokyo, Society of Mining Geologists of Japan, Mining Geology Special Issue no. 8, p. 29-38.

Wang Shangwen, ed., 1983, The petroleum geology of China: Beijing, Petroleum Press, 348 p. (in Chinese)

Wang Yunliang, Hughes, S.S., Tong Chunhang, Ziong Shunhua, Li Juchu, Zhou Rongsheng, and Li Jianlin, 1989, Geochemistry and petrology of Emeishan basalts and subcontinental mantle evolution in southwestern China: Chinese Journal of Geochemistry, v. 8, no. 1, p. 37-53.

Wanless, R.K., Stevens, R.D., Lachance, G.R., and Edmonds, C.M., 1968, Age determinations and geological studies. K-Ar isotopic ages, report 8: Geological Survey of Canada Paper 67-2, pt. A, p. 140-141.

Wannesson, J., Icart, J.C., and Ravat, J., 1991, Structure and evolution of two adjoining segments of the West African margin from deep seismic profiling, *in* Meissner, R., Brown, L., Dürbaum, H.-J., Franke, W., Fuchs, K., and Seifert, F., eds., Continental lithosphere: deep seismic reflections: Washington, D.C., American Geophysical Union Geodynamics Series, v. 22, p. 275-289.

Wark, D.A., 1991, Oligocene ash flow volcanism, northern Sierra Madre Occidental: role of mafic and intermediate-composition magmas in rhyolite genesis: Journal of Geophysical Research, v. 96, no. B-8, p. 13,389-13,411.

Wark, D.A., Kempter, K.A., and McDowell, F.W., 1990, Evolution of waning, subduction-related magmatism, northern Sierra Madre Oriental: Geological Society of America Bulletin, v. 102, no. 11, p. 1555-1564.

Wasserburg, G.J., and DePaolo, D.J., 1979, Models of earth structure inferred from neodymium and strontium isotopic abundances: Washington, National Academy of Sciences Proceedings, v. 76, p. 3591-3598.

Watanabe, T., Langseth, M.G., and Anderson, R.N., 1977, Heat flow in back-arc basins of the western Pacific, *in* Talwani, M., and Pitman, W.C., III, eds., Island arcs, deep sea trenches, and back-arc basins: American Geophysical Union Maurice Ewing Series 1, p. 137-161.

Waters, A.C., 1962, Basalt magma types and their tectonic associations: Pacific northwest of the United States, *in* MacDonald, G.A., and Kuno, H., eds., The crust of the Pacific basin: American Geophysical Union, Geophysical Monograph, no. 6, p. 158-170.

Watkins, N.D., and Baksi, A.K., 1974, Magnetostratigraphy and oroclinal folding of the Columbia River, Steens, and Owyhee basalts in Oregon, Wahington, and Idaho: erratum: American Journal of Science, v. 274, no. 7, p. 832.

Watters, B.R., Krynauw, J.R., and Hunter, D.R., 1991, Volcanic rocks of the Proterozoic Jutulstraumen Group in western Dronning Maud Land, Antarctica, *in* Thomson, M.R.A., Crame, J.A., and thomson, J.W., eds., Geological evolution of Antarctica: Cambridge, Cambridge University Press, p. 41-46.

Watts, A.B., Cochran, J.R., Patriat, P., and Doucoure, M. 1985, A bathymetry and altimetry profile across the Southwest Indian Ridge crest at 31 degrees S latititude: Earth and Planetary Science Letters, v. 73, no. 1, p. 129-139.

Watts, A.B., Weissel, J.K., Duncan, R.A., and Larson, R.L., 1988, Origin of the Louisville Ridge and its relationship to the Eltanin fracture system: Journal of Geophysical Research, v. 93, no. B4, p. 3051-3077.

Weaver, B.L., and Tarney, J., 1983, Chemistry of the sub-continental mantle: inferences from Archaean and Proterozoic dykes and continental flood basalts, *in* Hawkesworth, C.J., and Norry, M.J., eds., Continental basalts and mantle zenoliths: Nantwich (U.K.), Shiva Publishing Limited, p. 209-229.

Webster's Third New International Dictionary, 1971, Gove, P.B., ed.: Springfield, MA, G. & C. Merriam Company, 2662 pp.

Weertman, J., 1971, Theory of water-filled crevasses in glaciers applied to vertical magma transport beneath the oceanic ridges: Journal of Geophysical Research, v. 76, no. 5, p. 1171-1183.

Weertman, J., and Weertman, J.R., 1964, Elementary dislocation theory: New York, Macmillan and Company, 213 p.

Wegener, A., 1912, Die Entstehung der Kontinente: Geologrsche Rundschau, v. 3, no. 276-292.

Wegener, A., 1915, Die Entstehung der Kontinente und Ozeane: Sammlung Vieweg, no. 23, 94 p.

Wegmann, C.E., 1930, Über Diapirismus (besonders im Grundgebirge): Commission Géologique de Finlande Bulletin, v. 92, p. 58-76.

Wegmann, C.E., 1935, Zur Deutung der Migmatite: Geologische Rundschau, Bd. 26, Hft. 5, p. 305-350.

*Weippert, D., Wittekindt, H., and Wolfart, R., 1970, Zur geologischen Entwicklung von Zentral- und Südafghanistan: Hannover, Beihefte zum Geologischen Jahrbuch, Hft. 92, 99 p.

Weissel, J.K., 1981, Magnetic lineations in marginal basins of the western Pacific, *in* Vine, F. J., and Smith, A.G., organizers, Extensional tectonics associated with convergent plate boundaries: Royal Society of London Philosophical Transactions, ser. A, v. 300, no. 1454, p. 223-245.

*Weissel, J.K., Childers, V.A., and Karner, G.D., 1992, Extensional and compressional deformation of the lithosphere in the light of ODP drilling in the Indian Ocean, *in* Duncan, R.A., Rea, D.K., Kidd, R.B., Rad, U. von, and Weissel, J.D., eds., Synthesis of results from scientific drilling in the Indian Ocean: Washington, D.C., American Geophysical Union Geophysical Monograph 70, p. 127-156.

Wellman, H.W., 1966, Active wrench faults of Iran, Afghanistan and Pakistan: Geologische Rundschau, v. 55, no. 2, p. 716-735.

Wellman, P., 1979, Bouger gravity anomaly map, *in* BMR earth science atlas: Canberra, Bureau of Mineral Resources, 2 sheets (1 sheet of text + map sheet), scale, 1:10,000,000.

*Wellman, P., 1989, Upper mantle, crust, and geophysical volcanology of eastern Australia, *in* Johnson, R.W., ed., Intraplate volcanism in eastern Australia and New Zealand: Cambridge, Cambridge University Press, p. 29-38.

Wentworth, C.M., 1987, Implications for crustal structure in the western Coast Ranges, California, from studies along their eastern margin (Abstract): EOS, v. 68, no. 44, p. 1366.

Wentworth, C.M., and Zoback, M.D., 1989, The style of late Cenozoic deformation at the eastern front of the California Coast Ranges: Tectonics, v. 8, no. 2, p. 237-246.

Wentworth, C.M., Blake, M.C., Jr., Jones, D.L., Walter, A.W., and Zoback, M.D., 1984, Tectonic wedging associated with emplacement of the Franciscan assemblage, *in* Blake, M.C., Jr., ed., Franciscan geology of northern California: Society of Economic Paleontologists and Mineralogists, Pacific Section, Book 43, p. 163-173.

West, W.D., 1985, The Deccan trap and other flood eruptions; a comparative study: Proceedings of the Indiation National Science Academy, Part A, Physical Sciences, v. 51, no. 3, p. 465-494.

Westaway, R., 1990, Block rotation in western Turkey, 1, observational evidence: Journal of Geophysical Research, v. 95, no. B-12, p. 19857-19884.

Westercamp, D., 1979, Diversité, contrôle structural et origines du volcanisme récent dans l'arc insulaire des Petites Antilles: Bureau de Recherches Géologiques et Minières Bulletin, deuxième série, no. 3/5, p. 211-226.

Westercamp, D., and Tomblin, J.F., 1979, Le volcanisme rècent et les éruptions historiques dans la partié centrale de l'arc insulaire des Petites Antilles: Bureau de Recherches Géologiques et Minières Bulletin, deuxieme série, no. 3/5, p. 293-321.

Weyl, R., 1980, Geology of Central America (completely revised second edition): Berlin, Gebrüder Borntraeger, 371 p.

*Wezel, F.-C., 1988, A young Jura-type fold belt within the central Indian Ocean?: Bollettino di Oceanologia Teorica ed Applicata, v. 6, no. 2, p. 75-90.

White, R.S., 1984, Atlantic Oceanic crust: seismic structure of a slow-spreading ridge, *in* Gass, I.G., Lippard, S.J., and Shelton, A.W., eds., Ophiolites and oceanic lithosphere: Geological Society of London Special Publication no. 13, p. 101-111.

White, R.S., 1989, Asthenospheric control on magmatism in the ocean basins, *in* Saunders, A. D., and Norry, M. J., eds., Magmatism in the ocean basins: London, Geological Society of London Special Publication no. 42, p. 17-27.

White, R.S., and McKenzie, D., 1989, Magmatism at rift zones: the generation of volcanic continental margins and flood basalts: Journal of Geophysical Research, v. 94, no. B-6, p. 7685-7729.

Wicander, R., and Monroe, J.S., 1989, Historical geology. Evolution of the earth and life through time: St. Paul (MI), West Publishing Company, 578 p.

Wilkinson, J.F.G., 1981, Continential and mid-ocean ridge tholeiites; some similarities and contracts, *in* Subbarao, K.V., Sukheswala, R.N., eds., Deccan volcanism and related basalt provinces in other parts of the world: Memoir Geological Society of India, v. 3, p. 340-361.

Williams, H., 1978, Tectonic lithofacies map of the Appalachians: St. John's, Newfoundland, Memorial University of Newfoundland Map no. 1, scale, 1:1,000,000, 2 sheets.

*Williams, P.R., and Harahap, B.H., 1987, Preliminary geochemical and age data from postsubduction intrusive rocks, northwest Borneo: Australian Journal of Earth Sciences, v. 34, no. 4, p. 405-415.

*Williams, P.R., Johnston, C.R., Almond, R.A., and Sinamora, W.H., 1988, Late Cretaceous to early Tertiary structural elements of West Kalimantan: Tectonophysics, v. 148, no. 3-4, p. 279-297.

Wilson, J.T., 1954, The development and structure of the crust, *in* Kuiper, G. P., ed., The earth as a planet: University of Chicago Press, p. 138-214.

Wilson, J.T., 1965, A new class of faults and their bearing on continental drift: Nature, v. 207, no. 4995, p. 343-347.

Wilson, J.T., 1966, Did the Atlantic close and then re-open?: Nature, v. 211, no. 5050, p. 676-681.

Wilson, J.T., 1967, Theories of building contents, *in* Gaskell, T. F., ed., The earth's mantle: London, Academic Press, p. 445-473.

*Wilson, J.T., 1990, On the building and classification of mountains: Journal of Geophysical Research, v. 95, no. B-5, p. 6611-6628.

Winterer, E.L., Atwater, T.M., and Decker, R.W., 1989, The northeast Pacific Ocean and Hawaii, *in* Bally, A. W., and

Palmer, A. R., eds., The geology of North America; an overview: Boulder, CO, Geological Society of America, Decade of North American Geology, v. A, p. 265-297.

*Wolfart, R., and Wittekindt, H., 1980, Geologie von Afghanistan: Berlin, Gebrüder Borntraeger, 500 p.

Wood, B.L., 1978, The Otago Schist megaculmination: its possible origins and tectonic significance in the Rangitata orogen of New Zealand: Tectonophysics, v. 47, no. 3/4, p. 339-368.

Wood, C., 1983, Continental rift jumps: Tectonophysics, v. 94, no. 1-4, p. 529-540.

Woodhouse, J.H., and Dziewonski, A.M., 1984, Mapping the upper mantle: three dimensional model of earth structure by inversion of seismic wave forms: Journal of Geophysical Research, v. 89, no. B-7, p. 5953-5986, 6295-6297.

*Workman, D.R., 1977, Geology of Laos, Cambodia, south Vietnam and the eastern part of Thailand: London, Institute of Geological Sciences, Overseas Geology and Mineral Resources no. 50, 33 p.

*Worthing, M.A., Midobatu, C.K., and Nixon, P.H., 1992, Structural setting, petrology and emplacement of serpentinites in the Koki fault zone, Port Moresby, Papua New Guinea: Journal of Southeast Asian Earth Sciences, v. 7, no. 2-3, p. 147-158.

Worzel, J.L., 1965, Discussion, *in* Blackett, P.M.S., Bullard, E., and Runcorn, S.K., organizers, A symposium on continental drift: Royal Society of London Philosophical Transactions, ser. A, v. 258, no. 1088, p. 137-139.

Wright, J.B., Hastings, D.A., Jones, W.B., and Williams, H.R., 1985, Geology and mineral resources of West Africa: London, George Allen & Unwin, 187 p.

Wrucke, C.T., 1989, The middle Proterozoic Apache Group, Troy Quartzite, and associated diabase of Arizona, *in* Jenney, J.P., and Reynolds, S.J., eds., Geologic evolution of Arizona: Arizona Geological Society Digest, v. 17, p. 239-258.

*Wu Gongjian, Gao Rui, Yu Qinfan, Meng Lingshun, Dong Xuebin, Chui Zuozhou, Yin Ahouxun, Shen Xianjie, and Zhou Yaoxiu, 1991, Integrated investigations of the Qinghaie-Tibet Plateau along the Yadon-Golmud geoscience transect: Acta Geophysica Sinica, v. 34, no. 5, p. 552-562.

Wust, S.L., 1986, Regional correlation of extension directions in Cordilleran metamorphic core complexes: Geology, v. 14, no. 10, p. 828-830.

Wyllie, P.J., 1971, The dynamic earth: New York, John Wiley & Sons, Inc., 416 p.

*Xinh, L.T., 1986, Metallogeny of the Hoang Lien Son subduction zone: Geological Society of Malaysia Bulletin no. 20, p. 525-535.

*Xu Jiawei, Tong Weixing, Zhu Guang, Lin Shoufa, and Ma Goufeng, 1989, An outline of the pre-Jurassic tectonic framework in east Asia: Journal of Southeast Asian Earth Sciences, v. 3, no. 1-4, p. 29-45.

Yamagishi, H., 1985, Growth of pillow lobes--evidence from pillow lavas of Hokkaido, Japan, and North Island, New Zealand: Geology, v. 13, no. 7, p. 499-502.

Yamaji, A., 1991, Periodic hot-spot distribution on Io: Science, v. 254, no. 5028, p. 89-91.

Yan Zhide, 1987, Characteristics of strong earthquakes and tectonic relations in north-south seismic belt of China: Scientia, Sinica, series B, v. 30, no. 4, p. 438-448.

Yang Jingsui, Chai Yaochu, and Feng Binggui, 1990, Plume-type mid-ocean ridge basalt in the Bangong Lake ophiolite: geochemistry evidence, *in* Li Guangcen, Zhou Weiqin, and Nicolas, A., eds., Geology of the Himalayas, v. 2. Papers on geology: Beijing, Geological Publishing House, p. 477-491.

Yang Zunyi, Cheng Yuqi, and Wang Hongzhen, 1986, The geology of China: Oxford, Clarendon Press, 303 p.

Yao Huijin, 1988, Evolution of plate structure and evaluation of oil-bearing [prospects] in Turpan-Hami basin: Petroleum Exploration and Development, v. 15, no. 6, p. 12-18.

*Yap, F.L., 1986, Age determination on the Kuantan granite and dolerite dykes: Geological Society of Malaysia Bulletin, no. 20, p. 415-421.

Ye Hong, Zhang Botao, and Mao Fungying, 1987, The Cenozoic tectonic evolution of the great North China: two types of rifting and crustal necking in the great North China and their tectonic implications: Tectonophysics, v. 133, no. 3/4, p. 217-227.

Yeats, R.S., 1968, Rifting and rafting in the southern California borderland, *in* Dickinson, W.R., and Grantz, A., eds., Proceedings of conference on geologic problems of San Andreas fault system: Stanford University Publication Geological Sciences, v. 11, p. 307-322.

*Yin Jixiang, Xu Juntao, Liu Chengjie, and Li Huan, 1988, The Tibetan Plateau: regional stratigraphic context and previous work, *in* Chang Chengfa, Shackleton, R. M., Dewey, J. F., and Yin Jixiang, coordinators, The geological evolution of Tibet: Royal Society of London Philosophical Transactions, series A, v. 327, no. 1594, p. 5-52.

Yin Xiuhua, Shi Zhihong, Liu Zhanpo, and Zhang Yumei, 1980, The basic features or regional gravity field in [the] Chinese continent: Seismology and Geology, v. 3, no. 4, p. 69-75.

Yoder, H.S., Jr., 1976, Generation of basaltic magma: Washington, D.C., National Academy of Sciences, 265 p.

Yoder, H.S., Jr., 1988, The great basaltic 'floods': South African Journal of Geology, v. 91, no. 2, p. 139-155.

Yoder, H.S., Jr., and Tilley, C.E., 1962, Origin of basalt magmas: an experimental study of natural and synthetic rock systems: Journal of Petrology, v. 3, no. 3, p. 342-532.

York, J.E., and Helmberger, D.V., 1973, Low-velocity zone variations in the southwestern United States: Journal of Geophysical Research, v. 78, no. 11, p. 1883-1888.

*Yoshida, M., and Vitanage, P. W., 1993, A review of the Precambrian geology of Sri Lanka and its comparison with Antarctica, *in* Findlay, R. H., Unrug, U., Banks, M. R., and Veevers, J. J., eds., Gondwana Eight: assembly, evolution and dispersal: Rotterdam, A. A. Balkema, p. 97-109.

Yoshii, T., 1983, Cross sections of some geophysical data around the Japanese Islands, *in* Hilde, T.W.C., and Uyeda, S., eds., Geodynamics of the western Pacific-Indonesian region: American Geophysical Union and Geological Society of America Geodynamics Series no. 11, p. 343-354.

*Yu Zhihong, Liu Zhongpin, Wan Defang, and Fu Zijie, 1981, Tectonic map of the linear structures of the territory of China (by using the satellite images); Institute of Geology of Mineral Resources, Chinese Academy of Geological Sciences: Beijing, Cartographic Publishing House, 19 p. + 1 sheet, scale, 1:6,000,000.

*Zagruzina, I.A., Avdeyeva, O.I., Ge, G.G., Pavshukov, V.V., Fedorova, I.V., Yakovleva, L. V., and Nguyen Van Can, 1978, Novyye dannyye o vozraste granitoidov V'yetnama: Novosibirsk, Geologiya i Geofizika, no. 2, p. 142-147.

Zakharova, T.L., 1980, Izostaziya Baykal'skoy riftovoy zony: Novosibirsk, Geologiya i Geofizika, no. 5, p. 79-85.

Zaleski, S., and Julien, P., 1992, Numerical simulation of Rayleigh-Taylor instability for single and multiple salt diapirs: Tectonophysics, v. 206, no. 1-2, p. 55-69.

Zanchi, Z., and Angelier, J., 1993, Seismotectonics of western Anatolia; regional stress orientation from geophysical and geological data: Tectonophysics, v. 222, no. 2, p. 259-274.

Zandt, G., 1981, Seismic images of the deep structure of the San Andreas fault system, central Coast Ranges, California: Journal of Geophysical Research, v. 86, no. B6, p. 5039-5052.

Zandt, G., and Furlong, K.P., 1982, Evolution and thickness of the lithosphere beneath coastal California: Geology, v. 10, no. 7, p. 376-381.

Zapivalov, N.P., Serdyuk, Z.Ya., Zalazayeva, L.V., and Yashina, S.M., 1978, Neftegazonosnost' karbonatnykh porod silura-nizhnego karbona Zapadnoy Sibiri: Geologiya Nefti i Gaza, no. 1, p. 25-32.

*Zaytsev, N.S., and Luchitskiy, I.V., eds., 1979, Geologiya i magmatizm Mongolii, Sovmestnaya Sovetsko-Mongol'skaya Nauchno-Issledovatel'skaya Geologicheskaya Ekspeditsiya, Trudy vyp. 30: Moscow, Izdatel'stvo "Nauka," 235 p.

Zeng Zuoxun, 1990, Mechanical research on binucleus-type vortex structures: Acta Geologica Sinica (English Edition), v. 3, no. 4, p. 345-362.

*Zhai Mingguo, Cong Bolin, and Zhang Ruyuan, 1990, Two volcanic rock series in the Lancang Group, Yunnan Province, southwestern China: distinguishing and geological implication, *in* Ichikawa, K., Bunopas, S., Khoo, T. T., and Hada, S., convenors, Pre-Jurassic geologic evolution of eastern continental margin of Asia: Osaka (Japan), IGCP Project 224, Pre-Jurassic evolution of eastern Asia, Rept. no. 5, p. 75-80.

Zhang Buchun, Jia Sanfa, Wang Tonghe, and Zheng Binghua, 1985, Intraplate seismotectonic features of North China: Tectonophysics, v. 117, no. 1-2, p. 177-191.

*Zhang Ruyuan, Cong Bolin, and Li Yonggang, 1990, Blueschists in western Yunnan Province, China: petrology and its [sic] tectonic significance, *in* Ichikawa, K., Bunopas, S., Khoo, T. T., and Hada, S., convenors, Pre-Jurassic geologic evolution of eastern continental margin of Asia: Osaka (Japan), IGCP Project 224, Pre-Jurassic evolution of eastern Asia, Rept. no. 5, p. 71-73.

Zhang Yunxiang, Luo Yaonan, and Yang Chongxi, 1990, Panxi rift and its geodynamics: Beijing, Geological Publishing House, 415 p.

*Zheng Jianzhong, Xu Wenyue, Liu Jin, and Zou Ying, 1987, Some characteristics of the seismicity and the stress field in the Panxi Rift zone: Tectonophysics, v. 133, no. 3-4, p. 235-241.

Zhu Meixiang, 1986, Hydrothermal clay minerals in Rehai thermal field, Tengehong, Yunnan Province: Scientia Sinica, series B, v. 29, no. 4, p. 430-440.

Zhu Ming, 1989, The research on the isotopic ages of Mesozoic porphyries and porphyritic deposits in eastern China: Scientia Geologica Sinica, no. 2, p. 190-200.

Zhu Ying, 1979, Geotectonics of the North China massif and regularity of distribution of Anshan-type iron deposits based on preliminary analysis of aeromagnetic surveys: Geophysical and Geochemical Exploration, no. 1, p. 3-13.

Zhu Zhicheng, Ji Kecheng, and Fan Guangming, 1982, Preliminary research on the thrust wedges in some red basins of South China: Earth Science Journal, Wuhan College of Geology, v. 16, no. 3, p. 77-86.

Zhuravlev, Ye.G., 1986, Trappovaya formatsiya Zapadno-Sibirskoy plity: Akademii Nauk SSSR Izvestiya, Seriya Geologicheskaya, no. 7, p. 26-32.

Zietz, I., comp., 1982, Composite magnetic anomaly map of the United States. Part A: conterminous United States:

Reston, VA, U. S. Geological Survey Geophysical Investigations Map GP-954-A, scale, 1:2,500,000, 2 sheets.

Zindler, A., and Hart, S., 1986, Chemical geodynamics: Annual Review of Earth and Planetary Sciences, v. 14, p. 493-571.

Zoback, M.D., Zoback, M.L., Mount, V.S., Suppe, J., Eaton, J.P., Healy, J.H., Oppenheimer, D., Reasenberg, D., Jones, L., Raleigh, C.B., Wong, I.G., Scotti, O., and Wentworth, C., 1987, New evidence on the state of stress of the San Andreas fault system: Science, v. 238, no. 4830, p. 1105-1111.

Zolotukhin, V.V., and Al'mukhamedov, A.I., 1988, Traps of the Siberian Platform, *in* Macdougall, J.D., ed., Continental flood basalts: Dordrecht, Kluwer Academic Publishers, p. 273-310.

Zonenshain, L.P., Kuzmin, M.I., and Natapov, L.M., 1990, Geology of the USSR: a plate-tectonic synthesis: Washington, D.C., American Geophysical Union Geodynamics Series, v. 21, 242 p.

Zorin, Yu.A., and Osokina, S.V., 1984, Model of the transient temperature field of the Baikal rift lithosphere: Tectonophysics, v. 103, no. 1-4, p. 193-204.

SUBJECT INDEX

Solid Earth Sciences Library

Publications:

1. E.I. Galperin: *The Polarization Method of Seismic Exploration.* 1983 ISBN 90-277-1555-6

2. S. Hastenrath: *The Glaciers of Equatorial East Africa.* 1984 ISBN 90-277-1572-6

3. R.K. Verma: *Gravity Field, Seismicity and Tectonics of the Indian Peninsula and the Himalayas.* 1985 ISBN 90-277-1864-4

4. R. Haenel, L. Rybach and L. Stegena (eds.): *Handbook of Terrestrial Heat-Flow Density Determination.* With Guidelines and Recommendations of the International Heat-Flow Commission. 1988 ISBN 90-277-2589-6

5. E. Roth and B. Poty (eds.): *Nuclear Methods of Dating.* 1989 ISBN 0-7923-0188-9

6. G. Wagner and P. Van den haute: *Fission-Track Dating.* 1992 ISBN 0-7923-1624-X

7. J. Lima-de-Faria: *Structural Mineralogy.* An Introduction. 1994 ISBN 0-7923-2821-3

8. I.M. Varentsov: *Manganese Ores of Supergene Zone: Geochemistry of Formation.* 1996 ISBN 0-7923-3906-1

9. A.A. Meyerhoff, I. Taner, A.E.L. Morris, W.B. Agocs, M. Kamen-Kaye, M.I. Bhat, N.C. Smoot and Dong R. Choi; D. Meyerhoff Hull (ed.): *Surge Tectonics: A New Hypothesis of Global Geodynamics.* 1996 ISBN 0-7923-4156-2

Kluwer Academic Publishers – Dordrecht / Boston / London